Dragons and Tigers

A Geography of South, East and Southeast Asia

Second Edition

Barbara Weightman

California State University, Fullerton

WILEY

John Wiley & Sons, Inc.

D0166112

EXECUTIVE EDITOR   Ryan Flahive
ASSISTANT EDITOR   Deepa Chungi
PRODUCTION EDITOR   Barbara Russiello
EXECUTIVE MARKETING MANAGER   Jeff Rucker
SENIOR DESIGNER   Kevin Murphy
SENIOR ILLUSTRATION EDITOR   Anna Melhorn
PHOTO EDITOR   Tara Sanford
NEW MEDIA EDITOR   Tom Kulesa
FRONT COVER PHOTOS   Tom Van Sant/Geosphere Project/Photo Researchers, Inc.; Valerie Long/Lonely
                     Planet Images (bus)
BACK COVER PHOTO   Barbara A. Weightman

This book is printed on acid free paper. ◯

To order books or for customer service please, call 1-800-CALL WILEY (225-5945).

ISBN-13 978-0–471-63084-5
ISBN-10 0–471-63084-5

10 9 8 7 6 5 4 3 2 1

Preface

I have designed this book to meet the need for a comprehensive text on the geography of Asia. Knowledge of Asian landscapes is fundamental to academic and practical learning in the twenty-first century. Virtually half of the world's people live in Asia, 40 percent in China and India alone. With some of the fastest growing economies in the world, Asian nations, notably of the Pacific Rim, are critical components of North American and other countries' economic and political strategies. Australia, Canada, and the United States are key destinations for Asian migrants whose presence intensifies networks of trans-Pacific interrelationships. Some Asian nations are global leaders in environmental damage, and several are riddled with internecine strife. I think that learning about this dynamic and multicultural region is imperative.

Dragons and Tigers explores and illustrates conditions, events, problems, and trends of both larger regions and individual nations. I use a cross-disciplinary approach to discuss evolving physical and cultural landscapes. Nature-society interrelationships provide the foundation for my emphasis on social, economic, political, and environmental problems.

I have also attempted to provide plenty of material for instructor and student use. There are several new maps and photos in this second edition. The chapter bibliographies are multidisciplinary, referencing various approaches to the subject. I offer sufficient historical background for the grounding of current conditions and for the understanding of contemporary issues. Main point summaries are intended to sharpen the focus of the reader. Furthermore, I use close up, focus boxes to spotlight topics of interest and offer tables of demographic and quality of life data for comparative purposes. A glossary of terms is provided at the end of the text. I have tried to write a book rich and fascinating in content. My hope is that both instructors and students will want to read it.

Acknowledgments

I owe a debt of gratitude to several individuals. I especially thank my parents, Claudia Lowe, Louise Scheffer, and Eva Voigt, for their undying support over the years. I also thank the geography faculty at the University of British Columbia and the University of Washington for showing me the way to success in my academic endeavors.

I gratefully acknowledge the contributions of those who reviewed various segments of the book, and most especially Cindy Fan, *UCLA,* and Roman Cybriwsky, *Temple University,* who contributed to the chapters on China and Japan. Thanks to:

Patrick Buckley, *Western Washington University*
George Demko, *Dartmouth College*
Christine Drake, *Old Dominion University*
Philippe Foret, *University of Oklahoma*
Melinda Meade, *University of North Carolina, Chapel Hill*
Norman Moline, *Augustana College*
David Nemeth, *University of Toledo*
Mary Lee Nolan, *Oregon State University*
Bimal K. Paul, *Kansas State University*
Yda Schreuder, *University of Delaware*
Richard Ulack, *University of Kentucky*
Jean-Paul Rodrigue, *Hofstra University*
Gil Schmidt, *University of Northern Colorado*
Susan M. Walcott, *Georgia State University*

Any errors or deficiencies that remain in the book are solely my responsibility.

The people at John Wiley & Sons, Inc., also deserve credit. Special thanks go to Nanette Kauffman for seeing potential in my proposal and encouraging me to pursue it. Other individuals worthy of mention are Jay O'Callaghan, Publisher; Ryan Flahive, Executive Editor; Deepa Chungi, Assistant Editor; Kevin Murphy, Senior Designer; Tara Sanford, Photo Editor; Anna Melhorn, Senior Illustration Editor; Barbara Russiello, Production Editor; and Adriane Ruggiero, copy editor and proofreader.

Contents

Chapter 11

Japan: A Tiger Risen   280

Chapter 12

Korea and Taiwan: Tigers Rising   303

Chapter 1

✦

The Big Picture: Major Influences

"The idea of tolerance and mutual
concession is based on admitting
the compatibility of many different
philosophical views of the world."

HAJIME NAKAMURA (1964)

Introduction

Welcome to South, East, and Southeast Asia, a region of unparalleled diversity that is home to more than half of humankind. Close to 3.5 billion people inhabit a land area of about 8 million square miles (20,761,000 sq km). Here arose some of the world's most admirable and powerful civilizations. Here, too, are the world's highest mountains and several of its longest rivers. Vast, densely populated plains contrast with expanses of sparsely populated mountains, deserts, and tropical forests. Hundreds of ethnic groups, religions, and languages blend in a kaleidoscopic array of lifestyles and cultural landscapes. Democracies exist alongside socialist and militarist states. The region contains some of the globe's poorest countries as well as the ascendant **dragon and tiger economies** to which we are increasingly connected.

Think of the number of Japanese, Korean, or Taiwanese cars and electronic gadgets that you have. Check your clothing labels; where were they

assembled? Thailand? Indonesia? Do you eat Chinese, Indian, or Malaysian food? Is there a Buddhist or Hindu temple, or Muslim mosque near your community? We are all linked in some fashion to South, East, and Southeast Asia.

Diversity makes the pursuit of multiple themes necessary for exploring and appreciating this part of the world. However, in light of contemporary geographic thought, some themes will be more apparent than others. An overarching theme will be **landscape,** a concept that goes far beyond that which we observe. It encompasses our entire sensory experience; it is the content of our lived world. Individuals or groups do not function independently of other people or the forces of nature. Moreover, political, economic, social, and environmental systems are increasingly intertwined from local to global scales. Under this rubric, we will investigate both cultural landscapes and physical landscapes, in context of their interrelationships.

I will stress the **social construction** of landscapes as well as **quality of life** issues. How have internal

1

and external forces affected ways of life? How well are people's needs being met? Who determines what these needs are? I will also focus on **environmental transformation**. How have local and externally imposed values, attitudes, policies, and practices altered physical, biological, and zoological realms? What are the ramifications for human beings? In addition, I will explore the meanings and processes of development in the context of **globalism** or **globalization** and **regionalism**. Is the world becoming increasingly similar? Or, is the world becoming more dissimilar? How global is the global economy? Is it becoming more fragmentary with the formation of competitive regional organizations?

Dragons and Tigers

Dragons and tigers represent the dynamic forces of power and change. They are symbols long associated with South, East, and Southeast Asia. In China's *I Ching,* or *Book of Changes,* the light and dark aspects of *yin* and *yang* are often expressed as dragon and tiger. In Daoism, *yang* or goodness is controlled by a green dragon; *yin* or evil is controlled by a white tiger. Many a sage sought the dragon-tiger elixir as a key to immortality. The Chinese Emperor wore *dragon robes* and sat on the *dragon throne.* The dragon is used as a metaphor for mountains and rivers—to cut a pass or install water-control works is to "tame the dragon." Dragons bring luck and protection. They continue to be incorporated as architectural detail in temples, hotels, houses, and many other structures.

The tiger, aside from being a real, although increasingly rare, animal, is fabled in stories throughout these realms. In Korea, the tiger is a protective figure, watching over the land. Indonesia is rich in tiger lore with legends of tiger spirits, families, and villages. Tiger teeth and claws are said to induce prowess and potency. In South Asia, tiger images appeared on seals used by early Indus Valley civilizations more than 4,000 years ago. Betrothed couples in India wear red and yellow shawls that symbolize the tiger's stripes of fertility.

Today, dragon or tiger status is ascribed to growing economies. Japan, Singapore, South Korea, and Taiwan have all been described as *dragon economies* or

economic tigers, although today Japan and Singapore are described as developed economies. India is struggling for "tigerhood." Vietnam is "Asia's baby tiger."

These beasts are frequently seen in service and product names. Dragonair flies out of Hong Kong. Nine Dragon Beer is sold in China. Singapore's Tiger Beer is exported around the world. Tiger Balm ointment is a trusted remedy for headaches and insect bites. Lastly, there is a "dragon, phoenix, and tiger soup" The dragon is snake, the phoenix is chicken, and the tiger is cat.

Dragons and tigers have pervaded Asian landscapes in both subtle and obvious ways. That is why they have been given a place in the title of this book.

Defining the Region

Asia's expanse and complexity require its division into more manageable and meaningful areas of study. This text will concentrate on three geographic realms of Asia: South, East, and Southeast Asia (Figure 1–1).

According to geographer Harm de Blij (1997), **geographic realms** have three sets of spatial criteria. First, physical and cultural characteristics are the largest units into which we can divide the **ecumene** (i.e., the inhabited world). Second, geographic realms are founded on similarities of **functional interactions** between people and their natural environments. These are expressed in farmscapes, transportation networks, cityscapes, environmental damage, and countless other landscape features. Third, geographic realms focus on and incorporate the world's major population clusters. South, East, and Southeast Asia encompass the two largest clusters in India and China and the fourth largest in Indonesia. You will learn about the growth and distribution of these populations in a later chapter.

South Asia comprises India, Pakistan, and Bangladesh; the Himalayan states of Nepal and Bhutan; and the island nations of Sri Lanka and the Maldives. East Asia includes China, Mongolia, North and South Korea, Taiwan, and Japan. Southeast Asia encompasses the mainland countries of Myanmar, Thailand, Lao (Laos), Cambodia, and Vietnam, in addition to the maritime states of

Figure 1–1

Geographic realms. These three realms are based on similarities of functional interrelationships between people and their natural environments.

Malaysia, Singapore, Brunei, Indonesia, East Timor, and the Philippines.

"ASIA" AS A MYTH

These countries and their respective realms are parts of a larger entity called *Asia*. While we might think of Asia as a definable geographic region, actually it is more of an idea than a reality. More importantly, the evolution of the *idea of Asia* in Europe and the West shaped the nature and tenor of perceptions and interactions between Asian and non-Asian cultures throughout history. To understand this, we must explore what sometimes is called the **myth of Asia**.

There is no cultural or historical entity that equals Asia. It is more of a literary and psychological construct imposed from the outside than a geographical reality. Designations of "the Orient" and "the Far East" are also associated with the idea of Asia. Where do these terms come from, and what do they mean? What is Asia, and how did it come to be? How can it be a myth if it actually exists?

Asia, in fact, was defined by Europeans. From the time of ancient Greece, Europeans visualized the world as being divided into two or three continental parts: Asia, Europe, and sometimes Africa. Religiously inspired *mappaemundi* (maps of the world), portraying the earth as flat, often depicted Asia as half the known world! This *understood world* could also be divided into east and west: **Orient** meaning *east,* derived from sunrise, and **Occident,** meaning *west,* from sunset. For Europeans, *Orient* meant east and Asia. However, the initially blurry concept of Asia was filled in over time and concurrently expanded spatially.

CONCEPTUALIZING ASIA

The term *Asia* came from the Greek word for sunrise and originally referred to the Anatolian peninsula (Turkey) in the seventh century B.C., a region later known as Asia Minor. By the fifth century B.C., Asia had acquired continental meaning. Throughout this evolutionary process, comparisons and contrasts were drawn between east and west.

Centuries later, Karl Marx (1818–1883) drew distinctions between Asia and Western Europe. He subscribed to the concept of "An Asiatic Mode of Production." In this exploitative, economic system, the urban rich and powerful sucked all profits from an undifferentiated rural peasantry laboring in virtual enslavement.

Meanwhile, Asia's colonizers had developed their own frameworks of thought. You will learn more details in later chapters. A general model is provided by geographer James L. Blaut (1994), who refers to a **colonizer's model** of the world: civilization and progress diffuse from the European center to the culturally barren periphery. This is an interesting perception considering that Asian civilizations were much older and more advanced and complex than those of the colonizers.

Two things are important here. First, since classical times, Western writers have commonly contrasted their own civilizations with those of Asia. Asia has been perceived as a single entity, its internal diversity homogenized in false unity, its characteristics and institutions portrayed as alien and inferior to those in Occidental realms. The term "inscrutable" was often used to describe the frustrated Western perception of Asia. Recent research on the brain indicates that

Asians and Europeans may be "wired" to see the world differently. As the British writer Rudyard Kipling (1865–1936) said in the first line of his poem "The Ballad of East and West": "Oh, East is East, and West is West, and never the twain shall meet."

Second, environmental determinism colored historic and geographical accounts well into the twentieth century. Ellen Churchill Semple (1863–1932), an American geographer who taught at the University of Chicago and Clark University, was complimentary to China and Japan, yet described the people of India as ignorant and superstitious. She saw tropical peoples as servile and inferior due to the high temperatures in which they lived. According to Semple, anyone who went to tropical areas would soon be reduced to the lethargic state characteristic of those residing in these debilitating regions.

Derogatory assessments not only generated **xenophobia** (fear of foreigners) but also contributed to **ethnocentrism** and the **we-they syndrome.** Ethnocentric people judge foreign cultures by their own standards and often speak in *we* (better) versus *they* (not as good) terms. Economically developed, democratic Western nations have viewed themselves as superior to those in less developed, politically alternate Eastern ones. As you will see, this notion of cultural superiority continues to pervade East-West relations. Furthermore, the history of cartography reveals a long-established practice of centering the world on Europe. This promoted a **Eurocentric** worldview, the antithesis of China's **Sinocentric** (China centered) view.

Recently, I purchased a school notebook printed in Vietnam. On the back cover is a Eurocentered world map captioned by the words, "The World of [the] Future." What sort of maps are in your classrooms and academic books? Examine some world geography, world history, or world civilization textbooks and note how many pages are devoted to Eastern versus Western civilizations. What does all this say about Western perceptions of Asia? And what does it say about European influence in Asia?

European notions of Asia were also colored by religious beliefs. After all, the Garden of Eden was there, the Three Wise Men came from the east, and the sun, the light of salvation, rises in the east. Prester (Presbyter) John, a wise Christian king, was believed to have ruled somewhere in Asia. In the Middle Ages, countless pilgrimages were launched to find his kingdom. In that vein of thought, the Orient was a source of light and goodness and, conversely, the Occident was dark and evil.

Beyond these scripturally-inspired conceptions, Asia was portrayed as a mysterious place of dog-headed men and other freakish beings, as well as the heathen tribes of Gog and Magog, poised to pounce upon the unwary. More significantly, the Orient was thought to possess unbounded wealth in spices, jewels, incense, and other exotica—an idea underpinning numerous land and sea explorations.

It is important to recognize that this collective, conflicting imagery of an ill-defined Asia derived from both real and imagined experiences with a region extending eastward from Egypt into Ukraine and the Russian grasslands and southward into India. At the time, China was merely a fabled land on the periphery of this vast expanse called Asia. As exploration fostered increasing contacts between East and West, more places were added to the European knowledge map. Asia, as a conceived region, ultimately stretched eastward and southward into the Pacific Ocean, incorporating Japan and such far-flung islands as those in Indonesia and the Philippines.

By the nineteenth century, European powers were not only entrenched in many Asian countries for political and economic gain but were also struggling with defining their own boundaries. A series of wars, the rise and demise of assorted political powers, and ever-shifting national boundaries necessitated the revision of European frontiers. This was accomplished at the **Congress of Vienna** in 1815. The division between Europe and Asia was demarcated from Russia's Ural Mountains, through the Caspian and Black seas, and along the eastern end of the Mediterranean Sea to the African continent.

Today, Asia is recognized as part of the geographical landmass known as **Eurasia**. And so, there is a continental Asia in a Eurasian context. However, cultural, political, economic, and environmental regions, as we know them today, are no more coincident with these boundaries than with spatial constructs of earlier times. This is why it is preferable to consider smaller, more cohesive realms. *Although South, East,*

and Southeast Asia are the regions designated for discussion, I will use the term Asia *when referring to these collectively, or the* Asian continent *in general.*

Reading Place Names

There are many ways to spell **toponyms** or place names, especially in the case of Asia. This is, in part, because of varying transliterations of regional scripts into the Roman alphabet. Some of the spellings for the Chinese capital city of Beijing are: Peiping, Pei-ching, and Peking. Westerners' inability to pronounce indigenous place names led to alternate spellings. Singapore is the English version of Singha Pura that was formerly known as Temasek. Different ethnic groups may have their own versions of toponyms. For example, Urumchi in northwestern China is a Uyghur name. The Chinese call it Wu-lu-mu-chi. Sometimes places disappear and reappear as something else. For instance, East Pakistan became Bangladesh in 1971. Furthermore, place names are frequently changed. Tokyo was once known as Edo. Others are newer, as with the restoration of traditional names. Bombay is now called Mumbai, and Rangoon, Burma, is currently Yangon, Myanmar. Be forewarned: If you send a letter to "Bombay," it will likely be returned to you marked "No such place."

When speaking of a place in a historical context, I will spell it as it was referred to at that time, followed by the current spelling in parentheses. For example, "In old Siam (Thailand) . . . ." In modern context, current names will stand alone, for example, "Myanmar." This might seem confusing, but, as you will find out, different maps and atlases employ different spellings. More importantly, residents of places frequently continue to use the name that has been in use the longest, or that they have a preference for. For example, in the 1990s, Calcutta was renamed Kolkata, but many people continue to use the name Calcutta.

Religion and Belief Systems

Now let us turn to an aspect of culture of paramount significance: **religion**. Religion, or a set of sacred beliefs, pervades everyone's life, whether consciously or unconsciously. Aside from being evident in the cultural landscape in the form of sacred structures and other phenomena, religion functions in a variety of ways in society. Religions define male and female roles; affect political systems and policies; influence agricultural and dietary practices; shape environmental values; impact economic development; and, frequently, are used by politicians to create dissension and strife.

Hinduism, Buddhism, and Islam are the major religions in Asia, but Christianity and other belief systems such as Sikhism, Confucianism, Daoism, and animism are important in specific regional settings. At this juncture, I will discuss the origin, **diffusion** or spread, and key tenets of four religions in Asia: Hinduism, Buddhism, Christianity, and Islam. Others will be discussed in their regional contexts in subsequent chapters.

All four religions have their **hearth areas,** or places of origin, in continental Asia. Each of these diffused by one or more means: conquest, word of mouth (contact diffusion), migration of believers (relocation diffusion), and the work of missionaries (proselytization). Merchants and traders played significant roles in spreading religion, especially in the case of Buddhism and Islam (see Figure 1–2).

Figure 1–2

Origin and diffusion of major religions. Religions spread from hearth areas by word of mouth, migration, and proselytization.

HINDUISM

Hinduism is the oldest of the world's major religions. Its origins reach beyond the limits of documented history. Without an apparent founder, but with a substratum of belief from Indus Valley civilization, it emerged in northwestern India at least 6,000 years ago. From there, it diffused southward across the Indian peninsula. Much later, it spread eastward with the diffusion of Indian influences as early as the third century A.D. and with subsequent Indian migrations through the colonial era. Hindu communities still thrive in Malaysia, Singapore, and on the islands of Bali and Lombok in Indonesia.

Hinduism is a **polytheistic** religion, hosting more than 333,000 (meaning countless) gods. The most important of these, in terms of daily worship and rituals, are *Shiva* and *Vishnu* (Figure 1–3). However, all deities are manifestations of a one ultimate reality known as **Brahman**. There are, in essence, as many ways to practice Hinduism as there are gods. Although there is no set creed, adherents may draw upon an enormous body of literature. Earliest are the **Vedas**, from the Sanskrit word *veda,* meaning "knowledge". Another critical work is the **Upanishads**. This means "sitting down near" one's spiritual advisor or **guru**.

Hindus believe in the **transmigration of souls** based on accumulated *karma.* All beings possess a soul and are positioned within a hierarchy. Where one is, on what might be called a status ladder, is conditioned by the state of one's karma. Karma is the sum of an individual's deeds. Good deeds can move one up the ladder and bad deeds result in demotion. In fact, it may take numerous lifetimes to achieve the common goal of escaping the hierarchy and having one's soul unite with the One Ultimate Reality of Brahman.

Critical to the Hindu faith is the concept of **reincarnation**: the notion that a person is reborn into another cycle of existence depending on the karma of the previous life. If a woman deliberately harms or kills another person or an animal such as a lizard, she might be reborn as that lizard in her next life. To quote a 1960s saying: "Keep your karma clean!"

Hinduism is entrenched in all aspects of Indian culture. You will learn more about its impact on society and development in the chapters on South Asia.

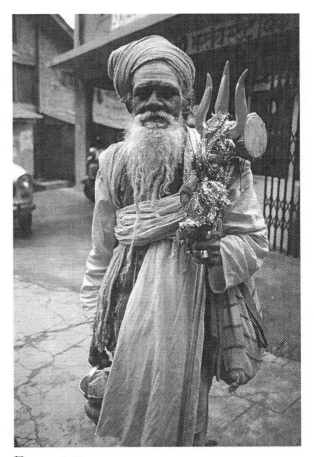

Figure 1–3

This Indian man is a saddhu—*a holy man who has divested himself of worldly possessions to seek the One Ultimate Reality. His trident indicates that he is a devotee of Shiva. Saddhus migrate among sacred sites and live off donations from the public. As a true* saddhu *(not a beggar), this man asked for nothing. He was merely present. Yes, I did make a donation.* Photograph courtesy of B.A. Weightman.

BUDDHISM

Buddhism also originated in northern India, in part as a reaction to the excesses of Hinduism prevalent by the sixth century B.C. From here, it diffused with missionary monks and traders via trade routes into China, and from there, across the Korean peninsula to Japan. It also spread southward and became established in Sri Lanka from where it spread seaward to Southeast Asia.

Adapted to local circumstances, Buddhism diversified and evolved into many schools of thought. An extensive body of literature was created, and many temples and monasteries accumulated rich collections of sacred books, tablets, and scrolls.

Buddhism and its various schools evolved from the teachings of Prince Siddhartha Gautama, born around 563 B.C. in what today is Nepal. After years in search of an explanation of life's harsh realities, he awakened to a profound conclusion: The cause of suffering is desire. This realization was the key to his enlightenment as the **Buddha.**

That the world is filled with suffering and that suffering is caused by desire became the first two of the **Four Noble Truths.** The last two are that it is possible to overcome desire by following the **Eightfold Path.** By choosing this right path of behavior (e.g., right knowledge, aspiration, and livelihood), one can achieve *nirvana.* As inscribed on the wall of a temple in Myanmar: "*Nibbana* is not situated in any *Bhumis* [stages of being] nor is it heaven. It is a state which is dependent upon ourselves." Nirvana is an indescribable state of desirelessness and peacefulness. It is absolute bliss.

Buddha advocated withdrawal from the material world of desire into a contemplative mode, making meditation a key part of Buddhist practice. He also emphasized being generous, kind, and vowing not to harm people or animals. Consequently, vegetarianism is widely practiced in Buddhism.

As Buddhism diversified with expansion, two main schools of thought developed. **Mahayana Buddhism** is practiced in China, Korea, Japan, and, because of Chinese influence, in Vietnam. This school holds that salvation can be achieved through the intercession of divine beings. Mahayana Buddhists believe the Buddha to be their savior but may commune with other deities as well.

One of the most popular of these is the **Bodhisattva** Avalokitesvara (in Sanskrit, the lord who looks in every direction). A Bodhisattva is one who postpones entry into nirvana in order to save other beings. Avalokitesvara is most popular in his female form and is enshrined in virtually all Buddhist temples. She is known as *Kuan-Yin* in China; *Kannon* in Japan; and *Kwanum* in Korea.

Figure 1–4
It was sunrise in the mountain town of Mamyo, Myanmar, where I watched Buddhist monks walk past devotees who put rice and other food into their "begging bowls." This scene exists on a daily basis throughout the world of Theravada Buddhism. Photograph courtesy of B.A. Weightman.

Theravada Buddhism developed in Sri Lanka as the monastic expression of Buddha's teachings. It places responsibility on the individual for salvation through the accumulation of merit by good behavior and religious activity. Monastic orders, or *sangha,* are very important, and monks and nuns are highly respected. It is Theravada Buddhism that diffused to Southeast Asia, where it is very much a part of the cultural landscapes of Cambodia, Laos, Myanmar, and Thailand (see Figure 1–4).

What happened to Buddhism in India? It rose to its zenith under Emperor Asoka in the third century B.C. But Hinduism eventually reasserted itself, and Buddhist adherents became a rarity. Aside from a few practitioners and scattered temple relics, Buddhism has faded from the Indian landscape. However, in its Himalayan refuge situation, Bhutan remains a Buddhist kingdom. Buddhism is also important in Nepal even though the Nepalese people are predominantly Hindu.

Buddhist temples and other landscape elements are increasingly common in Europe, North America, and Australia. This is a result of relocation diffusion as Chinese, Laotians, Vietnamese, and others continue to migrate to other parts of the world (see Figure 1–5).

Figure 1–5

This is one of the great halls of the Hsi Lai Temple in Hacienda Heights, California. It is specifically for bodhisattvas, persons who resist going to nirvana in order to help their fellow humans on earth. The plaza of rectangles replicates rice paddies and is called the Field of Merit. Hsi Lai Temple is similar to one in Kaohsiung, Taiwan. It appeals to multiple nationalities and accepts variation in Buddhist practices. Photograph courtesy of B.A. Weightman.

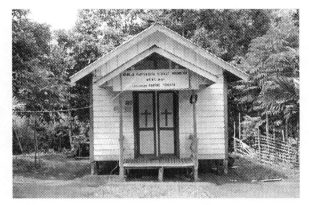

Figure 1–6

I came across this Pentecostal mission in the remote Dayak community of Mencimai in Kalimantan on the island of Borneo. Most of the people of this region follow folk belief systems or are Muslim. Even Christian converts retain elements of their traditional beliefs. There are not many converts in Muslim Indonesia. The church is built on stilts because of the tropical rains and surface flooding. Photograph courtesy of B.A. Weightman.

CHRISTIANITY

The spread of Christianity from its hearth region in Israel/Palestine is closely tied to European imperialism. Carried by colonizers, administrators, soldiers, sailors, traders, and missionaries, Christian doctrine arrived in Asia especially after the thirteenth century. It flourished in context of colonization that began in earnest in the sixteenth century, but competition with already established religions diminished its impact. Although Christians can be found throughout Asia, the most significant concentrations are in East Timor, the Philippines, eastern Indonesia, South Korea, and Vietnam.

Christianity is a **monotheistic** faith proclaiming the idea of a single deity called *God*. A Jewish man known as **Jesus Christ** is believed to be the Son of God, the Messiah that had been promised roughly 4,000 years earlier. The Messiah would restore the relationship between God and humankind. The teachings of Jesus and his disciples spread, as the Word of God, via contact conversion. Numerous versions of a book called the **Holy Bible** offer both a history of Judaism and early Christianity in addition to the teachings of Jesus and his followers. Many Jews and Christians believe that the Bible is the Word of God. Proselytization by missionaries, using the Bible, was and continues to be a powerful force in the diffusion of Christianity (Figure 1–6).

Christians believe in salvation through faith in Jesus Christ and adherence to the teachings of the Bible. Historic circumstance has seen the continual division of Christianity. Two major branches, Roman Catholicism and Protestantism, with its multiple denominations, have made the greatest headway in Asia. Christianity has grown to the point that it is self-perpetuating. Jesus' birthday (Christmas) is celebrated by close to 300 million Christians in Asia with people of other faiths often joining in the celebrations.

ISLAM

Islam, which translates as *submission to God,* is the youngest of the major religions. It originated around the subsequently holy cities of Mecca (Makkah) and Medina (al Madinah) in Arabia in the sixth century A.D. Islam swept east across Asia with successive waves of conquerors, traders, and missionaries. It replaced earlier religions in many areas. For example, Buddhism and Hinduism were once important in Indonesia but were virtually eradicated under the fervor of Islam. Islam is the dominant religion in Pakistan, Bangladesh, Malaysia, Indonesia, and Brunei. It is also a potent force in several other regions such as northern India and the southern Philippines. You will learn more about Islam in its regional contexts in later chapters.

Coming from the same core area, Islam has links to both Christianity and Judaism. It, too, is a monotheistic faith. Following a series of divine revelations, **Muhammad** (571–632) began preaching in Mecca and later Medina after being forced to flee there by his enemies. His teachings about a new way of life provided welcome structure in what was then a chaotic Arab world, and converts quickly accumulated. **Muslims** (those who submit to God) formed armies to invade and conquer the lands of **infidels,** or non-believers. Clashes with other established faiths resulted in the death of millions on all sides. Unfortunately, religious conflicts associated with Islam versus other belief systems still exist.

Muslims believe in one God, **Allah,** and recognize Muhammed as His prophet. They also view Jesus Christ as a prophet. Like the Jews, they avoid pork, perceiving it as unclean. To Muslims, the will of Allah is absolute. The Word of Allah is contained in the **Koran,** or *Qur'an,* which is Arabic for "recitation." Pivotal to being a good Muslim are the **Five Pillars of Islam.** These are: acknowledgment of Allah as God and Muhammed as His Prophet; prayer five times a day; a month of daytime fasting (known as **Ramadan**); the giving of alms; and at least one pilgrimage to Mecca. Held in the last month of the lunar calendar, this yearly event is known as the *hajj* and involves the convergence of millions of pilgrims in the holy cities of Mecca and Medina (Figure 1–7).

Dissension over succession to the Caliphate (ruling) authority eventually caused division in Islam, resulting in two major branches and several other sects. The majority of Muslims are orthodox Sunnis. The breakaway Shiites account for less than 15 percent. Shiism (*Shiah*) became rooted in Persia (Iran) in the sixteenth century and subsequently spread via Iraq and Afghanistan to Pakistan. It has even influenced Sunni Islam in India. Shiites are generally more fervent in their beliefs and more emotional in their practices. To them, religious leaders, or *Imams,* are infallible, as are their interpretations of the Koran and other religious literature. Shiites believe that Imams should rule while Sunnis believe that religious leaders should advise rulers.

Islam proscribes certain life-roles, modes of dress, and social behavior. However, the degree of stricture

Figure 1–7

This significant mosque is at Shah Alam, the new capital of Selangor Province west of Kuala Lumpur, Malaysia. Named for the Sultan Salahuddin Abdul Aziz Masjid, it was built in 1988 and holds 16,000 people. The mosque is oriented to face the holy city of Mecca. Loudspeakers on the minarets announce the call to prayer five times a day. Photograph courtesy of B.A. Weightman.

varies greatly from place to place, especially for women. Islam is also an influential force in governments of countries like Malaysia and Pakistan, where Muslims make up the majority population.

Islamism

In recent years there has been a resurgence of religious fundamentalism in the Islamic world. Muslims refer to this phenomenon as **religious revivalism**. Many Muslims disapprove of what they perceive as the erosion of traditional values in the wake of Westernization and secularization. They see Western culture as extravagant and immoral, and Godless governments as blasphemous. Both private property and profit-making are incompatible with basic values promoted by the Koran. As long as economic conditions were improving, dissatisfaction remained muted. However, as populations grew, poverty increased, and unemployment rose, fundamental Islamic ways became more appealing.

Hatred of the West was epitomized by the devastation of New York's World Trade Center on September 11, 2001, by Islamic terrorists. This and other atrocities are linked to an organization named al-Qaeda which is dedicated to punishing the perceived enemies of Islam.

Acts of terrorism have become all too common around the world including in several Asian countries. Pakistan is embroiled in fighting Islamic terrorists along its frontier with Afghanistan. It is also attempting to thwart fundamentalism within its own government. Terrorists blew up two night clubs in Bali, Indonesia in 2003 killing more than 200 people, mostly Australian tourists. Thailand is facing insurgency in its predominantly Muslim southern provinces. And the Philippines is trying to cope with Islamic militants in its southern island of Mindanao.

While these and other heinous acts have been fostered by Islamic fundamentalists, it is important to note that relatively few Muslims are fundamentalists and not all fundamentalists are terrorists. Nevertheless, it is true that increasing numbers of Muslims are calling for a return to the precepts of the Koran as interpreted by religious leaders.

FAITH, EXPLORATION, AND TRADE

Trade had long existed between Asia and Europe, dating back to the Roman Empire. But Arab Muslims laid the basis for a world trade system in the Mediterranean and Indian Ocean regions. As Islam is a way of life, there is no separation between religion and commerce. In fact, Islam is an ethical faith, providing even the rules of commerce. Believers who worked with each other understood rules of fair-trading practices. Non-believers who wanted to enter Muslim trading networks found conversion the easiest avenue. Each trading community served as a vehicle for the expansion of Islam.

In the eighth century, Arab Muslims conquered the Sind region of the Indus River delta. Until then, Buddhists had dominated trade, while Hindus focused on agriculture. Buddhist traders found it expedient to convert to Islam. Eventually, Islamic trading posts were established along India's west coast from Cambay to Calicut on the Malabar Coast (Figure 1–8). When the Muslim Mughal Empire in India expanded to the Bay of Bengal (1574–1592), access was gained to both the Coromandel coast and adjacent coasts of Southeast Asia. Muslims thereby achieved hegemony over the region's sea trade.

Prior to the appearance of Arab traders, China had been using South Asian merchants to carry their cargo. Steel had been made in China since the eighth century and was in great demand for weapons and farm implements. As a result of their overland and sea expansion, Muslims became active in the China trade and remained instrumental well into the twelfth century. Muslim trading communities existed in several Chinese cities such as Canton (Guangzhou) on the southern coast.

An international world system of commerce was now functioning via a network of trading ports

Figure 1–8

Trading centers and sea routes. Note that the Chinese explorer Cheng Ho navigated these waters long before the Europeans. The first European trading port was established by the Portuguese at Malacca (Melaka) in 1511.

and shipping lanes. By the fifteenth century, Muslim enclaves dotted the northern coasts of Java and Sumatra. Muslim-controlled Banten (Bantam) was also the focal point of Chinese trade in Southeast Asia. Makassar in Celebes (Sulawesi) was the key to the Spice Islands (Maluku). Achin (Aceh), Sumatra was situated at the northern entry to the Straits of Malacca. Of great importance was the founding of a transshipment point or **entrepôt** at Malacca (Melaka). Indians, Javanese, Chinese, and others migrated to these nodes of perceived opportunity where many of their descendants remain to this day.

Why this much interest? What was to be gained? Quite simply: riches. Demand was soon created for sandalwood and other aromatic woods; teak

and other hardwoods; silk and porcelain; slaves and exotic animals; and countless other products. But pepper, along with cinnamon, cloves, nutmeg and a host of other aromatic and edible spices became the most desired (Figure 1–9). Arabs conjured tales of gross birds harvesting cinnamon from ferocious bat- and snake-guarded swamps to dissuade others, especially the Europeans, from discovering their spice sources.

Neither Arabs nor Europeans were first in the spice trade. Indigenous states and empires were trading such commodities long before outsiders arrived. The Roman historian Pliny told of Indonesians in outrigger canoes driven by the prevailing winds unloading cargoes of cinnamon and cassia on the east coast of Africa in the first century A.D. The great

Figure 1-9

I had always wondered what mace looked like in the plant. Finally, I got a chance to see it in a market in Pinang, Malaysia. Mace is the stringy stuff covering the thin shell of the nut. It's amazing to think that these products changed the face of the world. Photograph courtesy of B.A. Weightman.

sea-borne empire of Srivijaya, centered in current-day Palembang near the east coast of Sumatra, controlled the Malacca and Sunda straits dominating east-west sea routes in the seventh century.

China was also involved in ocean trading. The talented Admiral Cheng Ho, in a series of voyages from 1405 to 1433, touched base at virtually every port in the region, from southern China to the east coast of Africa (Figure 1–8). He discovered that Chinese merchants were already well established in many of these places. Europeans truly were latecomers to this dynamic, lucrative world.

Europeans: Exploration, Commerce, and Conquest

By the late 1400s and early 1500s, Europeans became familiar with what Arabs had long been aware of: the annual reversal of winds associated with Asia's monsoon climatic system (see Chapter 2). "Monsoon" comes from *mausim*, Arabic for "season." By sailing west along the coast of the Atlantic Ocean and then south, around Africa's Cape of Good Hope and following the prevailing winds across the Indian Ocean between June and September,

and doing the reverse from December to March, sailing vessels could make the round trip in less than a year.

This was the **Age of Exploration** for Europeans on a zealous mission to convert the heathen to Christianity. But the real driving force was economic and thus similar to the aims of earlier Islamic traders. And so the Europeans ventured across the globe in quest of spices, gold, and other riches. In 1488, Bartolomeu Dias of Portugal rounded the Cape of Good Hope and sailed as far as the Persian Gulf. Upon returning home he announced that the sea route to India was open! Then in 1498, Vasco da Gama sailed as far as Calicut on India's Malabar Coast and later entered Lisbon's harbor with the first cargo brought from the "Far East" to Europe by sea. Ferdinand Magellan pioneered the westward route around South America's Cape Horn in 1519. Reaching the Philippines in 1522, he initiated Spain's entry into the region. Holland's (The Netherlands') 300-year involvement in the East Indies (Indonesia) commenced with the 1595 voyage of Cornelius de Houtman, who returned to the Netherlands with three shiploads of pepper and nutmeg.

England's Sir Francis Drake obtained a cargo of spices, plus booty plundered from a Portuguese ship, in his circumnavigation of the world (1577–1580). This accomplishment, in concert with Britain's defeat of the Spanish Armada in 1588, was a pivotal event in the sanctioning of further British expeditions. The race for territory was on, and it was inevitable that swords would cross.

Territorial gain and sea-lane control became paramount concerns. The importance of control over the Straits of Malacca cannot be underestimated. A plaque in the garden of the sultan's former palace in Malacca reads: "Whoever is lord of Malacca has his hand on the throat of Venice." The Italian city-state of Venice dominated international trade from the 1200s to the mid-1400s.

The Portuguese established trading ports from Africa to China, and by 1520 they dominated the seas. Intolerant of Islam and greedy for profit, they tried to force out the Muslims who remained along the coasts of Indonesia's islands and used Islam as a

Figure 1–10

This is the Porto de Santiago, main entry to the Portuguese fort of A Famosa in Melaka. The Portuguese established their fort and colony in 1511. The gate overlooks the Straits of Malacca (Melaka). Photograph courtesy of B.A. Weightman.

rallying point against the European intruders. The Portuguese moved into Cochin, Diu, Goa, and created a new port at Bombay (Mumbai) in India. They took the critical port of Malacca in 1511, and acquired control of Macao in 1557 (Figure 1–10). The Portuguese kept Ternate in the Spice Islands until 1570, when it was wrested away by the Dutch. However, the Portuguese remained in East Timor until 1975.

Both the Dutch and the British were anxious to oust the Portuguese from South and Southeast Asia. Backed by stockholders, with monopoly privileges and the right to raise armies and navies, their trading companies aggressively intruded into the area. **Factories,** or trading posts, were quickly set up throughout the region. The Dutch took Ternate in 1607 and founded Batavia on Java in 1619. Malacca fell into their hands in 1641, and Banten (by then British) was acquired in 1682. When the Dutch attacked the Johor port in 1784, the British responded.

Britain had initially concentrated on India but became intrigued with the economic potential of Southeast Asia, where they vied with the Dutch. They seized Malacca in 1796 and gained the island of Singapore in 1819. France entered the picture and established the first of its settlements at Pondicherry,

India, in 1674. However, the French were more interested in their African colonies and were not as aggressive as their contemporaries. And so the era of European colonialism was launched.

Colonialism and Imperialism

With colonization rampant in the fifteenth and sixteenth centuries, an extensive web of ports and shipping lanes fostered the evolution of an Atlantic economy as well. Silver and other resources from the New World became integral to economic exchange between Atlantic and Pacific realms. A world economy was emerging.

The number of imperial powers increased from two (Spain and Portugal) in the sixteenth century to five in the seventeenth century with the addition of Britain, France, and Holland. Collections of colonies worldwide were absorbed into the web of economic interchange. The only place not colonized in Southeast Asia was Siam (Thailand). It became a **buffer state** between British and French interests in mainland Southeast Asia (Figure 1–11).

Competition was fierce during this mercantilist period during which raw materials were brought to the colonial power, where factories produced manufactured goods to be shipped back to the colonies. Under **mercantilism,** colonies were not allowed to manufacture such goods. Neither were they allowed to trade with other than their own colonial overlord. The Seven Years' War (1756–1763) between Britain and France ended French ambitions in India and North America and allowed Britain to become the world's leading power. With colonies around the globe, it was said that the sun never set on the British Empire.

Britain expanded further into India, and France reached into Indochina. Holland tightened its grip on the Dutch East Indies (Indonesia). In order to control their monopoly on the spice trade and to jack up prices, the Dutch began destroying all the spice plants on some of the islands. Fearing instant poverty, the islanders resisted, with terrible consequences. For instance, almost the entire population of the Banda Islands was exterminated. By attacking and blockading ships owned by indigenous peoples

Figure 1–11

*Most of Asia was either colonized or strongly influenced by foreign powers, with the exception
of Thailand, which served as a buffer between English and French interests in Southeast Asia.
The Philippines was a Spanish colony from 1565 to 1898 and an American one thereafter.
Formosa (Taiwan) was a Japanese colony from 1895 to 1945. Although China was never a
colony, many places such as Canton (Guangzhou) and Shanghai had foreign concessions.*

and the Chinese, the Dutch managed to impoverish
other islands as well.

In 1869, the British- and French-owned Suez
Canal was opened, ushering in the **Age of Imperialism**
(1870–1914). By reducing the sea journey to Asia by
thousands of miles, interactions between Europe
and Asia intensified. With increased territorial gains
came friction. Rivalries, once confined to the European
continent, spilled over into the colonies. Territories
were exchanged between the British and the Dutch
and between the British and the French with no
consultation with the indigenous people.

Those colonized did not accept their inequitable
situation without protest. Resistance movements were

countered by imperialist wars of oppression. In
England these were known as Queen Victoria's Little
Wars. In British India, French Indochina, and Dutch
Indonesia, at least 70,000 died in battle before the
turn of the nineteenth century. Suppression of resistance in Burma alone (1823–1826) resulted in
15,000 lives lost.

World War I (1914–1918) was a cataclysmic
event in history. It spearheaded a new era for the
colonial world. The Treaty of Versailles in 1919
spurred nationalistic claims on the part of the
colonies, and reassigned them to the victorious
Western powers. But things could not be as they
were; the so-called *Belle Epoch* was over.

The Colonial City

Colonial cities were intended to serve the interests of colonial powers. They were guarded outposts that functioned as military, economic, and administrative centers. Architecture was grand, meant to impress the masses with the power and glory of the empire (Figure 1–12). Symbols of colonial rule were expressed in the form of architectural styles, statues, and monumental buildings with Western crests. Tudor-style apartment buildings still exist in the former British Quarter of Shanghai and Dutch churches can be found in Jakarta.

The colonial language was employed for naming places and streets and was the language of rule for administrative and educational purposes. Street signs such as Orchard Road and Rue Louis Pasteur remain in many former English and French colonial cities, and these languages continue to be spoken by educated elites.

To preserve race and class distinctions, separation was imposed on both people and landscape. Europeans built their own residential quarters, segregated from the indigenous people by physical barriers such as rail lines or waterways, or by military cantonments. Locals could enter the foreign quarters only with permission, and customarily did so in a servant capacity. Many sources refer to signs posted in China's and India's city parks stating, "No Chinese (Indians) Allowed."

Some cities were designed with a standardized plan such as set down in Spain's 1573 *Laws of the Indies.* This called for a grid-style street layout, centered on a main plaza surrounded by government buildings and a church. Similar but smaller plazas, built as the grid expanded, were intended to ensure the dissemination of order and godliness. This influence is particularly evident in the Philippines. Britain and France also liked the grid system because it represented order and regularity. Additionally, straight streets facilitated speedy access to trouble spots.

The colonial style city differed widely from the traditional city with its narrow, winding streets and bustling crowds. The colonial city had one focal point, but the traditional city had several, each focused on a government, religious, or market center. To the occupiers, the planned colonial city symbolized order within chaos.

As colonial power strengthened, city architecture and all that it stood for became even more imposing. Local people were relegated even further into the socioeconomic background. In Singapore and Kuala Lumpur, the British, in concert with Chinese and Indian migrants, dominated practically all activity. Indigenous Malays faded from the scene. This was also true in the Dutch East Indies. To overcome such injustices, Malaysia currently has special policies favoring Malays.

Figure 1–12

This building reflects both Dutch and Malay architecture. Eventually it was used by the British as the exclusive Melaka Club. Now it is an Independence Memorial for Malaysia.
Photograph courtesy of B.A. Weightman.

Imperialist Japan and the Rise of Nationalism

After centuries of feudalistic rule under powerful military leaders, or shoguns, the Emperor Meiji was restored to the throne in 1868. Young reformers

Figure 1–13

Japan's Greater East Asia Co-Prosperity Sphere. Japan won Formosa from China as a prize of war in 1895, annexed Korea in 1910, and established Manchukuo in 1932. It pressed for its goal to become "Britain of the Pacific" until its defeat by Allied Forces in 1945.

were determined to bring Japan into the Industrial Era. The Industrial Revolution had been initiated around 1750 in England and was in full swing elsewhere. Some saw potential for the island nation as the "Britain of the Pacific." Why could Japan not have a colonial empire? After all, it needed raw materials for industrialization and space for expansion. Japan flexed its muscles, attacking and defeating both China and Russia at the turn of the century. China's island of Taiwan (1895) and footholds on the mainland were the rewards of victory. Japan's empire was taking shape. Korea was annexed in

1910. Japan's territorial ambitions were fervently pursued in China, and by the mid-1930s, Japan dominated a vast empire (Figure 1–13).

Japan's ambition was to establish a **Greater East Asia Co-Prosperity Sphere (GEACPS)**, incorporating Asians minus the European imperialists. As they defeated both indigenous and colonial forces, they propagated the slogan "Asia for the Asians." Their grandiose plans were foiled when they drew the United States into the Pacific theater of war by bombing Pearl Harbor, Hawaii, on December 7, 1941. By 1942, Japan controlled all of the northern

Pacific west of the international date line. Japan then dominated 100 million people!

After its defeat in 1945, Japan was relieved of all its colonies but the GEACPS had left a legacy: Japan had been the vanguard of Asian nationalism. "Asia for the Asians" became a cornerstone of independence struggles. But new socioeconomic-political theories were on the horizon to shape the entire process.

Independence and Nationalism

European imperialism began to crumble under the weight of independence movements before and after the World War. After World War II (1939–1945), it crashed. During the process of **decolonization**, independence was both negotiated and fought for. Countless lives were sacrificed. France, which was the most resistant to giving up its colonies, lost 95,000 persons between 1946 and 1954, trying to keep its hold on Indochina. However, this significant loss was overshadowed by the deaths of well over half a million Vietnamese.

Asian independence leaders, schooled in the ideals of communism or democracy, were determined to build their own independent states characterized by dignity, equality, democracy, and self-determination. China's Sun Yat-Sen (Sun Zhong-shian) (1866–1925) was regarded as the first great Asian nationalist. While China as a whole was not a colony, some coastal areas were colonized, and foreign powers strongly influenced its government. The name Sun Yat-sen, along with others such as Ho Chi-Minh (1890–1969) in Vietnam and Sukarno (1901–1970) in Indonesia, soon became familiar to newsreaders and listeners. Such leaders instilled a new sense of national consciousness among their respective peoples.

Marxism and Leninism were attractive to some independence leaders because they provided explanations for what was wrong with a world under imperialism and capitalism, told how to get rid of the old order, and offered a model to put in its place. Marx based his theories on the plight of an urbanizing and industrializing Europe. Lenin recognized the potential power of both urban workers and the rural peasantry. Asians witnessed the success of the Russian Revolution of 1917, when the imperialistic Tsarist regime was overthrown and large estates were confiscated by the state in an attempt to equalize the use of land. Lenin's calls for revolution hit home with anti-imperialist, pro-nationalist groups. It is important to understand that communism was not the driving force behind independence uprisings in Asia. Nationalism and patriotism were.

Social movements of unprecedented scale and devastation brought independence, although it was achieved sporadically (refer back to Figure 1–11). Each country has its own story to be spelled out in later pages. A patchwork quilt of political systems emerged: democratic, communist, militarist. Whatever the end result, as the Chinese leader Mao Zedong (Mao Tse-tung) noted in 1957, "The east wind prevails over the west wind."

Prevalence of east-wind political, economic, and social systems did not bring stability, though. Communism with its massification, lack of incentives, and mismanagement was unable to achieve its ideals. It eventually collapsed in the Soviet Union in 1991. This momentous event, coupled with new economic realities of global proportions, spearheaded profound changes in Asia and the world.

New Global Order

Time-space convergence is an integral part of our changing world; more space can be covered in a shorter time span. In 1850, undersea communication cables were laid around the world. In that same year, an American-built clipper ship, the *Oriental*, made a record journey from Hong Kong to London in 97 days. In 1935, a freighter from Los Angeles, struggling in a winter storm, took 19 days to reach Tokyo. Sea-mail took weeks as well. Today, we can e-mail across the globe in mere seconds.

The number of Internet hosts has multiplied to more than 170 million. World Wide Web sites in Asia are supported and influenced by governmental and/or private organizations. Malaysia has the fastest-growing Internet system in Asia. Where political circumstances bar open communication (e.g., in Tibet or Myanmar), sites are sponsored by overseas communities.

Since 1980, telecommunication traffic has averaged an increase of 20 percent a year. By 1995, the number of e-mails exceeded the number of letters sent! The vast, intertwined, and overlapping channels of communication produced by flows of technology, hardware, and software disseminated by transnational or multinational corporations, organizations, and agencies are called **technoscapes**. Dramatic changes in technology have spurred far-reaching alterations in both national and international political, economic, and social systems. This **restructuring** is manifesting profound changes for people and the environment (Figure 1–14).

As the world's single superpower, the United States currently sits in the middle of an extensive and complex global network of flows of goods and information. The United States now exports more goods to Asia than to Europe. Money is transferred electronically among leading stock exchanges in London, New York, Hong Kong, and Tokyo. German BMWs are sold in the United States, yet they are assembled in Vietnam. Hewlett-Packard's global headquarters for microwave components and hard disk drives is in Penang, Malaysia. Designers Donna Karan and Ralph Lauren source their apparel to Taiwan, South Korea, and Singapore for manufacture. Clothing sold at The Gap, Saks Fifth Avenue, and many other stores are produced in South Asia.

The year 1997 was Canada's "Year of the Asia-Pacific." After the United States and Britain, the leading investors in Canada are Japan, Hong Kong, Australia, Singapore, and South Korea. Canada is not only tied economically with other Commonwealth countries such as India and New Zealand, but also with its Pacific Rim partners. Vancouver, British Columbia, is an important port on the West Coast of North America, particularly for Canadian grain and wood products destined for Asia. It hosted the Asia Pacific Economic Cooperation Conference in 1997.

Australia is also becoming increasingly tied to this Asia-Pacific trade network. A country that shunned immigrants from Asia until the 1970s now has engagement with Asia as a major component of its foreign policy. Half of Australia's immigrants come from Asia. Close to three-quarters of Australia's exports went to Asia in 1998, and it has defense agreements with several Southeast Asian countries. Australia is interested in being perceived geographically as a part of Asia. According to former Prime Minister Paul Keating, "Asia is no longer the 'Far East.' It is the 'near north.'" However, this notion has not readily been accepted in Asian circles.

The new world order engine was fueled by the disintegration of the Soviet Union because this event allowed capitalism to dominate. Communist nations, such as the People's Republic of China and the Socialist Republic of Lao, are experimenting with privatization of ownership and entrepreneurial profit-making. Moreover, as these countries become increasingly linked with global, capitalist enterprises and economic systems, non-participation in these will be virtually impossible. This is the essence of **globalization**.

The new global order has many dimensions. Change in one aspect has a ripple effect throughout all the others. With globalization, a decision made in London, Toronto, Seoul, or Sydney can alter the lives of thousands. The bulk of world business is conducted by **multinational** or **transnational enterprises** controlling financial, manufacturing, and distribution

Figure 1–14

I was on a river trip in Borneo and was amazed at the number of satellite dishes marking river and even interior communities. This scene is from Samarinda, a port on the Mahakam River in Indonesia. Televisions are often shared among houses, and American programming such as MTV is available. Note the stilt houses that allow for tidal changes and rising waters in the wet monsoon. Photograph courtesy of B.A. Weightman.

systems. Conglomerates typically have headquarters in Japan, North America, or Europe. By the end of 2000, over 150 of the world's top 500 transnationals had invested in China. Germany's electronics firm, Siemens, is established in 14 Asian nations from India to the Philippines. In China alone, it has 40 joint-venture projects and 21,000 employees.

Even foreign aid is dispensed more by international organizations than by individual nations. The International Monetary Fund (IMF), headquartered in the United States, is one powerful shaper of economic and other landscapes. Aid agreements are typically tied to the opening up of the recipient's economy to external investment and foreign goods. As you will discover, these aid projects can have dire social and environmental consequences.

But globalization is only part of the picture. A counter movement is **regionalism**. Regional organizations are concerned with such things as defense, cultural, environmental, and economic issues. Although globalism and regionalism appear contradictory, they are integrated phenomena operating at varying spatial scales and settings. Regionalism is a response to loss of self-determination through incorporation into multinational behemoths.

Regional trading blocks, such as the European Union (EU) or the North American Free Trade Association (NAFTA), regulate economic interchange in the interests of reducing foreign competition and to maintain the flexibility of extensive markets. The **Asia Pacific Economic Cooperation (APEC)**, is a similar organization.

APEC was organized in 1989 by then Australian Prime Minister Bob Hawke. This organization relies on *ad hoc* task forces and groups of business people, economists, and government officials to chart and achieve its goals. APEC's main goal is to achieve free trade in the Asia-Pacific region by 2020.

In 2003, APEC's enterprises involved more than a third of the world's population and generated more than half of the world's gross domestic product. The original membership of 12 has increased to 21, including East and Southeast Asian nations along with six non-Asian nations such as Canada, Mexico, the United States, and Australia. In other words, APEC is a Pacific Rim phenomenon.

The **Pacific Rim** is an economic concept of the late twentieth century. It refers to all the countries situated around the Pacific Ocean including those of North and South America; East and Southeast Asia; and Australia and New Zealand. **Asia-Pacific** is another, broader term. These countries' primary trading partners are one another.

While regional organizations set new paradigms for international negotiations, they are frequently criticized as being insensitive to concerns such as air pollution, toxic waste, environmental health, labor exploitation, and cultural survival.

Americanization

Consider the following facts. The cover of the July 4, 1997 issue of *Asiaweek* magazine, published in Hong Kong, shows Ronald McDonald's picture over the Gate of Heavenly Peace in Beijing's Forbidden City. The story is about "McChina." The Marlboro Man is pictured riding across the back cover of the magazine. There are Coca-Cola museums in China, Japan, and the Philippines, and Taiwan boasts the world's largest inflatable Coke bottle. Vietnamese versions of rap music boom in taxis and from music shops in Hanoi and Saigon. In China, Nike ads feature the "Flier Qiao Dan"—Michael Jordan. Once I witnessed a group of elderly women squatting around a TV on the floor of a communal longhouse in a remote region of Indonesian Borneo. They were laughing, pointing, and shaking their heads at the apparent silliness in front of them. It was MTV.

These are all examples of American **popular culture**, which is diffusing via transnationals and media into the remotest corners of Asia (Figure 1–15). What impact does this trend have on lives and landscapes?

In places such as Taiwan, Malaysia, and Thailand, some people think that American popular culture is detrimental to traditional values, especially those concerning youth behavior and familial relationships and responsibilities. In India, cooked-chicken sellers protested against Kentucky Fried Chicken for ruining their businesses. Elsewhere, small, family-run shops are being replaced by franchise chains such as

Figure 1–15

This Kentucky Fried Chicken on the island of Java in Indonesia shows cultural integration with the inclusion of Ayam Goreng—Fried Chicken with specific Indonesian spices. It is delicious. Photograph courtesy of B.A. Weightman.

7-Eleven, and home-made ice cream parlors by such franchises as Baskin-Robbins. Avon, Amway, and other direct marketers are changing long-held relationships between customers and local merchants.

Americanization is controversial. Although it is true that such enterprises provide employment, where do most of the profits go? Televisions are becoming ubiquitous in Asia. Although television can be a way of teaching and promoting public health or family planning, it also increases expectations and promotes consumerism. This may be good for business, but is it good for society in general? How does this medium and American programming such as CNN, MTV, and various comedy, crime, and drama series influence perception and behavior? In South Korea and China, for example, leaders are concerned about rising levels of obesity related to the increased availability of junk food. Malaysians and Taiwanese worry about inappropriate youth behavior. As young people move away from their families to pursue careers, older people worry about whether traditional networks of care will be available for them in their old age.

Some Asian leaders such as Lee Kuan Yew of Singapore and Muhammed Mahathir of Malaysia see popular culture as contributing to secularism and liberalism. Both men pursue modernization and change but are convinced that their own national cultures should transcend such potentially harmful Western elements. They urge their people to "turn east" for guidance.

🌐 Main Point Summary

1. More than half of humanity inhabits South, East, and Southeast Asia. These geographic realms are part of a larger entity called Asia.

2. Asia is characterized by tremendous cultural and physical diversity. Interrelationships between people and environments generate experiential landscapes (i.e., lived worlds or lifeworlds).

3. In the Western world, perceptions of Asia tend to be Eurocentric. Misconceptions of Asia have led to negative stereotyping, off-base broad generalizations, and a sense of superiority in attitudes and behavior.

4. Religion and belief systems pervade the lives of most people in Asia. Hinduism, Buddhism, and Islam have the most adherents, but missionary work is drawing increasing numbers to Christianity.

5. Arab Muslim traders were instrumental in charting the way for subsequent European ventures into the Indian Ocean and South China Sea. These waters were not "discovered" by Europeans, as they were already familiar to the Chinese and other sea traders. However, with God and gold in their minds, Europeans quickly gained footholds in the region.

6. Although an array of products was in demand, spices captured the European imagination in the early years of their entry into the region. As more opportunities arose for economic gain, serious competition ensued.

7. What initially were trading posts became colonial possessions, and what started out as colonies became imperial outposts of grand empires. Mercantilist policies bolstered homeland industries

but depleted resources, impoverished and en-slaved people, and wrought immeasurable and indelible social change on Asian cultures.

8. Two world wars and occupation by Japan al-tered the path of events. Indigenous, foreign-educated individuals emerged with a renewed sense of nationalism and patriotism to lead self-determination and independence movements. After significant bloodshed, the colonial powers withdrew from most of Asia.

9. In the aftermath of decolonization, a variety of political structures were established, with democracy vying with authoritarianism, and the concept of a free-market economy competing with communism for dominance. The collapse of the Soviet Union in 1991 ushered in an era of significant change.

10. The new world order comprises globalization on one hand and regionalization on another. With increasing power of multinational con-glomerates, individual nations band together to protect their own regional interests.

11. Americanization via business and media, while providing economic and other opportunities, is having a negative effect on health, behavior, and traditional cultural and social relationships.

⊕ Bibliography

Blaut, J. 1994. *The Colonizer's Model of the World: Geographical Diffusionism and Eurocentric History.* New York: Guilford.

Borthwick, Mark. 1992. *Pacific Century: The Emergence of Modern Pacific Asia.* Boulder: Westview.

Brierley, Joanna Hall. 1994. *Spices: The Story of Indonesia's Spice Trade.* Kuala Lumpur: Oxford.

Chapman, G. P., and K. M. Baker eds. 1992. *The Changing Geography of Asia.* New York: Routledge.

Cressey, George B. 1963. *Asia's Lands and Peoples.* 3rd edition. New York: McGraw-Hill.

De Blij, Harm J., and Peter O. Muller. 2004. *Geography: Realms, Regions and Concepts.* 11th edition. New York: Wiley.

Drakakis-Smith, David. 1993. "That Was Then, This Is Now: Forty Years of Social and Economic Change in the Tropical Third World." *Singapore Journal of Tropical Geography* 14: 81–102.

Georgacas, Demetrius J. 1969. "The Name Asia for the Continent; Its History and Origin." *Names* 17: 32–37.

Ginsburg, Norton. 1988. "Geography and the Pacific Century." *Asian Geographer* 7:1–11.

Hodder, R. 1992. *The West Pacific Rim: An Introduction.* London: Belhaven.

Jones, E., L. Frost, and C. White. 1993. *Coming Full Circle: An Economic History of the Pacific Rim.* Boulder: Westview.

Kitagawa, Joseph M. ed. 1989. *The Religious Traditions of Asia.* New York: Macmillan.

R. J. Johnston, Peter J. Taylor, and Michael J. Watts, eds. 2004. *Geographies of Global Change.* Oxford: Blackwell.

Lent, J. A. 1995. *Asian Popular Culture.* Boulder: Westview.

Levathes, Louise. 1994. *When China Ruled the Seas.* New York: Oxford.

Lewis, Martin W., and Karen E. Wigen. 1997. *The Myth of Continents: A Critique of Metageography.* Berkeley: University of California.

Michalak, Wieslaw, and Richard Gibb. 1997. "Trading Blocs and Multilateralism in the World Economy." *Annals of the Association of American Geographers* 87: 264–279.

Nakamura, Hajime. (1964) 1997. *Ways of Thinking of Eastern Peoples: India, China, Tibet, Japan.* London: Kegan Paul.

O'Meara, Molly. 1997. "Riding the Dragon." *World Watch.* 10: 8–18.

Risso, Patricia. 1995. *Merchants and Faith: Muslim Commerce and Culture in the Pacific Ocean.* Boulder: Westview.

Ryan, Paul R. 1989–1990. "Pacific Century Dead Ahead!" *Oceanus* 32: 5–12.

Spencer, Joseph E., and William L. Thomas. 1971. *Asia, East by South: A Cultural Geography.* 2nd edition. New York: Wiley.

Spencer, Joseph E. 1973. *Oriental Asia: Themes Toward a Geography.* Foundations of World Regional Geography Series. New York: Prentice-Hall.

Steadman, John M. 1969. *The Myth of Asia.* New York: Simon and Schuster.

Unwin, Tim. ed. 1996. *Atlas of World Development.* New York: Wiley.

Watanabe, Toshio. 1992. *Asia's Growth and Agony.* Honolulu: East West Center.

Chapter 2

✦

Environments and People

"Those who would take over the earth/And shape it to their will/ Never, I notice, succeed."

TAO TE CHING (C. 3000 B.C.)

Major Topographical Features

South, East, and Southeast Asia exhibit a grand and complex display of mountains, plateaus, drainage basins, and river valleys. All of these have been modified and have influenced people's actions and interactions since the beginning of human time.

The foundation for Asia's **topography** (landforms) is an extensive zone of ancient rocks that, in early geologic time, were folded, faulted, and intruded on countless occasions. Ongoing **hydrological** (water) and **aeolian** (wind) **erosion** and **deposition** have combined to generate Asia's physical geography as we know it today (Figure 2–1). Unlike North America and Europe, Asia was not covered with massive ice sheets, with the exception of high mountain systems such as the Himalayas. In fact, **glaciation** continues to play a critical role in sculpting mountainous regions.

The most imposing topographic features are the Himalayan ranges and adjacent Tibetan Plateau. The Greater or highest Himalayas extend 1,500 miles (2,415 km) from west to east with an impressive number of the world's highest peaks (check your atlas for locations). No fewer than fifteen peaks top 25,000 feet (7,000 m). Annapurna at 26,535 feet (8,090 m), Dhaulagiri at 26,689 feet (8,137 m), Kanchenjunga at 28,208 feet (8,598 m), and Everest, the highest at 29,028 feet (8,848 m), are the best

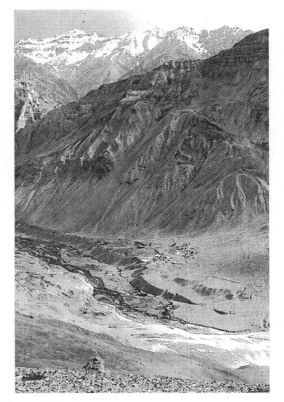

Figure 2–1

The Spiti Valley, in the Indian Himalayas, clearly illustrates the impact of water, wind, and ice in physical landscape formation. A small community with its barley fields clings to the base of the alluvial fan. The Spiti River is a tributary of the Indus River system. Photograph courtesy of B.A. Weightman.

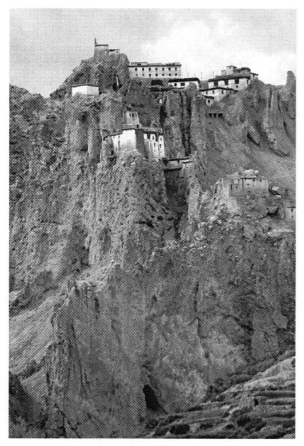

Figure 2-2

Many Tibetans live in the Himalayas of India's Himachal Pradesh. This gompa (monastery), at almost 13,000 feet (3,890 m), was built in the sixteenth century. Photograph courtesy of B.A. Weightman.

known mountains. Imagine a fractured rock wall around 4 miles (6 km) high and 200 miles (322 km) across. This is the majesty of the Himalayas.

Although cut by numerous passes, the Himalayan ranges have posed a formidable barrier between Indian and Chinese realms demarcating racial, linguistic, and other cultural differences. The mountains themselves form a region of **cultural convergence** where lifeways and belief systems blend in splendid landscapes (Figure 2–2).

The history of the Himalayas began 200 million years ago with the process of **continental drift** when the landmass known as Pangaea broke into pieces. At that time, Gondwanaland (Africa) and what we know now as India proceeded toward another, larger landmass called Laurasia. When the collision occurred 45 million years ago, mountains emerged in a colossal upheaval. The Tethys Sea vanished, consumed in the process. When trekking near the Tibetan border, I was delighted to find two **ammonites**: flat, spiral fossil shells, evidence of life in the ancient watery world of the Tethys Sea.

The Himalayan ranges are a product of **plate tectonics**. The theory of plate tectonics postulates that the upper portion of the lithosphere (the outer part of the solid earth composed of rock) comprises a patchwork of rigid plates. These are rooted in the underlying, more plastic upper layer of the earth's mantle. Movement along the fracture zones between the plates is responsible for a legion of topographic features and seismic phenomena.

Where plates diverge or move apart, molten material from the earth's core wells up into the space. Where plates converge or come together, spectacular mountains, pyroclastic events, and deep ocean trenches appear.

Did you know that Asia is the world's most disaster-prone region?

The Pacific Ring of Fire

Half of all major natural disasters are in the Asia–Pacific region. Because of the abundance and intensity of seismic activity, the region is often referred to as the *Pacific Ring of Fire* (Figure 2–3).

Volcanic eruptions and earthquakes have wreaked havoc on humanity for millennia. The most powerful eruption in recorded history occurred on the Indonesian island of Sumbawa in 1815 when Mount Tambora (9,348 feet, 2,850 m) blew up 100 billion tons of volcanic rock and debris. The death toll was 92,000.

Figure 2-3

The Pacific Ring of Fire. Do you live in one of these major fault zones? How would you prepare for a seismic event?

China's Tangshan earthquake in 1976 killed more than 240,000 people, and a 1993 quake in western India's Maharashtra state took 12,000 lives. The Philippines' Mount Pinatubo eruption in 1991 and Japan's Kobe quake in 1995 registered a relatively low death toll but caused considerable damage. In 2001, a 6.9 quake in Gujarat, India, killed more than 20,000 people.

Tsunamis, or seismic sea waves, can be even more damaging than the seismic event. The speed of a tsunami is proportional to the depth of the water; the deeper the water, the faster the wave. As the water approaches a shallow area, the fast-moving wave is slowed, causing the water to pile up on itself. Tsunamis have caused the deaths of hundreds of thousands of coastal and island people. For example, in 1976, a tsunami drowned 8,000 people in the Philippines' Moro Bay.

On the morning of December 26, 2004, a massive 9.0 earthquake struck off the western coast of Sumatra, Indonesia, setting off ocean shock waves that were felt more than 3,000 miles away on the coast of East Africa. The earthquake occurred when stress built up between the Burma and Indian plates, causing several hundred miles of the Burma plate to snap. This forced a massive displacement of water in the Indian Ocean. The resulting waves spread in all directions, moving as fast as 500 miles per hour. In the deep ocean, the waves may be imperceptible, but they slow down and gain height as they hit shallower water near shore. The worst-hit places were Aceh province on the island of Sumatra where more than 80,000 were killed, and Sri Lanka where more than 29,000 died. Thailand lost at least 5,000 people. In addition, thousands of people disappeared, their bodies swept out to sea. The tsunami also devastated infrastructures and ruined regional economies.

Although the outpouring of international aid was unprecedented for this disaster, the most affected areas such as Aceh province are not expected to recover for over a decade. And even as the recovery efforts proceeded, another earthquake hit on March 28, 2005, also of great magnitude.

In studying the December earthquake, scientists revised its scale from magnitude 9 to 9.3. Although this does not seem to be a large increase, 9.3 represents a $2\frac{1}{2}$ increase in scale. The December earthquake is now the second largest ever recorded in over 100 years of record keeping. The seafloor rupture of the earthquake was 750 miles, also the largest ever recorded, and three times larger than originally thought.

The Pacific Ring of Fire also includes the world's deepest ocean trench: the Mariana Trench, which descends 36,198 feet (11,033 m) below sea level. Another is the Indian Ocean's Java Trench at 23,376 feet (7,125 m) below sea level.

NORTH OF THE HIMALAYAS

If you were to cross the Himalayas northward you would be standing at an elevation of approximately 15,000 feet (4,500 m) at the edge of the stupendous Tibetan Plateau. You would be gasping for air and gaping in awe at more than 400,000 square miles (1,036,001 km^2) of rugged basins and ranges. Many of the rock-strewn basins contain permanent lakes. Mountain passes permit relatively easy access from one basin to another, but many of these passes are 17,000 feet (5,100 m) and higher. Here is where South Asia's four great rivers originate: the Sutlej, Indus, Ganges, and Brahmaputra (Tsang Po). China's major rivers, the Chang Jiang (Central River or Yangzi) and the Huang He (Yellow River) also originate here. The northern perimeter of the Tibetan Plateau is bounded by other massive ranges such as the Kunlun and Karakoram. The 28,250 feet (8,611 m) peak K2 rises in the Karakoram.

Northward again are the arid basins and mountains of northwestern China. The Tarim Basin is largest in expanse and the Turpan (Turfan) Basin lowest in elevation. The Heavenly Mountains—the Tien Shan—stand as one of this region's major divides. Although the Tarim and Turpan and related basins are not important in terms of population density, they are of critical importance economically and politically.

EAST AND SOUTHEAST OF THE TIEN SHAN

Eastward is a mosaic of rugged mountains and plateaus, along with plains and more basins. The Qinling Range, Loess Plateau, North China Plain, and Sichuan Basin are only four of many significant elements of China's physical landscape. These regions are dominated by the mighty Huang He and the Chang Jiang or Yangzi. Millions of people make a livelihood in these and other river valleys, basins, and plains. Even further east are the craggy Korean peninsula and Japanese islands.

SOUTH AND SOUTHEAST OF THE HIMALAYAS

South of the Himalayas, the North Indian or Indo-Gangetic Plain sweeps from Pakistan, across India, to Bangladesh. This subsiding trough of 300,000 square miles (777,001 km^2) is filled with fertile **alluvial soils** deposited by the river systems of the Indus, Ganges, and Brahmaputra. These water-deposited sediments, in concert with extensive river networks and relatively flat land, have helped make the Indo-Gangetic Plain one of the most densely populated regions on earth.

South of the Indo-Gangetic Plain rises India's dissected Deccan Plateau. Covering most of peninsular India, the Deccan is bordered by narrow coastal plains along the western Malabar and eastern Coromandel coasts. The tear-shaped island of Sri Lanka lies about 20 miles (32 km) to the southeast of the Coromandel coast.

Mainland Southeast Asia is a washboard of mountains and **fluvial** (river-related) landforms. Mountain ranges, with elevations as high as 18,000 feet (5,400 m), are generally aligned in a north-to-south direction. One exception is the S-shaped Annamite Cordillera that follows practically the entire coast of Vietnam from southern China. Mountain chains are arranged closely in the north with virtual slots of lowland in between. Further south, the slots widen into V-shaped valleys and eventually into open plains, basins, and deltas.

Mountains have created cultural barriers for the inhabitants of interstitial, riverine zones and they have encouraged ethno-linguistic diversity as each of hundreds of different culture groups maintained its own way of life in relative isolation from others for centuries.

Rivers are of paramount importance in mainland Southeast Asia. The Ayeyarwady (Irrawaddy), Chao Praya, Mekong, and Red River valleys have their headwaters on the Tibetan Plateau. So tortuous is the surrounding mountain terrain that the source of the Mekong was not found until 1994. River valleys foster support of millions of inhabitants and provide excellent sites for the growth and development of major cities, including most capitals.

Peninsular Southeast Asia extends 900 miles (1,449 km) from the Gulf of Thailand to the island of Singapore. The narrowest portion, the Isthmus of Kra, is roughly 40 miles (64 km) across. It widens to the south to form the Malay Peninsula.

ISLAND REALMS

South, East, and Southeast Asia are ringed by islands. Some, such as Sri Lanka and Singapore, are politically independent states. Others are part of island chains called **archipelagoes.** Japan, the Philippines, and Indonesia are all archipelagoes. Each comprises one or more large islands and hundreds, or even thousands, of smaller ones. Japan is dominated by the large island of Honshu, but there are three other sizable islands (Hokkaido, Kyushu, and Shikoku) and hundreds of lesser ones. The Philippines incorporates more than 7,000 islands along with the larger islands of Luzon and Mindanao. However, the biggest is not necessarily the most important. For instance, the Indonesian archipelago totals more than 17,000 islands, dominated in size by Sumatra, Sulawesi (Celebes), and Kalimantan (Borneo). But Indonesians and their political and economic centers are most heavily concentrated on the much smaller island of Java.

New Guinea, the world's second largest island (after Greenland), is politically divided between Papua New Guinea in the east and West Papua in the west. West Papua is part of Indonesia. This island is the focus of one of Indonesia's many political and environmental controversies (Chapter 15).

There are more island groups in the Indian Ocean. Strung between Myanmar and Sumatra is a 600-mile (966 km) chain of more than 200 islands. These are known as the Andaman and Nicobar Islands and belong to India. The majority are merely exposed tops of mountains protruding from the ocean floor. Southwest of India lie the Maldives, 115 square miles (298 km^2) of islands, banks, and reefs. Only about 10 percent of the roughly 2,000 of these are inhabited.

LIMESTONE AND CORAL ENVIRONMENTS

Chemically weathered limestone is the essence of **karst topography.** Fractured limestone formations, with at least 80 percent calcium carbonate, are subject to solution processes. Subsurface drainage

Figure 2-4
This is an example of tower karst topography in Guilin, China.
Photograph courtesy of B.A. Weightman.

channels are formed as water seeps through joints in the otherwise impermeable limestone. Chemical weathering takes place, resulting in pitted and lumpy surface forms and labyrinths of caves below (Figure 2–4). Karst regions in Asia are found in northwestern India, southern China, and in various parts of Southeast Asia. These formations are popular tourist attractions, drawing sightseers, photographers, rock climbers, and spelunkers from around the world.

Are you aware that the world's best dive sites are in the Asia-Pacific region?

Coral Reefs

Thriving in warm, tropical waters, coralline formations are abundant in Asia. It is estimated that more than half of the world's **coral reefs** are in the Indian Ocean, occurring along continental shelves and around islands.

Biological process is responsible for the formation of coral reefs. A coral is a small marine animal related to other marine invertebrates such as anemones and jellyfish. Corals excrete calcium carbonate from the lower half of their polyp bodies, forming a hard, calcified skeleton. Reefs are formed by colonies of corals that live in a symbiotic relationship with algae. Over centuries, skeletons accumulate on raised submarine features. This biological formation makes up a coral reef.

Coral reefs protect coasts from wave erosion and contribute to the process of sandy beach formation. They provide breeding grounds and nutrients for subsistence and commercial fish, and they are habitats for other food sources like mollusks and crustaceans. Also, they are invaluable tourism assets.

Coral reefs are the marine equivalent of rain forests, the richest ecosystem on earth. Indonesia is home to some of the world's most spectacular reefs. Their biological and economic importance cannot be underestimated. Indonesian waters contain more than 5,000 species of fish, 30 species of sea snakes, and 7 varieties of turtles, among countless other creatures. However, reef populations have been significantly reduced.

Look at Sulawesi, where scientists believe that tropical marine life began. Each year, thousands of pounds of pear oysters, mother-of-pearl shells, and *trepang*—sea cucumbers—are exported. Shells such as the helmet conch and giant clams are sold as well.

Failing to recognize the intrinsic value of coral reefs, people have destroyed and endangered them in a variety of ways. Reefs have been mined for construction materials such as lime. Corals and tropical fish have been harvested for the aquarium trade. Dredging for sand for cement has damaged them and disrupted the marine community. Land reclamation projects and fish ponds have eliminated them. Reefs are ruined by heavy oil spills and the dumping of industrial waste. Speedboats, mechanical fishing devices, and fish blasting have been extremely detrimental. Fish blasters use explosives to kill large numbers of fish easily.

Since the 1970s, many nations have recognized the significance as well as the severity of loss of their coral reefs and have taken steps to protect them. Action has also been spurred by the potential for tourism dollars generated by dive sites and ancillary facilities. Nevertheless, abuses continue and now include unlimited and uncontrolled tourist activities. Unfortunately, coral reefs continue to be threatened around the world. As they are part of a larger marine ecosystem, destruction of reef formations is detrimental to many other life forms as well.

Physical geography can also affect political geography. We know that rivers, mountain ridges, and other physical features are often used to demarcate political boundaries. Although some are disputed, most of these natural boundaries are stable. But bodies of water are problematic to define, presenting many opportunities for territorial disagreements.

Ocean Disputes and Conflicts

Oceans and seas are problematic because it is difficult to demarcate, in any practical sense, who owns which territory or mineral rights, or even who owns fish. In 1982, the United Nations issued The Law of the Sea Treaty, which allows each coastal state to claim a 200-nautical-mile **Exclusive Economic Zone (EEZ)** in which it controls resource rights. Countries are authorized to sell or lease rights in the EEZ to multinational corporations and can control fishing and pollution. However, problems arise when fishers stray over boundaries or pollution comes from another country's EEZ.

Designation of EEZs has not solved all problems. The 200-nautical-mile EEZ does not work where distance between states is less than 400 nautical miles. To solve this dilemma, a median line is used. This line divides the waters equidistantly between the opposing countries. For example, Malaysia and Indonesia share the Straits of Malacca, the width of which ranges from 8 to 126 nautical miles. The EEZ of each country ranges from 4 to 63 nautical miles. Nevertheless, disputes are common in such narrow passages partly because of the multitude of islands found in these waterways.

The case of the **Timor Gap** provides an example. In 1989, Australia and Indonesia struck a deal to explore and exploit the seabed between them. Here is thought to be one of the world's largest reserves of hydrocarbons. This perceived breakthrough followed years of controversy. In 1972, Australia and Indonesia agreed to formalize their sea boundaries along the deep Timor Trough. But East Timor was still controlled by Portugal. Both Australia and Portugal made conflicting claims to the seabed, issuing overlapping exploration licenses to oil companies. This disputed section became known as the Timor Gap. After East Timor became independent in 2002, it signed a deal with Australia allowing East Timor 90 percent of royalties from the shared zone of exploration.

Another point of controversy is the amount of territory that islands can claim. A small island can claim rights to fish and minerals such as manganese and oil over a disproportionate sea area. This makes possession of islands, no matter how small, critical. Jurisdiction over islands fuels tensions and conflicts in the Indian and Pacific oceans.

The Law of the Sea Treaty also guarantees ships *innocent passage* while passing through the waters of one country to the next. Passages such as the Straits

Figure 2–5

Conflicts at sea. Conflicts arise over ocean territories such as straits and islands, and over rights to resources such as fish and minerals.

of Malacca are vital to world shipping. To close these narrow waterways would be a relatively simple task in a time of conflict.

There are those who proclaim oceans and seas to be humanity's *common heritage,* a concept introduced by the Dutch as *mare liberum* in the sixteenth century. Unfortunately, this ideal is impractical in a commercial world, and the "scramble for the seas" has intensified (Figure 2–5).

Monsoon Asia

South, East, and Southeast Asia have been termed **Monsoon Asia** by geographers and other scientists. This is because these realms are affected by the **monsoon**—intense, seasonally shifting wind systems that involve an annual cycle of wet and dry seasons. Beyond equatorial regions, summers are wet and winters are dry (Figure 2–6).

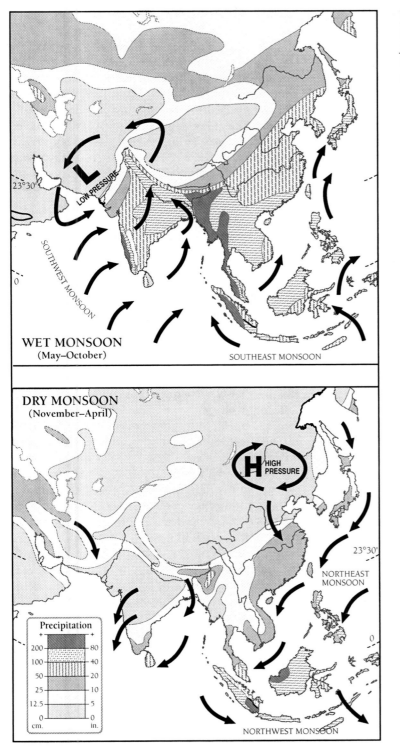

WET MONSOON
(May–October)

SOUTHEAST MONSOON

DRY MONSOON
(November–April)

NORTHEAST
MONSOON

NORTHWEST MONSOON

Precipitation

200	80
100	40
50	20
25	10
12.5	5
0	0
cm.	in.

Figure 2–6

Monsoon seasons. Remember that winds blow from high to low (pressure) and are named by the direction from which they came. All of Asia is influenced by the monsoon.

Equatorial regions experience year-round rainfall, although there are "wet" and "less wet" periods. This depends on the position of the sun, prevailing winds at the time, and topography. As wind systems shift north and south, different islands and even parts of islands receive more or less precipitation. Examine both January and July wind and precipitation patterns on a more detailed map in an atlas. This information is critical when planning a trip. You probably do not want to arrive in Bali in the wet period.

Another feature of equatorial regions is **convectional rainfall**. The combination of solar energy and abundant moisture availability in oceans and biotic communities causes rapid evaporation. Warm, tropical air rises, cools, and condenses. The inability of cool air to hold as much moisture as warm air produces rain. This convectional rainfall occurs nearly every day in the hotter, wetter months. In the drier months when evaporation rates are lower, several days might pass without a deluge.

Prior to rain, the humidity builds to an intolerable level. There is so much moisture in the air that 78°F (26°C) feels like 100°F (38°C). As the day wears on, storm clouds fill the sky, and in late afternoon, torrential rains wash the landscape and relieve the humidity. This is one reason why people work their farms in the early morning. Midday is a time to seek relief from the sun's intensity.

Latitudinal position is also significant, affecting both moisture availability and temperatures. In general, moisture decreases and temperature ranges increase inland and away from the equator. For example, average yearly precipitation at Guangzhou, on China's southern coast, is about 27 inches (686 mm). At Hohhot, capital of Inner Mongolia in China's interior, it is 17 inches (432 mm). Guangzhou's yearly temperature range is 56°F to 83°F (13°C to 28°C). Hohhot's is 8°F to 71°F (−13°C to 22°C).

Both rainfall and temperatures are critical criteria for agriculture because they determine the need for irrigation and the length of **growing season**. A growing season is the number of consecutive, frost-free days in a year.

What causes the monsoon seasons? The size of the Asian landmass or **continentality**, combined with its proximity to the Indian Ocean, are keys to explaining variations in air pressure and related upper-air circulation.

The interior of Asia is far removed from modifying maritime influences. Therefore, in winter, it is dominated by a high-pressure anticyclone while the central area of the Indian Ocean is dominated by the equatorial low-pressure trough known as the **Intertropical Convergence Zone (ITCZ)**. Winds blow from zones of high-pressure to zones of relatively lower pressure. Consequently, cold, dry winds blow from the continental interior to the periphery, dessicating (drying) the landscape in the process. Temperatures average between 60° and 68°F (16° and 20°C) at lower elevations.

Cold increases with elevation. The normal lapse rate is 3.5°F/1,000 ft. (6.4°C/1,000 m). The impact of cold is worsened with the **wind-chill factor** whereby body heat is lost more rapidly with stronger winds. Further, because of diminished atmospheric density at higher altitudes, the air's ability to absorb and radiate heat is reduced. Exposure to the sun's ultra-violet energy is intensified. While high elevations can produce oxygen deprivation and sunburn, exposure to extreme cold can induce *hypothermia*— frostbite with subsequent loss of body parts, or freezing to death.

Extremely cold temperatures, while found in the mountains and near the poles, can be experienced in lowland areas as well. One January, I was in a New Delhi train station. The platforms were overflowing with hundreds of homeless individuals and families huddled under blankets seeking escape from the bitter cold. A newspaper reported that 17 people had frozen to death in the streets overnight.

With desiccating winds and the coming of summer, the period of March to June becomes very hot. The winds begin to shift, and people pray and prepare for the coming of the wet monsoon. Ideally, the wet season occurs between June and September, but one sure thing about the monsoon is its unpredictability. If it is early, seeds can be washed away; if it is late, crops can wither and die. This explains the many ceremonies and festivals at this time of year to propitiate various gods and spirits associated with these essential rains.

At the time of the wet monsoon, continentality causes a low-pressure zone in the interior. The ITCZ shifts northward to the Tropic of Cancer. Solar radiation heats the landmass, leaving the surrounding waters relatively cooler. A subtropical high-pressure cell emerges over the Indian Ocean. As hot air sweeps outward across the ocean's surface, it quickly evaporates water. This huge, thunderous, sodden air mass moves into the lower-pressure ICTZ, bringing voluminous rains to the area south of the Himalayas. Precipitation is highest in coastal areas and on the windward slopes of mountains. As air masses move inland, available moisture diminishes and less precipitation is experienced.

Mountains and other areas of raised relief are highly influential as **topographic barriers.** The Himalayas, for example, provide a physical barrier to incoming air masses, channeling them westward across the Indo-Gangetic Plain. As moisture-loaded air meets this topographic barrier, it is forced to rise, cool, and condense, providing **orographic precipitation.** As a result, the windward slopes of the Himalayas receive abundant rainfall. The subsequently dry air masses flow over the crest and down the leeward slope, warming and drying the landscape in its path. Consequently, protected leeward slopes, or **rainshadow** areas, are comparatively dry.

Cherrapunji, India, situated at 4,309 feet (1,313 m) in the foothills of the Himalayas (just north of Bangladesh), is the record holder for most precipitation in a single year. Toronto, Canada, averages 30.1 inches (76 cm); Atlanta, Georgia, 50.4 inches (128 cm); and Singapore 89.3 inches (227 cm). In contrast, Cherrapunji averages 37.5 feet (1,143 cm) of precipitation a year. Such torrential rains, sweeping against the mountains from the Bay of Bengal, have caused the deaths of hundreds of thousands of people.

Another messenger of destruction is the **tropical cyclone,** termed a **typhoon** in the Pacific. Tropical cyclones typically develop between the latitudes of 8°; and 15°; when sea temperatures rise above 80°F (28°C). Beginning as a weak, low-pressure cell, the typhoon funnels as a deep, circular low. Moving with prevailing wind systems, it intensifies further and delivers violent storms. Depending on the location, typhoons generally strike between May and September.

In 1997, Typhoon Tina struck the coast of southern Vietnam, destroying everything in its path. Tina then moved on to wreak havoc in parts of Thailand and Bangladesh, where it degenerated into a tropical storm. In Vietnam, nearly 3,000 boats were sunk or swamped, and an estimated 4,000 people, mostly fishermen, perished. Tropical cyclones also occur in the Bay of Bengal causing major flooding in Bangladesh. Each year, tens of thousands of people are affected.

REGIONAL CLIMATIC PATTERNS

Although the monsoon affects the entire Asian region, smaller regions have their own climate characteristics. If you are interested in more detail regarding temperature ranges and precipitation amounts, be sure to consult your atlas. I will simplify here.

Climates in South, East, and Southeast Asia can be generalized into five categories: tropical or equatorial; arid; humid subtropical; humid continental; and highland (Figure 2–7). To understand this pattern, you can ask the following questions. How near or far from the equator is a place? Does it experience continental extremes or moderating maritime influences? What is the elevation? Is the place influenced by the presence of a topographic barrier? Answering these questions will help you understand regional similarities and differences in climate.

CLIMATE AND PEOPLE

Quality of life in the various climatic categories depends very much on what coping mechanisms are available. Flood control works, irrigation systems, water storage facilities, healthy work animals or functional machinery, electricity, sealed (paved) roads, insurance against crop loss, and emergency food supplies are only a few.

Millions of people in Asia have minimal means to cope with disaster. An erratic monsoon, a drought, a tropical cyclone, or a lengthy cold spell can mean ruination and total despair for those living on the edge of survival. Here is an example. In 1998, rains and pests destroyed the crops in a small area in southeastern India. Farmers unknowingly had purchased illegally diluted fertilizers and pesticides. Deeply in debt and sensing nothing but hopelessness, hundreds of farmers ingested their remaining chemicals. They believed their deaths would bring a government stipend to their wives, thereby saving their land from seizure by the moneylenders. Government payment

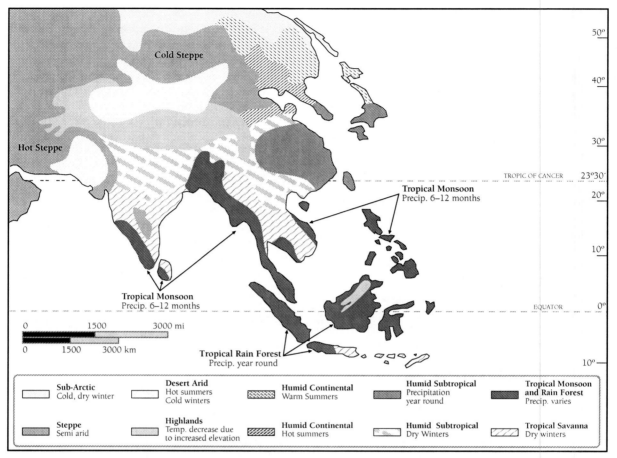

Figure 2–7

Climatic zones. Climate is a year-round phenomenon that varies with latitude, elevation, prevailing winds, and distance from moderating effects of water.

was sporadic and insufficient, and many women and their children are now landless and destitute. Those of us living in a high-tech world probably find it difficult to grasp this level of desperation. But for millions in Asia and elsewhere, desperation is a fact of life.

Natural Vegetation

Natural vegetation implies undisturbed, mature plant communities. Because of thousands of years of occupation, an undisturbed biome is a rarity. However, examination of natural, model biomes helps us to

understand original vegetation and reveals the measure of human impacts. The following biomes are coincident with the climate zones of Figure 2–7.

TROPICAL RAIN FORESTS

Tropical rain forests are most prominent in Southeast Asia where equatorial conditions provide plentiful year-round moisture and consistently warm temperatures. Here is the earth's most diverse biome. Of 250,000 flowering plant species in the world, 170,000 are tropical. Of these, 40,000 are in Asia, with 25,000 in Malaysia and Indonesia. Moreover, this region has tremendous **species diversity**. Western Europe, west of Russia and north of the Alps, has 50 indigenous species. Eastern North America has 171. In contrast, a single 123-acre research plot on the Malay peninsula exhibited 830 indigenous species.

Think of rain forests as having layers within which every plant competes for sunlight and moisture. Each layer has a different microclimate, with the upper layer having the most light, heat, and wind. The forest's biomass is in the **canopy**, high above the forest floor. Leaves appear on tree trunks at approximately 15 feet (5 m) above ground, and this begins the canopy that reaches upward to 100 feet (30 m) or more. Poking out of the canopy are the tallest trees called **emergents**. **Dipterocarps**, with their winged seeds, dominate the upper canopy. These are trees of grand stature, housing a wealth of flora and fauna. It is said that dipterocarps are the apex of evolution. The middle level of the canopy is densest. It has a spectacular array of mammal, bird, and other species. **Lianas** (woody climbers and vines) loop around and between trees. These are the bane of foresters because they can so entangle a tree that even when completely cut, it remains dangling, tied to adjacent trees. Some lianas are 8 inches (20 cm) in diameter. Orchids, ferns, and other **epiphytes** inhabit this zone as well. Epiphytes are physically, but not nutritionally, supported by other plant life.

Beneath this exceptionally rich environment is a more humid, darker understory and the forest floor. Ferns, bamboo, shrubs, and herbs shade a relatively open floor that is strewn with decomposing plant and animal litter and fungi. This rather thin layer provides some nutrients to vegetation. However, most tropical soils are deceivingly infertile, leached of their nutrients by copious rainfall and high temperatures. At lower layers, aluminum and iron oxides form rock-like layers called **laterite**. Since antiquity, laterite has been cut and used as building blocks for temples and other structures. Even giant trees have shallow roots to tap the litter layer for food. Some of these have supportive bases called **buttresses** (Figure 2–8).

Where light penetrates, dense growth can be found at floor level. Dense vegetation, or **jungle**, exists along rivers, or in areas that have been cleared

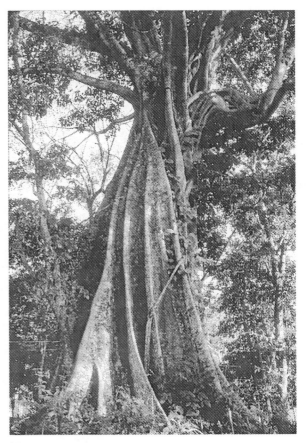

Figure 2–8

Here is a dramatic example of a buttressed tree in the tropical rain forest. Since soil fertility is at the surface, trees develop sprawling surface rather than tap-root systems. Some large species of trees develop blade-like supports like these. Photograph courtesy of B.A. Weightman.

and abandoned. These seemingly impenetrable areas belie the more open rainforest in their vicinity.

TROPICAL MONSOON FORESTS

Found primarily in mainland Southeast Asia and the Indian subcontinent, the monsoon forest is a transitional biome between tropical rain forests and more arid plant realms. Unlike tropical rain forests, monsoon forests must adapt to distinct wet and dry seasons. Many trees are **semi-deciduous**; they lose part of their foliage in the dry months (Figure 2–9).

Figure 2-9
This monsoon forest lines the Rapti River in southern Nepal. It is winter, dry season, and many of the trees have lost all or at least some of their leaves to compensate for the water shortage. Most of these trees are semi-deciduous. The Rapti is part of the Ganges River system and consequently is a sacred river.
Photograph courtesy of B.A. Weightman.

Tropical monsoon forests lack a continuous canopy. Their trees average only 50 feet (15 m) in height. Often cut for firewood, they make poor lumber. Teak, a highly valued hardwood for carpentry, is an exception. Trees also produce waxes, gums, resins, and lacquer. Monsoon forests possess a rich collection of wildlife.

Do your shoes have rubber soles? Do you use a palm oil-based soap? Have you used lacquer or latex paint? Have you used bamboo skewers or wooden chopsticks? Is there a coconut fiber mat at your front door? How much of the tropical forest do you use?

Riches of the Tropical Forest

Tropical forests are rife with desirable products such as fruit, tubers, plants, and other food crops; rattans and fibers; oils and resins, in addition to wood. Traditional ways of gathering forest products have little impact compared to modern logging activity. Logging is a prime reason why more than 40 percent of Asia's tropical forests have been destroyed. Foresters seek certain types of trees to cut for local use and export. I will describe two of these here: teak (*tectona grandis*) and meranti (*shorea spp*).

Teak is regarded as a fine, medium-weight hardwood. It is noted for its strength, stability, and durability. Termite resistant, it is valued for constructing housing and furniture. It used to be in great demand for ships and is still used in the construction of yachts. Teak is native to India, Vietnam, Burma, and Thailand, but is grown commercially on Java and in Sri Lanka. Logging for export has eliminated teak forests from Thailand.

There are several species of meranti, called *seraya* in Malaysia, an equatorial forest wood. These are giant, buttressed trees achieving heights of over 200 feet (61 m). Meranti are among the most important export woods of Asia. Logs are sent to Australia, Europe, and Japan for use in construction and in the manufacture of plywood.

Rattans and fibers are also in demand. Used for furniture, matting, and basketry, these are collected and assembled for local use and export. Some of the commercial rattans are climbing palms that wind their way around the forest for lengths of 250 feet (76 m) and more. Rattans have been known to grow to 600 feet (183 m). The main suppliers are the Malay Peninsula, Indonesia, and China. Kapok, a waterproof fiber harvested from ripe tree pods, and coconut fiber called *coir,* serve multiple uses.

Bamboo is among the oldest materials used by humanity. Seeds and shoots are edible, as are the worms that live inside. Water pipes, containers, spears, house frames, and construction scaffolding are only some of its uses. Bamboo is a grass related to corn and wheat. A woody perennial, it can grow to 60 feet (18 m). Some fast-growing varieties can grow a foot in one day. There are at least 1,000 species of the grass, and it grows around the world in most environments. However, commercial species are tropical, and 60 percent of these are in Asia.

Tropical forests are also rich in essential oils. Vietnam, for instance, produces around 400 of these, including anise (black licorice or anisette). In India, Sri Lanka, and East Timor, aromatic sandalwood has a recorded history of 4,000 years and is one of the oldest known perfumery ingredients. Sandalwood oil is commercially distilled from trees at least 30 years old. Extracted from plant leaves, patchouli oil was used to scent the carpets and rugs of India. Now cultivated commercially, especially in Malaysia, it is a key ingredient in French perfumes.

Resins, known as copals and dammars, are **exudates**; they ooze out of trees as a result of injury. These can be collected in soft or hard form. Old, fossilized, hard lumps, buried beneath the ground for ages, are highly prized. Copals are used in paints and resins. Dammar is the Malay word for resin. Softer than copal and soluble, dammars are used for such things as varnishes and wallpaper finishes. Gum resins and gum turpentine are drawn from pine forests in the Himalayan foothills and elsewhere.

Forest products are primary sources of pharmaceuticals. Strychnine is but one example. Indigenous to the monsoon forests of western, coastal India, it is still widely employed there and in Southeast Asia as a stimulant, tonic, or emetic. Strychnine is also used for spear and arrow hunting and to poison vermin. You have probably read about it in mystery novels as a means of murder. Prepared from the seed of the snakewood tree, large doses of strychnine are definitely deadly.

The World of Mangroves

Mangroves are littoral plant formations of tropical and subtropical coasts where trees and shrubs have adapted to inundated and saline conditions. At least 50 million years old, they originated in Malaysia and diffused around the globe. The world's largest remaining mangrove concentration is the Sunderbans in the Bay of Bengal. The Bengal tiger makes its home in this forested area. Mangroves comprise habitats for hundreds of living organisms from mammals, birds, crustaceans and fish, to algae, fungi, and bacteria. The entire community is called **mangal**.

Mangroves are significant for several reasons. They stabilize shorelines and decrease coastal erosion. Capable of generating over three tons of organic matter per acre per year, they build bottom sediment with their multiple roots. Animals such as water buffalo graze on them. And they act as windbreaks against coastal storms.

People find mangroves economically useful. Their wood is used in construction and provides a rich charcoal. Sugar from the nypah palm can be turned into alcohol and machinery fuel. Prawns, crabs, fish, and other delicacies that live among mangrove roots are harvested for food. Unfortunately, mangroves are disappearing rapidly with population increase, urbanization, industrialization, and the creation of commercial fish ponds.

Population pressure, along with government-sponsored resettlement schemes, have instigated the clearance of vast areas of mangal. Urbanization and industrialization have seen mangroves used as waste dumps and ruined by factory effluents and oil spills. International companies cut mangroves for pulp and particleboard. They are removed to make way for plantation agriculture, fish farming, and salt evaporation vats. Grandiose damming and water-diversion schemes alter the fresh–salt water balance essential to the survival of both mangroves and the creatures within.

Mangroves are capable of rapid growth. A yearly growth of 540 feet (164 m) has been recorded in Java. Palembang, Sumatra, a coastal port in the thirteenth century, is now 31 miles (50 km) inland. However, when the mangals are cleared *en masse*, they seldom regenerate. Consequently, with the twenty-first century perception of mangroves as wastelands, they have been radically reduced.

Tropical Savanna

Arid parts of India, such as the eastern Deccan, are characterized by grassland with scattered trees. These tropical savannas are transitional zones between monsoon forests and even drier regimes. Vegetation is typically **xerophytic**, or drought-resistant.

Spontaneous or set fires function to maintain the savanna. Early in the dry season, "cool" fires stimulate new growth. Late-occurring "hot" fires rage out of control, killing trees, seeds, and every living thing. Soils are more fertile and better drained, and savanna lands are widely used for grazing and crops such as grains and groundnuts (peanuts).

Desert and Steppe Environments

Xerophytic and often thorny shrubs and succulents are found in **deserts,** which are defined as regions receiving less than 10 inches (254 mm) of precipitation a year. Some plants bloom only sporadically, waiting perhaps years for rain. These are called **ephemerals.** Some areas of sandy or stony desert are devoid of vegetation.

Asia's deserts are primarily cold deserts. Winters can be extremely cold and summers unbearably hot. Because of sparse vegetation and openness, heat loss is rapid and nights are relatively cooler. Deserts can be made agriculturally productive with irrigation although this can be very costly.

In the plateau regions beyond the Tibetan Plateau, such as China's Nei Mongol (Inner Mongolia), there are vast areas covered with short grasses (Figure 2–10).

Figure 2–10

Here in Inner Mongolia, China, cattle graze on short grass steppe. Note the temporary holding pen in the background. After a few days, the cattle will be moved to a fresh area. Photograph courtesy of B.A. Weightman.

Referred to as **steppe**, these grasslands are the traditional home of nomads who move their herds from one pasture site to another in seasonal cycles. Steppes are China's most important grazing lands, but they are being used increasingly for cropping.

Mid-Latitude Forests

Many parts of China, Japan, and Korea possess broadleaf, **deciduous forests** characteristic of middle latitude locations. Deciduous trees drop their leaves in the cold season, using winter dormancy to survive until the warm days of spring. Oak, birch, and elm are three of the more familiar varieties. Needle trees such as pine, fir, and spruce exist in cooler northern regions and at higher elevations. There are nearly 2,000 woody plants in China's forests alone. These forests are important for both wood and non-wood products. Ginkgo, chestnuts, walnuts, tallow oil, tung oil, and birch juice are some of these. The vast majority of mid-latitude forests have been eliminated or damaged for the benefit of large human populations in these parts of the world.

Mountain Vegetation

Mountain plant communities vary vertically because of climatic change with elevation. This is known as **altitudinal zonation** of vegetation. Everything from tropical to mixed deciduous forests, to needle-leaf forests and grasslands exist with increases in altitude. Forested areas give way to alpine meadows, stunted trees, and the **tree line** where neither growing season nor soil depth will support tree life. Here the sparse, stunted trees give way to the low-lying flora, mosses, and lichens of sub-arctic **tundra** (Figure 2–11). Even higher is barren rock and ice. As with other biomes, population pressure is having deleterious impacts on mountain environments, especially in the Himalayas.

Wealth of Wildlife Lost

The existence of two broad zoogeographical divisions—Gondwanan and Laurasian—can be explained by plate tectonics. About 15 million years ago, the continental plate of Antarctica, Australia, and New Guinea split apart, leaving Antarctica behind. The

Figure 2–11

High in the Indian Himalayas— 15,000 feet (4,500 m)—a horse grazes on vegetation far above the tree line. Rich in grasses and flowers in summer, this terrain is covered with ice and snow in the winter months. Photograph courtesy of B.A. Weightman.

Australia-New Guinea section collided with south-eastern Laurasia, thereby creating the Malaysian and Indonesian archipelagoes. The collision took place just west of Sulawesi. This undersea fracture marks the division between Laurasian and Gondwanan faunal types. As mentioned earlier in this chapter, India, once part of Gondwanaland (Africa), crashed into Laurasia as well. In the process, India served as a raft carrying Gondwanan fauna to Laurasia.

Western and eastern **Malesia** (Malaysia and Indonesia) have strikingly different animal communities. Plant differences are not as distinct because plants more readily disperse across water. A combination of Gondwanan and Laurasian fauna inhabits the western region, while Laurasian live in the east. Here, animals such as the tree kangaroo and pygmy glider have similarities with Australian species.

The British naturalist Alfred Russel Wallace explored both South American and Southeast Asian rain forests. During his stint in Asia (1854–1862), Wallace discovered these two distinct zoogeographic regimes. He described their fauna as having "endless eccentricities of form and extreme richness of colour." The line of demarcation between the two zones is known as the **Wallace Line**.

Thousands of wildlife species live in Asia, many not yet identified. Many are **endemic**—they occur in a given area only. Having a restricted range, these are easily wiped out. Indonesia, the Philippines, China, India, and Japan have the most endemic species of flora and fauna, and these are seriously threatened.

Since international focus is usually on tropical regions, other climes are often ignored or forgotten. Think about this: half of Japan's native mammals are in danger of being eliminated across the archipelago. Two kinds of wolf and three types of bat are now extinct. Another 22 species, including the Japanese river otter, have not been seen in the wild for a decade. More than 2,000 species of terrestrial vertebrates live in China. Many are at risk. Endemic species such as the lesser and giant panda, the golden monkey, and the Yangzi alligator are seriously endangered. Przewalski's Horse, the earth's last truly wild horse, has not been seen in the wild since 1966.

The diverse environments of South Asia provide habitats for a plethora of animals, birds, and other

Figure 2–12

I spotted this elephant getting a drink of water in Kerala State in southern India. The elephant had just completed a day's work hauling logs in the forests of the Western Ghats. Elephants, as work animals, are rapidly being replaced by vehicles that are responsible for much greater environmental damage and pollution. Photograph courtesy of B.A. Weightman.

living things. However, Asian elephants, one-horned rhinoceros, sloth bear, and other species have sharply reduced populations (Figure 2–12). This is attributable to human pressure for land, modification or destruction of habitats, and poaching. Asian lions, once widespread and avidly sought by hunters, are now confined to India's Gir National Park in the western state of Gujarat.

Some Indian wildlife have adapted to living in human settlements. Bats, birds, mongooses, and monkeys

Figure 2–13

I discovered this poor creature in a small park/zoo in Banjarmassin, Kalimantan, Indonesia. It is an illegally caged proboscis monkey, known for its large nose. Given the rapid rate of deforestation in this region, the proboscis monkey is a threatened species. In another tiny cage was a baby orangutan that never stopped crying. Rescue efforts cannot keep up with the illegal poaching of wild animals. Photograph courtesy of B.A. Weightman.

subsist on the refuse produced by the inhabitats of cities, towns, and villages. The common langur and rhesus macaque are the most prominent, especially around temples. Here they can feed off food offerings and neither Hindus nor Buddhists will harm them.

Indonesia is the most mammal-rich, with close to 700 species of higher vertebrates, but it has more threatened species of mammals and birds than any other country (Figure 2–13). Nevertheless, in its forests can be found wildlife inhabiting every layer: from proboscis monkeys, gibbons, orangutans, flying foxes (bats), gliding snakes and frogs in the canopy; to tree kangaroos and other arboreals of the understory; to clouded leopards, tigers, and other cats of the lower understory and floor; to tapirs, rhinos, and wild pigs, foraging on the floor. Birds and insects are in countless varieties throughout. Unfortunately, many of these species are disappearing as forests are cut down for lumber and settlement.

Figure 2–14 shows the number of threatened species of mammals, birds, and reptiles in Asia. Threats come primarily from people who destroy habitats for settlement; seek economic gain from resource exploitation such as mining, forestry, and plantation agriculture; emit air and water pollution through industrialization and transportation; and hunt and trade in exotic live beasts, their skins, feathers, claws, and other body parts.

Recently, Indonesia exported thousands of live reptiles and more than 16,000 live primates. In 1997, the United States imported 1.8 million live reptiles worth US $7 million. This trade was mostly illegal. Countless birds are exported as well. Some are destined for medical research, some for zoos, and some for pet stores. Because they are not properly looked after, many creatures die in transit or in captivity, or are discarded when people tire of them.

Animal trade, much of it illegal, has increased with greater demand from industrialized countries. From tropical forests around the globe, the United States purchased 266,750 live reptiles and 19,323 live primates in 1989. Japan and South Korea continue to be major importers. Thailand, where it is possible to purchase virtually anything, no matter how rare, is a key source region.

Animal parts also generate big profits. For example, a Chinese alligator costs US $15,000. From 1980–1989, thousands of cat skins were exported from China, South Korea, and Mongolia. The largest buyer in Asia was Japan, with a total of more than 160,000 skins! Thousands more went to European states. The international trade in primates is increasing in the twenty-first century, even as they are nearing extinction. Most countries now have biome protection programs, national parks, and bans or limits on wildlife trade. However, illegal

Figure 2–14

Threatened mammals, birds, and reptiles. How many species will become extinct in the twenty-first century? What are humans doing to accelerate this process? What can be done to stop or slow it down? Source: World Resources Institute. 1996. World Resources 1996–97. *New York: Oxford University Press.*

trade continues, and the number of endangered species is even greater. Species destruction threatens the earth's biodiversity.

Biodiversity refers to the range of variation of living organisms and is usually defined in terms of genes, species, and ecosystems. It is generally accepted that we should strive to preserve biodiversity for several reasons: present and potential uses of biological resources especially in medicine; to maintain a biosphere supportive of healthy human life;

and for its own sake based on a concern for the well-being of all living things.

Recent findings reveal that tropical medicinal plants are more commonly found in "disturbed habitats"—where hunting and gathering or slash-and-burn agriculture is practiced, and that the virgin forest is quite devoid of them. But the knowledge of these plants and their uses is dying as traditional populations modernize and become more urbanized.

Does this mean that we should save the people and disturbed habitats and not worry about pristine rain forest? Remember that there are unknown, potential uses for rain forest flora and that this forest, with its own complex of ecosystems, is integral to the larger biosphere in which we live.

Some environmentalists speak of **environmental criticality** when referring to environmental degradation and associated socio-economic deterioration. Removal of forests, disturbed or not, is removal of habitat for all living creatures. Ways of life are destroyed and associated knowledge is lost. Thoughtless development reduces biodiversity of flora, fauna, and humans. Quality of life for all is surely related to the health of habitat.

Do the medicines you use derive from tropical plants or animals?

Plants, Animals, and Medicine

The United States imports more than $20 million worth of rain forest plants a year. In fact, 40 percent of medications sold in North America derive from plants. These are either harvested from forests by local people or are commercially grown. Cinchona trees, originally from South America, are grown in India to produce the anti-malarial quinine. But new, quinine-resistant strains of the disease have developed, prompting researchers to investigate anti-malarial potential of new plants.

Both modern and traditional medicine impact biotic communities. In India, for example, 2,500 plants are either used by traditional healers or exported. Nepal exports 90 percent of its 700 medicinal plants and herbs. These countries, along with Sri Lanka, are noted for traditional Ayurvedic medical practice. Grounded in Hinduism, Ayurvedic healing relies on natural products.

Traditional medicine from India, China, and elsewhere has an enormous pharmacopia, some of which has proved invaluable to modern medical practitioners (Figure 2–15). For example, reserpine, the source of modern tranquilizers, is extracted from the serpentine root shrub. The National Cancer Institute and other organizations, intent on finding cures for cancer and HIV-related diseases, are actively collecting tropical leaves, twigs, roots, and rhizomes from both forests and traditional

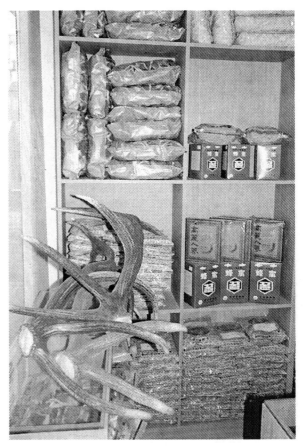

Figure 2–15

This is a traditional medicine pharmacy in South Korea, where animal and plant products are essential. Here you can see deer antlers and packages of ginseng roots. Photograph courtesy of B.A. Weightman.

healers. Thousands of samples have been collected yet these account for only seven to eight percent of the thousands of flowering plants in Asia. Many of these are vanishing in the wake of development.

Also useful in medicine are animal products. Snake venom derivatives can relieve phlebitis. And they can duplicate symptoms of muscular dystrophy and myasthenia gravis, enabling scientists to understand these diseases better.

Animal products are widely used in traditional practice, but Western scientists put little faith in the curative powers of most of them. A more important point is that the majority of Asian people believe in their efficacy. And, through experiments with placebos, scientists have shown that belief in a medicine's effectiveness can bring about actual improvement.

Faith in the mysterious workings of faunal cures drives the demand for faunal parts. In China, powdered deer antlers are employed to counter aging, and ground dried gecko is taken for respiratory problems. In Vietnam, boiled monkey bones are believed to improve circulation. Indonesia's trade in *jammu* or traditional medicinals involves millions of dollars a year. While many, better-educated urbanites are turning to Western imports, increasingly available in markets and pharmacies, rural and older populations continue to rely on long-trusted remedies.

One of the most exploited animals is the rhinoceros. Only a few hundred survive in India, Nepal, and Indonesia. A mere 70 remain on the island Java. Rhino parts are craved for their imagined powers as analgesics and aphrodisiacs. Its droppings are said to cure leech bites, its rawhide skin afflictions. Its "horn" (actually hard, agglutinated hair) guarantees the seller instant wealth. Unfortunately, rhino horn remains in demand even with public education and legal restraints on its trade.

Thinking About the Natural Environment

Environment is a multidimensional concept, and definitions range from anthropocentric to bioethical: the environment exists to support humanity versus it has value in its own right. The former leans toward the idea that there is a sharp division between people and nature, the latter toward the notion that human and natural worlds are entwined in an all-encompassing life-world. Those who view environment in terms of its use value tend to think that problems result from mismanagement and that better management of resources will fix things. The bioethical perspective is that nonhuman phenomena have intrinsic moral significance apart from their utility. Therefore, humans have a moral responsibility to care for and protect all aspects of nature. Between these "use it" and "save it" perspectives lie numerous other interpretations.

There is a common perception that Asian culture groups, at least historically, lived closer to and had more regard for nature. Nature must be important because both East and Southeast Asians employ natural metaphors regarding social position and morality. For example, reason (*li*) is fundamental to the history of Chinese philosophy. Originally, the character for *li* meant "well distributed veins on minerals or precious stones." Later, it came to mean "principle" and then, "universal principle." Besides, nature is intrinsic to religions and belief systems. However, there is no single Asian concept of nature.

Did you know that almost-extinct tigers are descendants of the extinct sabre-tooth tigers of Siberia?

Fate of the Tiger

At the turn of the twentieth century, an estimated 100,000 tigers roamed from Siberia to Indonesia. In the dawn of the twenty-first century, there are perhaps 2,500 remaining. Big game hunting in the colonial era, modern poaching, and loss of habitat are the major causes of this catastrophe.

In India, only about half of the funds designated for "Project Tiger" ever reach it. Poaching continues because impoverished people can earn US $100 for a dead tiger. Poisoning water holes with cyanide is one method, killing tigers and all other beasts who drink. Brokers make thousands on the sale of tiger products. An Indian tiger skin can bring as much as US $15,000 in Arab nations. Ground tiger bone can earn US $500 a gram. A 24-pound (11 kg) bag of tiger bones is literally worth a fortune. From 1975 to 1992, South Korea imported nearly 4,000 kg (8,818 pounds) of dried tiger bone from Sumatra—the equivalent of 338 to 620 tigers. There are only about 400 tigers alive in Sumatra today. Ground tiger bone is believed to assuage rheumatism and paralysis.

In southern China, an estimated 20 to 30 tigers eke out an existence in remote patches of forest. Massive forest clearing and the eradication of tigers as "pests" and "vermin" in the 1950s decimated the population. Although conservation programs now exist, these few tigers are in a precarious situation.

In mainland Southeast Asia, a few hundred tigers are rapidly meeting their demise. Aside from loss of habitat to logging concessions, plantations, and the like, tigers are starving because hunters have virtually eliminated entire game populations from range areas.

Conservation programs are largely ineffective. Lack of interest or funds to pay conservationists and game wardens, corruption, and poverty contribute to failure. Some efforts are being made to enlist the aid of local farmers in protecting "their" tigers, but the prognosis for the survival of the tiger is poor.

Although it is true that there is less of a distinction drawn between human and natural worlds in Asia, the range of attitudes and behaviors toward it is as varied as elsewhere.

Although culture is frequently seen in opposition to natural forces, these remain connected in a holistic, contextual existence. While Westerners assume a direct connection between people's perception of nature and their treatment of it, this is not necessarily the case in Asia, where contextualism and pragmatism are far more important. Nature can be worshipped and harmed simultaneously; one can pay homage to the tree prior to felling it.

Certainly, Asian perceptions of nature have not stood in the way of its destruction. Yes, Western impacts, foreign companies, and foreign investment all play their roles. However, political and economic elites, the formulators of national policies, facilitate, contribute to, and benefit financially from environmentally-damaging development schemes. With all the rhetoric about saving endangered beasts and habitats, the juggernaut of development continues to affect the health and well-being of nature and humanity alike.

Current thinking does not dichotomize use or development from environment. Since humanity cannot survive without natural resources and these cannot exist if depleted or destroyed, development and environment are inexorably linked. **Sustainable development** is essential in light of human impacts on air, water, flora, fauna, and other natural systems. It is critical for the survival and well-being of future generations.

Conceptually, sustainable development sounds like the right thing to do. But putting these ideas into practice unearths such questions as these: What is the precise meaning of *sustainable,* and who sets the standards? Who or what should be sustained? Is a geomorphic phenomenon such as a prehistoric cave entitled to the same consideration as a tiger population? Are animals more important than fish or plants?

In the upcoming chapters, you will read more about environmental problems and efforts to contain them. Keep in mind that because of the ephemeral characteristic of nature, local problems become regional ones, and regional problems become global ones. Keep in mind, also, that it is the poor who suffer most directly from environmental degradation. Unlike the majority of those living in the industrialized nations of the northern hemisphere, the majority of Asians live close to the land, without the mediation of air conditioners, water purifiers, and so forth. Geographer W. M. Adams (1995) says, "Sustainable development is synthetic, and constructed within the confines of current convention in northern-dominated debate about development. . . ." While debate lingers among policymakers, the full force of modern development continues to transform whatever is left of the natural world. In Chapter 3, I will discuss the dynamics of population, agriculture and food supply, and related social inequity.

⊕ Main Point Summary

1. The physiography of Asia is exceedingly complex. Ancient, massive mountain ranges and extensive, rugged plateaus offer unique environments for habitation and exploitation. The Himalayas and other ranges have channeled migration, served as barriers to cultural diffusion, and continue to hinder accessibility.

2. Fertile, alluvial lowlands such as the Indo-Gangetic and North China Plains support dense populations. Populations are generally clustered in plains, basins, and coastal lowlands.

3. The Indus, Yangzi, Mekong, and other rivers offer both food supply and power potential. Rivers are critical for settlement and as transportation arteries.

4. Oceans and seas provide problems as well as possibilities. Strategic and resource-rich areas such as the Straits of Malacca or the South China Sea are focal points of political conflict.

5. The physical and cultural diversity of fragmented island realms fosters instability. Political control is especially problematic in extensive archipelagoes such as Indonesia and the Philippines.

6. South, East, and Southeast Asia are collectively known as Monsoon Asia. Activities are regulated by distinct wet and dry seasons, and productivity is frequently marred by climatic variance, unpredictability, and disaster.

7. Natural vegetation reflects climatic regimes. Biomes, increasingly destroyed or threatened by human presence, house ever-decreasing species types and numbers. Both species and genetic diversity are endangered.

8. Biomes are varied, ranging from tropical rainforests in Southeast Asia to temperate, mixed forests in East Asia; from steppe in Mongolia to deserts in Pakistan; and from mangal in Bangladesh to tundra in the high Himalayas. Both people and wildlife have wide-ranging adaptations to the environmental challenges of these different biomes.

9. Wildlife diversity is unparalleled in tropical forests where an array of useful wood and resinous substances, medicinals, and food products can be found. Population pressure, development, and illegal logging and hunting operations threaten biodiversity. These also induce environmental criticality whereby the integrity of both environmental and social life-worlds is jeopardized.

10. Environment is a multidimensional concept with definitions ranging from anthropocentric to bioethical. Current, idealized thinking addresses sustainable development. Unfortunately, ideas and practices are seldom convergent. Population increase, urbanization, and economic development have left few remains of the "natural world."

Bibliography

Adams, W. M. 1995. "Sustainable Development?" In *Geographies of Global Change,* eds. R. J. Johnston, Peter J. Taylor, and Michael J. Watts, pp. 354–373. Oxford: Blackwell.

Balik, Michael J., Elaine Elizabetsky, Sarah A. Laird eds. 1996. *Medicinal Resources of Tropical Forests.* New York: Columbia University.

Bayliss-Smith, Tim, and Susan Owens. 1994. "The Environmental Challenge." In *Human Geography: Society, Space, and Social Science,* eds. Derek Gregory, Ron Martin, and Graham Smith, pp. 113–145. Minneapolis: University of Minnesota.

Bruun, Ole, and Arne Kalland. 1995. *Asian Perceptions of Nature.* Richmond, Surrey, G. B.: Curzon.

Durst, Patrick B., Ward Ulrich, and M. Kashio, eds. 1993. *Non-Wood Forest Products of Asia.* New Delhi: Oxford and IBH.

Kasperson, Jeanne, Roger Kasperson, and B. L. Turner, eds. 1995. *Regions At Risk: Comparisons of Threatened Environments.* Tokyo: United Nations.

Macintosh, Donald J. 1983. "Riches Lie in Tropical Swamps." *The Geographical Magazine* 55: 184–188.

Matthiessen, Peter. 1997. "The Last Wild Tigers." *Audubon* 99: 54–63 & 122–125.

McNeely, Jeffrey A., and Paul S. Sochaczewski. 1988. *Soul of the Tiger.* New York: Doubleday.

Myers, Norman. 1984. *The Primary Source: Tropical Forests and Our Future.* New York: Norton.

Peissel, Michel. 1997. *The Last Barbarians: The Discovery of the Source of the Mekong in Tibet.* New York: Holt.

Saghal, Bittu. 1997. "Tigerland." *New Internationalist* 288: 24–25.

Scott, Geoffrey A. 1981. "Mangroves and Man in the Malay Archipelago." *Sea Frontiers* 27: 258–266.

Smith, Holly. 1997. *Adventuring in Indonesia.* San Francisco: Sierra Club.

Stone, David. 1994. *Biodiversity of Indonesia.* Singapore: Archipelago.

Tangsubkul, Phophat, and Frances Lai Fung-wai. 1983. "The New Law of the Sea and Development in Southeast Asia." *Asian Survey* 23: 858–878.

United Nations Environment Programme. 1997. *Global Environmental Outlook.* New York: Oxford.

Vesilind, P. J. 1984. "Monsoons." *National Geographic* 166: 712–747.

Ward, Geoffrey C. 1997. "Making Room for Wild Tigers." *National Geographic* 192: 2–45.

Whitmore, T. C. 1990. *An Introduction to Tropical Rainforests.* Oxford: Clarendon.

World Conservation Monitoring Centre (WCMC). 1992. *Global Diversity: Status of the Earth's Living Resources.* London: Chapman and Hall.

WCMC. 1993. *1994 IUCN Red List of Threatened Animals.* Cambridge, U. K.: IUCN.

Chapter 3

✦

Population, Agriculture, and Social Disparity

> *"The people turn in allegiance to
> Humanity, as surely as water flows
> downward or as a wild animal
> takes cover in the wilderness."*
>
> MENCIUS (390–305 B.C.)

Population Growth and Dispersal

Eight thousand years ago, a mere five to ten million people inhabited the earth. Today, the earth's population is more than 6.3 billion. Most of this growth has taken place since 1830, when the world population first reached one billion. In 2004, South, East, and Southeast Asians accounted for more than 3.5 billion, or about 60 percent of the global population. How and why did this happen? What are the ramifications?

Estimates of early Asian populations indicate that they were relatively small, with sporadic periods of increase and decrease. By the mid-nineteenth century, population numbers had jumped significantly. However, growth was not an even progression. Before the advent of modern medicine, population numbers fluctuated across the globe with environmental disasters, war, famine, and disease. For example, in 1279, the Mongol conquerors of China exterminated some 35 million peasants. A pandemic of plague from 1347–1351 killed 75 million of the then 300 million Eurasians.

Although worldwide high-mortality disasters have been largely quelled, regional catastrophes continue to make dents in population totals. For instance, 17,000 died in Bombay from plague in the late nineteenth century. A famine and government policies killed at least 30 million Chinese from 1959 to 1961. A million Bangladeshis perished in a hurricane in 1970. And, in one of the worst disasters in human history, at least two million Cambodians were exterminated by government policy from 1975 to 1978! Nevertheless, overall numbers remain sizable and provide large **base populations** to generate notable increments of increase even though population growth rates have been declining in recent years.

Figure 3–1 shows the distribution of Asian populations. What accounts for this pattern? A combination of historical and geographical factors provides the answer. Asian populations originated in and expanded from a series of core areas over a period of

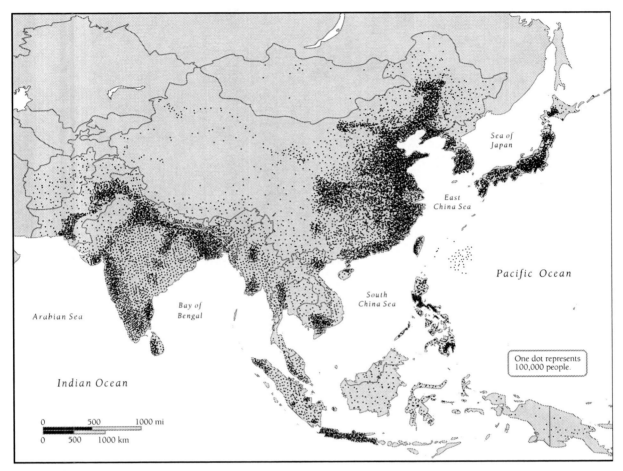

Figure 3–1

Population distribution. Compare this map with one of physical features. Note the relationship between population, plains and basins, and the significance of rivers. David L. Clawson and James Fisher, eds. World Regional Geography: A Development Approach, 8th edition, © 2004. Upper Saddle River, NJ: Prentice-Hall. Reprinted by permission.

more than 4,000 years (Figure 3–2). But where they cluster today reflects the availability of resources such as water supplies, fertile soils, minerals, and so forth.

CORE AREAS OF GROWTH AND DIFFUSION

Around 3000 B.C., there were two major clusters of population in East and South Asia: the North China Plain and the Indus River valley of northwestern India (now Pakistan). These regions possessed agricultural systems productive enough to permit growth and

stimulate expansion. A good measure of political control, social organization, and diversification of labor in rich environments allowed the development and maintenance of irrigation projects, as well as effective storage and distribution of agricultural surpluses. Population growth at this time derived from agrarian expansion.

Other smaller clusters existed around Japan's Inland Sea, in southern Korea, the Red River basin of northern Vietnam, Java, the upper Ganges River valley, and elsewhere in South Asia. By the year 1 A.D. regional populations had accumulated in these places.

Figure 3–2

Core areas and diffusion of development. Note how alluvial lowlands fostered settlement as well as the importance of rivers and coasts in colonial settlement.

By the fifth century, secondary population concentrations emerged in association with the upper Mekong River delta, Vietnam's central coast, central Myanmar, China's Szechuan basin, and southern India.

From the eleventh to the fourteenth centuries, periods of effective political control, accompanied by improvements in agricultural methods, fostered expansion into China's central Yangzi River region, Cambodia's Tonle Sap basin, the wet zone of Sri Lanka, and the lower Ganges River valley.

Growth burgeoned in the nineteenth century under the impact of European influences. European technology, coastally oriented trade, and imposed political structures combined to effect both population increases and change in distributions. New flood control technology with large-scale drainage canals facilitated the development of river deltas. Ports and their hinterlands became focal points of development and important migration destinations. Most of Asia's dominant cities grew dramatically during this period. Tokyo, Hong Kong, Shanghai, Calcutta (Kolkata), Bombay (Mumbai), Bangkok, Singapore, and Jakarta are some of these.

Core areas were critical to population growth and distribution in Asia. In fact, every country except for Laos is based on one or more core areas. However, there is more to the story. The landscape of population, like any other landscape, is socially constructed; landscapes are made by people in their particular life-world context. Social organization is especially significant in the development of Asian landscapes.

SIGNIFICANCE OF THE GROUP

Population growth and expansion were founded on intensive agriculture based on human labor. This is an undertaking requiring cooperative land management and production systems. Cooperation is encouraged because in Asia, the family is central and the group is more important than the individual. Established settlements were (and continue to be) **nucleated**; dwellings were clustered in villages (Figure 3–3). People and animals function in daily and seasonal routines of movement to and from their village and surrounding farmland. Cooperation is particularly important at planting and harvest times and in the maintenance of irrigation and flood-control systems. But this can only work in a situation of social and political stability. Consequently, long periods of stability coincided with population growth. Long-term chaos brought about population decline.

Figure 3–3

This aerial view over Pakistan shows two nucleated villages. People work the fields surrounding the villages. Much of this land is fragmented, meaning farmers do not own contiguous fields. Many farmers are tenant farmers working for landlords. Linear settlement patterns can be seen along roads and irrigation channels. Photograph courtesy of B.A. Weightman.

Where are you on the "I-we" spectrum of thought and action?

Thinking and Acting Collectively

In contrast to many societies where the individual is the locus of responsibility and action, Asian societies are characterized by collectivism. From birth, Asians are imbued with group consciousness extending far beyond familial and national levels to otherworldly existences. Identity derives from sets of expectations and responsibilities covering all aspects of life from emotional to economic. Long-established relationships among individuals, families, clans, work groups and so forth are regarded as both mutually beneficial and necessary to nourish and perpetuate family or organizational lineage. Thought and action revolve around "we" and collective agreement supersedes individual will.

The existence of different standards for different people is generally understood and accepted. The fact that royalty or elite groups exhibit abundant wealth, conspicuous consumption, or questionable behavior is tolerated as long as duties are performed correctly and moral responsibilities are met in ordering and providing for society at large. Authoritarian structure and order are considered prudent.

Religious beliefs frequently support a stratified status quo. Hierarchical concepts of order in the sacred realm bolster hierarchical relationships and consequential order in the secular world. *Carpe diem* (in Latin, "seize the day") and "you only live once" make little sense in context of profoundly enduring and seemingly endless continuities of collective existence.

If you ranked the five factors that shape everyday life, from the traditional Asian perspective, the ranking would be: fate, luck, **feng-shui** (in China; *vashtu* in India), virtue, and education, in that order. In contrast, Western perspectives rank education first, give only lip service to virtue, and denigrate fate, luck, and *feng-shui*. Furthermore, it seems ironic that Asian students these days are stereotyped in the West as being obsessed with education. To the extent that this is true, it demonstrates how rapidly globalization has undermined traditional Asian values and worldviews.

It is true that Western influences/philosophies espousing individual rights, freedom, and social equity are poking holes in the fabric of group identity in Asia. Change is most evident in places where values of individual achievement and entrepreneurial success are becoming prominent. Nevertheless, change does not mean erasure of tradition. Asians remain grounded, to varying degrees, in traditional collectivist consciousness.

Agricultural Foundations

Agriculture is understood as the cultivation of crops and rearing of livestock. Early humans were hunters and gatherers but at some point learned to grow their own food crops and raise their own meat supply. Plant and animal **domestication** most likely had several hearth areas. One of the most important was in Southeast Asia. Cultural geographers Joseph Spencer and William Thomas (1976) postulate that regional crop assemblages, such as taro, yams, and bananas in northern Thailand, encouraged agricultural developments (Figure 3–4).

Vegeculture most likely preceded seed cultivation; cultivating tuber and tree crops such as taro and bananas occurred prior to the domestication of rice and other grains. Land does not require clearing for root and tree crops, and a variety can be planted. These can be harvested at will; therefore, there is little need for storage facilities. Seed cultivation requires land clearance. Furthermore, grains have a specific growing season and must be harvested all at once, making storage necessary. Inadequate facilities invite losses to mold, insects, and rodents.

Primitive vegeculture developed in tropical regions around the world. Southern China, eastern India, and mainland and insular (island) Southeast Asia were such places. Vegeculture was practiced as early as 9000 B.C. in Thailand and was definitely a multiregional phenomenon by 7000 B.C. Evidence of vegeculture has turned up in many areas, including Taiwan and Japan.

In previous pages I noted that Asia possesses a wealth of indigenous plant species. In fact, many "familiar" foods originated there. However, crop assemblages were altered with increased interregional trade by land and sea. Plant diffusion was slow until the fifteenth century, but by the seventeenth century, botanical gardens had been established in Europe to foster plant exchanges around the world. By the nineteenth century, the industrial world demanded non-food, **industrial crops** such as cotton and rubber. Radical change derived from the introduction of plantation crops such as rubber and oil palm, especially in Southeast Asia. Commercial-variety rubber originated in

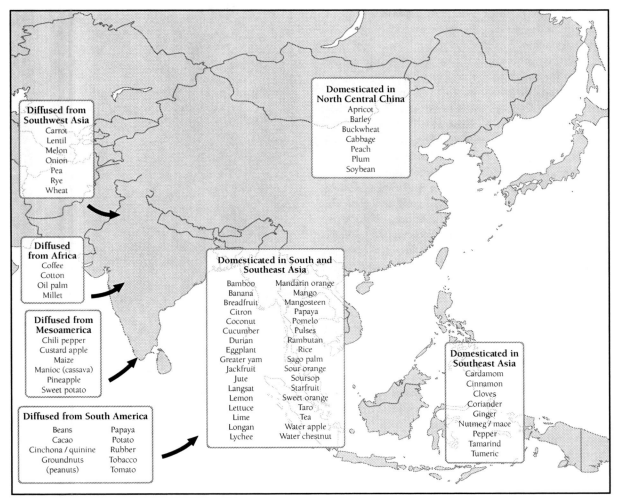

Figure 3–4

Crop assemblages, domestication, and introduction. Which of these grains, fruits, herbs, and spices are you familiar with?

South America and oil palms came from Africa. Today, plantation agriculture is transnational agribusiness, dependent upon the machinations of the global economy.

SHIFTING CULTIVATION

Early farming methods included **shifting cultivation,** a migratory, field rotation system found mainly in forested regions with relatively sparse populations.

An area is cleared by a **slash-and-burn** method—cutting down all but the largest trees and setting fire to the resultant debris. Burning brings nitrogen to the soil, which is then planted with the aid of a digging stick or hoe. Cropping is typically layered—below-ground tubers, surface plants, and above-ground plants and tree products. A field is used from one to three years depending on soil fertility. Then, it is abandoned and allowed to remain fallow for twenty to fifty years or more, during which it regenerates its natural flora. People move to a new area and begin again.

Shifting cultivation is an ecologically sound practice for maintaining soil fertility, reducing soil erosion, and encouraging forest rejuvenation. People participate in the ecosystem rather than superimposing their will upon it. However, this practice has drawn severe criticism as being destructive of forests, as well as being responsible for damaging fires.

Problems lie not so much with the method as with the impacts of population and deforestation. Dramatic population increases, combined with the removal of forests for plantations and other uses, mean that there is insufficient land available for shifting cultivators. For example, every year in Southeast Asia, anywhere from 98,800 to 1,976,000 acres (c. 400,000 to 800,000 hectares) of land formerly used for shifting agriculture is transformed to permanent agriculture. Consequently, shifting cultivators are unable to fallow their plots long enough. Replanting too soon produces a downward spiral from diminished yields to total abandonment.

Some shifting cultivation is still found in Southeast Asia, and in some areas of South Asia and southern China (Figure 3–5). It is known as **swidden** in Southeast Asia, but there are many local terms as well. In Indonesia, for instance, it is called *ladang*. While swidden plots are being clear-cut and permanently planted, the issue of soil fertility remains. Sustained crop yields in most tropical soils require the application of large quantities of fertilizer, essentially ensuring reliance on manufactured sources. Poorer farmers cannot afford chemical supplements, and many are forced to find subsistence elsewhere. Shifting cultivation is best described as a residual agricultural practice in Asia.

RICE CULTIVATION

Recent archaeological evidence from Zhejiang province in southern China indicates that rice was domesticated there around 5000 B.C. Scientists think that its hearth area was in Southeast Asia, possibly Thailand, some 9,000 years ago. Gujarat in western India was another site of domestication. Swidden plots housed initial rice cultivation that was not

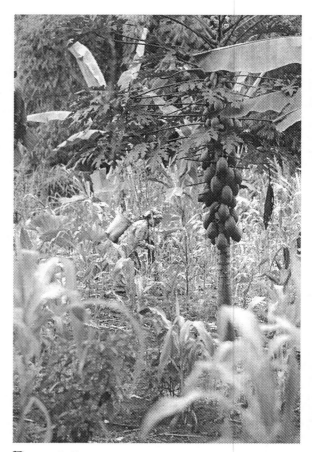

Figure 3–5

In this swidden plot in Kalimantan, Indonesia, the major crop is corn. A papaya tree stands in the foreground and a banana plant in the background. Eventually this plot will be abandoned for another. Photograph courtesy of B.A. Weightman.

irrigated. Wet-rice production did not become well established in China until the fourteenth century and in India even later. Rice cultivation diffused to Japan either via Korea or across the sea around 300 to 400 B.C. It was introduced into Java by sea from India, but as elsewhere, irrigated cultivation was not well organized until much later. Eventually rice replaced millets as the main food grain in the warmer parts of Asia.

Today rice is the staple food of more than half the world's population, and 90 percent of it is grown in Monsoon Asia. The inclusion of Asian colonies into the international economy stimulated population movements and the establishment of commercial, wet-rice systems in major river deltas. The Irrawaddy (Ayeyarwady), Menam (Chao Praya), and Mekong deltas were occupied between 1850 and 1930 and exported rice to Asia and the world.

"*Rice bowls,*" intensive rice-growing areas, were founded on environmental circumstances and human ingenuity regarding methodology and technology. Paddy preparation, water distribution systems, seedbed cultivation, sowing and/or transplanting techniques, and terracing and irrigating slopes are only some of the requirements (Figure 3–6). Farms are generally small, around 2 to 4 acres (1–2 ha). Plots are small as well, disallowing the use of large machines. Although the use of chemicals—fertilizers, pesticides, insecticides, and fungicides—are widespread, mechanization is not, except for some parts of Japan, South Korea, and Taiwan. Mechanization is increasing in parts of Southeast Asia such as Thailand and Indonesia.

In Asia, three types of rice are cultivated: *indica, japonica,* and *javanica. Indica* varieties have long grains that remain separated when boiled. *Japonicas* have shorter, stickier grains. *Javanicas* are somewhere in between. *Indicas* and *javanicas* dominate in tropical and subtropical areas while *japonicas* excel in cooler latitudes and at higher elevations. There are glutinous and non-glutinous varieties in all categories.

Farmers typically grow several varieties, depending on their own food preferences, market demand, and ceremonial requirements. They are also conscious of variation in maturation periods that can take from 90 to 260 days. Where temperatures permit, faster ripening varieties, like *indicas,* allow double or even triple cropping. Quick-ripening rice is critical where water supply is unstable.

Wet-rice cultivation relies on either rainfall or irrigation. In hilly areas, rice is grown on irrigated terraces. In deeply flooded areas like delta regions, floating rice is grown. It makes up a quarter of Bangladesh's rice output, for example. Elongated stems grow in concert with rising water in the wet season, keeping the grains above the water that may be several feet (meters) high. Floating rice is tended by wading or from boats.

Since the 1960s, farmers have been planting *high-yielding varieties* (HYVs) of rice, wheat, and other crops. These were developed at research institutions around the world in an effort to sustain burgeoning populations. High-yielding rice was developed at the International Rice Research Institute in the Philippines. These stronger, higher-yielding, pest-resistant seeds, along with appropriate management techniques, generated the **Green Revolution.** Tougher seeds enabled the use of hitherto vacant land. From the mid-1960s to the mid-1980s, land under rice production increased

Figure 3–6

Here in southern China a woman is bundling rice seedlings from the seed bed to the left. The seedlings will eventually be transplanted into a rice paddy.
Photograph courtesy of B.A. Weightman.

by 15 percent in Asia. But crop yields jumped by 74 percent. Scientists continue to search for a "super-rice" that will have even greater yields and be genetically protected against even more diseases and pests.

Innovations are generally adopted by the most innovative but not necessarily the biggest landholders. However, larger operators are likely to have more connections and, therefore, easier access to loans, farming cooperatives, irrigation systems, and so forth. Female-headed households and social minorities such as low class/caste or tribal groups seldom have equal access. Other impediments relate to environmental conditions or loss of cropland to urbanization. Nevertheless, the Green Revolution has generated quantum leaps in crop yields, lowered consumer prices, and made jobs available to landless unemployed. However, in some areas labor costs have increased to the point that farmers are forced to reduce labor inputs and use machinery and other cultivation methods. For instance, in Thailand, farmers are turning to broadcast seeding (using machines called seeders to scatter seeds over large areas) to counter the costs of hand-planting seedlings.

The *law of diminishing returns* dictates that increments of increase in crop yields will shrink relative to increases in inputs over time. Green Revolution agriculture calls for more knowledge in such areas as integrated pest management. And, in the case of rice, it demands more inputs, especially labor. With rising costs and reduction in cropland, food supply may not keep up with population growth. Geographer Jonathan Rigg (1994) notes that "the revolution is on-going," that social and other restraints shift and even disappear over time, and, "that one person's tradition-bound peasant is another's poor and exploited farmer."

"Without rice there is nothing doing."

Rice and Society

The Malay proverb above signifies the importance of rice in much of Asian society. In Chinese and some Southeast Asian languages,

"to eat" is the same as "to eat rice." Japan became known as *Mizumono Kuni*—the "Land of Luxurious Rice Crops." Rice is the foundation of culture. It is requisite for survival and integral to daily, weekly, monthly, and yearly routines and permeates both secular and sacred worlds.

Rice is highly digestible and nutritious when eaten with its bran intact and/or with vegetable and fish supplements. Claims of dietary deficiencies due to rice are simplistic. It is polished rice that has reduced protein. As part of modernization, rice mills are increasingly common, and rice is polished to a glaze. Thoroughly devoid of bran, the nutritional value of the polished grains is reduced.

Rice wine, beer, and liquor are widely consumed. Japanese saki is distilled from rice wine. In some places, a common container emphasizes the role of rice as a social locus. On several occasions, I have been invited to share the rice wine pot. Sipping out of one of several reed straws protruding from large clay containers, I experienced a sense of unity with my Lao and White Tai (in Vietnam) hosts.

Rice husks are fed to animals. In the Mekong delta, rice bran is fed to fish on fish farms. Rice bran oil is a basic ingredient of soap and many insecticides. Noodles and cosmetics are made from rice flour. The straw provides bedding for livestock and can serve as a low-grade feed. It also is the medium for cultivation of straw mushrooms.

Aside from its practical uses, rice permeates ceremonial and ritual activities. For instance, there is an annual event in Bangkok, attended by the Thai king, called *The Royal Ploughing Ceremony*, which calls for bounteous rice harvests. When Japanese Emperor Akihito was married in 1959, he carried a rice scoop to symbolize fertility. Folk dances still replicate stages in rice production, and rice motifs are frequent in weaving and basketry. Temples, goddesses, and gods are devoted to rice. It is integral to marriage, birth, and groundbreaking

> ceremonies. The importance of rice in Asian society can be summed up by a Balinese religious belief: rice is the "soul of man."

Population Increase and Expansion

Population numbers and distribution were tied to agricultural productivity. In stable and prosperous times, villages expanded and even coalesced. New farmlands were opened up and people moved into them. Where equal- or multiple-share inheritance was practiced, lands continued to be subdivided into ever-smaller plots. As populations grew and good lands filled, marginal lands became sought after. In-migration of one population forced the out-migration of another. Hunters and gatherers and swidden populations, many of whom were minority peoples, were seriously affected. Southeast Asia received countless non-Chinese ethnic groups as well as Chinese in context of China's southward expansion. Repeated invasions of Aryans and later Mongols into northwestern India drove extant populations eastward and southward there as well.

Population migrations did not end in ancient history; they are ongoing. As you will learn later, Chinese movement into Manchuria is a fairly recent phenomenon and the Chinese continue to migrate into their northern, western, and southwestern frontier regions. Vietnamese migration down the Annamite coast continues to press southward into the lower Mekong delta. Population processes and patterns are constantly changing in light of social and environmental circumstances.

SIGNIFICANCE OF CHILDREN

The key dynamic of population growth is the fact that people have children. This reality draws much comment and criticism, especially when the child-bearers appear not to be able to support their offspring. The necessity for having children does not rest solely on the concept of societal reproduction.

Children are the core of Asian family-group culture. Children are essential for the respect and continuity of family lineage. With strong bonds between people and their lands, children (usually sons) receive and are entrusted with guardianship of family property through generations. Children sustain the group, becoming a source of pride. Men and women are truly worthy when they produce children, more so with sons. Not having children is inconceivable to the majority of Asians.

There are also practical reasons for having children. In traditional societies, children contribute their services and labor very early. Four-year-olds look after two-year-olds. Seven-year-olds work at farm tasks or help in the family business. Children make important contributions toward their own and their families' survival. They are also expected to take care of their parents. Both boys and girls expect to look after their aged parents or parents-in-law. Where there is no old-age insurance or care facilities (as in most places), this cultural practice is critical.

But why do people have so many children? History demonstrates that life is uncertain. Natural disasters can strike at any time. Famine can quickly decimate a population. North Korea suffered a severe famine in 1998 although the extent of the famine's impact remains disputed. War has been all too common in the twentieth century, and millions continue to experience fighting as a way of life. Health care is available only to some and limited for all. In other words, if a family has only one or two children, the chances are not good that any will survive to carry out family responsibilities. In the War Museum of Hanoi, there is a picture of a woman who lost nine sons in wars fought in Vietnam from the 1950s to the 1970s. Who will care for her in her old age? People cannot be expected to have fewer children if many are needed to ensure an ongoing supply for labor, income, military duties, parental care, and the like. However, quality of life improvements can induce reduction in family size.

REGIONAL DYNAMICS

Table 3–1 lists statistical data used in traditional geo-demographic analysis. Traditional assessment of population trends is based on information about

Table 3-1 Selected Population Statistics[†]

2003	Total Pop. Million	Birth Rate %	Death Rate %	Nat. Inc. Rate %	% Change by 2050	% Urban	% −15/+65 Yrs. Old
Australia	19,900	1.3	0.7	0.6	48	85	20/13
Canada	31,600	1.1	0.7	0.3	16	79	18/13
U.K.	59,200	1.1	1.0	0.1	8	90	19/16
U.S.A.	291,500	1.4	0.9	0.6	45	79	21/13
South Asia							
Bangladesh	146,700	3.0	0.8	2.2	73	23	40/3
Bhutan	900	3.4	0.9	2.5	117	16	39/5
India	1,068,600	2.5	0.8	1.7	52	28	36/4
Maldives	300	2.4	0.4	2.0	77	27	39/4
Nepal	25,200	3.4	1.0	2.4	102	11	41/4
Pakistan	149,100	3.7	1.0	2.7	134	34	42/4
Sri Lanka	19,300	1.9	0.6	1.3	11	30	27/7
East Asia							
China	1,288,700	1.3	0.6	0.7	8	39	22/7
H. K. SAR*	6,800	0.7	0.5	0.2	10	100	16/11
Macao SAR*	400	0.7	0.3	0.4	76	99	22/7
Japan	127,500	0.9	0.8	0.1	−21	78	14/19
N. Korea	22,700	1.7	1.1	0.6	10	59	27/6
S. Korea	47,900	1.2	0.5	0.7	−8	79	21/8
Mongolia	2,500	1.8	0.8	1.0	45	57	36/5
Taiwan	22,600	1.1	0.6	0.5	−2	78	20/9
Southeast Asia							
Brunei	400	2.2	0.3	1.9	90	67	31/3
Cambodia	12,600	2.8	1.0	1.8	94	16	43/4
East Timor	800	2.6	1.3	1.3	84	8	44/5
Indonesia	220,500	2.2	0.6	1.6	43	40	31/5
Laos	5,600	3.6	1.3	2.3	102	17	43/4
Malaysia	25,100	2.6	0.4	2.2	86	57	34/4
Myanmar	49,500	2.5	1.1	1.4	30	27	33/5
Philippines	81,600	2.8	0.6	2.2	63	47	37/4
Singapore	4,200	1.1	0.4	0.7	6	100	21/7
Thailand	63,100	1.3	0.6	0.7	15	31	23/7
Vietnam	80,000	1.9	0.6	1.3	45	25	30/6
2003	Total Pop. Million	Birth Rate %	Death Rate %	Nat. Inc. Rate %	% Change by 2050	% Urban	% −15/+65 Yrs. Old

* Special Autonomous Region.
[†] These data indicate numerical trends. They are poor indicators of quality of life.
Source: Population Data Sheet 2003. Population Reference Bureau.

growth and change; density and distribution; and population composition. This information is gathered by various organizations from official sources based on censuses in individual countries.

Data are not necessarily reliable. Counts are not accurate because not all people are easily accessed. Moreover, people may lie about the number of children they have or the number of individuals in their household to avoid taxes, prohibitions on family size, or residency rules. Governments might skew figures for economic and political purposes. Whatever the case, look at these data not as absolute facts but rather as indicators of conditions and trends.

Growth and Change

Variation is a key word for understanding population dynamics in Asia. Table 3–1 demonstrates that some countries have relatively high birth rates (BRs). Bhutan, Maldives, Nepal, Pakistan, Cambodia, and Laos are among these. Nepal, Pakistan, Laos, and Myanmar have strikingly higher death rates (DRs) than other countries. The difference between these two measures is the natural increase rate (NIR). While Bhutan has a high BR (3.4%), it has a relatively low DR (.9%). This results in a high NIR (2.5%). Maldives has a similar situation: BR 2.4% − DR .4% = NIR 2.0% NIRs above replacement rate result in population increase. Too-low NIRs mean population decrease, a situation currently found in Japan.

Look at the "percent change" column. This indicates the anticipated percentage increase or decrease of a population from 2003 to 2050 given the current NIR. Pakistan, with its NIR of 2.7 percent, is expected to increase 134 percent. Notice that Japan is predicted to have fewer people in 2050 than it has now—a decrease of 21 percent. Having both a low BR and DR, Japan's NIR of 0.1 percent is insufficient to replace its population. Does this mean that the Japanese will ultimately disappear? Read more on this topic in Chapter 11. It is important to note that these are "snapshot" projections of current conditions. In fact, longer-term trends show a decline in NIRs.

Although India and China do not stand out in terms of NIRs, they are notable in terms of their population totals. Because of the size of these base populations, even a small percentage increase adds many people to the total. So, while India's NIR is only 1.7 percent, if current growth rates continue, it will add almost 546 million people by 2050. China, with its low NIR of 0.7 percent, nevertheless will grow by nearly 104 million people by 2050. I will discuss India's and China's population issues separately in the regional chapters.

Density and Distribution

Density and distribution data measure spatial characteristics of populations. Refer to Figure 3–1 and note that some areas are densely populated while others are sparsely populated. Environmental and historic circumstances contribute to these patterns. People tend to avoid extreme environments such as deserts and rugged, high-elevation terrain. Access to resources is crucial to survival. For example, arable land is found on plains and along waterways, and this is where most populations reside. Past circumstances have led to a phenomenon known as **historic inertia**—people migrate to and continue to live in places where other people are already established and opportunities are perceived to exist. Consequently, densely populated areas most likely will continue to be densely populated areas. As you will read in later chapters, some governments, like those of Indonesia and Malaysia, are trying to change their population distributions through programs encouraging migration from densely populated regions into sparsely populated regions.

Density figures are misleading. **Arithmetic density** indicates how many people exist per square mile (km) of land in a particular country. The Philippines, for instance, has a density of 638 persons per square mile. However, people are not evenly distributed throughout the islands. The northern island of Luzon is home to millions more than the southern island of Mindanao, and numerous islands have no inhabitants. Arithmetic densities reveal nothing about actual spatial distribution of people.

An alternative to arithmetic density is **nutritional density** that relates numbers of people to amount of cultivated land. Nutritional density assumes a direct relationship between the number of people and land

Figure 3–7
This is Kali Besar, one of Jakarta's several squatter settlements. Situated on a river near the sea, it is built on stilts to accommodate the tides. Photograph courtesy of B.A. Weightman.

productivity potential. This information is interesting but still misleading.

Density data provide only crude measures. Laos is a relatively sparsely populated country yet it is regarded as one of the poorest in Asia. Taiwan, one of the most densely populated, is seen as one of the richest. Densely populated Singapore, also one of the richest, has virtually no arable land. Density does not account for variation in human or land productivity, differences in dietary requirements, or quality of available diet, food imports, or unequal access to food supplies.

Another indicator of distribution is **percentage urbanized**. What proportion of a country's population lives in cities? Note that definitions of *city* and *urban* vary widely. Table 3–1 shows that most of Asia is not highly urbanized. China has at least 29 cities of more than one million people, but China is only 39 percent urban; most Chinese are rural dwellers. Thailand is often perceived as "Bangkok," but Thailand is only 31 percent urban. In fact, 16 of Asia's 24 countries are less than 50 percent urbanized. This emphasizes the fact that most people in Asia are farmers. Furthermore, urbanization does not necessarily equate better living standards. For example, North Korea is 59 percent urban but has very low living standards. Some of the worst poverty is found in cities where modern skyscrapers belie other realities of **squatter settlements** and homelessness (Figure 3–7). Look for more information on urbanization in the next chapter.

Population Composition

Population composition data reveal the sex and age structure of a particular population. **Population pyramids** are visual representations of relationships between males and females, and among different age cohorts (Figure 3–8). Most populations for Asian countries have wide bases, indicating high NIRs and youthful populations. Considering that it is female cohorts ages 15–45 that are potential child-bearers, this age group will increase dramatically as younger cohorts age. For example, a five-year-old girl in 2000 will be a potential mother in 2010 (or even earlier).

Table 3–1 lists percentages of populations under 15 and over 65 years old. These age cohorts, under traditional analysis, are perceived as being "dependent" on other age cohorts for survival. But in fact, significant numbers of these young and old people are contributors to their own survival, as well as that of other age groups. Moreover, these figures give an indication of the different kinds of challenges facing countries. For example, Japan has a problem with its growing elderly population in terms of providing services. It is almost the only

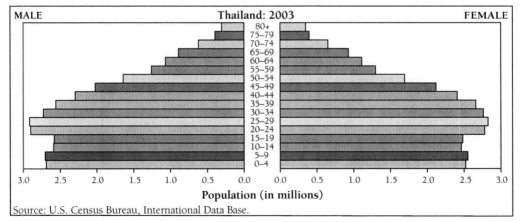

Figure 3–8

Compare these population pyramids. South Korea's narrowing pyramid base reflects its falling natural increase. Vietnam's broad-based pyramid indicates a relatively high natural increase until the 1990s when the natural increase rate started to decline. The impact of war from the 1950s through the 1970s is evident with the shortage of males in the 35-55 age cohorts. The narrowing base of Thailand's pyramid reflects the success of its family planning programs.

country in the world to have a higher percentage of its population over 65 than under 15. Most Asian countries face the opposite challenge of trying to provide services, such as education and employment, to their predominantly young populations.

Another measure of composition is **sex ratio**: the numerical ratio between men and women. Normally, populations have a balanced sex ratio. Although women outlive men, sex ratios at birth are biologically stable (assuming no intervention). The norm is 106 to 107 male births to every 100 females. But sex ratios in countries with long life expectancies reflect the fact that women live longer than men. In 2003, there were 97 men for every 100 women in the United States, for example.

In 2003, India and Bangladesh had sex ratios of 106, Pakistan 105, and China 106 males per 100 females. In China, among children age two or younger, there were 106 males for every 100 females. Demographers have discovered that millions of girls in these countries are "missing." In light of the importance of males in the extended and generational family, there is a strong preference for boys in Asian cultures. In China, for instance, a boy is called a "big happiness" while a girl is a "little happiness." Government-sponsored programs and personal efforts to limit family size can lead to **sex-selective abortion.** Modernization has brought sex-determination devices to hospitals and clinics, enabling women to "select" the sex of their offspring. In 1996, India outlawed these practices but they persist, nevertheless.

Female infanticide, the deliberate killing or neglect of girls, has a long history in Asia (and elsewhere, including Europe), particularly in times of strife. In such times, a girl might be perceived as a liability: she will leave the family and move into her husband's household, and often require a substantial dowry payment as well. A boy will remain, support, and care for his own multi-generational household. In tough times, men and boys eat first; women and girls often do without. Such cultural practices are not simple cruelty. They are strategies for survival in stressful life-world contexts. Improvements in quality of life witness reductions in female infanticide and infant mortality in general.

POPULATION AND QUALITY OF LIFE

Selected quality of life statistics are shown in Table 3–2. While life expectancies at birth (LE) have improved over time, only Japanese, Singaporeans, and Taiwanese can expect to live as long as people in Canada or the United States. Infant mortality rates (IMR) in the first year of life and child mortality rates (CMR) in the first five years of life are excellent indicators of living conditions. Cambodia, Laos, and Myanmar, among the poorest countries, exhibit very high rates. Remote Bhutan also suffers very high infant and child mortality rates.

The percentage of children under five years of age suffering from malnutrition, as represented by being underweight or underheight, is also high for many countries. In six Asian countries, 40 percent or more children are underweight. People in Japan, North Korea, and Singapore have the highest rates of access to decent water, while those of Cambodia and Laos have the lowest.

Access to health services and number of physicians per 100,000 people figures suggest survival potential with childbirth, disease, injury, or medical complications. Again there are dramatic differences between urban and rural settings. Since the majority of people in most countries are rural, health care delivery is often inadequate. In the United States, there are approximately 276 doctors for 100,000 people. Contrast this with 20 doctors per 100,000 people in Bangladesh or 16 doctors per 100,000 in Indonesia. China, Mongolia, and North Korea have relatively high numbers of doctors per 100,000 people. This calls the definition of *doctor* into question.

In China and elsewhere, "doctors" are not necessarily fully licensed physicians. They may be little more than medical aids with a few weeks' training. I met a young "doctor" in Xian, China, who told me he was a cardiologist. When I inquired about his training, he said that he had gone to "cardiology college" for four years after finishing high school. I have visited many clinics and hospitals in various parts of Asia and can assure you that conditions in most are grim. Several countries are extremely short of even minimally trained personnel, and sanitation

Table 3-2 Quality of Life Measures for Infants and Children

2003	Life Expect. at Birth		% Infant Mortality Rate (0–1)	% Child Mortality Rate (0–5)	% Low Birth Weight	% Children Under 5	
	M	F				Under Weight	Under Height
South Asia							
Bangladesh	59	59	5.1	7.7	30	48	45
Bhutan	66	66	6.1	9.5	15	19	40
India	62	63	6.6	9.3	26	47	46
Maldives	71	72	1.7	7.7	12	30	25
Nepal	58	59	7.7	9.1	21	48	51
Pakistan	60	60	9.1	10.9	21	38	—
Sri Lanka	70	72	1.7	1.9	17	29	14
East Asia							
China	69	73	3.2	3.9	6	10	17
H.K. SAR*	78	85	.26	—	—	—	—
Macao SAR*	75	80	0.4	—	—	—	—
Japan	78	81	0.3	0.5	7	—	—
N. Korea	61	66	4.5	5.5	—	60	60
S. Korea	72	76	0.8	0.5	—	—	—
Mongolia	63	68	3.0	7.6	6	13	25
Taiwan	73	76	0.6	—	—	—	—
Southeast Asia							
Brunei	74	79	0.6	0.6	—	—	—
Cambodia	54	58	9.5	13.8	9	45	45
East Timor	48	49	12.9	—	—	—	—
Indonesia	66	70	4.6	4.5	9	26	—
Laos	52	54	10.4	10.0	—	40	41
Malaysia	70	75	1.1	0.8	9	18	—
Myanmar	54	60	7.7	8.7	16	36	37
Philippines	77	81	2.6	3.8	18	28	30
Singapore	77	81	0.3	0.4	8	14	11
Thailand	68	75	2.0	2.8	7	19	16
Vietnam	70	73	2.6	3.8	9	33	36
2003	Life Expect. at Birth		% Infant Mortality Rate (0–1)	% Child Mortality Rate (0–5)	% Low Birth Weight	% Children Under 5	
	M	F				Under Weight	Under Height

* Special Autonomous Region

Sources: United Nations Development Project 2003. United Nations Environmental Program (UNEP) Geo Data 2003.

Table 3–3 Selected Quality of Life Indicators

2003	% Access to Safe Water	% Access to Improved Sanitation	# Doctors per 100,000 Persons	% Literate		% in Secondary School	
				M	F	M	F
South Asia							
Bangladesh	—	—	20	50	31	25	13
Bhutan	62	—	16	61	34	—	—
India	84	28	48	69	46	59	39
Maldives	100	56	40	97	97	67	71
Nepal	88	28	4	61	25	51	33
Pakistan	90	62	68	60	21	33	17
Sri Lanka	77	94	41	95	89	72	78
East Asia							
China	75	40	167	93	79	74	66
Japan	100	100	197	99	99	100	100
N. Korea	100	99	408	—	—	—	—
S. Korea	92	63	173	99	97	100	100
Mongolia	60	30	254	99	98	99	99
Taiwan	—	—	—	—	—	—	—
Southeast Asia							
Brunei	—	—	85	95	88	72	82
Cambodia	30	17	30	81	58	31	17
East Timor	—	—	—	—	—	—	—
Indonesia	78	55	16	92	83	55	48
Laos	37	30	61	64	33	34	23
Malaysia	—	—	68	91	84	59	69
Myanmar	72	64	30	89	81	29	30
Philippines	86	83	124	96	95	77	78
Singapore	100	100	135	96	89	74	70
Thailand	84	96	24	97	94	38	37
Vietnam	77	47	52	95	91	48	46
2003	% Access to Safe Water	% Access to Improved Sanitation	# Doctors per 100,000 Persons	% Literate		% in Secondary School	
				M	F	M	F

Sources: United Nations. 2004. Statistical Yearbook for Asia and the Pacific 2004.

conditions are poor. Moreover, both doctors and hospitals tend to be located in cities. This leaves rural people with poorly equipped and staffed clinics.

Table 3–3 also lists literacy rates and percentage of children in secondary school. This information is meaningful in context of population planning programs. South Asia, with the exception of Sri Lanka, has a poor record of children staying in school. In many places, girls attend school for only one or two years and are the most likely to leave to help with domestic chores and to look after younger siblings. This, along with historical disparities between males and females, explains the male/female differential in literacy rates (Figure 3–9).

Figure 3–9
This is a boys' religious school in Pakistan's Northwest Frontier. Here the boys spend a few hours each day learning to read the Koran along with some other basic education. Note the visiting girls standing to the right. What are they doing? How will the lives of these boys and girls differ? Photograph courtesy of B.A. Weightman.

Girls make up close to half the children attending school in China, but statistics do not show that they make up three-quarters of those enrolled but never attending. Some believe that education is wasted on girls, who will soon marry into another family's household. It is more practical that she help with chores at home as long as she lives there. It is well understood that, in more affluent circles, increasing educational achievement by women coincides with reduction in family size. An educated woman is more likely to understand the methods and reasons for family planning, and she is in a stronger position to make choices concerning her own life.

FAMILY PLANNING PROGRAMS

Controlling family size is a high priority for most nations of South, East, and Southeast Asia. Table 3–4 charts the progress of family planning in various countries. One measure of success is the **total fertility rate (TFR)**, which refers to the number of children a woman would have if age-specific birth rates remained constant throughout her childbearing years— age 15 to 45 (some sources use 15 to 49). Populations are growing despite the fact that TFRs have fallen.

Family planning programs were established in twelve Asian countries as early as 1970. These countries in particular have seen dramatic reductions in their TFRs. Malaysia, Thailand, and Indonesia are known for their success in reducing TFRs. Bangladesh has managed to reduce its TFR from an average of 7.0 in the mid-1970s to 3.6 in 2003.

Other countries such as Pakistan have not taken as well to family planning. Although programs have long been in place, the secondary position of women, with a mere 21 percent literacy rate, and rural poverty in general are not conducive to female decision-making regarding the number of children they bear. Only 28 percent of Pakistani women are using contraception, the second lowest number in Asia after Cambodia.

In light of political, social, economic, and cultural variation among and within nations, success in population reduction is not easily achieved, nor is the process the same for all. The following examples will illustrate.

Myanmar

The government of low-density Myanmar ignored controlling family size until it began to experience the consequences of rapid population growth. Programs stress the wider spacing of births. However, abortion is common as well. The quality of medical

Table 3–4 Family Planning and the Status of Women

	Family Plan. Prog. Estab. 1970s	Total Fertility Rate (1975)	Total Fertility Rate (2003)	% Married Women Using Contraception (2003)	Maternal Mortality per 100,000 Live Births (2003)	% Female Labor Force in		
						Agr.	Ind. (2001)	Serv.
South Asia								
Bangladesh	X	7.0	3.6	54	600	78	8	11
Bhutan		5.9	4.7	—	500	—	—	—
India	X	5.4	3.1	48	540	—	—	—
Maldives		7.0	3.7	33	390	—	—	—
Nepal		6.5	4.5	35	830	—	—	—
Pakistan	X	7.0	4.8	28	200	66	11	23
Sri Lanka	X	4.0	2.0	66	60	49	22	27
East Asia								
China	X	4.8	1.7	83	60	—	—	—
H.K. SAR*	X	—	0.9	86	100	—	—	—
Macao SAR*	—	—	0.9	—	—	—	—	—
Japan		2.1	1.3	56	12	6	22	73
N. Korea		5.7	2.0	53	35	—	—	—
S. Korea	X	4.1	1.3	77	20	13	19	68
Mongolia		5.8	2.7	60	65	—	—	—
Taiwan		—	1.3	71	—	—	—	—
Southeast Asia								
Brunei		5.4	2.3	—	22	—	—	—
Cambodia		5.5	4.0	24	590	—	—	—
East Timor		—	4.1	—	—	—	—	—
Laos		6.2	4.9	32	650	—	—	—
Malaysia	X	5.2	3.3	55	39	13	29	58
Myanmar		5.8	3.1	33	170	—	—	—
Philippines	X	5.5	3.5	49	240	27	13	61
Singapore	X	2.6	1.4	65	9	0	23	77
Thailand	X	5.0	1.7	72	44	47	17	36
Vietnam		5.8	2.3	77	95	—	—	—
	Family Plan. Prog. Estab. 1970s	Total Fertility Rate (1975)	Total Fertility Rate (2003)	% Married Women Using Contraception (2003)	Maternal Mortality per 100,000 Live Births (2003)	% Female Labor Force in		
						Agr.	Ind. (2001)	Serv.

* Special Autonomous Region

Sources: Population Data Sheet 2003. Population Reference Bureau. United Nations. 2004. The World's Women 1995. UNICEF.

care is indicated by the fact that 40 percent of maternal deaths result from botched abortions. Myanmar is a poor country with very limited infrastructure. Ethnic persecution and human rights violations are widespread. Contraception devices are difficult to get, especially in remote areas. Most are smuggled in from India and China.

Vietnam

Family planning has been promoted in Vietnam since the 1960s. Abortion is legal and free, making it the primary form of birth control. The government gives cash payments for sterilization or the use of IUDs but not for the use of birth control pills. In fact, only one percent of Vietnamese women use the pill, in contrast to 30 percent of Indonesian and Thai women. Misinformation about the efficacy and side effects of the pill has discouraged its use. Moreover, 60 percent of birth control providers believe that rural women would not remember to take it daily as prescribed.

Although family planning is widespread, it is not comprehensive. Men continue to reject male sterilization, and condoms are not readily available. In 1998, the government announced the establishment of a condom factory. However, infrastructural problems continue to make both supply and delivery unreliable.

Malaysia

Malaysia has considered reducing family size only since 1984 because it has been mainly interested in creating a large domestic market for the goods it produces. While programs appear successful in that TFRs have fallen from 5.2 in the 1970s to 3.3 in 2003, data mask ethnic group differences. For example, women of Chinese ancestry, who comprise 38 percent of the population, have a TFR of 2.0. Malay and other indigenous groups have a TFR of 4.0. Furthermore, tax breaks are provided for large families.

Bangladesh

Slowing population growth has been a priority in Bangladesh since the 1970s, when only 8 percent of women used contraceptives. A comprehensive program was developed that included training of field counselors. Regional clinics were set up throughout the nation, and 23,000 high-school-educated, female counselors, who themselves used birth control, addressed family planning issues with village women. The importance of family planning was stressed via radio, television, posters, and billboards. It costs the government US $62 to prevent one birth, but it saves US $615 on social services.

Today, 54 percent of Bangladeshi women use some form of birth control. The mean age at marriage has risen, the age at which women have their first child has increased, and more women stop having children earlier. Still, a large base population, in addition to a desired family size of three to four children, means that population growth remains high.

The case of Bangladesh is unique in Asia for two reasons. First, it shows that a high female literacy rate is not essential for reduced TFRs. Second, as geographer Bimal Kanti Paul (1997) points out, "This South Asian country disproves beliefs that birthrates drop only after economic conditions improve. It also refutes the contention that 'development is the best contraceptive'."

Basic population statistics should be interpreted as estimates of numerical change. They reveal nothing tangible about the well-being of people. On the other hand, projections and related measures are used to illustrate perceived population problems such as "overpopulation." The question of "too many people on earth" is a controversial one.

Are There Too Many People on Earth?

Fertility levels are down, but momentum remains for future growth as large, young cohorts move into childbearing range. Absolute numbers will continue to rise as long as the number of births per woman exceeds the replacement rate of 2.1. While fertility rates decline, the number of potential mothers increases. But, as Table 3–4 suggests, there are different rates of fertility decline. Higher fertility is typically associated with poorer people.

Scientists are reconsidering the Malthusian specter of famine in a world unable to sustain ever increasing numbers of people. Yearly reports from the Worldwatch Institute (WWI) point to an array of population growth impacts on sustainability of people and environments. One indicator for this dynamic is grain production.

World grain harvests are increasing at a much slower rate. Green Revolution improvements no longer bring about dramatic increments of increase in yields. Even though crop yields continue to increase, according to WWI's Lester Brown (1998), "There will eventually come a point in each country, with each grain, when farmers will not be able to sustain the rise in yields." However, many people disagree with Brown's neo-Malthusian predictions.

Soil erosion, scarcity of irrigation water, and land conversion are part of the picture. Loss of cropland to non-farm uses has been rising in urbanizing-industrializing regions (Figure 3–10). Factories, housing, shopping malls, and even golf courses gobble up cropland. More than 40 courses now occupy former rice paddies in the Xi delta region in southern China. Indonesia loses enough cropland each year on Java alone to feed nearly 400,000 individuals.

Such developments defile or reduce groundwater available for irrigation. According to the WWI, irrigation water in parts of India, Indonesia, and Malaysia will fall 15 to 30 percent short if urban needs are met. In coastal areas, saltwater is drawn into depleted aquifers, rendering the water saline and unusable.

Carryover grain stocks are also low, drawn down by unmet production needs. As supplies decline, prices rise, thereby reducing consumption by the poor. Concurrent rising affluence for others, in the wake of modernization, increases demand for such products as grain-fed meat, further reducing the supply of grain for the poor. From 1991 to 1995, incomes for China rose by half. In 1994, China had a net grain export of nine million tons. In 1995, rising prices led to the import of 16 million tons. In 2003, China faced a 40 million ton shortfall. Given projected population increases in Asia and worldwide, demand for grain is expected to exceed supply.

Increased demand for grain is not simply to put food directly into people's mouths. Rising affluence stimulates greater demand for meat, milk, and eggs, which requires more grain for livestock feeding. Asia's billions are moving up the food chain, consuming more grain-intensive products. Moreover, depletion

Figure 3–10

I took this picture from an expanse of rice paddies on the outskirts of Kyongju, South Korea. I was amazed at the urban encroachment on agricultural lands. This scene is ubiquitous in rapidly urbanizing South Korea. Photograph courtesy of B.A. Weightman.

of oceanic fish supplies is requiring additional grain to feed fish raised on land-based fish farms.

Asian countries spearhead demand for non-grain products. Yet, grain provides at least twice the amount of energy when consumed directly, as opposed to being eaten in non-grain form. In other words, grain-fed meat, fish, and poultry are not as sustenance-efficient as basic cereal.

China's demand for livestock products accounted for about two-thirds of world demand for grain in the 1990s. Lester Brown (1998) believes that China exemplifies why the **demographic transition model** (four sequential stages: high birthrate (BR) and death rate (DR); high BR and lowering DR; lowering BR and low DR; low BR and DR), based on centuries-long growth patterns in Europe, is not viable elsewhere because there are simply insufficient resources to sustain forecast numbers.

Skeptics of neo-Malthusian predictions of famine, disease, and pestilence trumpet technology as solving all. They argue that population pressure can lead to agricultural innovation and better productivity, as has happened in many places. Evidence also supports the fact that resource deprivation can engender self-imposed fertility control. Although this may be the case on a local scale, at the global scale, sustainability of both populations and resources remains problematic.

Does this mean that there is indeed a population crisis whereby masses of humanity face hunger, disease, and lives of misery? Geographer Alan Findlay (1995) argues that global economic structure, inequitable control over resources, and imbalanced population distributions are responsible for malnourishment and other social and environmental ills. In short, high-tech, rich nations use up far more than their fair share of resources.

According to Findlay, it is "structural forces which are producing uneven life opportunities for the world's population." The amount of food is not the problem—rather, it is access to it. Poverty and malnutrition have their roots in powerlessness. Political, social, and economic spaces vary greatly in their degree of power to exploit and control essential resources. Lack of power may derive from monetary, ethnic group, class, or gender inequities.

The rich, the politically important, the ethnic dominants, the upper classes, and males in patriarchal societies benefit over the rest, especially in stressful times.

Ironically, technology is largely responsible for both the cause and cure of population crises. Technology contributed to the dramatic lowering of death rates that spurred the population explosion in the first place. Today it enables industrial and urban land encroachment and is responsible for CO_2 emissions and chemical air pollution, water pollution, and other environmental hazards. Global networks of marketing and advertising by multinational corporations promote unwise application of new technologies, in addition to increasing expectations and encouraging mass consumption. Calories wasted, lost, or spoiled increase with affluence. Calorie loss, less than 10 percent in poor nations, is 20 to 40 percent in rich nations. There is increasing divergence between rich and poor, both among and within countries. Local misery arises, in good part, from global mismanagement.

Consumption in Asia

By virtually any measure, consumption of goods and services has risen steadily in industrial nations for decades, and it is growing rapidly in many developing nations. However, there is great disparity in consumption. The 12 percent of people living in North America and Western Europe account for 60 percent of global consumption. Nevertheless, consumption in Asia is increasing rapidly as incomes rise and populations grow. There is a global consumer class of 1.7 billion people, half in developing nations. Twenty percent are in India and China—362 million people.

China has had a reputation as the "land of bicycles." However, by 2002, there were 10 million private cars. In 2003, there were four million more. By 2015, there will be 150 million cars in China. The car industry provides jobs and stimulates other sectors of the economy but it also contributes to congestion and pollution.

China's income levels have quadrupled since 1980. Electricity demand has increased 400 percent, due to people buying and using electricity for lighting and appliances. Demand for oil is expected to double by 2025 making China the world's number two oil consumer and importer after the United States. Cellular phone use in China is booming with five million new subscribers a month. There are 180 million individuals currently using 250 million subscriptions because people get several for different cities. In 2004, India had 56 million cellular phone users and this number is expected to double by 2006.

Energy use in India and China is rising rapidly. Both rely heavily on coal and will account for two-thirds of the global increase in coal use by 2030.

Many countries have introduced credit cards to facilitate purchasing. There are 12 million MasterCard holders in Southeast Asia for example. These accounted for 7.9 billion dollars in sales in 2003.

Consumption is not a bad thing; one must consume to survive. However, consumption uses resources and generates pollution and waste material. One country to do something about this is Japan with its "Green Products" program.

Green products are manufactured and packaged with environmental concerns in mind. Japanese law requires local governments to have policies to buy green. By 2003, authorities in 47 prefectures and 12 major municipalities were buying green. Most progress has been made in the areas of paper, office supplies, information technology equipment, cars, and home appliances. In 2004, more than 2,000 businesses including Panasonic, Toyota, Sony, and Xerox were part of the Japan Green Purchasing Network.

Gender Inequality

Gender is a culturally constructed social institution. Gender roles are expected and assumed in cultural contexts, and society is organized around gender. Institutional behaviors and relationships are structured according to understood relative positioning and roles of men and women. In most cultures, gender inequity is the norm.

To understand gender inequity is to appreciate **power relationships.** Men and women are accorded unequal roles in virtually every society. This means that women do not have equal power to access education, earn comparable wages, make economic or political decisions, or even control their own destiny regarding inheritance, marriage, child-bearing, or career. Women do not have power over land or resources, or equal, if any, participation in public decision making.

Violence against women is a major risk factor for women in Asia. Gender-based violence—including selective abortion, infanticide, neglect, abuse, and domestic violence—affect millions of women. In India, at least 300,000 more girls than boys die each year, and one in four deaths of women between the ages of 16 and 24 is due to "accidental burns." These and other "accidents" result primarily from the failure of a bride's family to pay an original or newly demanded dowry, or from her "inability" to produce sons (a cruel irony since we know that males genetically determine the sex of a child). "Dowry deaths" are estimated between 5,000 and 12,000 a year in India.

Table 3–4 shows alarmingly high rates of maternal mortality. In part, these reflect access to health care. But they also reflect gender-based ill treatment. Particularly in Muslim cultures, women often are not allowed to be seen by a male physician. Too-early marriage and pregnancy, dietary deficiency, overwork, too-closely spaced births, and late-in-life pregnancies take an enormous toll. Births are frequently unassisted or aided only by a midwife. Complications typically mean death, and pre- and post-natal care is scant.

Given their sheltered and uneducated circumstances, many women in patriarchal societies may be unaware of their inferior status. They accept 12 or more hours of hard labor in and out of the home, eating last and least, constant pregnancy, forced sex, beatings, and even death as their lot in life. These women do not know that they have even the potential to make a decision or control anything. Even if they are aware of their lesser position, most often there is little they can do about it.

Are you aware that millions of women and children are sex slaves in Asia?

Women and Children Enslaved: Prostitution and AiDS

Millions of women and children under 18 work as prostitutes in Asia. Roughly 60,000 girls offer sex services in the Philippines. Thailand has as many as 800,000 children working in the "sex industry," and it is estimated that 10,000 girls are bought from Myanmar each year to replace those overworked, diseased, or dead. In Phnom Penh, Cambodia, a six-year-old can be acquired for as little as three dollars. As many as 50,000 Nepalese girls work as prostitutes in India. Their average age in 1994 was 10 to 14.

Commercial sex workers are in high demand throughout Asia. Places such as Bangkok, Manila, Mumbai, Ho Chi Minh City (Saigon), and Taipei are infamous sex cities. Sex tourism is partly responsible. Most "sex tourists" are from Japan, North America, Europe, and the Middle East. Most are heterosexual. However, prostitutes are also deemed essential for indigenous males. Mores in several countries restrict sexual behavior for females but not for males. Millions of men work away from their homes for weeks or months at a time, and prostitutes are seen as contributing to their well-being. Since local women are not readily available, imports are required.

Where do these girls come from? In Thailand, "labor agents" recruit girls from such places as Myanmar and Cambodia. Parents are paid a lump sum (this becomes a debt) and are assured that their daughter will be taken to work for a rich family, send money home, and attend school. In the face of abject poverty, sending a daughter off elsewhere might be seen as a blessing. The reality is that she will be detained in debt bondage—forced to work in a brothel until she has paid her "debt," plus the cost of food and clothing. Thailand has legal safeguards against such exploitation, but the law is rarely enforced by officials and police, who are almost entirely male and deeply entrenched in this money-making enterprise.

In India, labor agents scour the region for children of the impoverished. Nepal is a virtual gold mine in this regard. There is also a tradition of declaring one's daughter a *devadasi*, or Hindu temple servant. Dedicated to a particular goddess, a *devadasi* participated in ritual singing, dancing, and other religious practices. However, tradition has been corrupted, and today many of these girls are forced into prostitution by the unscrupulous. The practice has been outlawed, yet it still accounts for at least 20 percent of prostitution in India.

Boys are not immune to the sex industry. Boys often see prostitution as a way out of poverty, and many pursue the urban sex trade. Both boys and girls participate in the exceedingly lucrative pornographic movie and video business. The sex industry is closely tied to the drug trade, and both are associated with the spread of sexually transmitted diseases (STDs) (Figure 3–11).

According to the World Health Organization (WHO), there are at least 40 million people infected with HIV. In 2003, more than 14,000 new persons a day were infected around the world. In South and Southeast Asia, 90 percent of AIDS cases are in India, Myanmar, Cambodia, Thailand, and China.

In Thailand, HIV infection was virtually unknown prior to 1987. Now Thailand has five times the number of cases as India, which ranks second in the world for AIDS victims. It is estimated that 350,000 HIV-infected babies were born in 2000. There will be half a million

Figure 3–11

No translation is needed for this Vietnamese sign. AIDS comes from drug use. Like Vietnam, many Asian countries avoid the issue of sexual transmission in their anti-AIDS programs. Photograph courtesy of B.A. Weightman.

AIDS cases, more female than male, primarily in the 15–34 age cohorts. Thailand's life expectancy is predicted to fall as a result of AIDS-related deaths.

Prostitution and intravenous drug use are the main causes of HIV infection in Asia, and both are on the increase. High-risk groups are commercial sex workers and heterosexual men. International travel and tourism as well as the trucking industry are major contributors in the diffusion of STDs. Patronizing brothels on their journeys, the men bring HIV home to their unsuspecting wives and lovers.

The increase in HIV infection and AIDS cases has upped the demand for even younger sex workers. Men pay premium prices for virgins, believed to be disease-free. Once prostitutes are "too sick to work," they are left to the streets or "repatriated" to their home country to die. Some girls do manage to survive and even make money. Many of these return home as heroic "dutiful daughters" who have sacrificed themselves for their family. Their behavior is a model for the recruitment of younger sisters. However, the majority end up dead at a young age. Combating child prostitution is a major goal of women's movements in Asia.

Globalization is a contemporary and growing trend in the sex industry that has long been sustained by Europeans, North Americans, Australians, Middle Easterners, Japanese, and other groups. This flourishing enterprise has spawned a counterpart: "sex worker/ entertainer" migration streams. The Philippines, Thailand, China, and other Asian sources provide thousands of sex workers to such key destinations as Japan, Taiwan, and Saudi Arabia. Some 35,000 women leave the Philippines yearly to work as "female entertainers" in Japan. Terms such as *entertainer* or *artist* mask the reality of "body as commodity." Although some nations such as Thailand and the Philippines have enacted restrictions and regulations for employment abroad, they are largely ignored.

Prostitution is not only a woman's problem. Geographer Joni Seager (1997) states: "Prostitution is not a women's institution—it is controlled by men and sustained by violence." While local-level prostitutes might have a measure of control over their own destiny, "the global sex trade is almost entirely coercive, sustained by high levels of violence and predicated on the thorough subjugation of women."

WOMEN'S AND MEN'S FARMING

It is an accepted fact that women do most of the world's agricultural work. The prevalence of women in Asian agriculture is indicated in Table 3–4. However, male-reported statistics have long shorted and devalued female inputs. For example, women's contributions at planting and harvest times may be classified as "help" and not count as "work." Historically, in the realm of labor statistics, women have been rendered practically invisible.

Just as there are prescribed life-roles for males and females, so there are gender-appropriate animals and crops. Women's animals are small, of low value, and live within range of the house where they can be fed scraps. Chickens, goats, and pigs are considered women's animals. In fact, most poultry in Asia are raised by women. Men's animals are large, of high value, and require a spatially wide range for pasturage. Horses, cattle, and water buffalo are considered men's animals.

Crops are classified according to who controls their management and disposition, not who works them. Women's crops are usually vegetable and root crops suitable for subsistence or local consumption (Figure 3–12). These can be raised in spatially confined household gardens. Men's crops are typically grains, tree products, and cash crops such as tobacco, cotton, and jute designated for marketing and export, and cultivated in spatially scattered fields. These crop associations do not prevent both men and women from working in each other's domains however, although in most Asian societies women plant the rice seedlings.

Figure 3–12

A woman and her daughter sell fruit under a temple bell in Kathmandu, Nepal. The bell is rung by devotees to summon the deities. The woman is searching for lice in her daughter's hair. Mutual grooming by relatives and friends is an important mechanism of social bonding. Photograph courtesy of B.A. Weightman.

THE FEMINIZATION OF POVERTY

Gender inequity has drawn world attention since the 1990s. The United Nations and other organizations have begun to address this issue. Nevertheless, modern developments, contributing to the impoverishment of millions, continue unabated. The **global feminization of poverty** persists, with 70 percent of the world's poor being women.

The transition from subsistence farming to cash crop farming tends to favor men, as does increased commercialization. The significant role of women's gardens is downplayed or ignored, and these frequently disappear with commercialization. Studies in Sri Lanka have shown that this process has resulted in nutritional shortages. Loss of cropland to modern development forces women out of farming. Remaining prime land is generally assigned to cash crops, i.e.,men's crops. Moreover, men have greater access to capital, machines, water, and fertilizer, essential inputs that are controlled by men. Females are often employed as wage laborers on agricultural estates and in factories but do not receive equal remuneration. Men also tend not to share earnings with their entire family.

Mechanization can also be detrimental to women. In rice production for instance, tasks are often differentiated by gender. The introduction of rice planting machines in Taiwan took the business of transplanting seedlings out of the hands of women. Rice planting is now a man's job.

In Vietnam, men and women cooperate in rice production but women still do most of the work. Men prepare the paddies, women plant and weed. Both maintain irrigation and both do harvesting. Women hand-pound the grain before winnowing and drying it (Figure 3–13). Mechanized rice mills have recently displaced millions of women (and men) across Asia, hitting particularly hard in Indonesia, Bangladesh, and Vietnam. As rice mills eliminate perceived drudgery, they simultaneously eliminate what may be the only opportunity for women to earn income.

The increasingly intensive use of agricultural chemicals also takes its toll. Planting and weeding call for women to spend long hours in chemically polluted water, which causes a rise in related ailments. However, use of herbicides to kill weeds eliminates the job altogether. But the use of herbicides can contaminate water supplies causing more illnesses which are women's responsibility to deal with.

Because women's labor contributions are not officially apparent, women are marginalized and bypassed by development projects. Only a fraction of World Bank and other internationally funded projects target women directly. When funds are available, they are channeled to what are considered "women's activities." Yet it is known that women are more efficient money-handlers and more prudent in repaying loans. Furthermore, women, on average, are better at managing assets, generating additional income, and applying monetary increases to family well-being. Research in Malaysia and South Asia reveals that most programs favor the grooming of women as "housewives" in a nuclear family based on a perceived "American" model. These schemes emphasize female domesticity, feminine crafts, mothering and child-oriented activities, and their roles as "farmers' wives."

Gender discrimination also thrives in industrialized nations where segregation and subordination are

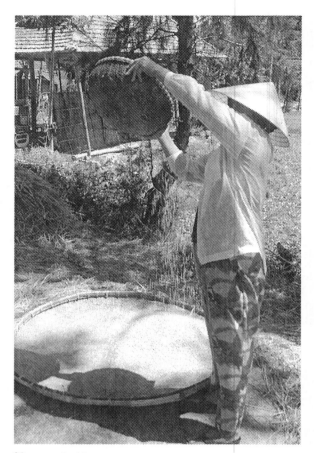

Figure 3–13

This Vietnamese woman is winnowing rice. By tossing it into the air, the chaff will blow away and the heavier rice will fall into the basket below. Winnowing is regarded as women's work. Photograph courtesy of B.A. Weightman.

intrinsic to patriarchal principles. In South Korea, women are clustered in certain occupations such as hotel workers and are paid less than men for the same position. Women are believed to have certain talents and weaknesses associated with their femininity: nimble-fingered and obedient, they lack skills required for managerial positions. Men do not like to work with or for women. In 2003, only 4.5 percent of women workers were in managerial positions in South Korea's 30 largest firms. Since every effort must be made to protect a woman's femininity, intense and complex business machinations and possible foreign travel in a "heartless marketplace" are viewed as detrimental to her welfare. Educational structure, content, and direction continue to foster these attitudes.

It might be assumed that increased education would broaden opportunity and boost income for women. But, this is not always the case. In fact, as women gain more education, male–female differentials may increase with persistent and pervasive discrimination and underutilization of women's talents. Moreover, educational systems continue to channel males into scientific and technological fields and women into the humanities. Textbooks portray perceived gender-appropriate roles.

In the wake of industrialization, a partnership between educational and industrial concerns has emerged. State-supported, cooperative programs have schools prepare students for employment as needed by the industrial sector. Factories recruit teenagers to work and even guarantee them a high school diploma. This is quite appealing to females whose parents are reluctant to support their schooling, especially beyond middle and high school levels.

With a deepening entrenchment of capitalism in Asian societies, the causes of and solutions to these different inequalities will be tied increasingly to foreign design and global enterprise. I am reminded of my recent introduction to the poverty-stricken landscape of Cambodia. Along the road from the airport to Phnom Penh, I was confronted with three large billboards: Coca-Cola, the Marlboro Man, and Godzilla. What does this say about foreign influence and modern development? The next chapter will look more closely at the idea and practice of development in Asia.

Main Point Summary

1. South, East, and Southeast Asia account for close to 60 percent of the world's people. A sporadic growth pattern was followed by rapid increase in the nineteenth century. Core areas of population emerged in valleys, plains, and coastal areas. Developments in water control and agriculture, dependent on cooperative human efforts, sustained and nourished growth and expansion.

2. Europeans, intent on harvesting and transporting Asian resources, focused their efforts in coastal areas. Ports became both focal points of economic activity and migration destinations. Many of Asia's largest cities owe their unprecedented growth to colonial influences.

3. Vegeculture was the precursor of agriculture in Asia, and numerous plants and animals were domesticated there.

4. Shifting cultivation, originally an ecologically sound practice requiring lengthy fallow periods, is declining in importance with increasing population pressure and forest removal.

5. Rice is perceived as integral to society in most of Asia. The foundation of culture, rice was domesticated in South and Southeast Asia and diffused throughout the region. Population increase on limited arable land has rendered farm plots very small in most places. Human and material inputs are intensive, and there is little room for machinery. Rice varieties are selected in relation to maturation period, personal preference, ceremonial needs, and market demand.

6. The Green Revolution fostered the use of high-yielding plant varieties. Yet, increasing demands and the law of diminishing returns means that food supply may not be able to meet population needs in the twenty-first century.

7. Children are at the core of family-group culture; there is a general preference for males. People continue to have several children for both personal and practical reasons.

8. Population growth and distribution are variable, with some countries having high natural increase rates and others low natural increase rates. Countries with large base populations, such as China and India, experience sizable increases even with lowered NIRs. Family planning programs also vary in practice and efficacy.

9. Asia has several world cities such as Tokyo, Seoul, and Shanghai. It also has many cities of over a million inhabitants, with the majority of these in India and China.

10. Sex ratios reveal that India and China continue to practice sex-selective abortion and female neglect and infanticide because of strong preferences for male offspring.

11. Statistics indicate that quality of life varies both within and between countries. Malnutrition and disease continue to plague countries such as Nepal, Laos, and Cambodia, and regions of China and India. Safe water, appropriate medical care, and basic education tend to be inadequate in all but the most developed countries such as Singapore and Japan and in more developed regions around capital cities.

12. While natural increase rates are falling, large, young cohorts entering child-bearing years are expected to put pressure on global food supply, especially grain stocks. Even with the Green Revolution, increments of crop yields will diminish and agricultural land will continue to shrink with urbanization and development.

13. Economic problems are essentially structural, centering on the control of production and distribution of resources. Power is inequitable and vested in international institutions and indigenous elites. Technological solutions often invite new environmental and social problems.

14. The culturally structured institution of gender prescribes roles, behaviors, and expectations that generally ensure the subservient position of women. Women are often segregated spatially, socially, and economically, and inequitable division of labor persists from home to work place. Development schemes may further disempower women and ensure their descent deeper into poverty. Young women are often sold into the sex industry. Some are desperate enough to voluntarily sell their services. In either case, they frequently succumb to violence and disease.

⊕ Bibliography

Asia Watch Women's Rights Project. 1993. *A Modern Form of Slavery: Trafficking of Burmese Women and Girls into Brothels in Thailand.* New York: Human Rights Watch.

Bender, William, and Margaret Smith. 1997. "Population, Food, and Nutrition." *Population Bulletin* 51/4. Washington, D.C.: Population Reference Bureau (PRB).

Bray, Francesca. 1986. *The Rice Economies: Technology and Development in Asian Societies.* Berkeley: University of California.

Brown, Lester. 1998. "The Future of Growth." In *State of the World 1998,* ed. L. R. Brown et al., pp. 3–20. World Watch Institute. New York: Norton.

———. 1998. "Struggling to Raise Cropland Productivity." In *State of the World 1998,* ed. L. R. Brown et al., pp. 79–95. World Watch Institute. New York: Norton.

Brown, Lester, et al. 2004. *Vital Signs 2004.* World Watch Institute. New York: Norton.

Cheng, Lucie, and Ping-chun Hsiung. 1998. "Engendering the 'Economic Miracle': the Labour Market in the Asia-Pacific." In *Economic Dynamism in the Asia-Pacific,* ed. Grahame Thompson, pp. 112–36. New York: Routledge.

Cho, Hyoung, and Pil-wha Chang, eds. 1994. *Gender Division of Labor in Korea.* Seoul: Ewha Woman's University.

Findlay, Alan. 1995. "Population Crises: Malthusian Specter?" In *Geographies of Global Change,* ed. R. J. Johnson, P. J. Taylor, and M. Watts, pp. 152–74. Oxford: Blackwell.

Gardner, Gary. 1996. "Asia is Losing Ground." *World Watch* 9: 19–27.

Gardner, G., E. Assadourian, and R. Sarin. 2004. "The State of Consumption Today." In *State of the World 2004,* ed. Linda Starke, pp. 3–21. New York: Norton.

Gore, Rick. 1997. "The Dawn of Humans." *National Geographic* 191: 84–109.

Huke, Robert E. 1985. "The Green Revolution." *Journal of Geography.* 84: 248–54.

Momsen, Janet H., and Vivian Kinnaird, eds. 1993. *Different Places, Different Voices: Gender and Development in Africa, Asia, and Latin America.* New York: Routledge.

Olshansky, S. Jay. 1997. "Infectious Diseases—New and Ancient Threats to World Health." *Population Bulletin* 52/2. Washington, DC: PRB.

Paul, Bimal Kanti. 1994. "AIDS in Asia." *The Geographical Review* 84: 367–79.

Paul, Bimal Kanti. 1997. "Changes in Reproductive Behavior in Bangladesh." *The Geographical Review* 87: 100–104.

Piper, Jacqueline. 1993. *Rice in South-East Asia: Cultures and Landscapes*. New York: Oxford.

Population Reference Bureau. 2003. *World Population Data Sheet*. Washington, D.C.: PRB.

Rigg, Jonathan. 1989. "The Green Revolution and Equity: Who Adopts the New Rice Varieties and Why?" *Geography* 74: 144–49.

Rigg, Jonathan. 1994. "Social Influences on the Adoption of New Rice Technology in Asia: Searching for Meaning." In *Atlas of World Development*, ed. Tim Unwin, pp. 142–43. New York: Wiley.

Riley, Nancy E. 1997. "Gender, Power, and Population Change." *Population Bulletin* 52/1. Washington, D.C.: PRB.

Sachs, Carolyn. 1996. *Gendered Fields: Rural Women, Agriculture, and Environment*. Boulder: Westview.

Seager, Joni. 1997. *The State of Women in the World Atlas* (2nd edition). New York: Penguin.

Starke, Linda, ed. 2004. *State of the World: The Consumer Society*. New York: World Watch.

UNDP. 2004. *Human Development Report*.

Chapter 4

Development, Urbanization, and Quality of Life

"If there were no contradictions and no struggle, there would be no world, no process, no life, and there would be nothing at all."

MAO ZEDONG (1893–1976)

There are many ways of defining and assessing development. No two countries are exactly alike in measurable aspects of development such as per person income, calorie intake, or number of doctors. More importantly, different countries have different ideas about what development is. For example, the strength of family ties may be more important than the availability of a medical doctor. Access to sufficient food may be more important than monetary income. Whatever the case, development is a concept widely used by governments and aid organizations in determining needs, investments, and project applications. In this chapter you will learn about various perceptions of development.

The Changing Meaning of Development

All nations have been assigned descriptors based on criteria such as per capita income (PCI) based on gross national or domestic product (GNP or GDP).

PCI is quite misleading, in part because it does not provide information on the buying power of a currency. A more contemporary measure of income is PCI based on purchasing power parity (PPP). PPP is based on GDP, accounts for price differentials among countries, and translates local currencies into international dollars. However, PCI measures do not recognize non-monetary transactions, such as bartering goods for services, that are very important in many Asian societies. Also, they assume an unrealistic equal (per person) distribution of money.

There is, in fact, a **rich-poor gap** not only between parts of the world, but also between countries and within individual nations (Figure 4–1). The global economy has grown seven-fold since 1950. The richest 20 percent consume 86 percent of the world's goods and services while the poorest 20 percent consume only 1.3 percent.

Assessment of a combination of development factors positions a country on a continuum of socioeconomic development. This development spectrum is based on misleading pairs of opposites, such as less

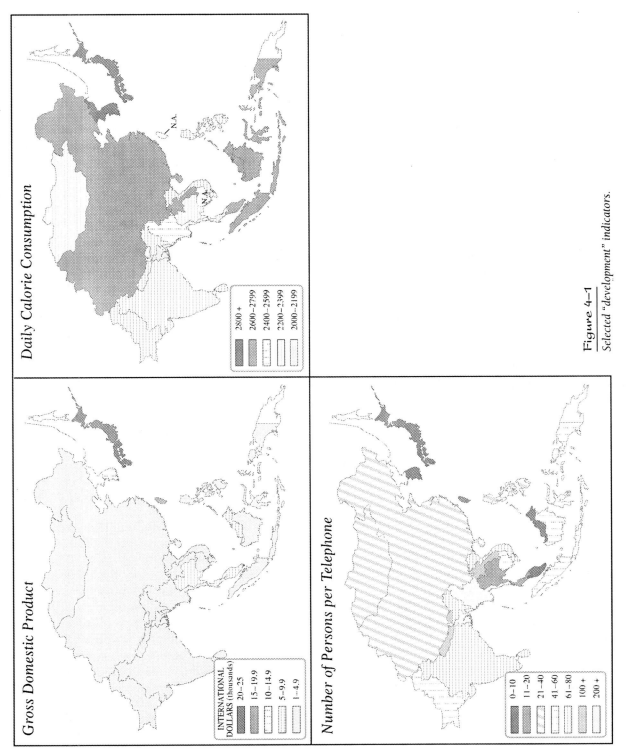

Daily Calorie Consumption

2800 +
2600–2799
2400–2599
2200–2399
2000–2199

Gross Domestic Product

INTERNATIONAL
DOLLARS (thousands)
20–25
15–19.9
10–14.9
5–9.9
1–4.9

Number of Persons per Telephone

0–10
11–20
21–40
41–60
61–80
100 +
200 +

Figure 4–1
Selected "development" indicators.

79

developed or more developed; overdeveloped or underdeveloped; and rich or poor. A term that is applied to several East Asian economies, including South Korea, Taiwan, and Hong Kong, is **newly industrializing countries (NICs)**. Others, such as India and Indonesia, are called **developing countries (DCs)**.

Through the latter part of the twentieth century, several schools of thought prevailed. For example, development trajectory proponents purported that all nations could proceed along an evolutionary path of economic advancement according to progressive stages already experienced in the West. Marxists stressed the ills of capitalism and class exploitation and the need to alter unequal power relationships within and between nations.

Another school of thought provided countries with ideologically based designations. Technologically developed, democratic nations formed the *First World,* the communist-bloc nations were called the *Second World,* and the remainder became the *Third World.* When the Soviet Union collapsed and China began adopting aspects of capitalism, and poorer countries such as Malaysia emerged as Economic Tigers, these assignations became obsolete.

A more recent theory argues that in an interdependent world, it is essential that balance be achieved to ensure mutual benefits through the control of trade between the rich, industrialized countries of the Northern Hemisphere and the poorer, former Third World countries of the Southern Hemisphere. This **North-South dichotomy** partitions the world into essentially superior and inferior realms and serves the notion of socioeconomic divergence between the haves and have-nots. It also connotes the notion of underdevelopment whereby the South's lack of development is at least partly a function of the North's development.

Regional inequities have always existed in Asia; they most certainly did not originate with Western impacts. But, ongoing disparities have been compounded by outside influences, especially in this century of global economic and political machinations.

Critics believe that most categorizations are pejorative, associated with paternalism and racism. Founded in "we-they" thinking, ideas of evolutionary progression along a path toward technological achievement and democratic institutions became moral justification for the application of Western social and economic principles to international relationships.

These principles have one thing in common: they begin with the notion of deprivation (judged by Western standards) and culminate with the concept of development incorporating measures of resource exploitation, urbanization, industrialization, and material consumption. Fraught with conflict and contradiction, such applications have long been contested by "recipient" nations.

Definitions and categorizations based on what people do *not* have and prescriptions as to how they *should* live, based on Western standards, have shaped perceptions and altered landscapes around the world. Moreover, students from NICs who study abroad learn non-indigenous concepts of planning and development and may return to apply these within their homelands.

Crowded, winding "unplanned" streets, which were once functioning, multifaceted communities, have been seen (through newly trained eyes) as inappropriate spatial geometries that must be rearranged for greater efficiency and visual appeal. Squatter settlements, home to millions, perceived as unhealthy warrens of poverty and criminality, have been razed. These are rapidly rebuilt; most governments now realize that it is more expedient to implement self-help programs within existing squatter settlements. The social costs of such upheaval cannot be measured.

The vision of nations speeding along a trajectory of technological advancement and material consumption has been shattered in many cases. Things that were supposed to happen under this rubric did not. China, for example, has initiated broad economic reform and has dramatically reduced its population growth, thereby ending the specter of hordes of Communist Chinese overrunning the planet. Singapore and Thailand have become regional economic powerhouses. Social and economic protest movements have been on the rise. Common issues such as global resource depletion and environmental crises are drawing international attention. Asian leaders speak of preserving Asian values and charting their own paths. As you will learn in the following chapters, Asian development models are rapidly and dramatically reshaping landscapes.

Cores and Peripheries

Despite indigenous development models, inequity between and within countries remains apparent and is often fueled by foreign investment. These landscape differences are better clarified in a **core-periphery** framework. The core-periphery concept expresses the idea that there is a powerful center and a less-powerful edge, a mode of thinking applicable in any context and at any scale. For instance, the world could be portrayed as having a Northern core and a Southern periphery. Or, in terms of Japanese investment in Asia, Japan is a core while its collective recipient countries form a periphery. Geographer Debra Straussfogel (1997) explains that core and periphery bracket a continuum that "represents one horizontal level within the larger global structure." She describes development as a process "whereby one structural configuration is transformed into another." As development alters socioeconomic spaces, restructuring takes place at multiple levels, thereby repositioning the country on the scale of "coreness" or "peripheryness."

An urbanized capital region is usually the core of its own state, with rural areas comprising the periphery. Some countries have several regional cores. For example, China's Beijing, Shanghai, and Guangzhou are all cores with surrounding peripheries. At another scale, these and other developing coastal cities can be conceived as a single core with the rest of China as the less-developed periphery. Cities also house core elite groups who hold relatively more power than the peripheral rural population. However, it is important to recognize that although effective power lies with the urban elite, potential power lies with the rural masses.

MULTINATIONAL INVESTMENT

Established core-periphery relationships have global impacts. Multinational corporations and capital investment by the United States, European Union, and Japan place recipient states in a position of dependency. Movement of labor-intensive manufacturing activities to peripheral locations to take advantage of cheap labor has been a hallmark of economic restructuring in the late twentieth century, often with unfavorable consequences. An assembly line might provide hundreds of low-wage jobs, but a downsizing or relocation decision in the core country can create instant joblessness. Many foreign firms in Asia are criticized for their exploitation of labor and resources and their cavalier attitudes concerning environmental quality.

Did you know that most American and British produced cigarettes are now sold overseas?

Smoking Asia

The WHO estimates that the twenty-first century will initiate 10 million tobacco-related deaths a year with more than two-thirds in developing countries. Developing countries already accounted for three-quarters of global tobacco production by 1980, and will consume 70 percent of the world's cigarettes by 2020. Tobacco use and production has declined in the industrialized world. However, exports have risen by 250 percent and Asia has become a prime marketing target.

Seven companies control about half of global sales. British-American Tobacco (BAT), along with the American companies Philip Morris (PM) and R. J. Reynolds (RJR), are the largest. In 1981, PM, RJR, and Brown and Williamson (a BAT subsidiary in the U.S.) formed the U.S. Cigarette Export Association. Flooding the Latin American market during the 1980s, they have now turned to Asia.

Convincing U.S. officials to claim that foreign restrictions on tobacco product advertising were, in effect, trade barriers, tobacco companies were exempted from national controls in Taiwan, Japan, and South Korea. U.S. market shares jumped accordingly. Advertising bans elsewhere are circumvented through sponsorship of sporting and other crowd-attracting events. In Malaysia, tobacco companies held free rock concerts and distributed free cigarettes to attendees. Promotion is furthered through hats, T-shirts, lighters, and other paraphernalia. Such strategies are aimed at both selling cigarettes to existing smokers and expanding the smoking pool.

Particular target groups are individuals with rising incomes, women seeking greater autonomy, teens, and even children and the poor. Billboards appeal to status, power, and independent thinking. A 1991 study revealed that 24 percent of teenage girls smoked while only eight percent of their mothers did. A 2004 study indicated that 25 percent of women smoke. Cigarettes are often sold individually. This makes Marlboro, Salem, and other brands affordable and ensures that the buyer will not see any warning label. This strategy also allows children to buy cigarettes, and there are increasing numbers of child smokers in the five to nine-year-old age group.

Cigarette smuggling is rampant, with one-third of all internationally traded tobacco landing in the black market. For instance, two out of every three packets sold in Vietnam have been illegally brought from Cambodia, Laos, Thailand, or China. Selling for "introductory" prices makes these more accessible and feeds the new smoker pool. Millions of cartons confiscated by the Vietnamese authorities are subsequently exported.

The tobacco industry can be important economically, contributing significantly, via taxes, to many national budgets as in Sri Lanka and Pakistan. The industry may also be a major employer. For example, in Vietnam, it provides more than 100,000 jobs and generates up to six percent of the national budget. One-third of the world's cigarettes are smoked in China where tobacco revenues are significant to the economy. China's domestic tobacco industry employs around five million people.

Some countries now have anti-smoking programs. Both Indonesia and South Korea target a portion of their tobacco revenues for anti-smoking campaigns. In Singapore, smoking is prohibited in public buildings. There are some no-smoking restaurants and cafes in Japan such as Starbucks. However, people there often ignore the no-smoking rule.

Although they know that cigarette smoking is harmful, tobacco companies produce 1,000 cigarettes every year for each man, woman, and child on earth. Currently, 3 million people die each year from tobacco-related diseases. Costs of medical care and lost productivity are in the billions of dollars. Will WHO's prediction of 10 million deaths annually materialize? Should there be curbs on the activities of American or other multinational tobacco companies overseas?

There is a World Tobacco Conference (held in Beijing in 1997) and there is an International Tobacco Control Convention endorsed by a majority of WHO members. The Convention seeks to control smuggling, regulate prices and advertising, and disseminate information. Members are in agreement that tobacco is addictive and that smoking is a type of social disease. Unfortunately, this deadly product is legal, and most limiting or controlling efforts have been thwarted by the tobacco companies.

Anne Platt McGinn (1997), a Worldwatch Institute researcher, believes that our familiarity with tobacco masks its threat. This makes the challenge a cultural one. How can public perceptions of smoking be changed? How can public health be made into a paramount value? What are your thoughts on this controversial issue?

The debatable concept of dependency has given rise to **dependency theory**. Dependency theory claims that cores can only grow through exploitation of their dependent peripheries. In order for cores to prosper, spatial imbalances must be maintained.

And, to understand a region's low level of development, it must be viewed in context of both local and global core-periphery relationships. From this perspective, trade is not mutually beneficial.

People in the periphery compete to work for very low wages, which drives down the price of goods. However, price reductions are not passed on to the consumer because producers in the core have a common desire to keep prices high. As cores and peripheries become increasingly integrated on international scales, regional economies are structured into situations of dependency on the demands of the cores and are unable to develop their own economies. But, geographies are constantly changing.

In global cores, financial, investment, and assistance rationales and applications are in question. In peripheries, economic independence and local development progress have become paramount concerns. Non-governmental organizations and privatization have taken on stronger roles in development. Such changes mean that dependency relationships are constantly shifting and increasingly complex.

Foreign Direct Investment (FDI) in Asia has increased with quantum leaps, notably in the case of China. Investment strategies are founded both in the needs of the investor and in the economic and political climate of the recipient country. Change in one area propels change in another. For example, India saw its foreign investment soar after its economic reforms in 1991. In North Korea, investment flows stopped after a 1993 crisis over its nuclear weapons program. Huge investments in China by the Association of Southeast Asian Nations (ASEAN) countries and the United States have been stimulated by the revision of China's tax code and its reassessment of incentives for foreign investment.

Any sign of instability can induce investors to retreat. War and political chaos have kept many away from Myanmar, Laos, and Cambodia. Vietnam's mercurial investment policies and disregard for venture profitability have turned off many potential investors. Indonesia's 1994 deregulation improved the investment environment for small- and medium-sized enterprises, and FDI rose significantly. However, political crises in 1997 and 1998 saw dramatic pullbacks. Nuclear testing by India and Pakistan in 1998 reverberated across the investment world.

Financial crises in Japan, Thailand, and elsewhere in Asia at the end of the twentieth century had far-reaching ramifications. Reductions in FDI are expected to be countered by mergers and acquisitions, with a gain in assets owned by foreigners. While reduced, FDI is expected to remain China-centered because investors remain attracted to the size of its market potential. Still, the presence of a cheap labor force near Singapore keeps Malaysia attractive for investment in the electronics industry. And, with its skilled work force, an improved infrastructure around Manila, and a new government committed to economic liberalization, the Philippines has also become a positive FDI environment.

FDI can bring benefits and spatial transformation. Governments can use incentives to generate investment in less-developed regions away from the core. Thailand, for example, encourages investment in areas remote from Bangkok. Moreover, Asia is no longer simply a cheap production realm; it is the fastest-growing consumer market in the world. Thus, there is a changing emphasis from fleeting use of a low-cost labor force to long-term commitment, including transfer of knowledge, skills, and technology.

FDI also creates a paradox. While providing employment and raising levels of disposable income, localization of production and management, creation of subsidiaries, and linkages with established firms simultaneously strengthen the multinational's web of influence.

Motorola, based in the United States, entered South Korea in 1967 to make transistors. An initial investment of $6.7 million has generated $260 million in exports, and the industry has spawned a semiconductor and wireless communication enterprise employing nearly 2,000 workers. As the company continues to expand in South Korea, it has earned the title of "Motorola Korea Group."

Swiss-based Nestle's operational style is to integrate itself in local economies. It has sent experts to virtually every Asian nation (except North Korea and Vietnam) to encourage coffee cultivation and the dairy industry. Large investment in manufacturing and marketing complements local product development. Nestle gains billions from Asian sales, a significant portion from Japan.

Multinationals are powerful entities of global proportions. Ford Motors has yearly sales in excess of the GDP of the Philippines. Toyota's sales rival that of Singapore. Interacting with nation-states, such conglomerates are instrumental in the creation of new economic landscapes. According to geographer Brian Berry (1989), "There are different outcomes in different world regions because they play their games on the differentiating checkerboard of culture."

PRIVATE INVESTMENT

This postmodern era is witnessing a dramatic increase in the amount of private capital flowing into Asia. Over 60 percent of international bank financing to emerging economies goes to ten countries, four of which are in Asia (South Korea, China, Thailand, and Indonesia). Deregulation of capital markets in Japan, the United States, and Europe, along with telecommunication networks permitting instant cash transfers, have facilitated this trend. International bank lending rose to its highest levels in 1996, with more loans to Asia than any other region. And in that same year, EU banks surpassed Japanese banks as the paramount lenders in Asia. The two largest borrowers were South Korea and Thailand, followed by Indonesia, Malaysia, China, and the Philippines.

In recent decades, macro-financial flows have enabled some developing nations to reverse capital flows in their own favor and fostered the rearing of Economic Tigers. On the other hand, in several countries such as South Korea and Thailand, banks floated excessive loans for too many poorly evaluated development schemes. Hundreds of projects now stand incomplete or unoccupied, their loans outstanding. Furthermore, critics of such large-scale development argue that it serves to widen the rich-poor gap.

After the financial crisis of 1997–1998, loans to Asia were cut back and are only now recovering. Many emerging market borrowers in Asia are running current account surpluses. Imports are being held down by weak domestic consumption and investment and exports are benefiting from improved competitiveness. And in the early 2000s, a slowdown in the United States economy has made lenders wary.

But what about the so-called middle class? There is a growing segment of Asian society demanding quality consumer goods, sophisticated leisure activities, better educational opportunities, and greater access to information. Political input and stability is important to this group, whose skills and purchasing power contribute to national well-being. According to Australian Asia scholars Richard Robison and David Goodman (1996), "The middle classes represent a new set of social interests that regimes must take into account."

OVERSEAS CHINESE

Who are the major investors in Asia? While Japan, the United States, and the European Union (EU) are the current leaders, Asians more frequently are investing in themselves. For instance, most of China's FDI comes from Asia-Pacific nations. One pivotal and powerful group is the **Overseas Chinese**, whom China scholar Sterling Seagrave (1995) refers to as *Offshore China*.

With enormous wealth and tightly knit family, clan, club, organizational, and territorial loyalties, Offshore China intricately involves more than 55 million individuals and countless corporations. In Southeast Asia alone, more than a hundred conglomerates have recently appeared, all owned by Chinese living in Indonesia, Malaysia, Thailand, and elsewhere in Asia. Although they live outside of China, most Overseas Chinese have long-standing connections with regions in southern China. At least 2,000 years ago, some of their ancestors left the southern provinces to colonize Southeast Asia or to serve colonial interests demanding cheap labor, and tens of thousands have followed since.

Focusing on business and trade, they maintained, expanded, and intensified their connections with various power structures to the point of becoming gatekeepers for any form of economic interchange. After each of a series of upheavals in China, more would flee or at least channel their capital elsewhere. With the Communist victory in 1949, billions of dollars left the mainland with the outflux of Nationalist Chinese and their supporters. By prudent saving and investment, timely information, and fast action via closely guarded networks, the Overseas Chinese have come to dominate the economies of East and Southeast Asia, with the exceptions of Japan and North and South Korea.

FOREIGN AID

The **spatial duality** of core-periphery development is frequently further entrenched with **bilateral** and **multilateral aid** programs. Most aid is bilateral—government to government transfers of cash, low-interest loans, material goods, food, or medical supplies. Technology and training are also transferred. Technical training, while improving skill levels, lays the groundwork for further capital input and even greater reliance on outside assistance. Aid might also take the form of preferential tariffs, whereby the donor nation agrees to buy goods produced in the recipient nation. Multilateral aid derives from international organizations such as the Organization for Economic Cooperation and Development (OECD), the World Bank (WB), or agencies of the United Nations.

If aid is concentrated in a core area, it strengthens the core and distances it further from peripheral realities. For instance, a full-service hospital is most likely to be located in the core, while poorly supplied clinics are placed in peripheral regions. Transportation development provides another example. Modern road construction typically radiates outward from the capital city. Starting as a paved highway, it eventually degenerates into an unpaved road, and ultimately into a dirt track, often impassable in the wet season. Transportation and other infrastructural conditions decline as distance from a core increases (Figure 4–2).

Tied aid, binding the source and destination in a not-always mutually beneficial agreement, is a common practice. Much official development assistance (ODA) is tied to structural adjustment programs (SAPs). SAPs require privatization, increases in exports, and adherence to free market principles. ODA is frequently siphoned off to pet projects for political reasons or to mega-projects tied to purchases from donor nations. A country might be required to import capital-intensive equipment—road-grading machinery, for instance. Such equipment gets the job done faster but also replaces labor in circumstances of already high unemployment. Moreover, because of administrative costs, less than

Figure 4–2

This crew is repairing a road in northern China. Note that there are more workers with shovels than machines. Such dirt roads can become impassable in the wet monsoon. Photograph courtesy of B.A. Weightman.

a quarter of ODA ever reaches the poor people it is intended to help.

In northern Malaysia, the Pergau hydroelectric dam was financed by Britain with soft loans—subsidized, low-interest money. In 1993, an audit revealed that the British government financed the project even though aid officials considered it a waste of financial resources. Part of the deal was that British companies in Malaysia would be granted additional development contracts. Indeed, British companies have been major recipients of Malaysian contracts since 1988. Aid is rarely simple giving with no conditions attached. Moreover, it often promotes dependency relationships.

DUAL ECONOMIES

Another spatial duality, notably expressed in urban landscapes, is the **dual economy**. Developing nations are characterized by two levels of economic activity: a *formal sector,* functioning at a large scale in the realm of national and international trade, and an *informal sector,* functioning at a smaller scale at local levels. The formal economy exhibits permanent employment, sets hours of work and pay, and regulates other benefits. The informal sector is characterized by family labor, small enterprise, and self-employment (Figure 4–3). It includes the hawking of goods and

begging. These two sectors are not necessarily mutually exclusive and should not imply relative superiority and inferiority.

A significant element of the informal sector in Asian cities concerns the selling of street foods (Figure 4–4). Professor Irene Tinker, one of the foremost authorities on women and world development, has written a book titled *Street Foods* (1997), much of which is devoted to food vendors in Asia. She refers to street vending as *microenterprise* and illustrates its vitality with case studies.

In Bogor, Indonesia, a city of 270,000, there is one food vendor for every 14 inhabitants. Street foods are vital to workers, families, and school children, and they often absorb less important agricultural products such as peanuts or corn. Some governments and planners view food stalls and mobile *becak* (pedicab) vendors as detrimental to traffic flow and hazardous to health. However, eradication efforts are rarely effective.

So vital are street foods in local economies that an array of local and external support systems has arisen to protect and enhance this informal service industry. Observers note the value of informal activities such as itinerant trash collecting, knife-sharpening, barbering, and the like for personal survival, sense of self-worth, social networking, and community service. But protection and enhancement ideals conflict with

Figure 4–3

Paper and wood umbrellas are made at this family enterprise in Myanmar. Three generations work in this small factory. Even the waterproofed paper is made on the premises. Photograph courtesy of B.A. Weightman.

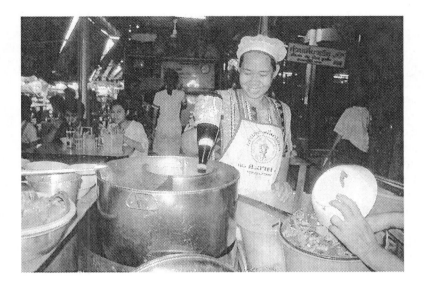

Figure 4–4
The cook at this Thai food stall is adding soy sauce to her specialty dish. Photograph courtesy of B.A. Weightman.

larger reality. Internationalization and newly affluent consumer demand have stimulated construction of multiple retail and service, Western-style malls that also have food courts, fast-food chains, and packaged snacks. These and other modernization phenomena are displacing street vendors who cannot afford to meet higher costs and new regulation requirements (Figure 4–5).

THE SHADOW ECONOMY

Increasing globalization masks a growing **shadow economy** of illegal or at least shady activities. Millions of people are involved in the black market—unregulated, untaxed, often unseen business. Smuggling of "protected" wildlife, contraband, arms, drugs, and people is widespread, especially in Southeast Asia. Crossing the border from Thailand to Laos,

Figure 4–5
Lot 10 is a popular mall in downtown Kuala Lumpur, Malaysia. Inside, escalators take you to several floors of shops and restaurants. Many stores are the same as those in your malls at home. In the basement is a fast-food court and a grocery store. Photograph courtesy of B.A. Weightman.

I was accompanied by a crowd of women, each with a few bars of soap, cigarettes, and other Thai and imported products, which they hid on their person or beneath other bundles under the vehicle's benches. Unavailable in Laos, these smuggled items were to be sold at roadside stands or local market stalls.

Increasing permeability of borders, official collusion, and poverty encourage these and other survival strategies. According to Ed Ayres, editor of *World Watch* (1996), "The growth of the shadow economy can be seen as pervasive evidence of human needs not being met by traditional institutions."

Growth of unregulated and illegal activities is also related to accountability. Not permanently based at a particular place, they are difficult, if not virtually impossible, to trace. Unregistered settlements, occupations, anarchic groups, and migratory populations transcend and escape the capabilities of extant legal strictures. Uncontrolled movement, especially rural to urban migration, is a major contributor to this situation. Every day, tens of thousands of individuals and families leave the hopelessness of their rural origins seeking opportunities perceived to exist in urban destinations. Lacking requisite skills, they swell the populations of squatter settlements and slums and bolster the ranks of the unemployed and the shadow economy.

Did you know that many products we buy are made by children, and that in Asia child labor is often necessary for survival?

Working Children

Tens of millions of children work in Asia. At least 15 percent of all 10- to 14-year-olds labor in fields and factories. Millions more, older and even younger, work 10- or 12-hour days in dangerous, medieval conditions in glass, match, and carpet factories, gravel pits, and brick kilns. They tote bricks for construction, carry baskets of raw sewage out of drains, and pick through garbage for anything salable. For the "boss" or factory owner, paying child labor a pittance is better than buying machinery for a fortune. Long hours and lack of nutrition, combined with virtually no safety standards, mean that accidents are commonplace. Dust, fibers, and chemicals cause respiratory damage and blindness. Severe injury can be fatal.

Most people condemn these practices and decry the cruelty of child labor no matter what the circumstances. Sanctions and boycotts might be successful in removing significant numbers of children from deplorable work environments, but the story is more complex.

Realities often make child labor essential for family survival. Many children of poverty have been sold into servitude by desperate parents. Many have crippled or diseased parents, or none at all. Without income they will literally starve to death. Working children are facts of life in many parts of Asia.

Enforcing child labor laws is problematic given disguised factories, hidden children, corrupt officials, and dire poverty in certain regions. Furthermore, not all work environments are terrible. Significant numbers of children work as contributors to the family farm or business in relatively decent conditions. Many go to school as well.

Education works against child labor. In fact, where schooling is compulsory, child labor rates decline. Still, families in poverty must make a decision as to whether long-term gain, via education, is worth the wait in terms of their immediate needs. Education calls for expenditures for books, uniforms, and food. This money and time spent must be balanced against lost wages.

Official and other organizations, while making efforts to end enslavement, brutality, and unsafe and unhealthy work environments, try not to further impoverish families or children by removing them from the work force altogether. Instead, they attempt to get the children and keep them in school and improve working conditions. Half-day classes are common in such countries as Myanmar, and children work before or after school. In Bangladesh, garment workers' unions actually run schools.

Child labor may be unacceptable in societies flush with palliative programs for the less fortunate, but in societies characterized by large numbers of desperate poor and limited or no safety nets, a single working child might be the only solution to survival for many.

Urbanization

Asian urbanization has skyrocketed in the twentieth century (Figure 4–6). In 1950, only 6 Asian cities were among the world's 30 largest. This number increased to 11 by 1970, and to 15 by 2004. Population and the spatial extent of cities grow by rural-urban migration, natural increase, and accretion of adjacent settlements. Some urban populations have risen dramatically only because of adminis-

trative boundary expansion. For example, Tokyo grew by 9 million people in the 1990s. But 6 million of these were added by a boundary expansion that incorporated 87 surrounding towns and cities. Moreover, forecasts of urban growth concentrating only in existing giant cities have not always materialized. In the case of China and India, it is medium-sized cities that are experiencing higher growth rates. Whatever the case, variation in city definition makes comparative urban data indicative rather than predictive.

Although cities grow significantly with incoming migration streams, growth from natural increase (NI) is even greater. These two factors are related as rural migrants have more traditional ideas about preferred family size. Consequently, NI rates in slums and squatter settlements are significantly higher than those in middle-class and elite districts. It is these poor city dwellers that many elite and middle-class groups consider to be the "population problem."

Today, most of the world's largest cities are in developing countries (DCs), with increasing numbers of these in Asia. More than half of Asia's largest cities are in India and China. With more than 30 cities of at least a million inhabitants, China has more large cities than any country on earth. Some countries show a marked degree of **urban primacy**, whereby the population of the largest city is greater than the combined

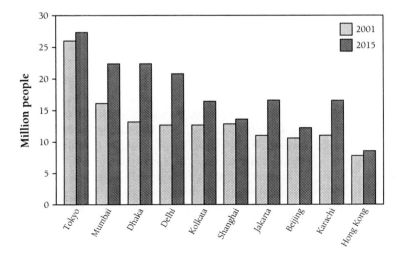

Figure 4–6

Urban growth. Cities grow in three ways: natural increase, in-migration, and expansion of boundaries. Source: United Nations Statistics—Urban Agglomerations 2004.

Figure 4–7
Cities of 1 million plus. India and China both have many cities of a million or more inhabitants, but most of their people are farmers.

populations of the next three largest cities (Figure 4–7). The largest city is then referred to as a **primate city**. Countries exhibit varying degrees of primacy depending on the population differential between the largest and next largest cities. Primate cities house relatively large proportions of national populations. Greater Seoul, for instance, houses 35 percent of South Korea's urban population, while more than a quarter of city-dwelling Japanese live in the Tokyo metropolitan region. Expansion and accretion absorb increasing shares of national populations into the urban world.

Primacy is not tied to a country's degree of urbanization nor to number of cities. Both South Korea and Japan are highly urbanized. Thailand is not highly urbanized, yet Bangkok is a primate city dominating the Thai economy. Shanghai, China's largest city, is not a primate city at the national level because China has several similarly sized cities. However, Shanghai might be seen as a primate city in a regional context, as most of China's economic development is concentrated in the eastern provinces. Primacy indicates spatial dominance at varying scales of observation.

MEGA-URBAN LANDSCAPES

In the fifth century B.C., the Chinese philosopher and sage Confucius postulated 3,000 persons as an ideal city size. In the twenty-first century, the Beijing-Tianjin metropolitan region is expected to house 37 million persons. In the early twentieth century, a mere 26 million lived in India's cities. At the turn of the twenty-first century, India's urban dwellers totaled more than 222 million. Such dense accumulations of humanity have reached unparalleled proportions with entanglements of human and environmental problems beyond the scope of simple solutions.

Polarization of development has created **mega-cities,** giant cities often incorporating several cities in a cluster. Clusters of cities, or **urban agglomerations,** have evolved into **mega-urban regions** that extend beyond metropolitan boundaries, incorporating towns, villages, and transportation corridors. These are complex fields of urban and rural interaction reaching 62 miles (100 km) beyond urban nodes.

Geographer T. G. McGee (1989) coined the term *kotadesasi,* joining the Bahasa Indonesia words *kota* (town) and *desa* (village), to represent the coincidence of urban and rural activity in the same spatial territory. This concept varies from *urbanization,* which implies a central urban system dominated by urban phenomena, processes, and values, and harboring only remnant rural elements and traditions.

Kotadesasi zones are characterized by an increase of diverse nonagricultural activities in predominantly agricultural areas. Members of the same household participate in sundry occupations in both realms, making economic linkages within the *kotadesasi* zone as important as those with the dominant city. Larger numbers of women work in non-agricultural activities. Proliferation of cheap transport such as bicycles, motorbikes, and scooters, along with public buses and trucks, allows for great mobility of people and goods within and between these and other zones (Figure 4–8).

These mega-urban, *kotadesasi* landscapes are kaleidoscopes of activity, including rice paddies, water buffalo-, bullock- and horse-carts, factories, golf courses, and traffic-jammed multilane highways. Industries locate in and subsequently pollute surrounding agricultural land. Agricultural production shifts from mono-crop cultivation (principally rice) to diverse production of vegetables, livestock, and fruit for regional and even international consumption.

In this ambiguously rural and urban environment, governed by overlapping jurisdictions, management of infrastructure can only be a nightmare. However, since the definitions of "urban" and "rural" are highly variable and frequently change, *kotadesasi* zones are sort of

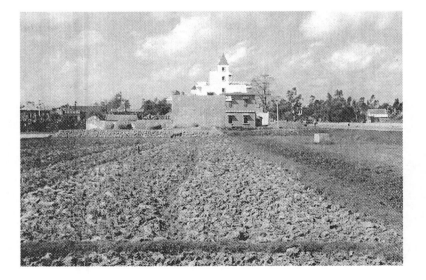

Figure 4–8
Agricultural fields and commercial establishments show mixed land use outside of Hanoi in Vietnam. In this town ceramic factories exist amid the rice paddies. The main service road to the port of Haiphong is to the right.
Photograph courtesy of B.A. Weightman.

gray territories where urban and industrial regulations may not apply. Small-scale and informal sector operators along with squatters, who find it difficult if not impossible to conform to city regulations, see these as spaces of opportunity. The result, according to McGee (1989), is that "there is no clear-cut division between rural and urban relations; rather, activities in the two sectors are fused and complementary."

URBANIZATION AND EMPLOYMENT

Cities and megacities are part of an all-encompassing international urban hierarchy topped by **world cities** such as London, New York, and Tokyo. The entire structure—from world cities, to powerful international economic centers such as Singapore, to regional market-oriented capitals such as Kuala Lumpur, through smaller regional and local urban centers—is interconnected by political and economic transactions conducted across earthspace and cyberspace. Information and its technology disseminate from **technopoles** such as The Netherlands' "Randstad" region, England's "M-4 Corridor," Taiwan's "silicon valley," centered on Hsinchu, or India's "silicon plateau" centered on Bangalore. From such world cities and technopoles, global corporations manage their global networks and subsets of urban and rural production and processing sites.

Research has shown that higher levels of urbanization are frequently related to degrees of dependency as measured by FDI, which has traditionally focused in cities. Urban job creation is a gravitational pull on the unemployed in the rural hinterland. Rural to urban migration draws less educated and unskilled people into the urban employment realm. More people require additional infrastructure, which generates more jobs, which stimulates more in-migration. This ongoing process generally results in a permanent excess of workers over jobs.

Underemployment is also boosted with urban growth in developing countries (DCs). Underemployment is a Western notion referring to the underutilization of labor. It might take the form of jobs available only on a cyclical basis such as agriculture, construction, and tourism that operate on daily, weekly, or seasonal schedules. Or, it might mean that a person takes ten hours to do a job that normally could be done in four hours. Another type of underemployment is the hiring of several individuals to do the job ordinarily held by one person. Also, there is *hidden underemployment*, based on group or family solidarity, where everyone participates in the work whether they are needed or not.

While unemployment denies means of livelihood, underemployment provides both income and sense of self worth (Figure 4–9). Homeless women

Figure 4–9

I took this picture at about 14,000 feet (4,200 m) in the Indian Himalayas. No machines were available for the repair of this main road. Instead, three women operated a shovel. In this system, one digs the shovel into the dirt, then the other two pull the rope that pulls the shovel to move the dirt. Some shovel-rope teams comprise only two people. While this system may not seem very efficient, it provides more employment than a machine. Moreover, the shovel-rope system is appropriate to the terrain. Photograph courtesy of B.A. Weightman.

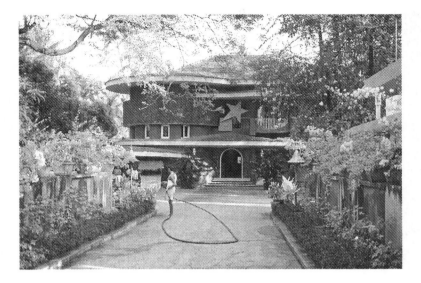

Figure 4-10

This beautiful house in India shows how the upper classes live. Note the gardener watering the plants. I took this picture on Christmas Day, and the star above the doorway indicates that this is a Christian household. The star is lit at night. Photograph courtesy of B.A. Weightman.

in Mumbai chop rocks to make gravel, carry bricks, mix cement, and do other arduous tasks. Meanwhile, they can live within the project they are building. When it is finished, they will move on to another construction site. Children work alongside their parents. Economically, at the national level, this is not an ideal situation because the cost of services for surplus labor takes away from potentially broader expenditures. But, individually, at the local level, underemployment is preferable to no employment.

With disproportionate investment and capital concentration, cities exhibit dramatically disparate landscapes. Landscapes of power and privilege contrast with those of powerlessness and denial. Yet, many poor are employed by their rich counterparts as watchmen/guards, house servants, gardeners, drivers, and the like (Figure 4–10).

Another vital urban phenomenon is the **transborder development region.** Trans-border development regions include economic development zones in adjacent political units. A region may include several countries, as in the case of the "development triangle" involving Malaysia, Singapore, and Indonesia. Or, it might involve several provinces, as in the case of southern China. Development zones are integrated across borders by transportation corridors and telecommunication links into the global system of production, commerce, finance, distribution, and consumption.

The pace of urbanization and landscape change in general is exacerbated with global economic interchange that, in essence, ignores political boundaries, human costs, and environmental concerns. Fortunately, there is growing awareness that we are running a race on a deteriorating environmental track. Development, whether indigenous or international in scope, is consuming the social and natural capital upon which it depends.

An important dictum is that "everything is related to everything else." Have you thought about the connections between information in this chapter and that of previous chapters?

Tale of the Tiger Prawn

Prawn fever in Asia has been triggered by consumer demand in such countries as Japan, Singapore, Canada, and the United States. Tiger prawn (large shrimp) farms made their initial appearance along the mangrove coasts of

China, Taiwan, and South Korea in the 1970s and have since diffused into South and Southeast Asia.

Investors convince rice farmers or fishers to sell their land or take up prawn farming for quick profits. Using their property as collateral, farmers take out high-interest-rate loans. However, what is promoted as a boon to the local economy favors only the "aquabusiness" investors, who also own cold-storage facilities and sell their own feed, chemicals, and pumping equipment. Companies are both national and transnational in scope, with an array of business interests. Mitsubishi of Japan and Aqua Star of the United States are two of these.

Coastal mangroves are cleared to make space for the prawn ponds. Fresh seawater must be brought in regularly, and the ponds must be pumped clean of prawn excrement and chemical additives. This wastewater further destroys coastal ecology, and ground water salinization ruins fields for rice. Finally, pollutants poison surrounding seawater, rendering it unsuitable to flush ponds. With increased contamination, prawn production declines, and ponds are ultimately abandoned.

Big business opens ponds elsewhere, but small entrepreneurs are doomed. No longer able to make loan payments, they become landless and unemployed. Developers take over the once productive rice fields for industrial, housing, and tourism projects. Those displaced provide a ready-made source of construction labor.

Thailand's eastern coast, a prawn production haven in the 1980s, has reaped this destruction. With mangroves gone and rice paddies abandoned, coastal communities have only tourism in their sights. Prawn companies have since moved to Thailand's western coast. The Thai government is loath to end the practice, not only because various officials are involved in the business, but also because it generates a significant portion of export earnings.

In Bangladesh, large aquabusiness investors made millions on prawns while ruining farmlands of thousands of already poor rice farmers. As prawn ponds were abandoned, investors purchased the land for next to nothing, and thousands of families migrated to nearby towns and cities.

The tale of the tiger prawn illustrates the interrelationships among numerous factors: big business, transnationals, foreign investment, foreign exchange, international trade, ecosystem destruction, environmental pollution, cropland abandonment, family ruination, displacement and unemployment, rural to urban migration, growth of cities, and expansion of urban-powered development into the countryside.

Postmodern Thinking

Geographers and others now recognize that *development is multidimensional* and that *there is no single appropriate development framework*. Geographer John Agnew (1987) emphasizes the return of cultural context to the concept of development, noting that every part of the world has its unique relationship with other places individually and collectively. New concerns, such as *meeting basic needs* and *economic reform*, have emerged. Fresh categorizations have been created as well.

The United Nations Development Programme (UNDP) published its first *Human Development Report* (HDP) in 1990. Although it argues the necessity for growth in national production, it also recognizes that some states have achieved high levels of human development with only modest per capita incomes. The HDP devised the Human Development Index (HDI), which ranks 175 countries according to a composite measure based on income, health, life expectancy, and education information. In 1995, a Gender Development Index (GDI), based on similar data, which indicates discrepancies between males and females, was compiled (Figure 4–11).

Intermediate technology has become fashionable. Beyond the digging stick and behind the tractor, such technology uses local human and

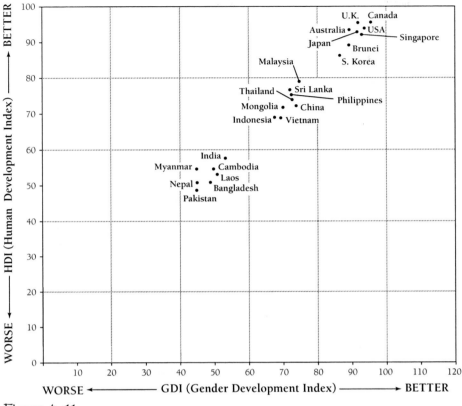

Figure 4–11

The human condition. Combining the Human Development Index with the Gender Development Index gives us a better idea of living conditions than measures such as per capita income or number of TVs per hundred people. Source: United Nations Human Development Report 2004.

material resources to effect positive change. For example, it uses local building materials rather than relying on foreign imports. Other examples include using solar energy or biogas for power. Such "appropriate" technology has become significant in many places in Asia and is tied largely to private organizations.

Non-Governmental Organizations (NGOs) reflect contemporary culture-based perceptions of the development process. All have an agenda, some political, some humanitarian, others religious. Nevertheless, NGOs tend to focus in peripheral settings such as urban slums and disadvantaged rural areas. Many NGOs are involved in health and human resource assistance programs. Ideally, culturally

derived development models and programs can be articulated with larger, international frameworks of change. Unfortunately, disparity between the ideal and the real seems unavoidable.

Because they address people's grievances and could become rallying points for political protest, NGOs are seen as potentially threatening by some governments. Indonesia, Malaysia, and other countries attempt to rein in NGO activities with tight controls and bureaucratic minutiae. Other countries, such as South Korea, have more comfortable relationships. Here, more than 20,000 NGOs maintain projects in areas ranging from environmental protection to political freedom and women's rights.

PRIVATIZATION

As development proceeds, demand for public services, infrastructure, and shelter rises. Not only do sizable debts and shortage of revenues constrain governments, but also corruption, patronage, and mismanagement frequently render state-owned enterprises money losers. Efficiency is precluded by labyrinthine regulations and bureaucracies. For example, it can take as long as 10 years to get a private phone installed in Pakistan. Community phones are the rule in most Asian countries, and call completion rates are generally low.

Growing dissatisfaction with central planning and government management of services has promoted *privatization* in various forms. Public assets or enterprises might be sold to private investors, as in Bangladesh. Or as in China, public agencies are *corporatized* and required to cover costs and manage operations more efficiently. More commonly, governments pursue policies encouraging private-sector participation in public service provision. Countries such as Thailand, Malaysia, and South Korea are promoting public service delivery by community, religious, or professional NGOs, cooperatives, and small operators. The Roman Catholic Church operates schools in the Philippines, and physicians' groups have opened clinics in crowded urban areas of Vietnam.

Privatization has affected the airline-passenger industry. State-owned Singapore Airlines and Malaysia Airlines were privatized in 1985, and Philippine Airlines and Thai Airways International were privatized in 1992. Deregulation has encouraged the development of new airlines such as EVA Airways and Asiana. Geographers John Bowen and Thomas Leinbach (1995) note the importance of airlines in the development trajectories of NICs. By increasing access to places, they open up new domestic and international markets. Asian airlines are now playing significant roles in global interchange. Even so, the retention of vested interests by many governments indicates their reluctance to relinquish full control over what they perceive to be a strategic element of infrastructure.

Housing is another sector in which governments have sought private help. Sites and services programs have gained significance since the 1970s. Government agencies install infrastructure and provide essential services on cleared land that is sold with low-interest loans. Poor families contract with private builders or construct their own dwelling with subsidized materials. In India, the government registers and assists housing cooperatives to purchase land and procure financing for low-cost housing. In several countries, private corporations are responsible for the development of complete communities inclusive of residential, commercial, educational, and recreational facilities. Despite these ambitious programs, most shelter for the poor still derives from the informal sector as squatter or temporary housing.

Privatization is not readily accepted everywhere. Individuals and constituencies with vested interests in state-controlled enterprises strongly resist privatization efforts. In Indonesia and Malaysia for instance, opposition is voiced by economic nationalists and Muslim fundamentalists, who detest Western consumerism and capitalism. Some people in these countries also fear that the already economically dominant Chinese will be the main beneficiaries. In Thailand, opposition comes from the military, which exercises significant control over public corporations to ensure employment for high-ranking retirees. Labor unions and other groups which depend on patronage, paybacks, and similar benefits also oppose privatization. In light of increasing urbanization, the provision of shelter and services will continue to challenge both private and public institutions.

CHANGING STRUCTURES OF CHANGE

A hallmark of postmodernism is rapid and on-going change. As views of development have changed, so have tools employed to interpret developmental processes. The **flying geese model** of economic development will illustrate.

The flying geese model, developed by a Japanese economist in the 1930s, was used in the 1960s to describe industrial life cycles in recently industrialized nations. Industry- and setting-specific at first, the model was advanced to examine changing industrial structures and international shifts in location.

According to the model, an industry starts with a product import or technology transfer from an industrialized nation. An imitation is domestically produced and substitutes for the original product import. As domestic production exceeds domestic consumption, imports decline and an export surplus is created. Subsequently, exports and production begin to decline. As domestic costs increase, industry looks elsewhere and becomes a net importer again. Industries are thereby cycled in an orderly fashion, akin to the structured pattern of flying geese. Over time, industries rise and fall, with the old being replaced by the new. A country that produced textiles in the 1920s might have focused on steel in the 1940s, chemicals in the 1960s, and automobiles in the 1980s. Each industry is increasingly sophisticated and technologically complex, illustrating the development process.

The "flying geese" analogy can also be applied to a sequence of countries as each moves along its flight path of industrial development. For example, in the 1950s, a labor-intensive manufacturing focus (textiles) characterized Japan as a NIC. This has been replaced by capital-intensive machine industries to the extent that Japan was classified as an industrialized country in the 1980s. From an international perspective, the "geese" of modernization are seen flying from more developed countries to less developed ones. At the same time, a sophisticated division of labor emerges among Asia-Pacific nations.

The 1990s witnessed criticism of the flying geese model of development. Perhaps applicable in the past with its vision of Japan leading the Asian flock down the road of progress, the model overemphasizes Japan's role and underplays that of the United States as an export market for Asian goods. Further, it does not account for the fact that imports might be from one's own international subsidiaries or joint ventures, or exports might be dominated by primary (non-manufactured) products.

Another criticism rests on the fact that industries do not pass *in toto* from one country to another in a linear pattern. Rather, change is manifested through sub-sector specialization. Each time a new technology is commercialized, a new pattern of clustered economies and division of labor will appear. Furthermore, postmodern-era production is interdependent, at once concentrated and dispersed across hierarchies transcending multiple borders. According to economists Mitchell Bernard and John Ravenhill (1995), firms no longer simply seek comparative advantage in location relative to raw materials or labor force for an entire production process. They assign location in terms of "the comparative and technological advantage of *particular* territorial sites for *particular* parts of the production process." Parts of the same product may be produced within the same industry at the same time in different countries.

As you will see in the following regional chapters, the meaning, practice, patterns, and consequences of development are in constant flux. What does *development* mean to you? How has it changed your community or city? Think about these things, because according to the Chinese philosopher Hsun-tzu (300–230 B.C.), "Through what is near, one understands what is far away."

⊕ Main Point Summary

1. The concept of *development* is difficult to define; its meaning varies spatially and temporally. Designations of *development* and measures of *progress* have been derived from Western notions and models. Western ideas have been superimposed on other landscapes, often with undesirable consequences.

2. Per capita income (PCI), a common indicator of level of development, has many shortcomings. The rich-poor gap and other variables are not revealed by this measure.

3. Asians are attempting to define their own culturally appropriate means and modes of development. This is no simple task in light of involvement in global technological and financial networks.

4. Foreign direct investment (FDI) has furthered core-periphery development patterns from local to regional and global scales. Multinational corporations have benefited at the expense of environmental and human health in newly industrializing countries (NICs). As cores and peripheries

become more integrated internationally, regional economies are structured into dependency relationships. Spatial patterns and networks of power relationships are in constant flux.

5. FDI in Asia has increased dramatically from the 1980s to the 2000s. In the late 1990s, political crises in Indonesia and monetary problems in Japan, Thailand, and elsewhere witnessed falling FDIs. China is of special interest to investors because of the size of its market potential; it now dominates the FDI scene.

6. FDI generates increased incomes and job opportunities but further entrenches local economies in multinational webs of subsidiaries and alien control mechanisms.

7. As a growing middle class of consumers demand quality goods and services, private investment is being sought to counter the hold of foreign-owned enterprises. Japan, South Korea, Singapore, and other Asian countries are investing more in Asia. Overseas Chinese are a powerful source of investment in East and Southeast Asia.

8. Most Asian countries have dual economies with formal and informal sectors. Informal activities such as food vending are essential for the survival of millions. Many rely on the hidden, often illegal activities of the shadow economy. Drug and cigarette smuggling, child labor, and prostitution are means of profiteering for some and survival for others.

9. Urbanization has skyrocketed, with more than half of the world's 30 largest cities in Asia. Asian cities are part of a hierarchical urban system topped by world cities such as Tokyo, New York, and London. This global, urban network is linked through technopoles via cyberspace.

10. Ongoing rural-urban migration and further population increase through natural increase (NI) place demands on city infrastructures and services that cannot be met. FDI in urban projects further attracts migrants and boosts unemployment and underemployment.

11. Megacities and mega-urban regions have emerged to include a disorganized mixture of urban and rural activities. *Kotadesasi* describes the coincidence of urban and rural land use in the same territory. Economic inefficiencies and jurisdictional confusion are problematic.

12. Recent development strategies include transborder development regions, NGO involvement with the urban and rural poor, government-supported self-help programs, especially in the realm of housing and services, and the implementation of intermediate technologies.

13. Some governments are releasing control over state-run enterprises to private groups. Services such as public transportation and housing provision have benefited significantly from privatization efforts.

14. Western domination has triggered a reaction to homogenization of landscapes and cultures. Greater emphasis is being placed on regional and cultural differentiation, which has the potential to backfire in factionalism. Consumerism and capitalism are being reassessed in context of both traditional and contemporary values and belief systems.

15. In this postmodern era, change is fast-paced and far-reaching. Commonly accepted models of development such as "flying geese" are no longer applicable, given the uncertain complexities of development from local to international scales.

⊕ Bibliography

Agnew, John. 1987. *Bringing Culture Back*. In: "Overcoming the Economic-Cultural Split in Development Studies." *Journal of Geography* 86: 276–81.

Ayers, Ed. 1996. "The Expanding Shadow Economy." *World Watch* 9: 10–23.

Bell, Morag. 1994. "Images, Myths and Alternative Geographies of the Third World." In *Human Geography: Society, Space, and Social Science,* ed. Derek Gregory et al., pp. 174–99. Minneapolis: University of Minnesota.

Bernard, Mitchell, and John Ravenhill. 1995. "Beyond Product Cycles and Flying Geese: Rationalization, Hierarchy, and the Industrialization of East Asia." *World Politics* 47: 171–209.

Berry, Brian. 1989. "Comparative Geographies of the Global Economy: Cultures, Corporations, and the Nation-State." *Economic Geography* 65: 1–18.

Bowen, John T. Jr., and Thomas R. Leinbach. 1995. "The State and Liberalization: The Airline Industry in the East Asian NICs." *Annals of the Association of American Geographers* 85: 468–93.

Cho, George. 1995. *Trade, Aid and Global Interdependence.* New York/London: Routledge

Douglass, Mike. 1995. "Global Interdependence and Urbanization: Planning for the Bangkok Mega-Urban Region." In *The Mega-Urban Regions of Southeast Asia,* ed. T. G. McGee and I. M. Robinson, pp. 45–77. Vancouver: University of British Columbia.

French, Hilary. 1997. "When Foreign Investors Pay for Development." *World Watch* 10: 8–17.

Grant, Richard, and Jan Nijman. 1997. "Historical Changes in U.S. and Japanese Foreign Aid to the Asia-Pacific Region." *Annals of the Association of American Geographers* 87: 32–51.

Lipton, Michael. 1977. *Why Poor People Stay Poor: Urban Bias in World Development.* Cambridge: Harvard University.

McGee, T. G. 1989. "Urbanisasi or Kotadesasi? Evolving Patterns of Urbanization in Asia." In *Urbanization in Asia: Spatial Dimensions and Policy Issues,* ed. Frank J. Costa, Ashok K. Dutt, Laurence J. Ma, and Allen G. Noble, pp. 93–108. Honolulu: University of Hawaii.

McGinn, Anne Platt. 1987. "Cigarette Traffickers Go Global: The Nicotine Cartel." *World Watch* 10: 18–27.

Quarto, Alfredo. 1993. "The Vanishing Mangroves." *Third World Resurgence* 39: 4–6.

Ransom, David. 1996. "The Poverty of Aid." *New Internationalist* 285: 7–10.

Robison, Richard, and David Goodman. 1996. "The New Rich in Asia: Economic Development, Social Status and Political Consciousness." In *The New Rich in Asia: Mobile Phones, McDonalds and Middle-Class Revolution,* ed. R. Robison and D. Goodman, pp. 1–16. New York: Routledge.

Rondinelli, Dennis A., and John D. Kasarda. 1993. "Privatization of Urban Services and Infrastructure in Developing Countries." In *Third World Cities: Problems, Policies, and Prospects,* ed. John D. Kasarda and Allen M. Parnell, pp. 134–60. Newbury Park: Sage.

Seagrave, Sterling. 1995. *Lords of the Rim.* London: Bantam.

Straussfogel, Debra. 1997. "Redefining Development as Humane and Sustainable." *Annals of the Association of American Geographers* 87: 280–305.

Tuan, Yi-fu. 1969. "Cultural Pluralism and Technology." *The Geographical Review* 79: 269–79.

Tinker, Irene. 1997. *Street Foods: Urban Food and Employment in Developing Countries.* New York: Oxford.

Weissman, Robert. 1997. "Move to Make U.S. Tobacco TNCs Accountable Internationally." *Third World Resurgence* 92: 7–10.

Chapter 5

South Asia: Creating Dilemmas of Diversity

"What is here is nowhere else;

what is not here, is nowhere."

THE *MAHABHARATA* (C. 1000 B.C.)

This chapter initiates the regional analysis that makes up the rest of this text. In it I will set the historical-geographical framework for subsequent discussions of specific countries in South Asia. For an understanding and appreciation of conditions, events, and swirling controversies in South Asia, historical knowledge is indispensable. Therefore, after a brief review of the physical setting, I will focus on South Asia's history up to and including the creation of India and Pakistan in 1947.

The Physical Setting

South Asia is a realm of spectacular highlands, vast deserts, rugged plateaus, fertile lowlands, and magnificent coastlines. Here mountains and coastlines combine to create one of the world's most sharply defined physiographic regions (Figure 5–1). Splaying southward from the Himalayan wall to the north, the Indian subcontinent is differentiated in the west by the Punjab, a region through which

flow the Indus River and its four major tributaries: the Jhelum, Chenab, Ravi, and Sutlej. Punjab means "five waters." South of the Punjab is the Indus River plain. This entire region is guarded by the Hindu Kush and other mountain ranges to the north and west, and the Great Indian or Thar Desert to the east. Farther east, the North Indian Plain extends to include the sacred Ganges River and the amazing Ganges-Brahmaputra delta. Southward, the Central Indian Plateau is separated from the Deccan Plateau by mountains and waterways such as the Vindhya Range and the Narmada River. These physical features have presented barriers to north-south movements throughout Indian history. Farther south is the highly dissected Deccan Plateau bordered by the Western and Eastern Ghats. The Deccan is cut by several rivers, most notably the Godavari, Krishna, and Cauvery. The island of Sri Lanka is located at the southeastern tip of peninsular India. It too is a troubled region. You will learn more about these regions in this and the following three chapters.

Figure 5–1

Landforms and waterways of South Asia. Compare this map with Figure 7–3 and note how topography influences settlement patterns.

Monsoon and Life Cycles

As I noted in Chapter 2, a large part of life including harvesting and planting, corresponds with the dry and wet monsoons. This is especially the case in South Asia, where the landscape evolves in a never-ending cycle of brown to green, sparse to lush, and lethargic to energetic. For millions, these cyclical events imitate the greater, overarching cycle of birth, decay, and rebirth. For countless others, they reveal the divine will or hand of God. Forgotten rituals or improper behaviors might trigger drought or flood with disastrous consequences.

In northeastern Hindu India, the great festival of Rathajatra at Puri celebrates the transition from the season of drought to the season of rain. A two-story-high vehicle, known as a juggernaut, is constructed to carry a likeness of Jagannath "Lord of the World," a form of Vishnu. To the clashing of cymbals and the chanting of priests, the behemoth is pulled toward the excited throng. As Jagannath nears, people joyously leap and sing and chant "Hari! Hari!" (God! God!). Each creaking and turning of the great wheels speaks of impending change and assures the coming of the wet monsoon. Participation in this and other seasonal events reaffirms rebirth and reasserts one's union with nature.

Archives of Time and Place

Separated from the Sino-Tibetan cultural realm by the Himalayas, the Indo-European culture realm has experienced its major cultural intrusions and contacts via mountain passes (through the Pamirs and Hindu Kush) and via Arab and other sea lanes in the Indian Ocean. Conquerors and adventurers, scholars and writers, soldiers and sailors, merchants and traders, men and women, rich and poor—all were captivated by the richly diverse and fascinating landscapes of what ultimately became the prize possession of the British Empire—India, Queen Victoria's "Jewel in the Crown."

When the British Empire reached its greatest extent in the nineteenth century, the map of South Asia was quite different. Pakistan and Bangladesh, for instance, did not exist. Actually, when ancient Indian empires were stunning the known world with their spectacular achievements, the British Empire was unheard of because it did not exist. A venture into South Asia's intriguing past is surely worth pursuing.

Geography plays a vital role in history; space and time are twisted together like a two-colored rope. Mountains such as the Himalayas and the Hindu Kush, and deserts such as the Thar (Great Indian), marked defensive frontiers. In the days when territory was traversed on foot or by animal, terrain conditions could serve as barriers or at least make distances impractical to cross. Control over strategic sites such as river fords and mountain passes could make or break empires. For instance, a dozen or so rulers chose Delhi as their capital because of its strategic situation. Delhi guards the western gateway into the **Doab** (meaning two rivers), the level and fertile land between the Ganges and its main tributary, the Yamuna (Figure 5–2). Much of India's history concerns mountain passes, especially those in the Hindu Kush. The Khyber Pass is one of the most fought-over passes in the world (Figure 5–3).

With slow and difficult travel, overextension could easily lead to territorial losses. The Deccan, for example, was a difficult region to conquer and keep, and it typically was subdivided among several rulers. Many an empire or state bordered the Vindhya Range, which forms the northern flank of the Narmada River, which runs westward into the Arabian Sea. Accessibility, or lack of it, played a key role in the relatively separate development of southern India.

Lack of consideration of the surrounding Indian Ocean blinded Indian rulers to the seriousness of maritime encroachment by European powers. Mere trading posts along the shores were not read as clues of impending conquest, as no one had ever conquered India by sea. The sea was seen as the black water, and its monsoons would certainly deter even the most daring intruders. Geographic perceptions

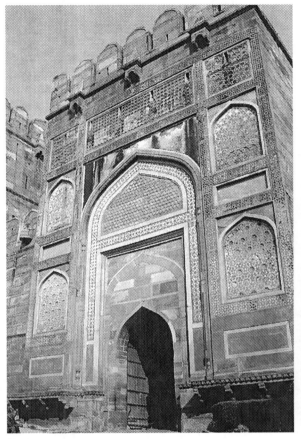

Figure 5-2

This is the Red Fort in Delhi, India. Built by the Mughal emperor Akbar between 1565 and 1573, its red sandstone exhibits Indo-Islamic architecture. It stands guard over the Yamuna River. Photograph courtesy of B.A. Weightman.

Figure 5-3

Here is the Pakistan side of the Khyber Pass. The road was built by the American Corps of Army Engineers. Note the hairpin turns, switchbacks, and tunnels. I remember traveling this route in a Russian car with a Pakistani driver who sped along the road, usually on the wrong side, with his hand constantly on the horn. We had several near misses with huge trucks headed for Afghanistan. Photograph courtesy of B.A. Weightman.

along with physical realities were clearly integral to historic developments in South Asia.

Many historians divide the subcontinent's history into Hindu, Islamic, and British periods. However, as landscape expressions of these transcend both time and space, I will use *ancient, medieval,* and *modern* to designate critical shifts in historic evolution. History is a progression through time, which makes necessary the inclusion of dates. Think of these as markers of time, not as something to be memorized. Reference to text and atlas maps will help you envision the upcoming story.

ANCIENT SOUTH ASIA

The discovery of crude stone tools in the Punjab shows that human life in India can be dated to the period following the Second Ice Age (400,000–200,000 B.C.). Cultivation of grain commenced around 7000 B.C. in South Asia, but major settlement did not occur until around 3000 B.C., when a notable increase in precipitation and water-control technology allowed large-scale, settled agriculture.

The Harappan or Indus Civilization emerged as the world's third major civilization after those in Mesopotamia and Egypt. It is thought that the people were dark-skinned, proto-Dravidians who spoke a precursor of Tamil, now an important language in southern India. With more than 100 settlements spread over a 1,000-mile (1,609 km) stretch of the Indus plain, the region also supported three major cities: Harappa, Ganweriwala, and Mohenjo-Daro (Figure 5–4).

When unearthed by archaeologists in the 1920s, the ruins of Harappa and Mohenjo-Daro revealed a phenomenal level of sophistication. Both cities conformed to a general model of ancient city planning found elsewhere. Constructed of baked bricks, each walled city had a fortified citadel and a large granary. Oriented to the cardinal directions, streets were grid-patterned. Courtyards shielded the two-story, dwellings of the elites from the smaller quarters of common laborers.

Amazingly, these cities had garbage collection, an elaborate system of running water, and sewage disposal. Elite homes had baths and "toilets," water channels to carry waste to underground sewers and septic tanks. Water was essential to the prosperity of these cities and apparently took on ritual/religious significance. The "Great Bath" at Mohenjo-Daro might have been used for ritual ablutions, suggesting the existence of a powerful priesthood. To this day, Hindu temples have rectangular water tanks for ritual washing and purification purposes.

Religious cults abounded, but the embryos of current beliefs were present. The Bo tree (a type of banyan), under which Buddha attained enlightenment centuries later, was sacred then as now. There were solar symbols, such as the wheel of light rays called *su-asti*, "well-being" in Sanskrit. In Germany in the twentieth century, this benign symbol was corrupted and misused by the Nazis as swastika, their emblem of Aryan purity. The *su-asti* remains a popular symbol in both Buddhism and Hinduism.

Achievements in urban development imply a sizable agricultural surplus and a well-organized, authoritarian control structure. Harappa alone supported around 40,000 people. Wheat, barley, and rice, along with domesticated sheep, goats, and fowl, plus wild game provided food surpluses that allowed people to indulge in other pursuits. Some settlements specialized in certain activities such as mining or pottery. Also, there was trade by land and sea with Mesopotamia. Indian crews navigated the Persian Gulf with the assistance of a "compass-bird"—a crow that would fly directly to the nearest point of land. There is evidence that Indic cotton was a valued trade item. To this day, cotton is India's most important export.

But the Harappans could not control all of nature's forces. Mohenjo-Daro was rebuilt at least ten times and apparently was abandoned in panic during a severe flood. Recent research shows that despite their mastery of water control, it was climatically changed water balances that were instrumental in the fluctuating fortunes of the Harappan civilization. Steep decline took place from 1800 to 1700 B.C., corresponding with a decrease in rainfall. Water problems and food shortages combined with other crises to drive the **Dravidians** southward.

Figure 5–4

Indus Civilization, c. 5000 B.C.

INVADERS FROM THE NORTHWEST

The Aryans

Around 1500 B.C., lighter-skinned, nomadic herders called **Aryans** appeared out of Central Asia and Afghanistan. The Sanskrit-speaking Aryans thrust through the Khyber and other passes of the Hindu Kush onto the plains of the northwest, where another wetter period nurtured pasturage suitable for their animals. It is important to note that while Aryans would have far-reaching influences in South Asian history, from the start they were largely incorporated into existing cultures. In other words, India was not "Aryanized." Rather, it became even more diverse and complex.

The Aryans brought cows and horses to India. In fact, cows were an early form of wealth and currency. Eventually these would become sacred in Hinduism. Apparently, horses were employed to pull chariots—the basis of cavalry warfare—and their use spread quickly throughout the continent. Horses made possible the expansion of empires, fostered the militarization of states, and enabled more expedient collection of taxes. As horse-breeding in India was difficult, they were brought from Arabia and Persia at great expense.

While copper and bronze were already in use, the discovery of iron near Patna, combined with Aryan knowledge of iron-working, were instrumental in territorial conquests. Used for weapons, tools, and plows, iron facilitated both forest removal and agricultural settlement.

Settlement along the Ganges had been limited to the Doab and the northern foothills and tributary valley. A possible reason is that rivers were more easily crossed there. Another reason is that dense jungles blocked access to the lowlands. Eventually another dry period narrowed the rivers and the jungle receded. Iron implements furthered settlement on the Ganges plain, and forest dwellers were either conquered or absorbed. In the *Mahabharata* (an epic poem of Hindu ideals and civilization), these non-Aryan forest dwellers were called *nagas* or "serpent people."

Unlike the Harappans, the Aryans did not have a script for Sanskrit, their spoken language. For centuries, the *Ramayana* (an epic poem about Rama, a prince of India) and other stories were memorized and passed on verbally. A script was not developed until the fourth century B.C., but religious taboos forbade the actual writing of religious tomes. Nevertheless, Sanskrit is rich in pejorative terms, and these were willingly used to denigrate non-Aryan peoples. Many were described as physical freaks, and these images became part of medieval Europe's perceptions of the East.

Seeds of Caste Division

The Aryans, whose own name means "noble" or "first born," used the term *dasa* for people with darker complexions. Initially, the term meant "enemy" and later, "subject." *Dasa* always implied dark-skinned people. Even today the preference for lighter skin persists. Marriage advertisements often list "wheaten complexion" as a desirable characteristic.

Social norms including social hierarchies were clarified in the *Shastras*, legal texts. These were the "Laws of Manu," Manu being the first man. Society became organized around four classes of people, or **varnas**, which means "color." The four classes were *brahmins* (priests); *kshatriyas* (warriors); *vaishyas* (peasant farmers); and *sudras* (serfs). Subjugated people, including the darker Dravidians, could only enter the system at the lowest level. Those who had closest contact with earth, fire, or water, such as artisans, smiths, and millers, were despised. The most skilled artisans in ancient India were Dravidian. *Sudras* were excluded from ritual affairs because they were seen as impure. In Hinduism, ritual purity was of the essence. This was the precursor of the **caste system**, which still endures despite the Indian constitution of 1949 that seemingly outlawed castes. The English word *caste* comes from the Portuguese word *casta*, used to designate people of different colors in Brazil.

By the sixth century B.C., regional rivalries had produced a string of states across the northern plains. A sophisticated commercial sector evolved, with all the necessary trappings such as coinage, taxes, and bureaucrats. Prosperity fostered economic specialization and *varna* subdivisions, known as *jatis*, multiplied.

A *jati* is typically defined and named by occupation. It is also an extended kinship group that practices group **endogamy**. This practice of marriage within the *jati* ensures continuity of its principles and practices. Caste and *jati* are at once divisive and unifying. While fragmenting society at large, they provide structured mutual exchange and support systems at communal levels.

Within a growing mercantile class, the *varnas* had to be further modified. Commercial dealers were taken into the *vaishya* class and farmers were demoted to the *sudra* class. Serfs and other low-order *sudras* became casteless (i.e., outcasts whose designation ultimately hardened into **untouchables**). Today, many untouchables prefer the term *dalits*, meaning the "oppressed."

After the establishment of Aryan supremacy in the Indus region and across the northern plains, an Indo-Aryan culture emerged. Large numbers of Dravidians were driven southward across the Deccan. There, they developed a very distinct civilization with their own languages. In fact, Dravidian languages are fundamentally different from Indo-Aryan languages. This difference has induced linguistic strife even into the twenty-first century.

BUDDHA AND MAHAVIRA

Two major religious events occurred in the fourth and fifth centuries B.C. First, **Buddha**, born Siddhartha Gautama in the Himalayan foothills of what is now Nepal, preached the paths and precepts of Buddhism across northern India. Buddha's "Middle Way" was appealing as it lay between the extremes of worldliness and asceticism. He founded *sanghas* (communities for monks) and *viharas* (wet-season refuges). Buddha himself was not worshipped for another hundred years; the first sculptures of him did not appear until the Christian era.

At first, his followers carved simple cells and halls out of mountain faces. As time went on, Indian stone carvers ultimately excelled in their craft. Many

carvings and wall paintings still survive in the wake of Buddhism's diffusion throughout Asia. Later, lands were provided for monasteries and a monastic code was devised. This code and the Buddhist scriptures were recorded in a script called *Pali*. Even today, ancient Pali prayer scrolls can be seen in Buddhist monasteries. After his death in 483 B.C., Buddha's relics were distributed and encased in structures called *stupas* (Figure 5–5).

The second event was the founding of **Jainism** by **Mahavira.** Both Buddha and Mahavira were

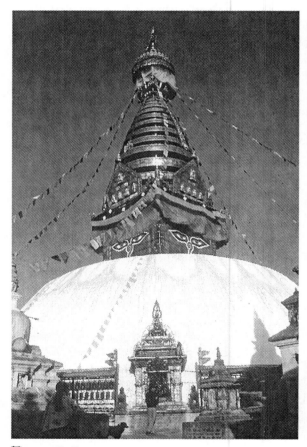

Figure 5–5

This impressive stupa *is Swayambunath in Kathmandu, Nepal. Sacred relics are confined inside the* stupa, *from which the eyes of Buddha see all things. Devotees move around the* stupa *in a clockwise direction, chanting and twirling their prayer wheels on the way.* Photograph courtesy of B.A. Weightman.

kshatriyas, both disdained social class distinctions, and both addressed the fears of never-ending rebirths. Mahavira was from the region of Bihar. He became a wandering ascetic at the age of 30. He preached in the Ganges valley until he died c. 612 B.C. In Jainism, the Buddhist injunction against killing became all inclusive: people, animals, insects, and even certain plants. Jainism was nonmaterialistic, but since at first it was only property holding that was forbidden, the religion became popular with the mercantile class. Cultivators and soldiers obviously failed to meet the nonviolence standards. Concentrated mainly in the state of Gujarat, Jains have managed to retain an unbroken tradition in India.

THE MAURYAN EMPIRE

Out of a period of uncertainty that followed an invasion and subsequent withdrawal by the Greeks, a new leader emerged: Chandragupta Maurya. By the end of the fourth century B.C., he had acquired an empire stretching from the Hindu Kush to the Bay of Bengal. This was the first time the political gap between northern India and southern India had been bridged. While not controlling the entire subcontinent, the Mauryas did control regions of the Deccan, important trade routes, and coastal regions. Great mercantile wealth was accumulated (Figure 5–6).

The Mauryan Empire is regarded as the first expression of the political notion of a state on the subcontinent. It was highly centralized, and many economic activities such as mining and land clearance were controlled by the state. The fact that death was imposed for defrauding revenues or killing an elephant (a capital asset) suggests that justice was harsh.

The most memorable Mauryan ruler was **Asoka**, who ruled from c. 268–233 B.C., the first fairly well-documented period in Indian history. Asoka is remembered not only because of the size of his empire but also because he adopted Buddhism as the state religion. He sent Buddhist missionaries, including his own son and daughter, to Sri Lanka. Although Asoka became a vegetarian, he tolerated those who were not. Thus, meat, including beef, was still available in the markets. The cow was not yet considered sacred. Asoka established hospitals, animal shelters, wells, and water tanks. His ruling principles and moral doctrines were inscribed on rocks and sandstone pillars around the country. The capital of the 40-foot (12 m) pillar of Samath, Gujarat, is a magnificent carving of four back-to-back lions. This official seal of Asoka was later adopted as India's national emblem.

THE GUPTAS AND INDIA'S GOLDEN AGE

When Asoka died, political conditions became chaotic. New Aryan migrants such as the Kushans, arrived from the northwest. The Kushans established their capital at Peshawar, gateway to the Khyber Pass. Around 320 A.D., northern India became united under the **Guptas**, under whom art and Sanskrit literature flowered. Derived from ancient sources, the *Puranas* were shaped and polished by the Guptas. These valued historic documents include Hindu myths, philosophical dialogues, ritual prescriptions, and genealogies. Art, science, and mathematics also flourished. The Gupta period marked **India's Golden Age** (Figure 5–7).

Under the Guptas, great strides were taken in mathematics. Numerals had been carved on some of Asoka's pillars, and these were further developed with the concepts of zero and nine numbers. The Arabs, who introduced them to the West, called them *Indisa* (Indian numbers). Europeans called them Arabic numerals. Aryabhata was the first to solve some basic problems of astronomy when he calculated pi to be 3.1416 and the length of the solar year as 365.358605 days. He knew that the Earth was round and that it rotated on its axis. He knew that an eclipse was caused by the moon's shadow blacking out the Earth. And all this was determined in India at least 1,500 years ago!

Fa-hsien, the first of three great Chinese Buddhist pilgrims to visit India, journeyed from China to India and Sri Lanka (399–412). He recorded that although the Gupta kings were Hindus, Buddhism flourished in the countryside. He commented on the well-being of the people and the fairness of the justice system. "Even those who plot treason only have their right hands cut off." He also noted that when certain people appeared, pieces of wood were struck to ensure avoidance of their evil presence. Fa-hsien had witnessed the caste system in action.

Figure 5–6

South Asia in the third century B.C. *The significance of the Indo-Gangetic Plain in Asoka's empire is unquestionable.*

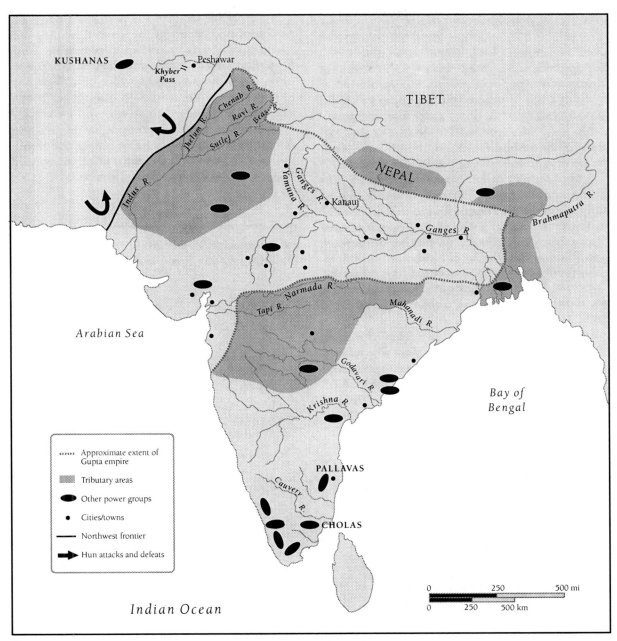

KUSHANAS

Peshawar

Khyber
Pass

TIBET

Chenab R.

Ravi R.

Jhelum R.

Beas R.

Sutlej R.

Indus R.

NEPAL

Ganges R.

Yamuna R.

Kanauj

Ganges R.

Brahmaputra R.

Arabian Sea

Tapi R.

Narmada R.

Mahanadi R.

Godavari R.

Krishna R.

Bay of
Bengal

PALLAVAS

Cauvery R.

CHOLAS

Indian Ocean

Approximate extent of
Gupta empire

Tributary areas

Other power groups

Cities/towns

Northwest frontier

Hun attacks and defeats

| 0 | 250 | 500 mi |
| 0 | 250 | 500 km |

Figure 5-7

*South Asia, c. 550 A.D. While the Indo-Gangetic Plain remained important in the north, river
and coastal regions fostered several powerful communities in the south.*

India's Middle Ages

North India was the focal point of Indian history until the beginning of the sixth century A.D. Then, regional kingdoms became dominant, only to be reabsorbed after being toppled by more powerful ones. During this period, central and southern India rose into prominence, and events occurred that would alter the geography of the entire Asian region. You will learn about Indian exploits in Southeast Asia in later chapters.

The Chinese pilgrim **Hsuan-tsang** spent 13 years in India (630–643) and reported to the domain of **King Harsha**, whose empire spread from the Punjab to Orissa and as far south as the Narmada River. Harsha was the last north Indian ruler to encroach south to the Narmada for 600 years. When Hsuan-tsang returned to China, he brought 20 horses carrying 657 Buddhist texts and 150 relics. He personally translated 74 of these texts, furthering the spread of Buddhism in China.

Harsha brought attention to influential western tribes such as the Rajputs of Rajasthan, who had settled around Jodpur. As *kshatriyas*, the Rajputs were highly militarized and exalted chivalry (Figure 5–8). They were also noted for their patronage of the arts, including the construction of the famous temples of Khajuraho.

The Rajputs honored and protected women and even educated some. At the same time they promoted child marriage and female infanticide. Being married to a military chief was a dangerous affair because wives were expected to die with their husbands on or off the battlefield. *Sati,* being burned alive on a husband's funeral pyre, was enforced. A low-class wife had the option of having her head shaved by a male untouchable, eating one meal a day, and enduring the degradation of widowhood. These practices diffused throughout India.

"Funeral flames dispel suspicion, honour lives when woman dies!" The Ramayana

Woman: Subjugation and Sati

Born into a male-dominated world, girls, if they survived the common practice of female infanticide, were worthy only as breeders of sons. No matter their caste, women were linked philosophically with *Sudras,* the lowest caste

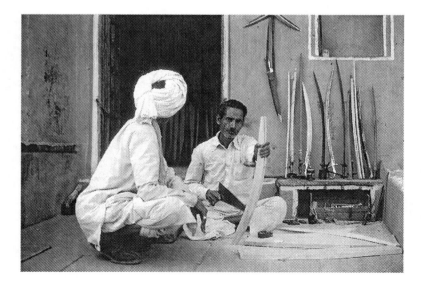

Figure 5–8

In Jaipur, India, these Rajputs are sword-makers. This occupation has been passed down through the family for generations. Photograph courtesy of B.A. Weightman.

group. Morally and intellectually, they were seen as weak. Chastity, loyalty, submission, and subservience were the watchwords of a woman's life. The essence of a woman's being was her relationship with her husband and his family. Without a husband, there was no point to her existence.

Child marriages were common. In the last census of the nineteenth century, there were 10,000 widows under the age of four, and over 50,000 between the ages of five and nine in and around Calcutta alone. In marriage, a girl "took away" wealth from her own family and was "gifted away" to the groom's household, where she was expected to be selfless and sacrificial (literally). A properly dutiful wife was expected to climb willingly onto the funeral pyre of her late husband and be burned alive to purge the sins of the entire family. A pregnant widow was expected to be burned on a separate pyre after giving birth. With this act of courage and devotion, a wife would earn the title of *sati* and be venerated and honored forever. *Suttee* is the English version of *sati*.

Sati comes from the word *sat* meaning "truth." A *sati* was a woman "true to her ideals." Since a woman's only value was to serve her husband, his death eliminated justification for her living. Those who survived widowhood suffered countless indignities and lived the remainder of their lives in degradation and disgrace. Remarriage was forbidden to higher castes.

The origins of *sati* are unclear, although references to such an act are made in some of the post-Vedic scriptures such as the *Mahabharata*. Increasingly prevalent after 400 A.D., actual dictates appeared around 700 A.D. By that time the life of a woman apart from her husband was declared sinful.

Sati soon became a "religious duty" for *kshatriyas* like princes and warriors, but soon spread to other castes. Although a wife was supposed to self-immolate voluntarily, most had to be forced onto the pyre. In situations where there was a harem, all the wives would burn, commit mass suicide, or be killed. Some were buried alive.

Initially, *brahmin* women were forbidden from seeking this sort of salvation. A *brahmin sati* would be guilty of suicide and would therefore disgrace herself and her husband. However, eventually they too adopted the practice.

In the eighth century, Muslim rulers attempted to check *sati* but were unsuccessful. Islam had already introduced the customs of veiling and **purdah**—the complete seclusion of women. By the 1200s, Muslim male and female social lives were completely separate. While the Hindu Laws of Manu instructed women to worship their husbands as gods, the Koran sanctioned similar behavior, and added the practice of polygamy. Hindu and Muslim customs regarding women became intertwined, especially in the north.

The British, who came to India in 1608, were unable to take action against *sati* for 200 years. From 1815 to 1828, there were close to 8,000 (reported) cases in Bengal alone. The British finally banned the practice in 1829. However, *sati* continued in most regions, including its stronghold in Rajputana (Rajasthan).

Wife-suicide or killing was not unique to India. It was practiced in China, Egypt, Russia, and elsewhere.

The caste system was well entrenched in this era. Caste taboos meant that scavengers and butchers were confined to mud quarters beyond the city wall. Moreover, they were compelled to walk on only the left side of the road.

Southern India was not entirely isolated from activities in the north. Cultural influences diffused from the northwest southward via the western highlands. Many Sanskrit words were incorporated into Dravidian languages. Contact with the Mauryas brought new ideas of statehood. Brahmin families were instrumental in the spread of Hinduism. Translation of sacred texts and orthodox practices strengthened the hierarchical structuring of society. Oppression of low castes and untouchables was frequently intense.

INVASIONS AND EMPIRES

Eighty years after the death of Muhammad in 632, an Arab empire extended from the Pyrenees in Spain to the Indus valley. The Arabs initiated their territorial gains in India when they established themselves in Sind at the mouth of the Indus in 712. Gujarati rulers tried to fend off conquest, in part by welcoming Zoroastrians (named after Zoroaster, c. 628–551 B.C.) who fled Persia and formed India's Parsi community, now concentrated in Mumbai (Bombay). Parsis, for whom fire is the essence of purity, would later make significant contributions in finance, trade, industry, and the arts. It is thought that the Persian water wheel was introduced at this time (Figure 5–9). The Arabs also introduced paper and gunpowder. By 871, they had established independent dynasties in the Sind and the Punjab, and in 1000, a holocaust began.

Mahmud (971–1030) of Ghazni (in Afghanistan), son of a Turkish slave, plundered India with 17 invasions over the next quarter century. Called "The Sword of Islam," Mahmud and his armies swarmed through the Khyber Pass during the dry monsoon. After devastating everything in their path, they sped back to Afghanistan before the wet monsoon could make the rivers of the Punjab impassable. In 1023, they sacked the temple at Sarnath and slaughtered 50,000 Hindus. Interestingly, Mahmud was never interested in acquiring territory. For him, India was merely a treasure trove.

Why were these invasions so successful? First, fragmentation in terms of social and political divisions precluded any united front against them. The Cholas, rulers in southern India, were far from the fray. The Rajputs, with their independent group structure, were never able to coordinate their efforts, despite their strong fighting abilities. The caste system was another hindrance to correlative action. In contrast, anyone could join Muslim armies. With ability and loyalty, even slaves could rise to top positions. Methods of warfare also differed. Indian armies fought with relatively cumbersome elephants. Muslim horse-archers could easily outrun them. With a rapid charge, they

Figure 5–9

This animal-powered Persian water wheel in Rajasthan, India, is evidence of Persian influences. Photograph courtesy of B.A. Weightman.

would suddenly swerve their horses sideways, release a hail of arrows, and race away.

Did you know that the game of chess was probably derived from elephant warfare?

Elephants and Chess

Around 500 B.C., the elephant arrived on the Indian battlefield. Elephant warfare strategy is faithfully represented in the game of chess, probably invented in India in the sixth and seventh century A.D.. The king directs the battle from atop his elephant. His moves are restricted to protect him because if he is killed, the army is vanquished. Generals, cavalry, and runners conduct the battle. Infantrymen (pawns), primitively armed, slow, and untrained, are valued only for their numbers. Flanks are guarded by elephants, who take front-line positions in the last stages of the battle.

The game diffused via Persia to Europe. The word *chess* is thought to be derived from *shah*, the Persian word for king. *Shah mat* approximates "checkmate," meaning "the king is dead."

Elephant warfare remained in place for 2,000 years, reaching its pinnacle under the Mughuls in the sixteenth century. By that time, firearms and artillery, brought from Europe by the Turks, had been introduced into India. The combination of raining arrows from mobile horse-archers, bullets from musket-armed sharpshooters, and explosive barrages from ranks of artillery brought the era of elephant warfare to a close.

THE DELHI SULTANATES

The Arab conquerors inaugurated Delhi as their capital in 1229 and for nearly 90 years struggled to maintain control of northern India. The second sultan, Raziyyat, ruled for a mere three years. A chronicler of the time described her as being a wise ruler and competent military leader. "But of what use were these qualities to her as fate had denied her the favour of being born as a man?" Raziyyat was ultimately deposed and executed.

After defeating the Mongols in 1279, the notoriously cruel Balban had thousands of Mongol prisoners trampled to death by elephants as entertainment for the Delhi court. Afterward, the victims' heads were piled as a pyramid outside the city gate. Balban had copied the standard Mongol style of revenge.

In 1309, Balban's son, Alauddin, invaded southern India. The loot, supposedly including the famous Koh-i-noor diamond, was hauled back by 1,000 camels. Another profitable campaign required 612 elephants to carry plundered treasure.

During Alauddin's rule, society became even more partitioned along religious lines. Cities and commerce were predominantly Muslim. Entire castes of artisans converted to Islam to escape their low-caste status. By and large, farmers were Hindus, represented by their village headmen. They could not accumulate property as a consequence of Alauddin's belief that "the Hindus will never become submissive and obedient till they are reduced to poverty."

During the tenure of the Delhi Sultanates and thereafter, many Arab Muslims married Indian women. The ruling class accumulated harems of inordinate size. Furthermore, they promoted the practice of *purdah*. In many countryside villages in northern India, even Hindu women adopted this practice of veiling and remaining secluded from nonfamily males and strangers in general.

Other changes occurred as well. The Delhi court favored the sophistications and fineries of Persian culture. Persian became the official language, influencing Sanskrit and maintaining its presence even in some parts of the south into the nineteenth century.

Persianization

For an 800-year period, Persian culture dominated elite culture in north and northwestern India as well as in other Muslim-controlled areas. The Persian connection was an ancient one. However, the process of Persianization was initiated with the Turkish invasions of the eleventh century and furthered by subsequent immigration of large numbers of Persian soldiers, administrators, scholars, and others into India. Urdu, cultivated with Islamic orientations and written in a Persianized, Arabic script, evolved as the national language of Indian Muslims.

Figure 5–10

South Asia, c. 1525. While the north was unified under the Delhi Sultanate, the west and south were fragmented among various rulers. The Portuguese were the first Europeans to set up trading posts, most of which were on the Arabian Sea.

> Even the Mughals, the last Indian court of empire, retained Persian dress, literary, artistic, and architectural styles, and the Persian language. Without the later introduction of English by the British, Persian could well have remained the language of political and cultural elites in South Asia.

Two significant events took place in the sixteenth century. First, the Portuguese appeared in the south, where they destroyed Hindu temples at Goa and Madras (Chennai). Goa fell to Portuguese control in 1510. Second, the Mughals rose to power and ushered in India's modern era.

India's medieval period ended with the arrival of the Mughals, who introduced the modern territorial state. Until this time, India was more a collection of structures than of territories (Figure 5–10). Cities, towns, villages, and temples; rulers, soldiers, and priests; merchants and farmers; all were ensconced in not always well-coordinated networks. There was a large degree of local autonomy. All this was to change in the era of Mughal authority and British hegemony.

India's Modern Era

In 1526, on the traditional battlefield of Panipat, 50 miles (80 km) from Delhi, thousands of elephants, horsemen, and footmen of the Sultan of Delhi, Ibrahim Lodi, amassed to face the impending terror, Babur (1483–1530). In only a few hours Lodi was dead, his massive army crushed by an onslaught of artillery, a cavalry attack from the rear, and elephants run amok. In succeeding encounters, Babur's mobile artillery overwhelmed army after army, including the Rajputs. After besieging Delhi and other fortified sites, he swept victoriously into Bengal. Babur, "The Tiger," direct descendent of Timur and Chinghis (Genghis) Khan, thus became the first Great Mughal. The name *Mughal* is derived from the Persian word for Mongol.

THE MUGHAL EMPIRE

Babur, while establishing order, treated the Indian people as subjects rather than prey and forbade his soldiers to maraud them. He and his son and successor, Humayun, initiated functional divisions in administration. Humayun died prematurely in a fall, leaving his young son Akbar to rule. Akbar (1542–1605) quickly consolidated his empire in further military campaigns, defeating the Rajputs and conquering Gujarat and Bengal (Figure 5–11).

Akbar was poisoned by his son, Jahangir the "World Seizer." Jahangir's beautiful Persian wife, Nur Jahan (Light of the World), became very powerful and was instrumental in furthering Persian culture and building in northern India. Politically astute, she arranged the marriage of her niece, Mumtaz, to the next truly Great Mughal: Shah Jahan (Ruler of the World).

Shah Jahan (ruled 1627–1657) extended Mughal rule into the south. Setting up his capital on the Yamuna at Shahjahanabad (Shah Jahan's city), he developed sites of architectural magnificence including Jama Masjid, India's largest mosque, and the Red Fort (in today's Old Delhi). Living in sumptuous splendor, Shah Jahan had 5,000 women in his harem. Nevertheless, when his beloved Mumtaz died, he had what has become India's most famous landmark—the Taj Mahal—erected in her honor (Figure 5-12).

For a while, the empire enjoyed relative peace and prosperity. Vassals of earlier regimes, tribal leaders, petty rulers and the like were absorbed into the larger empire, making it increasingly cumbersome. By recognizing these groups as *zamindars* (landlords), the Mughals bought their loyalty but encouraged peasant exploitation. The Mughals supported trade which in turn fostered urban growth. The cities were supported by the agrarian masses. As a result, Mughal culture became essentially urban and increasingly differentiated from the majority rural Hindu population. The Urdu language expanded, with contributions from Persian, the language favored by the elite. Arabic was employed by religious scholars. Hindi was spoken by the general populace. These linguistic differences were exacerbated elsewhere adding to and creating other divisions.

Figure 5–11

South Asia, c. 1700. Again the focus is the Indo-Gangetic Plain from which the Mughal Empire expanded southward. More Europeans including the British, Dutch, French, and Danes are on the scene. Note the concentration of activity near Bombay (Mumbai).

Figure 5–12

This is the magnificent Taj Mahal in Agra, India. It was built from 1632 to 1653 by the Mughal emperor Shah Jahan as a tomb for his wife Mumtaz who died bearing her fourteenth child. Thousands of workers and designers were brought from Europe and Asia to erect this monument of marble and semi-precious stone. As the costly edifice progressed, one of India's worst famines ravaged the countryside. Photograph courtesy of B.A. Weightman.

After the reign of Akbar, interest in science and technology waned in India. Court brilliance, architectural display, and military prowess took precedence. Lavish expenditure, constant warfare, and court intrigue peaked under Shah Jahan's son, Aurangzeb.

The Mughal empire reached its greatest extent under Aurangzeb (ruled 1658–1707). But Aurangzeb's cruel, anti-Hindu policies and uncontrolled expansionism promoted his empire's downfall. Non-Muslims—Hindus, Jains, Buddhists, Sikhs—were brutalized mercilessly. Non-Islamic structures were razed and conversions forced upon pain of torture and death.

Tax burdens in general became increasingly oppressive to finance court excesses and military ventures. Newly conquered areas of the Deccan yielded proportionately less revenue than the fertile northern plains. Revenue collection was stepped up and special taxes placed on Hindus. Discontent was rife, but when protesters gathered outside the Red Fort, Aurangzeb ordered them crushed by the imperial elephants.

THE SIKHS

Sikhism emerged in the fifteenth century. Its core area was the Punjab, a region of incessant Hindu-Muslim conflict. Its founder, **Guru Nanak** (1469–1539),

taught that "Man will be saved by his works alone . . . There are no Hindus and no Muslims. All are children of God." He opposed all forms of discrimination, including caste. His teachings, compiled by his disciple and successor Guru Angad, became the **Sikh** holy book, the *Adi Granth.* Guru Angad met with Akbar, who gave him a temple site at Amritsar near Lahore.

In the sixteenth century, some Sikhs became involved in political intrigue, and their leader was executed by the Mughul ruler Jahangir. Subsequently, the Sikhs began to take on a militaristic character and by 1650 claimed the Punjab as Sikh territory. Persecutions by Aurangzeb intensified their militarism. Under Guru Govind Singh, they vowed not to cut their hair, choosing instead to wear it knotted under a turban. They also swore to avoid tobacco and alcohol and renounced *purdah* and *sati.* Sikh women were accorded almost equal status with men. Sikhs also adopted the surname of *Singh* (meaning lion). There is a saying that "All Sikhs are Singhs but not all Singhs are Sikhs."

Sikhism grew as a force to be reckoned with, eventually becoming the last bastion of resistance to the British takeover in India. Only in 1849, after two British campaigns, were the Sikh armies defeated. Eventually Sikh regiments were incorporated into the British-Indian army. See the next chapter for more on the Sikhs and Punjab.

In order to alleviate the friction of distance that made central control difficult, Aurangzeb moved his capital to Aurangabad. There, despite an army of hundreds of thousands of elephants, guns, artillery, cavalry, and at least half a million camp followers, Aurangzeb was no match for the guerrilla-style tactics of Shivaji, "The Mountain Rat."

Shivaji (1627–1680), the zealous Hindu leader of the Marathas, had erected a complex of fortifications at defensible sites in the fractured western Deccan. Headquartered at Poona (Pune), behind Bombay (Mumbai), Shivaji was able to lash out, then quickly retreat to his mountain strongholds. The Marathas turned out to be the Mughal Empire's most formidable opponent. In 1644, Shivaji sacked the main Mughal port at Surat. In ensuing confrontations, Aurangzeb the Muslim and Shivaji the Hindu stood as protagonists of their respective faiths, deepening the division between believers.

As disenchantment with the Mughals grew, common gripes and shared concerns bonded individuals in competing groups. The Sikhs—bound by faith, the Jats—farmers bound by kinship, and the Marathas—bound by nationalistic aspirations, challenged the Mughal hegemony. Many revolts were led by the *zamindars*. Mughal retaliation was unforgiving. The situation worsened in 1702–1703 when famine and plague eradicated more than two million people from the Deccan.

The empire disintegrated in chaotic circumstances, with the strongest and cleverest manipulators and power-grabbers staking out their territories. And so arose the powerful **maharajas** (Great Rulers) of states such as Baroda, Gwalior, and Hyderabad. During this fray, the Persians invaded and sacked Delhi in 1739, carting off the jewel-encrusted Peacock Throne and other treasures. As Delhi declined, cities such as Lucknow and Hyderabad became centers of Mughal culture.

EUROPEAN SEA POWER

When Babur raged into India, the Portuguese already controlled the Indian Ocean with a monopoly on the pepper trade. Their fortified outposts at Goa and elsewhere served as customs stations where Asian merchants had to obtain letters of protection in order to avoid Portuguese attacks on the high sea. The land-based Mughals were interested in international trade insofar as it was a source of silver and gold. They remained unconcerned about the intricacies of its functioning as long as the precious metals kept flowing into their treasury.

The Indians responded to the Portuguese monopoly by welcoming European competitors with the intention of playing them off against each other. In London, the British East India Company was founded in 1600. Two years later, the Dutch East India Company was founded in Amsterdam. Indian textiles were desired products, and factories (storehouses) were erected in the foreign concessions. But, as the textile trade boomed, factories became involved in ordering, directing, and financing the industry. Soon they became instruments of foreign control.

By 1740, the British had eclipsed the Dutch, who then focused their attention on Indonesia. Starting with a collection of trading posts, the British fought the Mughals, forged alliances with local rulers, and eventually ousted their European competitors. By exploiting local rivalries, jealousies, Hindu-Muslim and other tensions, they were able to expand their influence. Three key commercial centers emerged.

On the Coromandel coast, Fort St. George developed as Madras. Madras soon flourished as an entrepôt.

With the dissolution of the Mughal empire, many merchant ships moved to Bombay. Its natural harbor was safely protected by the British fleet. As the empire further dissolved, Indian shipping networks floundered and interregional trade with Bengal and the Malabar Coast dissipated. The British in Bombay had their sights on trade with Europe.

Britain's Fort William, a swampy, unhealthy, deep-water anchorage 80 miles (129 km) up the Hooghly River, expanded to become Calcutta, the British colonial capital from 1772 to 1912. British Bengal trade boomed in the eighteenth century, and further inroads were made into the interior. There, factory agents established contact with indigenous spinners and weavers, and British artisans came to train them in the ways of British fashion. However, little investment was made in the means of production, and the workers remained mired in poverty. Farmers were forced to grow industrial, cash crops in order to meet their tax obligations. This caused food shortages, increased indebtedness, added millions to the ranks of landless and unemployed, and enriched the *zamindars,* tax collectors, and money lenders. Indian industry declined, and India became an importer of European manufactures and an exporter of raw materials such as raw silk, cotton, indigo, sugar, and opium. The British shipped the opium to China to satisfy their imperialist ambitions there.

By 1750, Indian distrust of the British was widespread. British racism marred the landscape. The city of Calcutta had grown to 100,000 people and was divided into "Blacktown" and "Whitetown." Attempts to get rid of the British were immediately thwarted. With Colonel Robert Clive's connived victory over a vastly larger force at the Battle of Plassey in 1757, the British achieved control over Bengal. British sea power was transformed into British land power.

For the next hundred years, the British struggled to expand their territorial control and suppress opposition on the subcontinent. Repeated wars were fought with various groups such as the Marathas and the Sikhs. In 1819, the British gained control of Rajasthan and became the largest land power on the subcontinent. By 1846, they had taken the Sikh capital at Lahore, the fertile lands of the Punjab, and the exquisite mountain vale of Kashmir. Then, in what seemed an insignificant move at the time, the British sold Kashmir to the highest bidder, a Hindu chieftain, Gulab Singh. By 1850, they controlled an empire extending from the Indus River to the Bay of Bengal and from Kashmir to Ceylon (Figure 5–13).

Colonial transformation was unstoppable. As the Indian landed aristocracy saw their lands appropriated, Western influences were instilling new ideas about development. Western missionaries were challenging Hindu beliefs and practices, especially the caste system. English replaced Persian as the official language of government and education. According to historian Lord Macaulay (1800–1859), "A single shelf of a good European library is worth the whole native literature of India and Arabia." Macaulay's ideas led the way in developing India. The British were to bring up a "class of persons Indian in blood and colour, but English in taste, in opinions, in morals, and in intellect." Schools and colleges were built to accomplish this task.

As governor general of India, Lord Dalhousie (1847–1856) launched an ambitious program of modernization. Expanding on the previous work of British engineers, he constructed the first 450 miles (724 km) of the Ganges Canal, which today provides hydroelectric power for a large region and irrigates two million acres of farmland. Under the Dalhousie administration, electric and telegraph services were installed, an efficient postal system developed, and the first 300 miles (483 km) of railroad track laid. While it is true that India ended up with Asia's best rail system, there were negative consequences.

On the one hand, railways permitted the relief of famine with grain shipments, while on the other hand they created famine in areas from which the grain was sent. Grain shortages drove prices beyond the reach of the impoverished masses, and millions starved. For example, 10 million in Bengal and Bihar starved in 1769–1770. At least 23 famines took their toll between the eighteenth and twentieth centuries. The last severe food crisis was in 1943, when 1.5 to 3 million died as a result of grain shortages, usurious prices, profiteering, and government negligence and incompetence.

Railroads also opened up India as a vast consumer market. But, instead of stimulating interior development, they favored coastal cities by giving rise to new cotton milling and shipping facilities. Local textile mills closed. Towns and cities bypassed by the rail line declined. By 1879, three-quarters of India's textile mills were in and around Bombay.

In the minds of Indians, infrastructural change did not have the impact of social change. The British had outlawed the practice of *sati* in 1829. In 1856, another law permitted the remarriage of widows, anathema to Hindus. By British law, Hindu princely states could not be inherited by any other than a blood heir. Any "heirless" properties reverted to the British Crown. Some states, even those loyal to the British,

Figure 5–13

South Asia, 1856. By this time Britain had control over most of the region. Other colonial possessions were reduced to small coastal colonies.

were annexed under the guise of mismanagement. Such attacks on Indian culture and Hinduism itself were cause for alarm, and in 1857 there was a mutiny.

The army of the East India Company comprised mainly British-trained Indian troops called **sepoys**. Intrigue sparked rumors that cartridges, which had to be bitten off to load the new Lee-Enfield rifles, were greased with pig and cow fat. With the pig being unclean to Muslims and the cow sacred to Hindus, a mutiny erupted. The Sepoy Rebellion of 1857 triggered nationwide violence. Whole communities of Europeans or Indians were slaughtered, and cities were reduced to rubble. Some states such as Hyderabad, Baroda, Rajasthan, and Punjab remained loyal to the British and the rebellion was ultimately crushed.

In 1857, the British government, horrified at the extent of atrocities, abolished the East India Company and formally took over India as a colony, a status it would retain for nearly a century. The era known as **the Raj**, meaning the "rule," had begun.

THE RAJ

Under British government control, social reforms were abandoned and attention was given to maintaining the status quo. The army was reorganized under British officers. The British legal system was introduced, and the Indian Civil Service was formed. This initiated the cumbersome bureaucracy that, like the road and railway system, would tie the country together. In all of this, Indians were deliberately excluded from positions of power. The country was measured, cataloged, mapped, organized, and centralized. The Grand Trunk Road stretched more than a thousand miles between Calcutta in Bengal and Peshawar near the Afghan border.

In 1877, Queen Victoria was declared Empress of India, and her representative in India became the viceroy. New treaties offering varying degrees of autonomy were made with the various princely states. There were more than 500 of these native states, occupying nearly 40 percent of the landmass. Larger states such Hyderabad, Punjab, Bengal, and Assam had governors or commissioners who reported directly to the viceroy. Smaller states had advisors. But these political alliances, with their incumbent privileges, contributed to even greater division of an already deeply divided region.

The British avoided contact with their Indian subjects. New towns, suburbs, and military stations were built across India. In cities, military cantonments and civilian districts known as **civil lines** were spatially and socially separate entities, complete with their own accommodations and services. These British mini-worlds in the lowlands had their counterparts in the highlands: **hill stations** (Figure 5–14).

Figure 5–14

The hill station of Simla clings to a mountainside in Himachal Pradesh. It served as the summer capital for the British in India from 1865 to 1939. It was much cooler than either Calcutta or Delhi. All government materials were hauled by train and caravan from Calcutta (Delhi after 1912) each year. Photograph courtesy of B.A. Weightman.

Hill Stations

Hill stations are mountain-amenity landscapes offering opportunities for rest and recreation at relatively cool elevations of 5,000 to 7,000 feet (1,500–2,200 m). They existed throughout the colonial world and remain important centers of commerce and tourism in many parts of Asia.

In India, hill stations were of great significance, and not just as social, economic, and political centers for Europeans. They symbolized the paramountcy of the Raj and revealed prevailing attitudes of nineteenth-century Europe: racism based on beliefs grounded in environmental determinism.

At that time, it was believed that white people could not withstand the rigors of torrid, foreign lowlands for extended periods. The combined onslaught of heat, dust, and exposure to the "natives" would surely damage European sensibilities and possibly cause offspring to be less than normal. It was essential that Europeans be restored in more familiar, temperate climes. As one Englishman remarked, "like meat, we keep better there."

Hill stations have been described as "island(s) of British atmosphere hung above the Indian plains," and "comforting little piece(s) of England." At the highest location, a protective nest of European architecture, trees, plants, and parks implanted British ideals of salubrious, picturesque, and restorative havens. In contrast, the "eyesores and nuisance" of the Indian quarters and bazaars on the slopes below were discreetly hidden behind "attractive" buildings and hedges.

Hill stations thrived as social centers. They glittered with banquets, pageants, and balls and catered to European tastes with afternoon teas, activities at the Club, and the frivolities of games and sport. Indian servants remained in the background, ears cocked for the next command. In later years, educated Indians were permitted to join the fun but remained fundamentally disliked by the British.

The British perceived India's climate to have three seasons: the cold, the hot, and "the rains." They adjusted their schedules to suit regional variations. In the north, for example, the cold lasted from October to March. Then, woolens and warming fires were needed, and British ladies remained on the plains with their husbands. By April or May, the plains were roasting and **amenity-migration** was made to the cooler hill stations, where thick-walled bungalows reduced temperatures even further. These small, one-story houses and the name **bungalow** diffused from India throughout the Western world.

Some hill stations became so popular that they were used as summer capitals by government functionaries. For instance, the governments of Bombay repaired to Poona (Pune); Madras to Ootacamund (known by the British as "Ooty") and Bengal to Darjeeling. In 1864, the Viceroy of India proclaimed Simla in the Himalayas as the empire's summer capital, and the entire government and all its files and documents were moved there from Calcutta each year from 1865 to 1939.

Hill stations did not boom until the opening of the Suez Canal in 1869 and railroad construction in the colonies. A narrow-gauge rail line carried its first passengers to Simla in 1903. Prior to that time, people and all their trappings were literally carried or carted up the mountain track by their Indian servants. As late as 1932, nearly a third of Simla's summer residents were porters or rickshaw pullers, most working to pay debts at home.

Increased population stimulated commerce. In 1830, Simla had only 30 houses. By 1881, there were over a thousand. The opening of the railway saw the population spiral to more than 30,000 by 1898. Market gardening contributed to increased commerce between hills and plains, and the town's commercial sector grew rapidly.

Indian areas frequently were targeted for improvements, which meant replacing their "unhealthy" warrens with environments suitable for European pursuits. Indians were forced by displacement to live in even more crowded conditions.

Simla, as the Raj capital for at least eight months a year, hosted important political events. Here in 1913–1914, the British, Chinese, and Tibetans met to fix their common boundaries. It was in Simla that the Indian National Congress was conceived and shaped and that the Muslim League had its beginnings. Here both organizations participated in discussions leading to India's independence and partition in 1947. And here, too, Prime Minister Indira Gandhi granted statehood to Himachal Pradesh in 1971. Simla, with its viceroy's mansion and British architecture crowning the teeming bazaars below, is now a popular destination for Indian upper- and middle-class tourists and is still a popular site for national and international negotiations.

India had been Britain's major trading partner since 1840. With the opening of the Suez Canal in 1869 and the expansion of railways, the route to Britain was significantly shorter and India was drawn increasingly into the international economy. British capital was invested in rail works, and coffee and tea estates (plantations). India exported these products, along with raw cotton, wheat, and jute (fiber). They

imported machinery and textile manufactures (made in England from Indian cotton). Eventually an industrial base was established.

India's industry rose and fell with British policies. Her economy was used to make up trading deficits with other industrialized nations. Once-renowned manufactures such as cotton, brocades, and silks were eliminated. However, raw material-based industries such as textile mills and jute, cement, and sugar factories were eventually built. Old trading or religious centers evolved as industrial centers, attracting landless peasants as a cheap labor supply. By 1900, Calcutta was awash with slum dwellers and squatters. British India's most famous author, Rudyard Kipling, described it in his story, "The City of Dreadful Night."

Indians were left to their own resources to develop other types of industry. J. N. Tata (1839–1904) was one of several Indian industrialists. His family founded the country's steel industry with the opening of the Tata Iron and Steel plant in Jamshedpur, Bihar, in 1907. The Tata family continues to reign as India's largest industrial dynasty.

Grand architecture of what became known as the Victorian-Mughal style housed train stations, post offices, and other important buildings. Bombay's Taj Mahal hotel was one of the grandest of all (Figure 5–15).

In 1911, Queen Victoria's grandson, King-Emperor George V and his wife, Queen-Empress

Figure 5–15

This Victorian-Mughal hotel was built by the great Indian industrialist Tata. Opening in 1903, it had granite staircases, electric fans, and 400 beds. It was the only luxury hotel in Bombay where Indians were permitted to stay.

Photograph courtesy of B.A. Weightman.

Figure 5–16
Old and New Delhi. Fascination with geometry and perspective, concern for order and accessibility, and desire for grand vistas and impressive layouts led to New Delhi's design. This plan stands in stark contrast to the narrow streets and winding passageways of the fortified old city. While New Delhi was planned, Old Delhi grew naturally.

Mary arrived in Bombay, the first British royal visit in history. At a key moment in the spectacular durbar (assembly) in Delhi, the monarchs announced from Shah Jahan's Red Fort that the imperial capital was to be transferred from Calcutta to Delhi, where a new city would be constructed.

Designed by British architects and finally inaugurated in 1931, New Delhi was designed with broad, straight avenues, well-suited for parades; prominent Victorian-Mughal government buildings, stately palaces for the viceroy and other notables; and spacious bungalows for the elite. European in plan, yet Indian in style, New Delhi was a contradictory counterpart to the traditional maze of Old Delhi (Figure 5–16).

THE ROAD TO INDEPENDENCE

The notion of self-rule had its roots in the nineteenth century, when there was a renewed interest in Indian history, culture, and religion. European scholars pursued translations of ancient works, and these were avidly read by growing numbers of educated Indians. At the same time, Indian religious scholars had reinvigorated Hinduism. Some even suggested modifications—such as the abolition of caste—more appropriate for a modern society.

The Indian National Congress, formed in 1865, provided a forum for nationalist sentiments and became the main instrument for the independence movement. The **Congress Party** continues to be one of the most influential in India. The British disdained or ignored expressions of desire for greater autonomy.

When the British divided the state of Bengal into Hindu-dominated and Muslim-dominated parts in 1905, without consulting any Indian leaders, Bengal erupted with huge demonstrations. Picketers were beaten; leaders were whisked away without trial. Four years later, the division of Bengal was revoked, but by then nationalistic mind-sets had hardened against the British. Nationalists went underground and launched a campaign of terror against British rule.

During the same period, Muslims were actively looking out for their own interests. The Muslim League was created in 1906. While Muslims made up more than 20 percent of the population of India, they were scattered and not well represented in positions of power. Certainly the Congress Party was run by brahmin Hindus. In 1916 it was agreed that Muslims and Hindus should have separate elections and representation.

World War I (1914–1918) was a significant event for India and Indians. About 1.2 million Indian troops served the empire by fighting on the side of the Allied forces. In the aftermath, British colonies such as Australia and Canada were granted significant control over their own affairs. There were those who thought that India would be rewarded in a similar manner. It was not. But imperialist ways would not prevail. Anything short of independence was unacceptable to the Indians. After all, the Chinese Revolution of 1911 and the Russian Revolution of 1917 demonstrated that popular uprisings could overthrow despotic regimes.

British opposition reached its lowest point in 1919 with a massacre of unarmed Indian demonstrators at Amritsar in the Punjab. The toll was 400 dead and 1,000 injured. The British Raj lost the loyalty and support of millions. Mohandas Gandhi (1869–1948), considered a "saintly fanatic" by the viceroy, was profoundly affected by this event.

Did you know that Dr. Martin Luther King, Jr., who led the civil rights movement for African Americans in the United States, was strongly influenced by the philosophy of Gandhi (Figure 5–17)?

Gandhi: Great Soul

Mohandas Gandhi (1869–1948), born in Gujarat, experienced discrimination as an

Figure 5–17

This Delhi sign shows Gandhi promoting vegetarianism. Note that the sign mentions ahimsa—the sanctity of all forms of life. Note also the use of both Hindi and English. Photograph courtesy of B.A. Weightman.

Indian practicing law in South Africa. There, on behalf of the Indian community, he developed the strategy of non-violent protest against injustice. Upon his return to India in 1914, he abandoned his European dress and adopted the lifestyle of a *saddhu*. Traveling far and wide, he used traditional Hindu values such as *ahimsa* to spread the concept of peaceful, mass protest against British oppression. He employed traditional symbols such as the spinning wheel to rebuild an Indian sense of nationalism.

Jawaharlal Nehru, in his autobiography (1942), said that Gandhi was "like a beam of light that pierced the darkness." Moreover, rather than direct from the top down, "He seemed to emerge from the millions in India . . . and the essence of his teaching was fearlessness and truth." Gandhi's notion of nonviolence was not simply the absence of violence; compassion and responsibility for the welfare of others were vital components of the concept.

Gandhi spoke out against Hindu-Muslim animosities as well as violence against women. He pointed to male chauvinism as being responsible for the genesis of *sati*. He abhorred the inequities of the caste system, saying that untouchables should be called *harijans*: children of God. Beloved by the masses and many political leaders, he was given the title *Mahatma*, meaning Great Soul. Mahatma Gandhi was instrumental in bringing about India's independence from Britain. He is often referred to as the "Father of Independent India."

Gandhi, who came to lead the Indian National Congress movement, launched three mass campaigns of civil disobedience in 1920, 1930, and 1942. These marches, sit-downs, boycotts, and blockades resulted in retaliatory violence and imprisonment for thousands. But the British did make concessions. In 1932, they began recruiting Indian military officers. In 1935, the Government of India Act and a new constitution put the power of eleven states in the hands of Indian representatives. Indians moved into higher ranks of the civil service. In sports, science, and literature, Indians began receiving world recognition such as Olympic gold medals and Nobel prizes.

In 1929, Gandhi nominated **Jawaharlal Nehru** (1889–1964) to succeed him as leader of the Congress movement. The dynamic, Western-educated Nehru, who could eloquently express his empathy for the Hindu masses, would eventually become India's first prime minister.

The only serious rival to Congress was the Muslim League, founded in the Bengali city of Dhaka. It was led by European-educated **Muhammed Ali Jinnah** (1876–1949), who believed that Muslims and Hindus had evolved as "two nations." As Nehru's Hindus gained power, thousands of Muslims joined the League. With Nehru's refusal to recognize Jinnah as the spokesperson for all Muslims, an attempt at power-sharing failed. Jinnah then turned to convincing Muslims that a Congress election victory would be a Hindu victory.

World War II (1939–1945) was the final catalyst in bringing about India's independence from Britain. As Britain had declared war on Germany without consultation, the Indian provincial representatives resigned. Being called upon to "Quit India," in 1942, the British arrested the entire Congress leadership and 60,000 activists. Then, they invited Jinnah into the government. With Congress power diminished, Jinnah was able to put the Muslim League on an equal footing. The British had agreed to Indian independence at the end of the war, but now it was too late. Muhammed Ali Jinnah demanded a separate Muslim state: Pakistan "Land of the Pure." Unable to speak Urdu, he remarked in English, "If not a divided India, then a destroyed India."

Hindu-Muslim violence was widespread, and on August 16, 1946, the "Great Calcutta Killing" event cost the lives of 5,000 Muslims and Hindus. Bloodshed swept across northern India, with violence and atrocities sparing no community. Faced with the prospect of complete anarchy, Nehru and other Hindu leaders became convinced that a Muslim state was indeed necessary.

A new viceroy, Lord Louis Mountbatten, was sent to India in 1946 to negotiate an independence agreement. He quickly saw that the division of India into Muslim and Hindu entities was inevitable. Pakistan was to encompass heavy Muslim concentrations in the northwest and northeast, meaning that the new nation would consist of two separate entities over 860 miles (1400 km) apart. Moreover, to give Muhammed Ali Jinnah his Pakistan, both Punjab and Bengal would have to be divided. The consequences were terrible.

THE PARTITION OF INDIA

As both Punjab and Bengal had mixed populations, no boundary line would be satisfactory to all. Punjab, in particular, was mixed with Hindu, Muslim, and Sikh communities interspersed. The decision was made to award 62 percent of the land and 55 percent of the population to Pakistan. As of midnight January 14, 1947, millions found themselves trapped in a country hostile to their faith.

The partition of India in 1947 spurred the largest migration in human history. Around 12 million people moved between West and East Pakistan and India seeking refuge in their own religious communities. The ensuing savagery was unparalleled. Entire villages were massacred, and trainloads of people were butchered and torched. At least a million people were killed in the wake of India's partition.

These events deepened Hindu-Muslim animosities. A horrified Gandhi took up residence in the Muslim quarter of Old Delhi and vowed to fast to death if the communal violence did not cease. Although calm returned several days later, many Hindus were enraged by his seeming support for the Muslims. On January 30, 1948, Gandhi was assassinated by a Hindu fanatic. A distraught Prime Minister Nehru spoke to his nation: "The light has gone out of our lives and there is darkness everywhere."

In that same year, Muhammed Ali Jinnah, Pakistan's first political leader, died as well. Spurning Muslim prohibitions against pork, alcohol, and smoking, he died of lung cancer.

The nearly 600 princely states were diverse geographically, socially, and politically, precluding any uniform policy regarding their status. With the creation of India and Pakistan, each state was given the option of joining either country. Most acquiesced to India but some, such as Hyderabad, had to be coerced by the Indian army. Kashmir presented a special case, its complexities defying easy solution.

The Maharaja of Kashmir was a Hindu, but the majority of his subjects were Muslim. While he wavered on his decision, hoping to remain independent, Pakistan tribesmen rebelled. The Maharaja then acceded to India and asked for military help. Indian troops were unable to oust the Pakistanis. With stalemate in 1949, a "line of control" became the *de facto* boundary. This resolution was not a solution, and Kashmir remains a region of conflict to this day.

Nationalism and Regionalism

Complex, diverse, and paradoxical—this aptly describes the cultural mosaic of South Asia today. With ancient roots and an amalgam of influences from within and without, empires have risen and fallen, leaving indelible legacies on the landscape. Despite continual religious, linguistic, and political conflict and restructuring, India and the newer nations of South Asia remain intact. Nevertheless, the interminable dynamics of nationalism and factionalism continue.

The idea of the territorial state was introduced by the Mughals, who developed an ordered bureaucracy. This merged into the British civil service, an all-encompassing, multi-leveled network of control and regional division. Yet, persistent national agitation did reap change. Constitutional reform ended with a British, parliamentary-style, democratic system. And the federalism that the British employed in their devolution of power was retained after independence. Highly centralized power of the Mughals and British fed nationalistic tendencies and was furthered by the partition of India. Yet federalism, which allows for regional differences and varying degrees of local autonomy, has been fundamental to the stamina and resilience of the Indian and other governments in the region. At the same time, it has facilitated regionalism, localism, and communalism. National and regional ambitions and inequities underlay countless examples of instability and strife marring the subcontinent into the twenty-first century.

⊕ Main Point Summary

1. Physical geography has affected Indian history. Mountain passes and river valleys invited settlement yet channeled invasions. Mountain ranges and rugged plateaus afforded protection and insulated from intrusion. Monsoon cycles gave rhythm to invasion strategies, seasonal migrations, and shipping schedules. Agricultural, religious, and social activities continue to be geared to the wet and the dry seasons.

2. Favorable site factors, exemplified by the fertile Indus valley, the Doab, and the larger Ganges plain, were conducive to the support of large populations. Natural harbors provided both shelter and accessibility, fostering the development of cities such as Bombay (Mumbai), Calcutta (Kolkata), and Madras (Chennai).

3. The Indus valley produced one of the world's greatest civilizations. Those who built Harappa and Mohenjo-daro were predecessors of the Dravidian people who now inhabit southern India. Pressed southward by the various Central Asian invaders, their languages remain distinct from those spoken in the north.

4. Around 1500 B.C., Aryans invaded via the Khyber Pass, introducing Sanskrit and the *Vedas* to northwestern India. The *Vedas* and subsequent writings became the foundation for all modes of thought and behavior.

5. The Aryans brought horses and cows to northern India. Cavalry warfare allowed for the effective dispatch of traditional elephant forces. Cows, initially significant as beasts of burden and sources of sustenance, eventually became sacred.

6. The division of society into caste—*varnas* and *jatis*—provided group cohesion and standardized inter-group relations. It also engendered social divisions that became even more splintered and rigid with time.

7. In the fourth and fifth centuries, Buddhism and Jainism emerged. Buddhism spread across Asia and resulted in different schools of thought such as Mahayana and Theravada in the process. With the arrival of Islam, Buddhism ultimately vanished from the land of its birth. In South Asia, only Sri Lanka and Bhutan have remained Buddhist. Jainism, with its strong emphasis on *ahimsa*, retained its foothold in Gujarat.

8. The Mauryan Empire of the third century was the first expression of a political state on the subcontinent. But India's Golden Age was in the fourth century, when, under the Guptas, great advances were made in literature, art, science, and mathematics.

9. Subsequent invasions from Central Asia destroyed the landscape and brought an end to Buddhism in India. The following centuries saw the rise and fall of numerous states and empires. The most extensive was that of King Harsha, who was the last Indian ruler to reach south of the Narmada River.

10. By the 600s A.D., caste was firmly entrenched, as were attitudes to women. *Sati* was widespread among upper castes, especially the warrior Rajputs. With further Islamic invasions bringing the customs of veiling and *purdah*, restrictions for women intensified.

11. Delhi was inaugurated as capital of the Delhi Sultanate in 1229. Under successive rulers, heavy taxation for military ventures forced millions into poverty. Hindus were widely persecuted. Meanwhile, the court became increasingly

Persianized, adding Persian refinements to upper class landscapes and lifestyles. The excesses of the Delhi Sultanates were halted by Mughal forces at the Battle of Panipat in 1526.

12. In the sixteenth century, the Portuguese and other Europeans established trading posts along the Malabar and Coromandel coasts. Ultimately, the British would dominate and change the Indian landscape forever.

13. The sixteenth century also saw Babur take power as the First Great Mughal. Under his successors, the Islamic Mughal empire was consolidated and controlled through a hierarchical structure of government. Taxation increased to provide the lavish expenditures required to construct architectural wonders and fund military exploits. Exploitation of an already overburdened and indebted peasantry allowed the accumulation of fortunes of urban-based landlords and money lenders. Rulers and ruled became distanced economically, politically, and socially. Finally, the Mughals were denied their invincibility by the Marathas and the empire collapsed into a mosaic of competing states.

14. European powers took advantage of internal dissension to advance their territorial ambitions. The British, with established settlements at Bombay, Calcutta, and Madras, consolidated their power through conquest, intrigue, and alliances with local potentates. They advanced their control with the construction of roads and railroads. Economic activities shifted to coastal locations and interior cities declined spurring labor migration to foreign-dominated port cities.

15. Attempts to oust the British were thwarted. With the Battle of Plassey in 1757, British control of India was essentially complete. Racist policies, cultural interference, and economic exploitation by the British East India Company fomented rebellion. Strife and famine impelled the British Government to take over India in 1857.

16. Under the Raj, the subcontinent was cataloged and mapped in detail and an elaborate bureaucracy put into place. However, Indians were excluded from positions of consequence. British settlement was encouraged by both the opening of the Suez Canal in 1869 and railroad expansion. Myths of environmental determinism helped to convince the British of their racial superiority and "necessary" social separation was epitomized by spatially divided cantonments and hill stations. The British-designed capital at New Delhi, which differed sharply from traditional Old Delhi, fixed perceived cultural differences on the landscape.

17. The nineteenth century witnessed a resurgence of Hindu and Muslim nationalism and anti-British sentiment. Western-educated leaders Jawaharlal Nehru, Muhammed Ali Jinnah, and "Mahatma" Gandhi led their followers in an independence movement that resulted in the Partition of India in 1947. This division into Hindu India and Muslim Pakistan unleashed terrible violence and triggered the largest migration in history.

18. While unified by British infrastructure and economic and political systems, the region was divided by religion, language, and caste. Moreover, Pakistan comprised two pieces, spatially and culturally distant. South Asia of 1947 was a complex and paradoxical mix of unifying and disintegrative forces that would wreak havoc throughout the twentieth century.

⊕ Bibliography

Allchin, F. R. ed. 1995. *The Archaeology of Early Historic South Asia: The Emergence of Cities and States*. Cambridge: Cambridge University Press.

Basham, A. L. 1968. *The Wonder That Was India*, 3rd ed. London: Macmillan.

Brown, J. 1977. *Gandhi and Civil Disobedience*. Cambridge: Cambridge University Press.

Choi, Dong Sull. 1997. "Origins and Realities of Suttee in Hinduism." *Comparative Civilizations Review* No. 36: 38–53.

Davies, C. Collin. 1963. *An Historical Atlas of the Indian Peninsula*. London: Oxford.

Dickason, David G. 1975. "The Indian Hill Station. Geographical Record." *The Geographical Review* 65: 115–116.

Dutt, Ashok K., and M. Margaret Geib. 1987. *Atlas of South Asia*. Boulder: Westview.

Farmer, B. H. 1993. *An Introduction to South Asia*. New York: Routledge.

Kenny, Judith T. 1995. "Climate, Race, and Imperial Authority: The Symbolic Landscape of the British Hill Station in India." *Annals of the Association of American Geographers* 85: 694–714.

Kulke, Hermann, and Kietman Rothermund. 1990. *A History of India* (revised edition). New York: Routledge.

Lapierre, Dominique, and Larry Collins. 1975. *Freedom At Midnight*. New York: Simon and Schuster.

MacMillan, Margaret. 1988. *Women of the Raj*. New York: Thames and Hudson.

Murphey, Rhoads. 1996. *A History of Asia*. New York: Harper Collins.

Narasimhan, Sakuntala. 1990. *Sati: A Study of Widow Burning in India*. New York: Penguin.

Nehru, Jaharwalal. 1942. *An Autobiography*. London: Chatto and Windus.

Pandian, Jacob. 1995. *The Making of India and Indian Traditions*. Englewood Cliffs: Prentice Hall.

Pubby, Vipin. 1996. *Shimla Then and Now* (2nd edition). New Delhi: Indus.

Schmidt, Car. 1995. *An Atlas and Survey of South Asian History*. Armonk: M.E. Sharpe.

Smith, B. 1995. *Classifying the Universe: The Ancient Indian Varna System and the Origin of Caste*. New York: Oxford.

Sopher, David. 1964. "Man and Nature in India: Landscapes and Seasons." *Landscape* 13: 14–19.

Spear, Percival. 1963. *A Study of the Social Life of the English in Eighteenth Century India*. London: Thames and Hudson.

Watson, Francis. 1979. *A Concise History of India*. London: Thames and Hudson.

Wolpert, Stanley. 1991. *India*. Berkeley: University of California.

Wolpert, Stanley. 1984. *Jinnah of Pakistan*. London: Oxford.

Chapter 6

✦

South Asia: Pakistan and the Himalayan States

"From your past emerges the

present, and from the present is

born the future."

MUHAMMAD IQBAL (1873–1938)

In this chapter, I will focus on the Indian sub-continent's northern regions and tell you about Pakistan, the disputed region of Kashmir, Nepal, and Bhutan (Figure 6–1). These mountain zones are both physically and culturally fractured. High altitudes and deserts have limited accessibility and foster cultural differentiation, communalism, and regionalism.

Within the formidable Himalayas and Karakoram are the tenuous and oft-disputed boundaries between Pakistan and Afghanistan, Pakistan and India, and Pakistan, India and China. Political instability has marked the evolution of virtually every state in the region. I will start with Pakistan at the time of India's partition in 1947 and explain the factors leading to the transformation of East Pakistan into Bangladesh in 1971. Then, I will describe current conditions and problems across the region, beginning with modern Pakistan.

Nationalism and Regionalism: The Devolution of Pakistan

In 1948, British geographer O. H. K. Spate wrote that he saw the creation of East and West Pakistan as "the expression of a new economic nationalism that has inevitably taken into its hands the immensely powerful weapon of immemorial religious and social differentiation." In 1947, Pakistan had two "wings," the eastern wing being only one-sixth the size of the western wing yet holding eight million more people. This was only one of a litany of differences between the two parts of this new nation—centrifugal forces that would ultimately result in its political demise.

For the split nation of Pakistan, unity could not be maintained with diversity. Contrasts were too numerous, and conflicts were too intense. West Pakistan is dry; access to water and drought is of critical concern. East Pakistan is humid; flooding is a yearly

Figure 6–1

Northern South Asia. The political complexity of this region is compounded by its rugged physical geography.

event. Aside from differences in size and population, the west possessed 80 percent of the cultivable land, and its share per farmer was more than double that of the east. West Pakistan had the country's capital and only port at Karachi. East Pakistan had far less infrastructure and was much poorer economically.

Cultural differences are exemplified in linguistic differences (Figure 6–2). There are several languages spoken in the west. These are from different language families and possess numerous dialects. Bengali, a branch of Sanskrit, is spoken in the former eastern wing of the country. In 1948, the Karachi government declared Urdu as the national language of Pakistan, enraging Bengalis and deepening the chasm between east and west. Linguistic

division would prove problematic in the years and decades ahead.

Language and Nationalism

Pakistan (formerly West Pakistan) has reaped the consequences of lacking linguistic cohesion. About two-thirds of the population speak Punjabi. Punjab is the most economically productive province and contains the nation's capital, Islamabad. Sindhi is spoken by about 12 percent, mostly in Pakistan's second

Figure 6–2

Major languages of Pakistan.

branch of the Indo-European language family. Sindhi, Urdu, and Punjabi belong to the Indic branch. Brahui is a Dravidian language. Balti is a Sino-Tibetan language.

Since language is the context for thought and action, and conceptualizations in one language do not necessarily exist in another, differences can act as centrifugal or divisive forces in any centripetal drive for national cohesion.

industrial province of Sind. Urdu is spoken by only 8 percent, yet it was proclaimed the national language. Along with English, it is the language of the elite and government. Urdu speakers are concentrated in cities, especially Karachi. Language riots broke out in Sind in the 1950s when university students were forbidden to write their essay responses in Sindhi. More riots occurred in 1965 when Urdu signs and radio programs replaced Sindhi ones in Hyderabad. In the 1970s, some of Karachi's Urdu-speaking Muhajirs, postpartition Muslim migrants from India, demanded equal representation in government even though they were a minority population. Others agitated for a separate state.

These incidents epitomize the sense of deprivation felt by Sindhi-speaking leaders and the emerging middle class. A Sindhi Language Authority was established in 1990. Language still fuels violence in Sind and elsewhere.

Several other languages are spoken, with the greatest diversity in the mountainous regions. Pashto and Baluchi belong to the Iranian

Divisions sharpened with the economic and political policies pursued by the government in West Pakistan. There were arguments over exploitation of natural resources, industrial development, acquisition and allocation of monetary resources, communal and regional representation, and the role of Islam in government.

Although both parts of Pakistan were essentially Muslim, there were differences in practice. A history of communal strife marred the relationship between Muslims, Hindus, and Sikhs in the west. Moreover, some ethnic groups (e.g., the Pathans) were extremely fervent in their application of the mores and laws of Islam in their daily lives. For Bengalis, religious interaction with Hindus had become routine. They attended each other's festivals and reciprocated in temple offerings.

The question of Islam's role in government remains a critical aspect of political instability in Pakistan. In 1956, under pressure from conservative Islamists, the government declared Pakistan an Islamic Republic. Government was to operate according to the Koran and **sunna**, the way of life prescribed in Islam, based on the teachings of Muhammad. Religious moderates, who had pushed for modernization and democracy, were sidelined.

Other problems centered on the rich-poor gap and other real and perceived inequities. After partition, many business-class Punjabis moved to Karachi's Sind Province, where they took over former Hindu businesses and came to dominate the economy. Added to the fray were the **Muhajirs**, Muslims who had previously lived in India. The urban-dwelling, Urdu-speaking Muhajirs, many of whom were educated and experienced in business and civil service, came to dominate administration. But, these Punjabi and Muhajir urban elites were resented by the Sindhis. As the government struggled to establish some degree of equity under the weight of vested interests and expected patronage, the rich-poor and urban-rural gaps widened. Urban riots and peasant revolts led to the imposition of martial law in 1958.

With ensuing economic and political instability, East Pakistan looked to a new political party—the Awami League led by **Mujibur Rahman (Mujib)**. Mujib would fight for greater political autonomy for the east. The following years of hectic east-west relations served to strengthen the resolve of the Awami League.

In 1965, Pakistan went to war with India over border demarcation in the Rann of Kutch as well as in Kashmir. The year-long war was costly. Since both border disputes were with West Pakistan, East Pakistan felt unduly burdened. More disturbances again drew military rule, which lasted until 1970 when elections were to be held.

In Karachi, the Pakistan's People's Party (PPP), under the leadership of Zulfikar Ali Bhutto, ran on the slogan "Islam our Faith, Democracy our Polity, Socialism our Economy." Bhutto also declared a "thousand years' war with India."

Elections were delayed because of a crippling cyclone in East Pakistan. Then, entirely dissatisfied with West Pakistan's relief efforts, the Bengalis swept the Awami League to victory, a pivotal event because the proposed system of "one person one vote" would give the more populous east a majority in the National Assembly. As it became clear that Mujib rather than Bhutto might rule Pakistan, the National Assembly was "postponed" by the military.

Withholding power from Mujibur Rahman and his Awami League was a fatal error. East Pakistan erupted with strikes, demonstrations, and insurgencies. The Bengalis refused to pay taxes. West Pakistan countered with brutal military action—complete with rampant murder, pillage, and rape. Immediately, 250,000 refugees poured into India. Ultimately 10 million would seek refuge across the border.

Faced with hordes of penniless refugees in the already poor state of West Bengal, India seized the opportunity to retaliate against its arch enemy in Karachi. Indian intervention brought the east-west war to an end, and in 1971, East Pakistan became the independent nation of Bangladesh, with Mujib as its prime minister. Bhutto became prime minister of West Pakistan, which was renamed Pakistan.

Pakistan: Land of the Pure

DIVERSE LANDSCAPES

Pakistan in 1971 was politically divided into four highly differentiated and unequal parts: the Northwest Frontier Province (NWFP); Punjab; Sind; and Baluchistan (Figure 6–3). To understand contemporary trends and problems we must look at the different characteristics of each region.

Northwest Frontier Province (NWFP)

The rugged hills of the NWFP are capped by a dramatic knot where the Karakoram, Pamir, and Hindu Kush have buckled together. Afghanistan's Wakhan Corridor, a 16,000- to 24,000-foot (4,880–7,315 m) buffer zone between Tajikistan (part of the former Soviet Union) and the former Indian Raj, lies in this region, extending to China in the east. Below glaciers and barren, rocky slopes are forests of pine, deodar, and evergreen oak. At lower and dryer elevations, these give way to acacia, dwarf palm, and coarse grasses. Wood and stone houses cling to steeply walled ravines, and finely constructed rock terraces support irrigated agriculture (Figures 6–4 and 6–5).

This dramatic and broken terrain is, in fact, a **shatter zone**, inhabited by a kaleidoscope of independent and feuding tribes. Many cross the border with Afghanistan. Most are Pathan, a general term for tribes such as the Afridis or Waziris. Most speak Pashto, an Iranian dialect. While some Pathans had accepted British rule and laws, others followed *pukhtunwali* (local Pathan law). Other groups such as the Chitralis are not Pathan.

Cutting through the mountains are the Kabul and Swat rivers. These converge in the Vale of

Figure 6-3

Physical geography and provinces of Pakistan. Compare this map with Figure 6–10. What geographic relationships do you see?

Figure 6-4

This scene shows Mount Trich Mir, at 25,260 feet (7,578 m). I took the picture overlooking the Chitral River in northern Pakistan. The alluvial terraces are irrigated by gravity flow of water. Rice, fruit, and vegetables are grown here. Photograph courtesy of B.A. Weightman.

Figure 6-5

Below, in the town of Bahrain, Pakistani houses cling to the sides of the rocky slopes. The structures are made of wood and stone with no mortar. The building in the foreground is a grain-grinding mill where water power turns the huge stone grinding wheels. Photograph courtesy of B.A. Weightman.

Peshawar, a lowland separated from the Indus River by a sandstone ridge. Under irrigation, the deep alluvial soils support wheat, corn, sugar beets, and tobacco (Figure 6-6). In Peshawar and other towns of the NWFP, Punjabis became prominent in business. Peshawar was also a budding industrial center.

Punjab

In 1947, Pakistan's portion of the Punjab region (formerly India) held 60 percent of the population. It also dominated the armed forces and produced more than half of the GNP. Punjab long had been the breadbasket of the subcontinent. Here flow the Indus, Jhelum, Chenab, Sutlej, and other vital rivers. Out of the mountains and foothills, streams and their *doabs* (lands between the rivers) slope downward with masses of alluvium fanning onto the Indus plain. Wheat, cotton, rice, oil seeds, and sugarcane thrive under irrigation. This region has great potential for development.

Sind

Sind, more than any other region, is wedded to the Indus. To the north and west of Sukkur lie the Kirthar Mountains. Alluvial fans permit cropping,

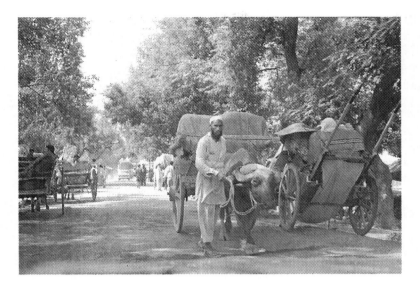

Figure 6–6

In Pakistan's Northwest Frontier, water buffalo pull carts of tobacco to the market at Mardan. The water buffalo are also suppliers of milk with a high fat content. Photograph courtesy of B.A. Weightman.

but this western valley is underlain with clay and therefore subject to waterlogging and *salinization,* the accumulation of salts. Northward, the land turns to clay-desert, called *pat.*

Centuries of seasonal flooding below Sukkur have produced a *doab* between the Indus and the Nava rivers. Each part of this eastern valley at some time has been traversed by an Indus stream. This fertile and productive region has invited water control works through the centuries. The *doab,* actually the ancient Indus delta, fades eastward into the Thar Desert (Figure 6–7).

Further south, small but significant limestone outcrops support the Sukkur water barrage and the city of Hyderabad. The delta of the Indus merges with the mud and salt flats of the Rann of Kutch.

Figure 6–7

This aerial photo shows irrigated lands along the edge of the Thar Desert in India. Photograph courtesy of B.A. Weightman.

Since the river distributaries have long been in flux, the coast is dotted with dead ports. When Indus floods converge with high ocean tides, the coast floods inland for up to 20 miles (32 km). Mangrove and tamarisk thickets fringe old and new beach-ridges and bands of clay-like silts.

Sind epitomizes heat and aridity. Summer temperatures can be well over 100°F (37.8°C) at midnight! It rarely rains, but when it does, it pours. Karachi, with an annual average rainfall of about 8 inches, has recorded 12 inches (305 mm) of rain in 24 hours.

Second to Punjab in size and wealth, Sind has great agricultural potential. In 1947 it was in the process of being colonized mainly by immigrants from Punjab. However, rural Sind was controlled by *zamindars* (landlords) such as Ali Bhutto. Tensions between zamindars and peasants were only compounded by Punjabi and Muhajir intrusions.

Baluchistan

Although twice the size of Punjab, rugged and arid Baluchistan possessed the smallest population and was the poorest province. Sedentary farming, pastoral nomadism, and related crafts and trade were ways of life among the various tribes of the craggy outcrops, plateaus, and mesas of the Suleiman and other ranges sweeping southward and westward to the Iranian border. The southern Makran coast has been described as "bizarre," comprising gigantic pillars and pedestals, layered and etched like old teeth. Northward to Afghanistan lie volcanoes, expanses of black pebbles called *dasht*, shifting red dunes, and salt-encrusted *playas* (dry lakes) called *hamun*. A local proverb refers to Baluchistan as "the dump where Allah shot the rubbish of creation."

Baluchistan's only notable urban center was Quetta, a hill station strategically situated between two passes into Afghanistan. In the hills above Quetta at 5,490 feet (1,650 m), juniper trees were cultivated. Oil of juniper berries gives gin (e.g., Bombay Gin) its distinct flavor. Coal reserves offered some economic potential but these have since proved to be of poor quality. This region was also known for its *karez*, underground irrigation

systems using hand-dug, gravity-flow tunnels to divert water from aquifers to surface streams and ponds. *Karez* systems still exist in arid lands from Iran to western China and Oman, even further west.

ROLLER COASTER DEVELOPMENT

Pakistan was saddled with an infrastructure purposefully laid to connect cantonments with strategic frontiers and tuned to economic and military needs of the former Raj. As a result, the bulk of the road and rail network remained in India. As goods historically moved overseas via Bombay, Karachi was a rather pathetic port. Pakistan gained only five percent of the Raj's industry. Although the British had constructed canal projects, new water power and irrigation development schemes were essential. In essence, Pakistan received only the leftovers of the vast British system.

Physical geography, cultural variation, and inadequately developed infrastructure did not foster the economic or political integration of Pakistan. In fact, its economic history has been turbulent. Wracked with political instability, policy shifts, and changing focus, Pakistan muddled its way from one crisis to the next into the twenty-first century.

Although Pakistan was an agricultural country, early stage planners, many of whom were urbanites from India, turned to industrialization. Cotton and jute mills were built for fiber processing previously done in Bombay and Calcutta. Agriculture received little attention. Farmers were poorly paid for their crops, and lack of incentive hindered their productivity. By 1950, this decline, combined with rapid population growth, made Pakistan a net food importer for the first time in its history.

The sluggish economy was revived in the 1950s when landlords returned to the political arena. With a decade of support from the *zamindars*, and benefits of the Green Revolution, Pakistan's agricultural yields rose to be among the world's highest.

Under another military government, a major geographic shift took place. Karachi, with its immigrant population, squatter settlements, and other unseemly aspects, was perceived as unsuitable as the capital of a new Islamic state. For this and more practical reasons, a new capital, Islamabad, was to be built adjacent to Rawalpindi (Pindi) in Punjab, the true heart of Pakistan. Aside from its governmental capacity, it would function as a **growth pole**. With the evolution of Islamabad, industry and commerce would be decentralized from Karachi, Punjab would become developed, and a presence would be maintained close to the unstable regions of NWFP and Kashmir. Islamabad, meaning "City of Islam," also exemplifies the concept of a **forward-thrust capital**.

Islamabad took on its function as capital in 1963. As there are restrictive laws as to who can live in the new capital, most workers actually live in Pindi and are bussed back and forth to their government jobs.

International relations also affected Pakistan's roller coaster development. In 1949, the long established economic interchange between Pakistan and India ground to a halt in a disagreement over monetary policy. But the Korean War (1950–1953) generated a market for leather, wool, cotton, and jute products, perfect timing to support farmers and the newly constructed processing mills. In 1954, Pakistan signed a Mutual Defense Agreement with the United States, and this brought substantial amounts of American investment. But this economic boom ended in 1965 with the outbreak of war with India over Kashmir.

The United States stopped military assistance to Pakistan and reduced its level of monetary assistance. With the breakaway of East Pakistan in 1971 and the perception of India as a "friend of the United States and enemy of Pakistan," Pakistan turned away from South Asia and toward the Islamic countries of the Middle East.

Beginning in 1974, hundreds of thousands of workers migrated to the oil-exporting countries of the Middle East to work in construction, oil, and other industries, and in services. Large numbers of these came from poor, *barani* (dry, non-irrigated) regions. Pakistan continues to receive large amounts of remittances from the Persian Gulf states and has recently incorporated Iran and adjacent Islamic states into its sphere of interest.

Relationships with China were also a factor. Border disputes were settled in the early 1960s and both countries pursued closer ties, taking a stand against India. In 1965, a joint-venture project was begun to link Pakistan and China by a highway through the Karakoram. Islamabad would be linked to Beijing via the NWFP and Kashgar, gateway to China's ancient Silk Route. Built largely with human labor (more than 500 died), the 500-mile (805 km) road through the Khunjerab Pass at 14,928 feet (4,478 m) and Hunza was officially opened in 1986. Environmental problems continue to plague this cold and wind-swept region. Not only is the pass closed in the winter months but also landslides frequently render the highway impassable.

Another important event occurred in 1979 when the Soviet Union invaded Afghanistan. Suddenly, for the United States, Pakistan became **a frontline state**, a strategic barricade against the spread of communism and Soviet access to the Arabian Sea and Persian Gulf. Aid was restored, and Pakistan became the third-largest recipient of American aid after Israel and Egypt.

The war in Afghanistan brought millions of refugees into Pakistan, where they joined with their ethnic kin, not only in refugee camps but also in economic pursuits. In cities, Afghans replaced workers who had migrated to the Middle East. Afghans who migrated to the United States sent remittances to their relatives in Pakistan. In addition, the war stimulated commerce in guns and drugs that supported new economic investments. However, vital structural problems were still not addressed.

Zulfikar Ali Bhutto was ousted in a military coup in 1977 and was later executed for conspiring to murder a political opponent. In 1988, when **Benazir Bhutto**, his daughter, became the youngest and first female leader of a Muslim state, Pakistan was essentially bankrupt. When Iraq invaded Kuwait in 1990, triggering the Gulf War, a sudden leap in oil prices, loss of remittances, and reduction of exports put Pakistan in a critical position. Assistance was provided by Saudi Arabia in return for the service of 6,000 Pakistani soldiers deployed as part of Operation Desert Shield, the name of the United States action in the Gulf. Even with additional support from the World Bank (WB), Pakistan has nowhere near the resources needed to overcome its economic problems. Moreover, these are compounded and exacerbated by an increasing rich-poor gap, rapid population growth (2.7 percent in 2003), increased religious conservatism, and political instability.

Years of military dictatorship, interspersed with short periods of unstable civilian government, have been characterized by mismanagement, patronage, nepotism, corruption, and violence. Leaders have been deposed, executed, or have died under mysterious circumstances. Benazir Bhutto has held office twice (1988–1990 and 1993–1997), only to fall each time under corruption charges. In recent years economic growth has faltered while communal strife intensifies.

The War On Terror

The western boundary of South Asia can be characterized as a frontier. Mountainous, tribal, and fractious, these *Tribal Areas* have long been a zone of contention. Here, no one, not even a Pakistani official, can enter without the permission of a tribal chief and/or **mullah** (religious leader).

From 1979–1989, the Soviet Union occupied Pakistan's western neighbor, Afghanistan. Even as the Afghans battled the Soviets, they fought each other. Conflict drove several million Pushtun and other Afghan refugees across the border into Pakistan's contentious and unstable Tribal Areas.

Within the context of the Cold War, the United States saw an opportunity to thwart the Soviets and sent a massive amount of weapons to Afghan resistance fighters. Even as the Soviets were being defeated, Usama bin Laden was financing the campaign and organizing his followers into a terrorist movement called al-Qaeda (Arabic for "The Base"). As the strict, fundamentalist Taliban took hold in Afghanistan, Islamic fundamentalism emerged across the border in Pakistan. Mosques and *madrassas*

(religious schools) expounded Islamic doctrine and hatred of America.

The Taliban regime brought stability to Afghanistan but at a dreadful social price. In this stable environment, the al-Qaeda movement flourished. Under the leadership of bin Laden it prepared to launch attacks against Western interests around the world. The most dramatic of these attacks took place on September 11, 2001 when al-Qaeda terrorists hijacked airplanes and flew them into the World Trade Center in New York City and the Pentagon in Washington, D.C. In response, the U.S. retaliated in Afghanistan. The Taliban were defeated and many al-Qaeda members were killed or apprehended. Others, including the al-Qaeda leadership, escaped, possibly into the Tribal Areas.

After the events of September 11, Pakistan's military leader, General Perez Musharraf, who took power in October 1999, decided to side with the United States in the war against terror. This was a complete reversal for Pakistan, a former supporter of the Taliban's extremism.

The Pakistan-Afghanistan border is South Asia's most critical region as it lies on the eastern edge of a vast Islamic realm extending as far west as Morocco. Pakistan's coherence and stability are essential in the war against terror. However, Pakistan's role in the struggle, as dictated by Musharraf, is not upheld by the majority of Pakistanis. They voice their disapproval by voting for fundamentalist politicians, voicing support for remnant Taliban, and giving sanctuary to al-Qaeda operatives. The future of Pakistan hangs in the balance.

Azad Kashmir/Jammu and Kashmir

Muslims first arrived in Kashmir in 711. By the sixteenth century the region was known as an "ornament of the Mughal Empire." After a period of Afghan and Sikh rule, the British, having defeated the Sikhs, sold the regions of Gilgit, Ladakh (now in India), and Kashmir to Gulab Singh, a Hindu, Dogra chieftain.

Muslim agitation began in the 1930s. With partition in 1947, Gilgit, Hunza, and other predominantly Muslim mountain states acceded to Pakistan. The British "encouraged" Kashmir's Hindu ruler to join India. That this mainly Muslim region would become part of an independent, mainly Hindu India was anathema to Pakistan. Geography also played a role in Pakistan's opposition. Jammu and Kashmir contain the five headwaters of the Indus River. Whoever controls these rivers controls Pakistan's fate (Figure 6–8).

With Muslims agitating to join Pakistan, NWFP tribesmen entered the scene to "liberate Jammu and Kashmir" from Hindu India's grip. A revolutionary government was formed in northern Kashmir, which became known as **Azad Kashmir**. This stimulated a significant migration of Hindu Kashmiris to India.

Fearing for the lives of his fellow Hindus, the Maharajah of Kashmir fled from Srinagar to Jammu and signed an agreement to join India. The Indian army took control and promised a plebiscite. In 1949, a United Nations' Cease Fire Line dividing Azad Kashmir from Jammu and Kashmir was accepted by both India and Pakistan. In 1956, Jammu and Kashmir formally entered the Indian Union. The promised plebiscite was never held.

In 1965, with accusations of infiltration, spying, and sabotage, another war broke out and lasted six months. The following year, in a meeting at Tashkent in the Soviet Union, Pakistan and India agreed to settle their disputes peacefully.

The promise of a peaceful solution was shattered in 1971 with the East-West Pakistan war and Indian intervention on East Pakistan's behalf. Fighting in Kashmir continued until 1972, when a new "Line of Control" was

Figure 6–8

Territorial disputes. The Siachen Glacier at 22,000 feet (6,600 m) is the highest glacier and battlefield in the world.

drawn. A Kashmiri Accord was signed by India and Pakistan in 1974.

Since that time, military fighting has been sporadic. Skirmishes take place mostly on the Siachen Glacier. This ice mass is strategically important because it commands a China-India-Pakistan triangle formed by China's 1962 annexation of India's Aksai Chin region.

Civilian conflict increased in the 1980s and 1990s. Muslim uprisings, election boycotting, and anti-Hindu terrorism intensified. In 1993, the Jammu and Kashmir Liberation Front and Hizbul Mujahideen united for the liberation of Kashmir. The Indian military has been ruthless in its suppression of insurgents. One result is that tourism, once a mainstay of Kashmir's economy, has picked up to some degree. Overall however, the industry is decimated.

Srinagar and the Vale of Kashmir are open for business even as inhabitants' vehicles, shops, and facilities remain damaged or closed for repairs. Unlike in the past, today's tourists come mainly from India and not from America and Europe. Many travelers see the potential for more violence and choose to stay away. After all, this is still disputed territory.

A beacon of light appeared in 2003 when India and Pakistan started to mend their relations. A cease-fire has been called and both countries are supposed to come to some agreement about Kashmir.

PEOPLE WORKING THE LAND

Supplying Food and Water

In the 1990s, almost half of Pakistan's labor force worked in agriculture. Only about 40 percent of Pakistan is suitable for cropping, and 90 percent of this land is on the Indus River plain in Punjab and Sind. Traditionally, crops fall into two seasonal categories: *rabi* and *kharif* (Figure 6–9). It is estimated that women do about 60 percent of the farm work in fields located close to the house. Since strict *purdah* (female seclusion) is the norm in the countryside, women work outside only when necessary and rarely venture far from the family compound (see Chapter 5 p. 111).

Agriculture plays an important role in the overall economy. Cotton apparel accounted for 35 percent of exports in 2003. Cotton acreage has been increased, and with new seed varieties, yields are above world averages. Maize, wheat, and rice are the major grains, and Pakistan specializes in the production of *basmati* rice. The Hindu word *basmati* means "fragrant" and refers to the nutlike flavor and aroma of this small but long-grained rice. Most wheat and rice is consumed domestically—increased yields are absorbed by a growing population. Even with Green Revolution and other improvements, grain and most other crop yields continue to fall short of world averages. Pakistan ranks fifth in the world in terms of acreage devoted to sugarcane but only sixty-ninth in productivity. Inefficiencies and inequities are systemic.

Since independence in 1947, Pakistan has increased its cultivated land by one-third, largely through the expansion of irrigation. Acquisition of water has always been a challenge—waterworks date back 4,000 years. Both the Mughals and the British constructed irrigation canals. In 1960, the **Indus Waters Treaty** gave Chenab, Jhelum, and Indus water rights to Pakistan and gave Ravi, Beas, and Sutlej water rights to India. Then, the World Bank-sponsored Indus Basin Development Fund was established to build new waterworks and link western rivers with the fertile soils of eastern Punjab. The giant Mangla Dam on the Jhelum and the Tarbela Dam on the Indus provided flood control, hydroelectric power, and irrigation. The 2-mile (3.2 km) wide Tarbela, completed in the 1970s, is the world's largest earth-filled dam. Today, Pakistan has the world's most extensive, continuous irrigation system.

Increasing water supply to croplands has a downside. Added supply raises the water table—the upper level of groundwater—and root systems can be inundated. In Punjab and Sind for instance, the Indus water table used to be 53 feet (16 m) below the surface. With increased irrigation, it rose to within 10 feet (3 m) of the surface. In addition, salts accumulate with rapid evaporation of moisture. To counter the loss of cropland from waterlogging and salinization, thousands of pumps, tube wells, and conduits had to be installed (mostly privately funded).

Traditional sociopolitical relationships influence the effectiveness of the irrigation system. Although the government controls the main power nodes and channels, intermediary and smaller channels are regulated by longstanding local power structures, relationships, and legal codes. Consequently, water is not necessarily delivered when crops most need it, and those at the system's extremities might not get enough, if any. The whole system was set back in September 1992, when floods destroyed countless irrigation networks and displaced as many as three million people.

Who Owns the Land?

Inequities remain apparent in farm ownership. Pakistan has struggled with land reform in efforts to reduce absentee landlordism, shrink large farms

KHARIF and RABI CROPS

KHARIF CROPS
- sown after the onset of the wet monsoon (June–July)
- harvested in autumn
- dominate southern Pakistan in Sind and Baluchistan
- floodplains and tank irrigation

RABI CROPS
- sown at the end of the wet monsoon
- harvested in spring
- dominate northern Pakistan in Punjab and NWFP
- doabs and uplands

TYPE OF CROP

MAIN PRODUCTION REGION
(B = Baluch; P = Punjab; S = Sind;
NWFP = Northwest Frontier Province;
IRV = throughout Indus River Valley)

KHARIF		
	RICE	P & S
	JOWAR (MILLET VARIETY)	P & IRV
	BAJRA (MILLET VARIETY)	P & IRV
	SESAMUM (SESAME)	S & P
	MAIZE (PULSE: LEGUMINOUS GRAIN)	NWFP
	GROUNDNUTS (PEANUTS)	P
	COTTON	IRV
	VEGETABLES	EVERYWHERE
	PEARS	B & NWFP
	PLUMS	B & NWFP
	GRAPES	B
	DATES	S

RABI		
	WHEAT	P
	JOWAR	P & IRV
	BARLEY	P & NWFP
	GRAM (PULSES—GRAM = CHICKPEAS	P & S
	—MASOOR = LENTILS	P
	—MATTAR = PEAS	S
	—MUNG = KINDEY BEANS)	P
	UNSEED (OIL SEED)	P
	RAPE (OIL SEED)	IRV
	MUSTARD (OIL SEED)	IRV
	TOBACCO	NWFP
	VEGETABLES	EVERYWHERE
	CHILIES	S
	APPLES	B & NWFP
	APRICOTS	NWFP
	BANANAS	S
	CITRUS	P
	GUAVA	P
	MANGO	S
	PEACHES	NWFP

NOTE: SUGARCANE IS GROWN IN RABI AND HARVESTED IN KHARIF IN PUNJAB.

Figure 6–9

Kharif and Rabi crops.

and enlarge small ones, consolidate fragmented holdings, and increase peasant ownership. However, these efforts have met with strong opposition from the landed rich, who also hold political power. In 1947, a mere 1 percent of farmers owned a quarter of the agricultural land, and 65 percent of farmers held only 15 percent of the land. While there has been some size reduction and an increase in the number of medium-sized farms, 34 percent of farms are still under 5 acres (2 ha).

About one-third of Pakistani farmers are tenant farmers, required to give 50 percent of their crop to the landowner. And, despite 1992 laws opposing the practice, **indentured servitude** is widespread, especially in Sind and parts of Punjab. Poor farmers, forced by circumstance to borrow from *zamindars*, become victims of usurious interest charges and thus remain eternally indebted to a land-owning family willing to exploit the servitude for generations.

Industrialization and the Question of Energy

With government backing, industrialization has progressed at a rapid pace. Even so, heavy industry is limited. Pakistan has only one integrated steel mill, a Soviet Union-funded venture near Karachi. This mill depends on imported iron and coal and still runs at only partial capacity. Moreover, industry is very much tied to agriculture. Manufactured goods account for about two-thirds of exports. These are mostly cotton products and processed foods. Pakistan's major trading partners are the United States, Japan, and Persian Gulf nations.

In recent years Pakistan has pursued trade relationships through supranational organizations. In the mid-1980s, the **South Asian Association for Regional Cooperation (SAARC)** was formed to disassociate political conflicts from economic cooperation. SAARC includes Bangladesh, Bhutan, India, Maldives, Nepal, Pakistan, and Sri Lanka. Many meetings have been held. The **Economic Cooperation Organization (ECO)**, established by Pakistan, Iran, and Turkey in 1985, now includes Afghanistan, Azerbaijan, Kyrgyzstan, Tajikistan, Turkmenistan, and Uzbekistan. In 1994, construction started on a highway connecting Pakistan with Turkey via Iran.

Energy has been of paramount concern both for industrial production and improvement of quality of life. Water-generated power is of great significance but alternative sources are being sought. Unfortunately, Pakistan's coal is relatively poor quality, but natural gas sources have been developed in Baluchistan and oil is produced in Punjab. Natural gas pipelines link Islamabad, Faisalabad, Peshawar, and Quetta, and another line from Iran via Afghanistan is under construction. Additional oil fields have been discovered in the south. At least 20 international companies are involved in their exploration. However, development has been limited by lack of security in both Sind and Baluchistan.

The fact that a third of Pakistan's power is still derived from animal dung, wood from already severely deforested areas, and *bagasse* (sugarcane fiber), implies that the country must continue to seek alternative sources for power production.

Nuclear power is one alternative. Pakistan's first heavy-water nuclear reactor was built near Karachi with Canadian aid in 1965. In 1974, when India exploded its first atomic device, Ali Bhutto pledged an equal response: Pakistan would build an "Islamic Bomb" if the people had to "eat grass" to do it. In the 1990s, another plant opened at Chashma on the Indus about 149 miles (240 km) south of Islamabad. Fuel is provided from a uranium enrichment plant at Kahuta, also near Islamabad.

Although North American and European nations have stopped providing nuclear technology to Pakistan for fear of its military intentions, China continues to do so. And in 1998, Pakistan exploded an atomic device. Soon after, India responded with its own nuclear event. Now that both nations have nuclear capabilities there is even more reason for them to mend relations and solve the problem of Kashmir.

Electrification of villages has been a high priority. Power in the countryside can light homes, pump water, and grind grain. The government is convinced that improving rural conditions will stem the ongoing rural-to-urban migration streams that have turned Pakistan's cities into unmanageable, crowded behemoths.

Gulf Gateway

Without a sound industrial base, Pakistan will be hard-pressed to survive in an ever-more competitive global order. However, Pakistan has a strategic, geographical advantage: it is a natural juncture for the proposed trans-regional gas and oil pipeline from Central Asia, Iran, and Qatar to world energy markets. While the pipeline is only in the planning stage, a recently developed deep-sea port at the coastal town of Gwadar in Baluchistan is about to change radically the trade patterns around the Arabian Sea.

The port in Gwadar, 419 miles (675 km) west of Karachi, will connect Afghanistan and Central Asian states with Europe. A deep-water channel more than nine miles (14 km) long will permit large vessels, such as container ships and oil carriers, to anchor. Pakistan hopes that this will draw shipping away from other less convenient ports in the region.

A master plan proposes a modern airport and a network of roads linking Gwadar with the country's main cities in the east and north. This plan also includes a tax-free export processing zone and an industrial complex site. The local population, mainly fishing folk, eagerly awaits the potential benefits of this development.

China agreed to partially finance the project in 2002. China's interest lies in the fact than the port facility offers a toehold in this energy-rich region as well as a shorter link to Xinjiang Autonomous Region in China's western frontier. Xinjiang is 2,790 miles (4,500 km) from China's east coast but only 1,550 miles (2,500 km) from Gwadar.

The ultimate success of the port hinges on its usage by the Central Asian states of the former Soviet Union. In this respect, cooperation with Afghanistan, which lies between Pakistan and Central Asia, is critical.

Although Pakistan disowned the Taliban to join the United States in the war on terror, it is still struggling to live down its legacy of involvement in Afghan affairs. Moreover, Afghanistan is a long way from being stable; skirmishes on its border with Pakistan keep international relations tentative at best. A July 2003 attack on Pakistan's embassy in the Afghan capital of Kabul and a general distrust of its intentions among the Afghan populace emphasize how far Pakistan is from realizing its ambition to be the center of a large Central Asian trading block.

HYPERURBANIZATION

The spectacular growth of cities in Pakistan can be summed up with the example of Karachi, which had a million people in 1950. This rose to 8 million by 1991 and 10.4 million by 2001. Lahore, the second largest city, had 850,000 in 1950, and now has more than 4 million people. City growth in Pakistan is fast enough to be called **hyperurbanization.**

City development has always been tied to the availability of water. Urban centers have the power structure to influence water projects. Understandably, patterns of urban distribution remain associated with rivers and canal systems, and administrative strength and core elites continue to be linked with urban places.

Under colonialism, cities were created and supported to serve colonial purposes. Along the Grand Trunk Road, towns were developed at river crossing points to function as centers of control, administration, collection, and distribution. With the introduction of railroads, urbanization gravitated to rail stations and lines. Bypassed by the railroads, small communities were isolated, and many became deserted. Expanded irrigation projects increased agricultural production, but the Green Revolution and land reform favored medium-sized farms. Very small farms were absorbed. Large-scale, rural-urban migration ensued, and towns became specialized in

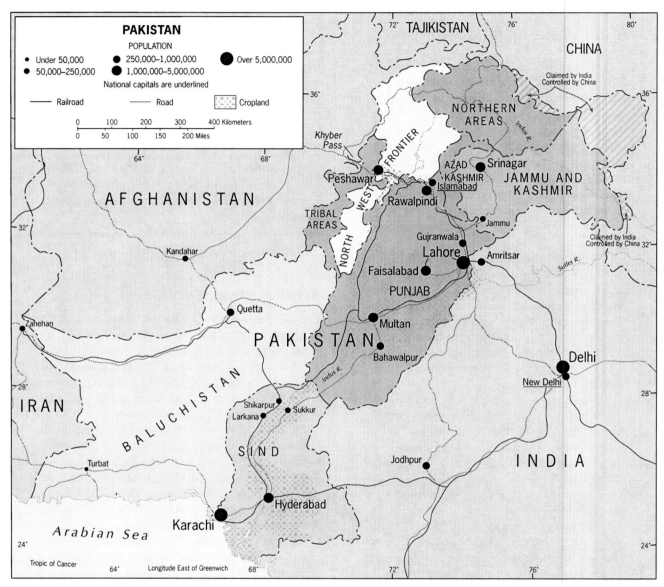

Figure 6–10

Pakistan's cities. From H. J. de Blij and P. O. Muller, *Concepts and Regions in Geography,* Second Edition, 2005, p. 278. Originally rendered in color. © H. J. de Blij and P. O. Muller. Reprinted with permission of John Wiley & Sons, Inc.

various agriculture-related processing and manufacturing activities.

Three-quarters of Pakistan's urban centers are situated in the Indus Basin (Figure 6–10). The remainder in the uplands of Baluchistan and NWFP are more like overgrown towns. In fact, the largest cities are experiencing the greatest growth. Karachi and Lahore grew more than 250 percent from 1951 to 1981. Most of this growth was from in-migration. Afghan refugees have been major contributors to rapid growth in Quetta and Peshawar.

There have been several Five-Year Plans and urban plans devised since independence. However, political instability, vested interests, and resistance to change have rendered the overall results disappointing. Diffusionary tactics, including growth-pole development (moving the center of power to decentralize activities), have floundered because new industry and ancillary activities tend to gravitate to existing agglomerations and infrastructures. Furthermore, in Pakistan rail rates favor locating industries as close as possible to the port of Karachi and the newer port at Gwadar.

Karachi, the country's primate city, continues to grow by accretion of *katchi abadis*—squatter settlements on the periphery. Efforts to decentralize people into new *metrovilles* outside the city have been largely unsuccessful. Land intended for the poor is purchased by speculators, who leave it vacant. Even some of the low-income participants who move into metroville housing, sell their properties to upper-income buyers for immediate monetary gain. The city government now concentrates on improving services and supporting self-help programs in the *katchi abadis* and other poor areas of the city. Meanwhile, the city's larger infrastructure is crumbling, resulting in contaminated water, accumulation of refuse, and power shortages for everyone. Frustrations arising from these problems are exacerbated by rivalry between the richer, Urdu-speaking Muhajirs and relatively poorer Sindhis, Pathans, and other ethnic groups.

Most residents of Pakistan's cities come from rural backgrounds. Consequently, urbanization is couched in rural tradition. Most work is in the manufacturing and service sectors and is labor-intensive. The informal sector of the dual economy is critical to the survival of millions.

Urbanization is obviously tied to population growth. Although Pakistan has had family planning since 1952, programs have not been very effective. The rural fertility rate is 6.6 children per woman (average TFR 4.8 in 2003); the average age at marriage is 17. With only 11 percent of women using any form of contraception, most are exposed to child bearing for 33 years! Needless to say, infant and

maternal mortality rates are high. The already-low sex ratio of 105 males per 100 females is decreasing, indicating better care and increased life expectancy for males versus female infanticide, neglect, and high infant mortality rates for girls.

Part of the problem is the repression of women and lack of education. Enforced seclusion and traditional attitudes prevent most females from attending school or even learning to read. Pakistan's literacy rate is 60 percent for men but only 21 percent for women. Although schools have been built for girls in several areas, less than 10 percent of girls ever attend. More girls go to school and attend university in urban centers, but life roles of "suitable positions," arranged marriage, and general subservience to men are still expected for most. The status of men among men is vital in South Asian, Muslim society and must be maintained in part through modest and deferential behavior of women, especially those in a man's immediate family.

Muhammed Ali Jinnah had been a strong supporter of rights for women. Gains were made such as women getting the right to own property. The election of Benazir Bhutto suggested that progressive strides would be made. But modernization has impacted primarily elite women. Very few women work in formal or informal sectors full-time. There are scores of women's groups such as the Women's Action Forum (started in Karachi in 1981), but opposition to women's rights has only grown stronger. With Islamization being carried forward in a relatively conservative fashion, women's movements (largely urban and elite) face tough obstacles. Increasing violence against women in recent years has put women's groups into a position of defending the few rights they have rather than acquiring new ones.

Pakistan's rich-poor gap has not narrowed significantly. In 1999, the poorest 20 percent had 8.8 percent of the wealth. The richest 20 percent possessed 42 percent. With nearly 80 percent of Pakistan's population over 25 having little or no schooling, this situation is unlikely to change.

Endemic poverty has spurred the informal economy. In recent years the drug and arms trade, strengthened by upper-level collusion and mafia-like operations, has become a critical issue within the

Figure 6–11

In Saidu Sharif, these boys are rolling balls of opium to be sold in local shops for processing elsewhere. Photograph courtesy of B.A. Weightman.

international community (Figure 6–11). Not only does this trade generate income but it allows Pakistan a measure of political pull in this and other regions of turmoil.

Today, Pakistan is beset with problems. Regional and structural economic imbalances; rapid population growth and urbanization; and sociopolitical conflicts riddle this turbulent land. When considering Pakistan's volatility, keep in mind its rocky beginnings and increasing complexity.

Himalayas: Abode of Ice and Snow

Now that you know something about the trials and tribulations of Pakistan and the turmoil in still-disputed Kashmir, I trust that you will be interested in learning about the often-called "exotic" countries of Nepal and Bhutan. Both are landlocked, isolated, and very mountainous domains. Both face dilemmas associated with modernization.

The Himalayas extends across eight nations: Afghanistan, Pakistan, India, China, Nepal, Bhutan, Bangladesh, and Myanmar. The relative isolation promoted by the world's highest peaks and deepest gorges has sustained both natural and human diversity amongst the 120 million who live here. Many people speak Sino-Tibetan languages and practice **Tantric** or Vajrayana **Buddhism** (Figure 6–12).

Figure 6–12

These are only two of a plethora of gods in Tantric Buddhism practiced in the Himalayan regions of South Asia. Photograph courtesy of B.A. Weightman.

The critical positioning of these various peoples between the political giants of India and China has deeply affected their ways of life and threatened or even eliminated their autonomy. Only Nepal and Bhutan remain as independent states in the Himalayas. Ladakh and Sikkim, once independent, are now part of India. Border conflicts and unstable political boundaries are facts of shatter-zone existence.

THE KINGDOM OF NEPAL

Landlocked Nepal is pinched between Tibet (China) and India. Nepal's history has encompassed both survival in context of environmental constraints and maintenance of autonomy and cultural identity within the clutches of competing world powers.

Ten Himalayan peaks mark Nepal's boundary with Tibet (Figure 6–13). Here glaciers and treeless tundra give way to **cloud forests** that coat the slopes from 8,000 to 13,000 feet (2,400–39,000 m). These mountain rainforests are blanketed year-round in fog and mist. Up to 140 inches (356 cm) of rain fall from June to September, but even in the dry season, the forest drips with moisture. Oaks, laurels, maples, and magnolias mix with ancient spruce, fir, and cedar trees. Orchids, morning glories, azaleas, giant ferns, and rhododendrons color the gray-green forest mass. Flowering stands of rhododendron, standing 60 feet tall, turn acres of land crimson. This once-pristine forest houses rare and unusual animals such as the red panda, sun bear, clouded leopard, fishing cat, and goat-antelope. Human encroachment has put this habitat in danger.

Southward from the high Himalayas, the land drops precipitously into lower mountains and foothills. This region contains the 12-by-15-mile (19 by 24 km) alluvial Kathmandu Valley. Here are Nepal's main urban centers, including the capital of Kathmandu. With more than a million people the capital city has grown 62 percent in a decade

Figure 6–13

Physical geography of Nepal. Nepal serves as a buffer state between China and India.

Figure 6–14

A Hindu temple in Kathmandu is backed by a modern office building. Notice the cloth market held on the temple steps. Photograph courtesy of B.A. Weightman.

is the Royal Chitwan National Park, with its tropical monsoon forests, giant sal trees, 12-foot (3.6 m) high elephant grass, and diminishing populations of tigers, one-horned rhinos, sloth bears, and an array of other animals and birds. Here too flows the sacred Rapti River, a tributary of the Ganges.

One of my most memorable journeys was floating silently in a dugout canoe along a narrow channel of the Rapti. Mugger crocodiles rested on the narrow banks, in the shadow of the towering elephant grass. A one-horned rhino stood belly deep in the water munching on river plants. As the sun rose higher, the crocodiles slithered into the water, right under my canoe! I was further startled by a ruckus of bird cries in the upper branches of the tall sal trees. Hundreds of parakeets soared overhead, streaking the blue sky with their brilliant green plumage.

Nepal, with its population of 25 million, is culturally and socially diverse. Nepali is an Indo-European language related to Hindi but infused with Tibeto-Burman. Another indicator of the country's crossroads situation is religion. Although most people are Hindu, Buddhism and Hinduism intermingle in everyday practice. Life is further complicated by the influence of the Hindu caste system. Here, people are differentiated not by occupation but by group customs. The caste system affirms group cohesion and its flexibility adds to the confusion.

Nepal is the world's only Hindu Kingdom. The king is the head of state. However, it is also a parliamentary democracy headed by a prime minister. Historically, the region was fragmented into a mosaic of competing states. In the eighteenth century the western Gurkhas defeated the ruling Newars of Kathmandu. In 1790, Nepal was unified. Border disputes with British India led to the Anglo-Nepalese war of 1814–1816 that was fought with the rhythm of the monsoon. In the end, Nepal had to accept a British "resident," thereby giving Britain a toehold in Nepal.

From 1816, Britain ruled Nepal as a buffer state. Nepalese troops—the **Gurkhas**—fought for the British against the Indian Mutiny of 1856 and made their mark on military history for their fighting ability and heroism. Gurkha regiments would fight for Britain in many battles, a factor in Britain's granting of independence to Nepal in 1924.

(Figure 6–14). About two-thirds of Nepal's people live in this hill region, terracing the slopes with rice and wheat. Yields are generally poor. Below lies the marshy plain known as the **Terai** at 900 feet (270 m). In recent years large numbers of migrants from the hills have descended into Terai to find jobs (Figure 6–15).

The Terai is an extension of the Indo-Gangetic plain. Approximately one-third of Nepal's population lives here. Many of these are high-caste Hindus of Rajput extraction from India. The Terai constitutes only 17 percent of the land but generates 60 percent of Nepal's GNP. Main occupations are in commercial agriculture and industrial concerns. Competing for space

Figure 6–15

This is a Tharu house in the Terai region of Nepal. The Tharu migrated to Nepal from India many years ago. The house is of wattle and daub (reeds and mud) with few openings to repel insects. The inside of the house is very dark, but this is not so important because most work is done out of doors. The roof is a convenient place to air bedding and dry clothes.
Photograph courtesy of B.A. Weightman.

Gurkha regiments stationed overseas have contributed significant remittances to Nepal. But, the return of Hong Kong to China in 1997 and pullbacks from other areas of the world have reduced British Gurkha troops by thousands. While they are still stationed in Singapore and elsewhere, most Gurkhas now fight for the Indian army. Remittances have been severely reduced and many older Gurkhas, dismissed without pension, live in poverty.

Nepal is one of the world's least developed countries. Its problems derive from both internal and external causes. It is landlocked, far from the ocean, and has few resources. Political instability, corruption, and favoritism have characterized the government throughout most of this and the previous century. Rule through a tiered *panchayat* (council) system has been largely ineffective at the village level. The established power structure controls funds and reaps the most benefits. Interestingly, a recent political party called itself *Panchayat*. It became very unpopular; to speak of "village *panchayats*" now is politically incorrect.

Political dissention continues to plague Nepal's government even after the overthrow of the king in 1990 and the assassination of the royal family in 2001 by Prince Dipendra. The prince was unhappy with the bride his family had chosen for him and slaughtered eleven family members including the king and queen.

Today, Maoist guerrillas roam the countryside killing at will. There have been bombings in the capital and the tourism industry has taken a significant blow. This situation remains unresolved.

Nepal's buffer situation places it in a difficult position. Large numbers of refugees have entered from Tibet, especially after China's takeover in 1959. The Chinese want this migration stream to be stopped. Nepal has bought arms from China, and India wants that activity to cease. India wants rights to purchase land in Nepal, first rights for power from Nepal's hydroelectric plants, and an end to border smuggling. Given the physical setting, a strong relationship with China is unlikely. Physical and cultural propinquity position India as the country with which Nepal has to deal.

Indians are already prominent in Nepal's industrial and commercial sectors. Town development in the southern Terai, dam building on at least five rivers, the beginning of a power grid shared with India, manufacturing enterprises of items such as clothing, carpets, cement, sugar, bricks, soap, and matches are largely Indian investments. Apparel and carpets are exported primarily to the United States and Germany. Imported manufactures come mainly from India.

Aside from politics, there are numerous, complex economic, social, and environmental problems. Recent research has demonstrated that multiple agencies working in specific areas have failed to recognize the interrelationships among all of these. Moreover, integrated and holistic approaches have been difficult to implement in context of unfair power relations. The rural poor with their "overpopulation" and "inappropriate" land practices were usually blamed. However, *blame the victim* philosophy is no longer in vogue.

Researchers are beginning to recognize that the people of Nepal (and elsewhere) managed to survive in their difficult environment for centuries prior to Western impacts and that modernization and commercialization have severely damaged both fragile environments and cultural integrity. Many "solutions" have backfired, primarily because they are couched in the industrialized world's "scientific expertise" and "development" (read economic growth and consumerism) mentality.

While Nepal is one of the largest recipients of foreign aid on a per capita basis, it ranks as one of the lowest in human development. Nearly half its population is below the age of 15. Prenatal care is free but only 24 percent of newborns receive it. Contraception use doubled from 1987 to 1997 to 29 percent. By 2003, the figure had risen to 35 percent. Overall agricultural production is declining, and about half of the labor force works only part-time. Much aid has been siphoned off into the coffers of bureaucrats and politicos. Unlike some other Himalayan regions, lack of road development keeps Nepal's hill regions isolated and in chronic poverty. Road building—by creating access to the more developed lowlands—could bring health care, education, and basic consumer goods. But lack of roads also prevents commercial forestry and other environmental impacts.

Because change is slow, attitudes persist even in light of new approaches to solutions. Even with grassroots participation and co-management programs, according to geographer Barbara Brower (1993), "the old ideology, though explicitly repudiated, lives on in the assumptions, goals, plans, and tactics of Western-trained professionals seeking to involve local people in programs of resource management."

Nepal's economy relies heavily on tourism, although this benefits primarily the hill and mountain regions. Strangely, Nepal, under the guidance of Western advisors, has invested little in this sector. When tourists do go to Nepal, their activities are largely uncontrolled. According to geographer Nigel Allan (1990), "Nepal has become a petting zoo for upscale, middle-class foreign tourists who are anxious for a holiday experience in the exotic and mysterious 'East'."

BHUTAN: LAND OF THE THUNDER DRAGON

Bhutan (also called **Drukyul**), is a Lamaistic Buddhist kingdom struggling to keep its culture and identity in the face of outside influences and modernization. Landlocked and isolated, it was virtually closed to the outside world until the 1960s. Like Nepal, Bhutan is one of the world's least developed countries.

This Switzerland-size kingdom once comprised monastic power bases centered on *dzongs*—fortified monasteries. Its deep valleys were linked by cultural and trade contacts facilitated by iron-chain suspension bridges constructed in the fourteenth century. Given the highly differentiated terrain and natural communities ranging from glaciers and tundra in the north to tropical rainforests and malarial swamps in the south, cultural variety is to be expected (Figure 6–16).

The Bhote tribes of northern Bhutan are originally from Tibet and adhere to the Drukpa Kagyudpa sect of Tibetan Buddhism. Commonly called **Drukpas**, they revere the king as their spiritual leader. The Nepalese of the south are primarily Hindu. Most people speak Dzongkha, a Tibetan dialect. Others speak Sharchop or Nepali or one of at least 18 other different languages or dialects.

In 1731, Tibet imposed suzerainty over Bhutan. Forty years later, in light of constant regional struggles and raids into Assam and Bengal, the British became involved. After a lengthy civil war one family triumphed and, in 1907, the first hereditary king was enthroned with British support. Four years later, Bhutan signed a treaty with the British to counter Chinese designs on the region. Later, Bhutan agreed to rely on the advice of India (Great Britain) for

Figure 6–16
The Kingdom of Bhutan.

action in foreign affairs. This agreement was formalized by India in 1949.

While Bhutan's current foreign policy leans toward isolationism, China's 1950 claims to Bhutan and its 1959 occupation of Tibet cemented Bhutan's already strong ties with India. The first road from India to the capital of Thimpu was opened in 1962. Now, several roads from India are furthering cultural intrusion and the trappings of development. For India, Bhutan is a convenient buffer state between it and China. It is a strategic training and staging ground for troops that could be quickly employed against China or separatist groups in Assam. Bhutan is also a source of hydroelectric power for India's industrial region west of Calcutta. India is active in dam construction and has negotiated power-sharing agreements. In fact, most of Bhutan's generated power is used by India. As most foreign aid and development assistance is also from India, Bhutan finds itself in an ever-tightening dependency relationship.

Modernization began in the 1950s. However, the country is too small to develop any economies of scale. For example, Druk Air now has two planes. Even the capital of Thimpu has only 46,000 people. But, as the king once pointed out, "gross national happiness" supersedes "gross national product." Although there is some small-scale industry in the

south, including chemicals, cement, and wood products, the majority of Bhutan's inhabitants work directly with the land.

Subsistence farmers cultivate grains and vegetables, and herders raise yaks and cattle and practice *transhumance* (moving livestock between the highlands in summer and the lowlands in winter). A mere six percent of the land is cultivated. Most cropping is confined to the flatter alluvial valley floors and narrow terraces. However, Bhutan is no longer self-sufficient in cereals. Efforts are being made to diversify agriculture from traditional maize and rice to fruit and vegetables, especially for export to India. One long-standing export crop is **cardamom,** a spice commonly used in South Asian cuisine. There is evidence that some commercial crops are culturally and environmentally inappropriate.

The Land of the Thunder Dragon is in trouble. Nepalese have resided in Bhutan since the nineteenth century. In the mid-twentieth century, many of Nepal's soaring population have illegally migrated across the border, settling in the southern lowlands of Bhutan alongside the existing Nepalese community. When Bhutan conducted a census in 1988, it discovered that ethnic Nepalese made up an unspecified but significant proportion of the total population. The former kingdom of Sikkim had already been overrun by Nepalese and, for protection, elected to join India. Bhutan's king did not want a repeat performance in his realm.

In an effort to salvage Bhutanese culture, the king launched *Drignam Namzha,* a "traditional values and etiquette" policy that promoted Dzongkha in the schools and required all people to wear traditional Bhutanese dress on certain occasions. Many Nepalese objected. But when the king declared 1958 as the cutoff date for legal immigration and asked for proof of citizenship, they were outraged. The Nepalese demanded political representation but were ultimately expelled from Bhutan. Currently, an estimated 80,000 to 100,000 Nepalese live in refugee camps in southeastern Nepal. Recently, Bhutan has softened its border policies and set up screening agencies for immigrants. The influx has slowed dramatically but the problem remains unsolved. Violent acts against refugees in Bhutan are increasingly common.

Bhutan's ruler and his Drukpa subjects are very concerned about the precarious state of their Buddhist culture and the endangered state of their physical environment in light of modernization. Consequently, tourists are limited to 3,000 a year and must be confined to guided tours. Sixty percent of the land is designated as forested, and sustainable forestry is national policy. Nevertheless, reforestation programs are not keeping up with deforestation. Large tracts have been set aside as floral and faunal preserves. With its array of biomes, Bhutan houses the world's only population of golden langur (a leaf monkey) and some of the few remaining populations of one-horned rhino, tiger, and Asian elephant. A greenbelt and elephant corridor have been created along the Indian border.

Even with sustainable development policies, Bhutan has numerous social and environmental problems. Poverty, disease, illiteracy, high infant mortality, and low life expectancy are universal. However, Bhutanese women have more freedom and power than in most other South Asian societies. Women are often the major bread winner in a family, and daughters inherit their parents' property. Environmental problems are increasingly widespread and are similar, if not as intense, as in other Himalayan communities.

MOUNTAIN DEVELOPMENT AND ECO-DESTRUCTION

Magnificent, awesome, challenging—such descriptors belie the reality of the Himalaya's geological instability and eco-vulnerability. In this oft-perceived "barren wasteland," millions struggle daily to eke out a living. Long subject to earthquakes, landslides, and flash floods, Himalayan populations more and more are confronted by problems related to environmental fragility and degradation.

There are three geo-ecological zones (Figure 6–13). The Outer Himalayas has an average elevation of between 3,280 and 6,560 feet (1,000–2,000 m). This lowest region extends almost unbroken from the Indus River to the Brahmaputra and includes numerous long, flat-bottomed valleys called *duns.* These are filled with gravelly alluvium (look again at Figure 2–1).

In between this and the Great Himalayas is the Middle Himalayas. About 129 miles (206 km) wide and between 9,840 and 16,400 feet (3,000–5,000 m) in altitude, this zone comprises the Himalayan foothills. Ranges run obliquely from the Great Range or stand disconnected. Most people live here, concentrated in such valleys as the Kathmandu Valley of Nepal and the Vale of Kashmir. This is the region of most critical environmental and human damage.

The Great Himalayas, the highest zone, is sparsely populated. But, in recent years people from the Middle zone have come into the region, placing additional stress on its limited resources. This is also the environment affected most by tourism. Mountain climbing is a major tourist activity.

Tibetan-influenced Buddhism predominates in the region of the Himalayas, although Hinduism and Islam are important in places closer to India. Monasteries, *chortens* (structures containing relics of a holy person), *mani* walls (walls of stones with prayers carved on them), prayer flags, and pilgrimage tracks are Buddhism's impress upon the Himalayan landscape. Dwarfed by Chomolungma, "Mother Goddess of the World" (Everest), Kanchenjunga, Dhaulagiri, and other pillars of the gods, it is little wonder that millions believe in the sanctity of Himalayan peaks.

Mount Kailas

Not the highest peak in the Himalayas, but the most significant in human terms, is 22,028 feet (6,608 m) Mt. Kailas. Believed by both Hindus and Buddhists to be the sacred center of the universe, Mt. Kailas is central to religious conceptions of the world known as *mandalas*. A *mandala* is a symbol that one contemplates during meditation. From this sacred pinnacle flow the four great rivers: the Sutlej, Indus, Ganges, and Brahmaputra. In reality, the sources of these important rivers lie within a 60-mile (96 km) radius of Mt. Kailas.

The mountain also figures prominently in pilgrim circulation. Thousands of pilgrims converge on this site each year to complete the 34-mile (55 km) circumnambulation (a walking meditation) upward to 18,600 feet (5,580 m). In 1900, Kawaguchi Ekai, a young Zen monk from Japan, walked across Nepal and traversed an east–west trail across the Tibetan Plateau. As recorded in Scott Berry's book *A Stranger in Tibet* (1989), the monk wrote of his first vision of Mt. Kailas and its sacred lakes: "Verily, verily it was a natural *mandala*. The hunger and thirst, the perils of dashing stream and freezing blizzard, the pain of writhing under heavy burdens, the anxiety of wandering over trackless wilds, the exhaustions and lacerations, all the troubles and sufferings I had just come through, seemed like dust, which was washed away. . . ."

Although the Himalayas were apparently remote and inaccessible, isolation was not complete. Traders plied precipitous routes, transferring goods such as salt and wool from Tibet and silk from China to eager merchants in India. Through trade and religious pilgrimage, connectivity was maintained within and without the region. Although contact with Tibet is now limited, yak caravans continue to trek remote areas, linking them with newly built truck routes. Aside from salt and other raw materials, the caravans and trucks now carry cans of cooking oil, tanks of propane, and other products. Pilgrimage continues to be a means of information and economic interchange. But the pilgrims now talk of how much their lives have changed, mentioning the new satellite dish atop the local monastery.

Roads and Dams

The colonial era brought British explorers to the Himalayas. Thereafter, the region was open to commerce with the densely peopled plains of India. Then imported mass-produced goods were introduced to mountain communities, and the region was exploited for resources such as minerals, timber, and its cheap labor supply by foreign power structures. Himalayan peoples soon became enmeshed in political and economic dependency relationships beyond their control.

Development schemes such as roads and dams are often based on the interests of outsiders. India has been very active in the region, understandably in light of its strategic interests regarding Pakistan, China, and Bangladesh.

After a border war between India and China in 1962, India built numerous road networks as high as 17,000 feet (5,100 m). Road construction—especially dynamiting—destabilized the hills and soil. The roads also gave lumber companies access to virgin forests with the resulting disappearence of forests and wildlife. The threat of tigers was replaced by the threat of landslides.

Road construction has also delivered sweeping cultural change. Truck routes funnel in new products and new ideas. They are also conductors for increased prostitution and the diffusion of AIDS. Commercialization, resource extraction, and mass tourism have effectively devastated natural and human landscapes to the point that the integrity of traditional human-environmental relationships has been destroyed.

The Himalayas function as the water tank of Asia. It has the highest runoff of any mountain range, providing water for one-tenth of the world's population. However, torrential rivers invite dam construction to generate hydropower for electricity and water for irrigation. Since the 1950s, dams have proliferated, particularly in the upper Ganges and Indus tributary areas. Dam construction destroys forests, degrades land, and displaces mountain and hill people, forcing them to move to other already stressed regions. Pakistan's Mangla Dam, completed in 1967, receives so much silt and debris from the Jhelum that its operational life has been cut from 100 to 50 years.

Soil erosion is accelerated by development projects and general overuse of natural resources. Rivers run brown with silt, precious topsoil needed to sustain crop yields. Some 300 cubic meters of topsoil annually wash from Nepal's Himalayan ranges to the plains of the Ganges and Brahmaputra below. The stain of Himalayan silt can be seen as far as 400 miles (645 km) from shore in the Bay of Bengal.

Logging

Forestry has exacted the greatest toll from the Himalayan ecosystem. Large-scale cutting commenced in the late nineteenth century, when the British created 6.5 million railway ties from one of the world's finest deodar forests. Since 1950, logging, agriculture, and urbanization have destroyed nearly half of the forest cover in the Indian Himalayas alone.

Throughout the Himalayas, original village forests are so degraded that people (mostly women) must spend hours every few days to find and gather firewood. Imagine carrying at least 50 pounds of wood on your back—secured by a single wood-fiber strap around your forehead—and then trudging barefoot to your home where you will cook a meager breakfast for the family, do household chores, feed the animals, and do farm work. Every year, a Nepalese family spends at least two months' worth of time gathering firewood.

In 1970, a huge storm raged over the great mountain Nanda Devi (25,650 feet or 7,695 m). Tons of earth, boulders, and trees crashed down slopes. A tributary of the Ganges rose more than 60 feet (18 m). Six road bridges, 24 buses, and some 600 homes were swept away, and nearly 200 people were killed. Mountain dwellers were stunned at the magnitude of this disaster.

Realizing that vast forest areas had been cleared by commercial companies, the mountain dwellers were determined to fight back and save their environment. In the north of India's Uttar Pradesh state, where leopards and tigers once prowled the forests, a group of women literally held on to trees as loggers wielded their axes. This grass roots movement became known as *Chipko Andolan,* meaning "the movement to embrace the trees." As women in other areas engaged in "tree-hugging" protests, the **Chipko Movement** drew international attention and became instrumental in the formulation of India's conservation policy. Although village preservation groups have become ubiquitous, population and other pressures limit their effectiveness.

Tourism

Tourism and other forms of commercialization have had far-reaching impacts. The annual tourist count on the slopes of Mt. Everest alone was 18,000 before the 2001 political crisis in Nepal reduced that number. Tourism is a double-edged sword. While it generates revenue and employs significant numbers of locals, tourist facilities and activities pollute the land and water, destroy delicate vegetation, and place even greater demand on wood supplies than local populations. Villagers strip forests to provide hot water for "needy" tourists. Tourists deface rocks and religious sites, and some even go so far as to steal the inscribed stones of sacred *mani* walls. Tourism also compromises cultures.

One people to be drastically affected by tourism are the **Sherpas** of Nepal. When Sir Edmund Hillary became the first Westerner to ascend to the top of Mt. Everest from Nepal in 1953, he was lauded throughout the world. Standing in the shadow of his glory was a Sherpa named Tenzing Norgay. The Hillary expedition is now known as the Hillary-Norgay expedition.

Although Sherpas are famed for their mountain-climbing abilities, they are not traditionally mountain climbers. Seeking new pastures for their yak herds, the ancestors of the Sherpas migrated into the Everest region from Tibet around 500 years ago. They began trading with people in lowland Nepal and India. When the Tibetan border was closed by China in the 1950s, Nepal was opened to foreign climbing expeditions.

Sherpas had already served as servants and porters for British climbers attempting to reach Everest's summit from Tibet. However, when Hillary and Norgay achieved the summit from Nepal, the word *Sherpa* was heard around the world.

Sherpas are now dependent on tourism. Although some lead expeditions to conquer peaks, others guide trekking groups at lower elevations. Some Western climbers admit that they pay Sherpas to take risks they are unwilling to take. On some occasions, injured Sherpas have been abandoned by climbers too anxious to reach a summit. There are other negative consequences. Earning money to be there in the first place and then dirtying the land is offensive enough to the gods but a terrible death in the snows even furthers the detriment to one's *karma* and diminishes chances of a good rebirth.

Master-servant relationships are now being transformed into client-guide relationships. Today, many climbers go lightweight and even without oxygen. Sherpas are employed to maintain base camps. However, media attention is usually given to the large-scale expeditions, and those guides who do climb seldom receive recognition equivalent to that of their clients. Have you ever heard of Sundare Sherpa, who has ascended Mt. Everest five times without oxygen? Or Kaji Sherpa, who in 1998 climbed the peak in 20 hours and 24 minutes, two hours faster than the previous record?

Figure 6–17
This is your friendly "rent-a-yak." Tourism is entering even the most remote regions of the Himalayan region.
Photograph courtesy of B.A. Weightman.

About Yaks

Imagine a one-ton, hairy ox standing 6 feet high (2 m) at its shoulder. Imagine a beast that is able to carry as much as 300 pounds (136 kg) for 20 or more days at an elevation of 18,000 feet (5,486 m). This is a yak, the most useful animal for residents of the high Himalayas (Figure 6–17). Aside from their packing abilities, they provide meat, milk, butter, cheese, wool, and leather. Yak products are used in traditional medicine and their tails make fine dusters for the sacred scrolls in Buddhist temples.

First domesticated in Tibet, yaks cannot survive well below 10,500 feet (3,500 m). A blood hemoglobin is thought to be responsible for a yak's ability to exist in thin air at high altitudes. It begins to lose its condition around 7,000 feet (2,100 m). Females, called *dimos*, are smaller than males and give about one quart of milk a day. *Dimos* produce a calf once every two years; populations do not reproduce quickly. Yaks can live 25 years.

There are around 12 million domesticated, black, brown, and gray yaks. Wild yaks are much larger and are always black. Wild yaks, however, are very rare; their populations diminished by illegal hunting and loss of habitat.

Yaks can be crossed with cattle to produce animals better-suited to lower elevations. This animal is called a *dzho* and looks like a cow with long hair.

Tourism is one of the fastest-growing industries in the Himalayan region. Nepal now earns one-quarter of its foreign exchange in this manner, but only 10 percent of this money ever reaches the rural poor whose environment is the main attraction. Sadly, mountain dwellers can now claim to live in the world's "loftiest garbage dump."

Who is to blame for this disaster? Some argue that the mountain people have ruined their own land through population increase, overfarming, overgrazing, and careless cutting of trees. Others target outsiders—Japan, India, the West—as being most responsible. This dichotomy can be clarified in light of the paradigm that everything is interrelated and that a holistic approach is essential.

PEOPLE AND THE MOUNTAINS

For thousands of years, Himalayan villagers have survived by subsistence farming and herding. For many, transhumance was a way of life. But all this has changed as the region becomes more entangled with the forces of modernity.

Populations increased with immigration of Indians from the twelfth century and again with eighteenth- and nineteenth-century food surpluses. However, road construction in the 1950s allowed the advent of modern, although limited, health care. This resulted in a drastic decline in death rates, an increase in life expectancy, and, inevitably, a population explosion. Land use intensified on already limited farming and grazing areas, resulting in reduced crop yields and ruined pastures. Greater firewood requirements meant that woodlands were rapidly depleted. Productive land use systems that sustained smaller Himalayan populations for centuries were unable to support the added numbers.

The average population growth for the entire Himalayan region is 2 percent. This, combined with a decrease in the area of cultivated land, and a mere 1 percent increase in farm yields means that local food production can no longer keep pace with human requirements. Because living conditions are deteriorating, women feel the need to have more children to help them.

With self-sufficiency destroyed, ecological migration began in earnest. People sought new land in higher and even more fragile environments, where too-steep slopes and thin soils cannot support intensive cropping or grazing. A second, mostly male ecological migration to lower elevations is an ongoing process. Cities and towns of the hills and lowlands swell with unemployed males, while women and children assume added burdens in their mountain villages.

Many outside experts, agencies, and governments have held the view that the people were "destroying their own environment" with their primitive and harmful methods. But their proposed solutions have been drawn from experiences elsewhere and often do not address the intricacies of the varied niche environments of mountain settings. Consequently, modernization efforts and palliative programs have worsened the situation and increased dependency relationships with outside political and economic forces.

Commercialization has transformed a barter economy to a cash economy. This, along with public intervention, improved accessibility, media exposure, and tourism have combined to raise expectations. Simultaneously, social and economic inequities have intensified, with little opportunity for improvement.

Widespread input substitution has upset the traditional balance. For example, following the decline of mountain pastures, sheep and goats were substituted for cattle. This has further destroyed pastures and exacerbated erosion because goats, lacking upper front teeth, tend to rip out plants. In the past, larger ruminants merely cut the vegetation, allowing it to eventually grow back.

Shortage of firewood has increased the use of animal manure (mixed with straw) for cooking fuel. Therefore, chemical fertilizers must be substituted for the lost manure. Farmers, now dependent on alien manufacturers, can be totally ruined if they lack the purchase price or improperly use the product.

With declining grain yields, farmers have been encouraged to switch to cash crops, meaning that people who traditionally produced their own barley, rice, corn, and other crops are compelled to buy imported grains. Some commercial crops are environmentally and culturally inappropriate. New technologies, such as hand tractors, have opened nutrient-poor submarginal lands, quickly making them unfit for production. Irrigation water shortages have been created by the diversion of water to industrial and other uses.

For development, local governments must promote links with external markets. Horticulture, hydroelectric power, irrigation and other resource extraction, and tourism are rarely sensitive to local needs. Moreover, mountain regions tend to interact with mainstream markets on unequal terms. External markets perceive mountains as cornucopias of resources ready for the taking, and very little revenue is ever plowed back.

Government efforts in land reform are strapped by delays and muddied by vested interests. Resource-centered programs such as afforestation (the planting of trees for commercial purposes) or pasture development focus on techniques rather than user perspectives. They are generally area-specific, not taking into account the potential impacts on related areas.

Research in the region of the Himalayas has made clear at least three important facts. First, a society's inability to live within the usable limits of a biophysical resource base generates responses that inevitably cause environmental alteration and degradation. Second, market forces can be singled out as the most significant factor underlying those responses. Third, policy makers and farmers perceive neither the problems nor the solutions in the same way. This derives from differing degrees of closeness to the problem and differing stakes in outcomes of both problem and solution.

While the situation is critical in many areas, especially environmental, for many people general well-being has improved. Land reform, health care, income subsidies, increased education, self-help programs, and so forth have exerted some positive impacts. However, most of these have bonded mountain communities to outside power structures more tightly than ever.

Clearly, this northern region of South Asia is rife with complex problems. Some of these arise from environmental constraints. Others emerge from transitional political settings. At the turn of the twenty-first century, however, two trends are apparent. International and global economic and strategic concerns are increasingly overriding the impacts of local geographies. And, global frameworks are counterbalanced by cultural and territorial identity-consciousness. This is an era of rampant and proliferating nationalism. Unfortunately, in many regions environmental integrity and social well-being have become lost in the shuffle.

⊕ Main Point Summary

1. Political instability has marked the evolution of practically every state in this region. After the partition of India, a split Pakistan could not withstand the centrifugal forces that ultimately brought about its political demise. Differences in size and population between west and east were compounded by economic imbalances and cultural contrasts. Language and religion were particularly potent forces in communal strife; the question of Islam's role in government remains an ongoing issue in Pakistan.

2. After partition, many business-class Punjabis moved into Sind and came to dominate the economy. Others, known as Muhajirs, came to dominate administration. These groups are resented by the Sindhis, and internecine strife persists.

3. Pakistan has fought India over border demarcations. Kashmir remains an unsolved issue.

4. When East Pakistan protested the draining of its resources for military forays against Indian and other expenditures in the West, West Pakistani troops countered with brutal military action. This stimulated a massive refugee movement into India's West Bengal. Indian intervention ensured the creation of an independent Bangladesh in 1971.

5. Pakistan is characterized by dramatic regional variation. Its rugged Northwest Frontier Province is a cultural shatter zone, difficult to control, and with ties to tribal groups in Afghanistan and Kashmir. Punjab, the land of five rivers, is the breadbasket of Pakistan. It is also important industrially and politically. Sind is tied to the Indus River and is second to Punjab in size and wealth. Pakistan's largest city and port—Karachi—is here. Baluchistan is a rugged and remote tribal area with little development.

6. Pakistan's infrastructure was built largely in response to the economic and military needs of the British Raj. This fact, combined with physical and cultural divisions and political instability, has foiled efforts at balanced economic development.

7. Pakistan's more recent roller coaster development is tied to its relationships with India, China, and the United States. China is viewed as a supporter. India is seen as a protagonist (especially in the case of Kashmir), and the United States has been perceived, at least until very recently, as a friend of India.

8. The Karakoram highway has permitted greater interchange between Pakistan and China. Increased traffic, especially of tourists, has brought about dramatic change in formerly isolated tribal areas.

9. Pakistan is a troubled country suffering decades of military rule and political instability. Rising Islamic fundamentalism, terrorism, and the question of jurisdiction over Kashmir are serious problems of international proportions that Pakistan and its neighbors must address.

10. Almost half of Pakistan's labor force works in agriculture. Maize, wheat, and rice are the major crops. Irrigation is critical to production, and giant dams have been constructed on the Jhelum and Indus for flood control, hydroelectric-power development, and irrigation expansion. But, additional surface water has caused waterlogging and subsequent salinization. Thousands of tube wells have been installed to correct this problem. However, unfair practices regarding control over various parts of the system mean that many farmers do not get water when they need it.

11. Inequities remain apparent regarding farm ownership. About one-third of Pakistanis are tenant farmers. Many are landless and spend their lives in servitude to landowners.

12. Industrialization is limited by resource availability, internal conflict, and corruption. Development of oil and gas fields by international companies is hindered by lack of security in Baluchistan and Sind. Proposed pipelines from the Middle East are hampered by political instability in Afghanistan.

13. Pakistan is a nuclear power.

14. Hyperurbanized, three-quarters of Pakistan's cities are in the Indus Basin with the largest cities continuing to grow even larger. Rural-to-urban migration is a serious problem, and squatter settlements accrete to city fringes.

15. Urbanization is tied to population growth. Fertility rates are high, and women have little say in their lives. The preference for males is reflected in a highly imbalanced sex ratio. Gender inequity is only one of a host of problems faced by Pakistan, where the majority live in poverty.

16. The Himalayas extends across eight nations. Relative isolation among the world's highest peaks has fostered cultural diversity. Nepal and Bhutan are the only independent states within the Himalayan region. Both must balance their lives between the powerful influences of China and India.

17. Fantastic forests cover mountain slopes below the treeless expanses of glaciers and tundra. Populated zones, including Kathmandu, are in

alluvial river valleys and foothills. Slopes are terraced to grow rice and wheat but yields are generally poor. About one-third of Nepal's population lives in the Terai, an extension of the Indo-Gangetic plain.

18. Nepal is culturally diverse, the world's only Hindu Kingdom, and one of the world's least-developed countries. Nepal's political instability and corruption have led to a widened rich-poor gap and environmental exploitation. India continues to play an influential role in Nepal's "development."

19. Nepal has one of the world's highest rates of foreign aid yet it remains one of the world's poorest in terms of human development. Lack of infrastructure prevents certain forms of development but also deters environmental exploitation.

20. Nepal relies heavily on tourism, but this impacts mainly the hill and mountain regions. Both environmental and cultural integrity are severely threatened.

21. Bhutan is a Lamaistic Buddhist kingdom struggling to keep its identity under the pressure of tourism and development.

22. Bhutan, like Nepal, serves as a buffer state between India and China. Bhutan leans toward isolationism but is strongly impacted by India.

23. Large numbers of Nepalese live in Bhutan, many illegally. This is a source of tension. Bhutan does not want to go down the road of Sikkim and lose its independence to India. The Bhutanese government has launched a "traditional values and etiquette" policy to ensure the maintenance of cultural integrity.

24. The Himalayas have three geo-ecological zones: the Outer; the Middle; and the Great. Environmental damage is most severe in the Middle zone.

25. Tibetan influences are apparent on the landscape. Mount Kailas is an important center of pilgrimage. Increased tourism is despoiling this and other sacred places.

26. Development schemes are based on the interests of outsiders. India is especially active in road and dam construction. Although the Himalayas has long been accessible by yak caravan, roads have opened the region to trucking and with it, pollution, prostitution, and AIDS. Human-environmental integrity has been compromised in many areas. Sherpas are examples of this predicament.

27. Overpopulation and development have led to ecological migration, which has caused even greater environmental damage. Outside agencies blamed the people for their troubles and imposed unsuitable "solutions." Commercialization, input substitution, environmental exploitation, external trade requirements, population increase, and other factors have combined to bring ruination to many parts of mountain regions. On the other hand, modernization has improved the standard of living, at least somewhat, for countless individuals and families.

⊕ Bibliography

Agrawal, Arun. 1996. "Forest Management in the Indian Himalayas: An Institutional Perspective." *Human Dimensions*. 1: 6–10.

Alexander, Brian. 1987. "A Beast For All Seasons." *International Wildlife*. 17: 45–50.

Alexander, David. 1984. "Saving the Trees, Saving Ourselves." *International Wildlife* 14: 18–22.

Allan, Nigel. J. 1990. "Household Food Supply in Hunza Valley, Pakistan." *The Geographical Review*, 80: 399–414.

Allan, Nigel, J. R. 1998. Review of P. R. Karan and H. Ishii. 1996. "Nepal: A Himalayan Kingdom in Transition." Tokyo: United Nations. *The Geographical Review*, 87: 421–22.

Berry, Scott. 1989. *A Stranger in Tibet: The Adventures of a Wandering Zen Monk*. Tokyo: Kodansha.

Bray, John. 1993. "Bhutan: The Dilemma of a Small State." *The World Today*, 49: 213–16.

Brook, Elaine. 1988. "Through Sherpa Eyes." *Geographical Magazine*, 60: 28–34.

Brower, Barbara. 1993. "Co-Management vs Co-Option: Reconciling Scientific Management with Local Needs, Values, and Expertise." *Himalayan Research Bulletin,* 13: 39–49.

Burki, Shahid Javed. 1991. *Pakistan: Continuing Search For Nationhood.* 2nd edition. Boulder: Westview.

Chopra, Pran. 1986. "The View From Bhutan: The Temptations and Traps of Development." *Ceres,* 19: 24–27.

De Blij, Harm J., and Peter O. Muller. 2004. *Geography: Realms, Regions and Concepts.* 11th edition. New York: Wiley.

Denniston, Derek. 1993. "Saving the Himalaya." *World Watch,* 6: 10–21.

Fisher, James. 1991. "Has Success Spoiled the Sherpas?" *Natural History,* 2: 38–45.

Flavin, Christopher. 1990. "Last Road to Shangri-La." *World Watch,* 3–4: 18–26.

Gerrard, A. John. 1990. *Mountain Environments: An Examination of the Physical Geography of Mountains.* Cambridge: MIT.

Hausler, Sabine. 1993. "Community Forestry: A Cultural Assessment: The Case of Nepal." *The Ecologist,* 23: 84–90.

Jodha, N. S. 1995. "The Nepal Middle Mountains." In *Regions At Risk: Comparisons of Threatened Environments,* eds. Jeanne X. Kasperson, Roger E. Kasperson, and B. L. Johnson, pp. 140–85. London: Heineman.

Khan, Ashraf. 2003. "Gulf Gateway." *Far Eastern Economic Review,* October 8: 54.

Karan, P. P. and Cotton Mather. 1985. "Tourism and Environment in the Mount Everest Region." *The Geographical Review,* 75: 93–95.

Karan, P. P., and Shigeru Iijima. 1985. "Environmental Stress in the Himalaya." *The Geographical Review,* 75: 71–92.

Mandelbaum, David G. 1993. *Women's Seclusion and Men's Honor: Sex Roles in North India, Bangladesh, and Pakistan.* Tucson: University of Arizona.

McColl, Robert W. 1987. "House and Field in the Karakorams: The Interaction of Environment and Culture." *Focus,* 37: 15–19.

Meier, Richard. L. 1985. "Islamabad Is Already Twenty-Five." *Ekistics,* 52: 212–16.

Metz, John, J. 1990. "Forest Product Use in Upland Nepal." *The Geographical Review,* 80: 279–87.

Rahman, Mushtaqur. 1996. *Divided Kashmir.* Boulder: Lynne Reinner.

Rahman, Tariq. 1995. "Language and Politics in a Pakistan Province: The Sindhi Language Movement." *Asian Survey,* 35: 1005–16.

Rauniyar, Durga, S. 1997. "Spotlight: Nepal." *Population Today,* 25:7.

Scarborough, Vernon, L. 1988. "Pakistani Water: 4,500 Years of Manipulation." *Focus,* 38: 12–17.

Searight, Sarah. 1987. "Priming the Pump." *The Geographical Magazine,* 64: 19–24.

Siddiqi, Akhtar Husain. 1989. "Urban Development in Pakistan." In *Urbanization in Asia: Spatial Dimensions and Policy Issues,* eds. Ashok K. Dutt, Laurence J. C. Ma, and Allen G. Noble, pp. 47–74. Honolulu: University of Hawaii.

Spate, O. H. K., 1948. "The Partition of India and the Prospects of Pakistan." *The Geographical Review,* 38: 5–29.

Spate, O. H. K., and A. T. A. Learmonth. 1967. *India and Pakistan: A General Regional Geography,* 3rd edition. Bungay, Suffolk: Richard Clay (Chaucer).

Stephenson, Bryan. 1994. "Bhutan: The Development Dilemma." *Geographical Magazine,* 66: 51–55.

Taylor, George C. Jr. 1965. "Water, History, and the Indus Plain." *Natural History,* 74: 40–49.

Thapa, Gopal, B., and Karl Weber. 1991. "Deforestation in the Upper Pokhara, Nepal." *Singapore Journal of Tropical Geography,* 12: 53–67.

United Nations. 1988. "*Population Growth in Mega-cities. Karachi.*" Population Policy Paper No. 13. New York: United Nations.

Weiss, Anita. M. 1998. "The Slow Yet Steady Path to Women's Empowerment in Pakistan." In *Islam, Gender, and Social Change,* eds. Yvonne Haddad and John Esposito, pp. 124–43. New York: Oxford.

Young, L. J. 1991. "Agriculture Changes in Bhutan: Some Environmental Questions." *Geographical Journal,* 157: 172–78.

Chapter 7

✦

South Asia: India, Giant of the Subcontinent

"We wear the dust of history on our foreheads, and the mud of the future on our feet."

SHASHI THAROOR (1997)

Scholar and writer Shashi Tharoor (1997) believes that "The central challenge of India as we enter the twenty-first century is the challenge of accommodating the aspirations of different groups in the Indian dream." The reality of a billion individuals, belonging to countless ethnic, linguistic, religious, caste, and other defined groups, makes India's continued existence as a federation amazing (Figure 7–1). *Federalism* is a governmental system whereby central and regional authorities are voluntarily tied in a mutually interdependent relationship. In this light, India is a classic paradox; it is a unity of disunity. It is multitextured, a cauldron of competing interests, a collection of centers and peripheries (Figure 7–2).

Independent India was conceived with diversity in mind. Its ability to change, redrawing boundaries along linguistic lines for instance, is a large part of its *raison d'etre*. Federalism thrives on the tension between cores and peripheries, as long as the links are not severed. With this in mind, India's Constitution grants extraordinary powers to federal authorities when unity and the integrity of the nation are threatened. Thus far, separatist movements have been

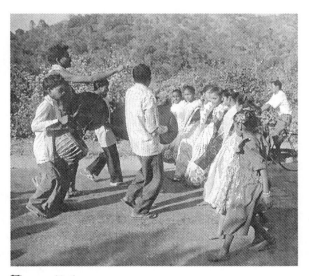

Figure 7–1

Khond dancers in Orissa State. I happened across this group on a road in the Eastern Ghats. They were celebrating the mother goddess, a yearly event. The dancing was accompanied by a huge feast and animal sacrifices. Some of the women almost danced themselves into a trance. Photograph courtesy of B.A. Weightman.

Figure 7-2

States and cities of India. From H. J. de Blij and P. O. Muller, *Concepts and Regions in Geography,* Second Edition, 2005, p. 281. Originally rendered in color. H. J. de Blij and P. O. Muller. Reprinted with permission of John Wiley & Sons, Inc.

Figure 7-3
Population density. David L. Clawson and James S. Fisher, eds., *World Regional Geography: A Development Approach,* 8th edition, 2004. Upper Saddle River, N. J: Prentice Hall.

quashed. However, after 50 years of independence, the voices of dissension are more frequent, much louder, and more powerful. These arise from social inequity, regional disparity, and political instability. These are some of India's challenges.

Demographic Challenges

Figure 7-3 shows that India is already densely populated. At its partition, India's population was 344 million. According to the 1991 census, it was 844 million. By 2003, it had reached more than one billion. Although India's NIR has fallen from 2.1 in 1978 to 1.7 in 2003, overall population growth is phenomenal. India adds at least 18 million a year—2,000 babies are born every hour! Its area is just half that of China, but it will probably exceed it in population in the twenty-first century. Moreover, food production is not keeping pace with population growth.

Fertility behavior is strongly influenced by prevailing social and economic conditions as well as by religious/cultural norms and expectations, such as early marriage. Of the 4.5 million marriages each

Figure 7–4

Population dynamics. From H. J. de Blij and P. O. Muller, *Concepts and Regions in Geography,* Second Edition, 2005, p. 275. Originally rendered in color. H. J. de Blij and P. O. Muller. Reprinted with permission of John Wiley & Sons, Inc.

year, an estimated 3 million brides are between 15 and 19 years old. India has a young population, with 36 percent under the age of 15.

Not all populations are growing at the same rate; regional variations are apparent (Figure 7–4). And, as I noted in Chapter 3, poorer families tend to have more children. Poorer women, who have less control of their bodies or have no access to birth control information or devices, also have higher NIRs (Figure 7–5).

Less than half of couples of reproductive age use contraception. The most popular form of birth control is surgical sterilization, but this is elected by couples who already have three or four children. India's TFR has fallen from 6.0 children in the 1950s to around 3.1 in the early 2000s, but periods of quickly falling fertility have been followed by periods

of stable levels. Whether India has reached another plateau is not clear.

In 1996, the Indian government eliminated national targets for acceptance of fertility control methods. The new "Target-Free Approach" puts family planning at the community level, where grassroots workers set targets according to individual needs. Acceptance of all forms of birth control have risen since that time. But, the effects are yet to be seen.

For every 1,000 men in India, there are 929 women—a figure on the decrease. This trend of decline has existed unabated since the first census in 1861. The gender gap reflects widespread discrimination against women and is more apparent in the north, where sex ratios are considerably lower than in the south. The states of Rajasthan, Haryana, and

Uttar Pradesh, for instance, have much lower ratios than Kerala or Tamil Nadu.

Caste differences play a significant role in these discrepancies. There are more higher caste women in the northern states, especially of the kshatriya or former military groups. These castes are noted for their seclusion of women from outside contacts, including education. Female illiteracy rates are high, and women are expected to produce sons. Female children are not usually killed at birth but rather neglected in their early years. Fed last and less than their male siblings, their ailments ignored, many girls never reach the age of five.

Another issue is health care. Some 300,000 children, more girls than boys, die yearly of diarrhea. This is attributable to lack of sanitation and neglect. However, this does not mean that health care has not improved overall. Indeed, death rates have been lowered all over the country. Killer diseases such as smallpox have been eradicated, and infant mortality rates have been lowered. The life expectancy of Indians has risen from age 32 in 1947 to age 63 in 2003. In fact, many more Indians are living longer and are capable of having more children.

India has a less-than-successful history of family planning in terms of programs and their impacts. In 1952, it became the first country in Asia to establish an official family planning program with the aim of curtailing population growth. Various national and state programs were instituted, including promotion of IUDs and sterilization. Three million men underwent vasectomies in 1970–1971.

Economic incentives were common; food rations were withheld from some families with more than three children; others lost a month's pay unless a worker was sterilized. Such programs are believed to have averted more than 85 million births from 1956 to 1986. Even so, progress was problematic.

Ethnic, religious, and caste rivalry fostered fears of persecution and genocide, and violent resistence became frequent. Capitalizing on these fears, politicians were quick to blame family planning for other social ills. Many organized programs have fallen by the wayside. I recall visiting a village where a "two-children" poster had been painted on the side of a school build-

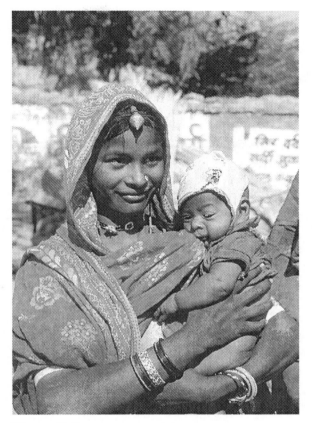

Figure 7–5

This young woman is approximately 15 years old. She is a member of the untouchable, sweeper caste. She lives in a temporary tent on the outskirts of Udaipur in Rajasthan. Photograph courtesy of B.A. Weightman.

ing. As the children crowded around me, I noticed that the poster was in a terrible state of disrepair. It was barely legible. One boy jumped up, pointed at it, and said in English: "Small family very funny." Everyone laughed, the teacher louder than anyone.

The 1991 census results shocked many into realizing that fertility rates must be reduced at all costs and that women must be empowered to have more control over their lives. But women remain last in line for education, job training, credit, medicine, and even food. Constantly changing governments cannot provide the stability needed to sustain long-term empowerment strategies. Yet, experts agree that this is the most effective mode of reducing population growth rates.

India's population growth will continue its spectacular rise, given its current size and youthfulness. Despite the use of family planning, more than 40 percent live in areas where crude birth rates (CBRs) are in the mid- and upper- 30s. World Bank predictions indicate that even if India achieves a two-child family by 2015, it will still have a population of 1.9 billion!

Kerala: A Woman's State?

Kerala, in southern India, is a small, not particularly prosperous state. Yet, it has the highest literacy rates in India, even higher than those for Delhi. Although women still lag behind men, more than 80 percent of both genders are able to read and write. Prior to India's partition, only 11 percent were literate. Clearly, this state has made great strides in educating its population (Figure 7–6). Why is this state so different?

Kerala's government spends 40 percent of its budget to maintain a vast network of schools and colleges. Education has long been valued in society, especially among higher castes. Since most higher castes were traditionally matrilineal, many women continue to enjoy freedoms unattainable elsewhere. During the Raj, most of Kerala was ruled by independent maharajas who promoted education in the Malayalam language. Literacy rates were much higher in Kerala than in states with English-based schooling. About 20 percent of Kerala's citizens are Christian. This group attaches more importance to equal education for males and females. Access to education at Christian schools was an attractive lure for the conversion of low-caste Hindus. In addition, knowledge of record-keeping was essential to maintain the status of Kerala's international traders.

In 1990, Ernakulum (Cochin) became the first administrative district to be officially declared fully literate. This was the result of a literacy drive entailing motivational strategies so appealing that the project's founder received commendation from the United Nations. Other communities followed suit with the support of various official agencies. Volunteer teachers were mobilized, two-thirds of them women. Local religious leaders were recruited to back up the media blitz. Interestingly, this was accomplished by a communist state government. Most impressive has been the progress achieved by tradition-bound Muslim women.

Figure 7–6

A middle-class family in Kerala. This family owns a lumber business. Note the difference in dress between the more traditional parents and the more modern children. Photograph courtesy of B.A. Weightman.

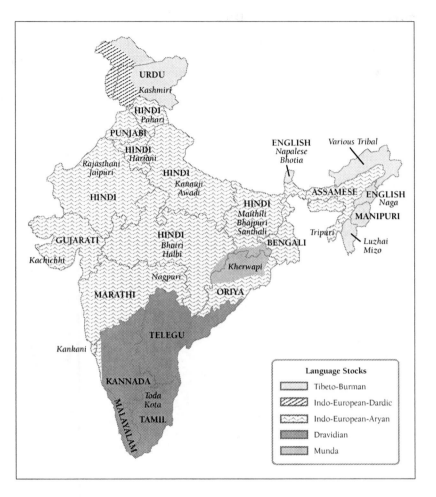

Figure 7–7
Language stocks. Note the division between Indo-European and Dravidian linguistic regions.

Language Stocks

	Tibeto-Burman
	Indo-European-Dardic
	Indo-European-Aryan
	Dravidian
	Munda

Linguistic Conflicts

With 15 national languages and more than 100 tribal tongues in India, it is no wonder that language has been a divisive force. This is especially significant because language is even more important than ethnicity or nationality for individual and group identity. Unlike Europe, where Latin long served as a medium of cultural diffusion, in India, Sanskrit was replaced by Persian as the court language. Persian was later replaced by English. Educated elites, versed in one or more of these languages, became divorced from the masses, who remained steeped in their own linguistic worlds.

Four language stocks, each with its own set of language families, produce a complex language map (Figure 7–7). The Munda and Tibeto-Burmese stocks are relegated to relatively isolated areas, including the Eastern Ghats and the Himalayas. The main cleavage is between the Indo-European and the Dravidian language stocks. The most widely spoken language is Hindi, a member of the Indo-European language family. Hindi speakers also occupy India's central states, a core ringed by people speaking unrelated tongues. Dravidian includes 4 main languages and 18 lesser ones. These occupy the southern states, having been pushed out of the Indus valley with historic migrations. The most well known Dravidian language is Tamil.

Hindi is one of India's official languages, a source of aggravation to Muslim Urdu speakers. In 1994, riots broke out in several states over the broadcast of a 10-minute news program in Urdu in Bangalore. Hindi derived from Hindustani, a language used in Mughal bazaars and military camps. It incorporates numerous Persian words, as this was the Mughal court language. There are several dialects, but Indian schools teach "Standard Hindi" written in the Devanagari (Sanskrit) script.

Urdu is similar to Hindi in grammar, but from a political perspective it is different. It borrows from Arabic and Persian and is written in a Persian script. Muslims in northern India see Urdu as their "special language."

Transcending this linguistic web is English, which was made the language of high school education in the 1830s. Contrary to Indian advice, the British ignored linguistic boundaries when establishing state boundaries. In 1953, there was a Telugu movement for a linguistic state. In 1956, there were language riots in Bombay between Marathi and Gujarati speakers. Boundaries had to be changed.

Bombay was moved to Maharashtra. In the south, Kerala and Tamil Nadu were formed to benefit Malayalam and Tamil speakers, respectively. Orissa was defined in the eastern tribal region, and Karnataka was created for Kannada speakers. These are only a few examples of boundary restructuring following partition.

Boundary alterations did not mask the fact that India's elites spoke and read in English—a "foreign tongue." Government efforts to force Hindi as India's only national language were met with strong resistance. It has spread, nevertheless, via media such as schools and television, and interregional migrations. Hindi is even a *lingua franca* (common language) for mixed migrant worker groups. English remains the language of private schools, colleges, and universities. Moreover, India's modern, urban social class is differentiated by English more than any other cultural force. Many intellectuals argue that an English education should be the means employed to "uplift" the poor masses.

Language issues remain significant. In Goa, recent attempts to force English speakers to learn Konkani caused mass demonstrations. Politicians representing Uttar Pradesh and Madhya Pradesh are calling for the eradication of English. Disagreements and conflicts have obvious impacts on political and economic development.

Caste Complexities

Even though there is a growing ambivalence toward caste in India's upper social strata, caste rules and regulations apply among virtually all people. Brahmins,

Figure 7–8

Bonda woman and child. Numbering about 6,000, the Bonda live in geographic isolation in the Eastern Ghats. They are hill cultivators, growing mangoes and jackfruit, but still hunt with bows and arrows. Theirs is a female-dominated society. The very independent Bonda speak Koui, an Austro-Asian language. Photograph courtesy of B.A. Weightman.

Rajputs, Marathas, and Jats are examples of regular castes. But more than two-thirds of Indians belong to so-called *scheduled castes and tribes*. Scheduled tribes are also called *adivasis* meaning "original inhabitants." These include such groups as the Nagas of Nagaland and the Bondas of Orissa (Figure 7–8). Another term employed by the government is *other backward classes* (OBCs) that include the bulk of the population—the "lower middle classes."

The Indian government has published lists including 405 scheduled castes, 255 scheduled tribes, and 3,500 OBCs. These groups are entitled to special social and political benefits. Discrimination in favor of the "backward" classes did not begin with independence. It began in the state of Mysore (Karnataka) in 1921. In this state, 92 percent of the people belong to the backward classes for which university admissions and jobs are reserved. In contrast to predictions at the time, the state did not collapse, and the capital of Bangalore has become the center of India's computer industry.

Tribal Troubles

India has 255 scheduled tribes plus numerous unofficial ones. Many of these groups live in the Eastern Ghats and the hills of Assam. In 1944, a little-known but decisive battle was fought in the hills and jungles around Kohima in Nagaland. Here, the British broke the Japanese advance from Burma, ending their plans to capture the Raj. This remote region has remained fractious, with insurgencies against the Indian government mixing with intertribal conflict. Seven states cover the upper basin of the Brahmaputra River. Incursions by Christian missionaries and Bengali Muslims add to the fray.

The Nagas, who became Christian in the nineteenth century, have been up in arms since 1946. They were granted statehood in 1963. In 1993, 316 people were killed in ethnic massacres between the Nagas and the Kuki tribes in the Manipur hills. In Meghalaya, the Khasis are rebelling against outsiders, and Assam has witnessed rioting between ethnic Assamese and Bengali Muslim migrants from India and Bangladesh. If this seems very confusing, be assured that this is only part of the story of tribal troubles.

Smuggling is another part of the picture. Teak, aromatic agar wood, jade, gems, and heroin are among the smuggled goods. Eastern India is a key route for the movement of heroin out of Myanmar. Drug lords cum politicians rule their territories with personal armies. Border towns are hotbeds of crime and prostitution. Burmese girls, retired from the Thai brothels because they are HIV positive, now work here. Addicts number anywhere from 15,000 to 40,000 in Manipur alone. The government claims that the HIV infection rate is more than 50 percent.

Why does the Indian government spend so much money on maintaining control of this region? After all, it is attached by only a narrow corridor of land at Siliguri. New Delhi's concern lies in the fact that the region is of strategic and economic importance. It borders on four nations: China, Bangladesh, Bhutan, and Myanmar. Further, it is rich in oil, timber, and tea. It also has great hydroelectric potential. Culturally, it is the meeting point between Hindu-based culture groups and the animist and Christian Tibeto-Burman people of the hills. Keep your eye on this region as one of South Asia's hot spots.

Caste as a form of kinship or marriage lineage circumscribes who can marry whom and delineates mutually supportive behaviors and responsibilities among members. In this context, hundreds of thousands of castes exist. As members and non-members are sharply defined, favoritism and nepotism to one group are expected by other groups.

Caste is not static. Castes' positions in relation to one another are constantly being renegotiated on the basis of fluctuating power, status, ritual behavior, political mobilization, education, and geographical location. Caste holds diverse peoples together via a hierarchical structure permitting flexibility within the caste. Upward mobility is potentially possible.

Caste may be associated with occupation—*jati*. While one's *jati* is fixed at birth, its position within the hierarchy is not. Moreover, modernization has confounded the association of birth assignation with practical occupation.

Abolition of discrimination has applied only to *untouchability*, not to caste. Those in low level or polluting *jatis*—the Dalits—seek recognition as regular members of Indian society. Whatever their origin, their subordination derives from perceived ritual impurity rooted in antiquity. Some scholars argue that untouchability was exacerbated by the British, who were overly concerned with caste inequities and thereby drew undue attention to them.

Today, Dalits demand their rights, and government programs have facilitated their move into many upper echelons, including politics. Education, in particular, has invested them with expectations for equal treatment. Indeed, "untouchables" now have many opportunities historically denied them. Nevertheless, stepping outside the bounds of untouchability is no easy task; deeply rooted social taboos are difficult to expunge.

Affirmative action policies have created privileged sections within underprivileged groups. An interesting point is that pro-action policies are part of the official Constitution, and they were written by a former untouchable. The situation that those classified as "backward" go forward by virtue of their "backwardness" is resented in many circles.

The progression of these formerly subordinated classes has political impacts. Many politicians avidly seek the loyalty of entire socially categorized populations. Appointments are made with an eye toward potential mobilization of caste followings. Just as significant is the fact that 85 of 545 seats in India's Parliament and 22.5 percent of government jobs are reserved for scheduled castes and tribes.

Caste then, in terms of establishing hierarchical relationships and interactions, is not disappearing, although it is exhibiting greater flexibility. The poor and lower-middle-class masses are seeking economic improvement, while the privileged are trying to retain the status quo. Brahmins find themselves taking orders from or eating food prepared by Dalits. Increasing numbers of marriage ads state "caste no bar." Although

quotas and special privileges for untouchables have propelled many into upper echelons, these have sometimes triggered violence. Most untouchables continue to exist at the very bottom of Indian society.

Dowry Deaths

In July 1996, Maju Singh was doused with kerosene by her mother-in-law and set afire by her husband. Before she died, she managed to tell police that her husband's family had been dissatisfied with the amount of dowry (money and goods) she had brought into the marriage. Maju's death was one of an estimated 6,000 to 15,000 **dowry deaths** a year in India.

The term *dowry death* refers to a newlywed bride who, upon moving into her husband's family home, is harassed over the goods and money she brought into the marriage, leading to her murder or suicide. As late as 1987, dowry deaths were concentrated in the corridor connecting Punjab, a traditionally patriarchal and violent part of northwest India, to Delhi and eastward to Uttar Pradesh. But by the mid-1990s, these atrocities had spread to more than half of India's states and territories.

Most Indian marriages are still arranged. Today, Indian brides and their families feel compelled to buy their way into a marriage alliance with "gifts" of cash, jewels, and consumer goods. The availability of a host of household appliances, as well as their increasing promotion on television, has created a desire to have things among the lower-middle-class and others who lack the purchasing power to buy what they are encouraged to want. As the groom's family uses dowry demands to acquire these items, the bride's family is stuck with the bills.

Problems centered on dowry might begin as early as at the wedding ceremony when the groom's family demands more than what has already been promised. If the bride's family cannot meet their obligations, the bride may be harassed, beaten, or burned alive "accidentally."

In general, Hindus face an imperative to get married and stay married no matter the circumstances. Furthermore, ancient laws insist that a wife suffer her husband's aggressions. A recent United Nations study showed 65 percent of Indian women to be subject to domestic abuse. Most marital breakdown is considered to be the wife's fault. A traditional Hindu parting to a newlywed daughter says, "We are sending your bridal palanquin (an enclosed litter) today. Let only your corpse come out of that house." Moreover, most parents are reluctant to take an aggrieved daughter back home as this would give the family a reputation as being "difficult" and could hinder the possibility of future marriages for siblings.

In the mid-1980s, the Indian government recognized dowry death as a quasi-manslaughter crime. However, as many Indians use unstable kerosene stoves for cooking, it is relatively easy to pass the event off as an accident. Unfortunately, this type of violence has only increased in the twenty-first century.

Religious Conflicts

In December 1992, a rampaging mob demolished an unused mosque in Ayodhya, Uttar Pradesh claimed that it had been built on the birthplace of Lord Rama. This inflamed the sensibilities of India's largest cultural minority: 130 million Muslims. Communal riots broke out nationwide, and hundreds were killed and injured. This incident was merely a spot on the tapestry of religious conflict in India's officially secular democracy. The overwhelming majority of Indians are Hindu. In fact, in some languages such as French and Persian, the word for "Indian" is *Hindu*. It once meant "people beyond the Indus," but history has rendered it, correctly or incorrectly, a term designating a nation and a religion. However, like India itself, Hinduism is a polyglot of practices and lifestyles. Unfortunately, in the twentieth-first century, this kaleidoscope has become blurred under the forces of socioeconomic inequities and political ambitions.

The unlikely notion of *Hindu chauvinism* has emerged in the political arena with the Bharatiya Janata Party (BJP). This fundamentalism is not simply religious. It incorporates the resentments and animosities of caste, belief, and regional sectarianisms. It cannot be simply religious, because Hinduism is not a singular faith divorced from all other faiths. Hinduism is not a monolith; it is inclusivistic in nature. But it is convenient to appeal to differences in practice or detail of belief for those with a particular political or economic agenda.

In the early 1990s, the BJP called for a ban on English, the promotion of Sanskrit, and a ban on proselytizing by other religions. By 1999, the BJP had achieved national power and Hindus were killing Christians in Gujarat, Orissa, and elsewhere. Gujarat is also a hotbed of Hindu-Muslim violence. Religious insurgencies continue to plague this vast and complex country.

Hindutva

Hindutva means "Hinduness"—a desire to make India a country where Hindu principles prevail. This is the guiding precept of the BJP and is expressed as Hindu nationalism, Hindu patriotism, and Hindu heritage. *Hindutva* believers want to foist a Hindu curriculum on schools, inhibit activities of non-Hindu religious proselytizers, change family law in ways that would upset Muslims, and render non-Hindus as outsiders.

This philosophy worries Muslims and other non-Hindus. It also concerns those who believe that India's secularism, its separation of religion and state, are indispensable to the survival of its democracy. Moderate Hindus and non-Hindus oppose these divisive ideas. However, they do acknowledge the appeal of the BJP which swept into power on a Hindu nationalist platform. They also realize that it is only the constraints of a coalition government that keeps the excesses of *Hindutva* from being put into practice.

Figure 7–9
Patterns of agriculture. What is the relationship between crop distribution and water availability? Note that millet can be grown in many different environments.

Agricultural Landscapes

Figure 7–9 illustrates the wide variety of crops cultivated in India. While crop yields have risen significantly in the latter half of the twentieth century, they do not equal those achieved elsewhere nor, because of inadequate distribution facilities, are they sufficient to meet population requirements. Human and animals continue to dominate labor even as the use of tractors and other machines becomes more common (Figure 7–10).

Water availability has increased via tube wells, storage tanks, and canal irrigation. More than 14 million pumps have been installed, and wells now are the major source of irrigation water. Farmers have been able to put more acreage under production and use high-yielding seeds. Some well owners have started water markets, selling their water to others to open up even more farm land.

Timely application of water can increase yields markedly, but supplies are overexploited in many areas. Other areas have never had enough. Management agencies lack coordination, and the entire issue of water, like land, is very political. According to Robert Repetto, writing for the World Resources Institute (1994), "In India, the rights to an immensely valuable resource [water] were distributed gratis in a pattern even more unequal than that of land distribution, reinforcing rural inequalities in income and wealth."

Outright famine has not plagued India for decades. However, food insufficiencies and malnutrition remain troublesome in many areas, especially when distribution networks fail or costs are prohibitive. A critical contributor to improved food supply in India was the 1966 Green Revolution, which began in Punjab with high-yielding varieties (HYV) of wheat and Tamil Nadu with HYV rice. But scientific improvements have not been equitably distributed, and they benefit different crops in different ways. In the case of wheat, for instance, Punjab, Haryana, and western Uttar Pradesh have benefited most because of suitable irrigation water supply. In the case of rice, Tamil Nadu, Kerala, Andra Pradesh, and Jammu and Kashmir have profited most. Other rice-growing states or regions with variable rainfall and inadequate water storage and distribution systems use a different approach. Here, farmers plant a *rabi* (dry season, winter) HYV wheat crop.

Uttar Pradesh, where one in six Indians reside, has seen variable consequences of the Green Revolution. Uttar Pradesh is India's largest grain producer, yet yields are only half those in Punjab. One government- and WB-supported countermeasure is the reclamation of **sodic land**—land where crusts of salt or other alkalis prevent nutrients and water from going below the surface. Application of gypsum and flushing with water can bring such soils into production. Farmers are given free gypsum, loans to buy water pumps, and counseling on crop and farm management.

Agricultural improvements have not erased food import requirements, which are projected at 14.1 million tons in 2010, with a further 2 percent increase each subsequent year. These requirements reflect population needs, deceleration of crop yield improvements, economic growth and increased consumption, and greater consumption of non-vegetarian foods. These factors pose challenges to Indian agriculture that does not make the best use of its land in terms of potential productivity.

Thirty percent of India's land is dry-cropped. With irrigation, these lands could become relatively affluent. In Maharashtra, experiments with water conservation, tree planting, and grazing control

Figure 7-10

Man and water buffalo harrowing a rice paddy in Orissa. This action helps to break up the lumps of clay-like soil. Rice cultivation relies heavily on human and animal labor. Photograph courtesy of B.A. Weightman.

Figure 7–11

Rice paddies and coconut estates line the coast of Kerala State. These are the first in a vertically zoned series of crops in the Western Ghats. Rubber, pepper, cardamom, and other crops grow at higher elevations. The coconuts provide copra (meat) and coir (fiber). The man is pedaling a bicycle rigged with paddles to move the water from one field to another. This is another example of the importance of human labor in agriculture. Photograph courtesy of B.A. Weightman.

have led to agricultural improvements and the creation of a viable market system (Figure 7–11).

India is now an international exporter of food but must continue to increase crop yields to feed its own ever-growing population. However, there are obstacles. How can India sustain increased yields? Will there be an endless supply of ever-better HYV seeds? Policies regarding the ownership and use of HYV seeds have become a global issue. How far can biotechnology go? Nearly 500 **transgenic plants** (human-made) were developed in 1994–1995. Will these be available to Indian farmers? Would Indians even want to eat these products? (Think of the beautiful tomato developed to be machine picked; it looks great but tastes like nothing.) Who would buy India's exported food of a transgenic nature? Several countries in Asia have already banned human-engineered foods.

Technical improvements may not be appropriate for all farmers. For example, large tractors or harvesters are unsuitable for small fields. Moreover, who can afford such machines? How will fuel be acquired and paid for? Are different religious or caste groups willing to cooperate in the acquisition and sharing of such equipment? Storage, packaging, and

shipping to avoid losses and disease are further issues to be grappled with.

Why Waste Waste?

Judicious use of animal and even human waste can improve the lives of millions. Cow dung, when fermented or digested in an enclosed, cement container, produces methane gas, carbon dioxide, and a nutrient-rich slurry. From this *biogas* plant, methane can be piped into homes and used as cooking gas, or it can fire a diesel engine to generate electricity or pump water. The slurry is an excellent fertilizer. Generating biogas is culturally appropriate as well, because all products of the cow are considered sacred.

Biogas was produced as early as the sixteenth century in Persia and has been used in India for almost 100 years. Now, more than two million small "digesters" and a thousand community-size plants have been put in place in rural, agrarian regions. Biogas provides light to millions who are not linked to the larger electrical grid and is a boon to the 80 percent of rural India who lack access to sufficient cooking fuel. The Indian government hopes to construct digesters for the 12 million households that have enough cattle to maintain a sufficient supply of dung.

Human waste can also be put to good use. Sewage-fed lagoons produce 10 percent of the fish consumed by Kolkata (Calcutta). A natural wetland east of the city has been turned into a 29,652-acre (12,000-ha) aquaculture operation. While the sewage provides nutrients for the algae that carp and other fish eat, the process dilutes the high concentration of fecal coliform bacteria in the sewage. The resulting effluent can be safely used for irrigation in urban gardens. City gardens in the rich, composted soils of old garbage dumps produce 150 to 300 tons of vegetables a day and employ around 20,000 people in East Calcutta.

Deceleration in crop yields is now occurring in key areas such as Kerala and Punjab. This can be attributed to several factors. Even with the widespread installation of tube wells, water management is often ineffective. Often, water control projects have been installed in regions where crops do not require irrigation. Fertilizers and pesticides may be adulterated. Seed multiplication farms are not operating efficiently. Roads have fallen into disrepair, disrupting distribution of both inputs and outputs. Rural electrification, expanded to all but 80,000 villages (out of 580,000) by 2004, plays a significant role in pumping and other aspects of water control. Unfortunately, mismanagement has resulted in intermittent power supply in some areas and weeks of blackouts in others.

Changes in agriculture also have social consequences. I will give you two case studies, but of course there are many more. The first involves caste relationships and the second centers on the changing role of women.

DEMISE OF THE *JAJMANI* SYSTEM

In Punjab the custom was for Harijan laborers to work for their Jat landlords in exchange for grain and fodder. This was called the *jajmani* **system**. Harijan laborers received one-twentieth of the wheat harvest. This was much more than an economic system. It was a centuries-old social contract rooted in caste relationships and practiced among Sikhs and non-Sikhs alike. It required a certain decorum and specific exchange relationships including water, tea, and food.

With the advent of HYV, the one-twentieth share became greater from the HYV fields than the traditional variety fields. Furthermore, some farmers could afford tractors and no longer needed Harijan labor on their large plots. With concomitant economic growth in towns and cities, and extended bus routes, many Harijans sought urban employment. New-found wealth meant that Harijans no longer had to accede to Jat requirements or even requests. Jats purchased motor scooters; Harijans bought bicycles. Activity spaces were expanded, and Harijan-Jat social distance was extended. The intergroup ceremony and camaraderie surrounding the harvest disappeared.

More tube wells were installed, and underground pipes reduced evaporation. Rice was grown in addition to wheat. With quick-maturing seeds (e.g., 60-day lentil), triple-cropping became possible. When labor was needed, migrant workers from the Gangetic plain were hired until that source dried up. Now many Jats do their own labor and must work longer hours to support all their essential equipment. There is a rising level of debt—some jats sell their land in fear of bankruptcy. Rich landowners buy these parcels, circumventing property holding limitations by registering titles among family members. Small farmers cannot compete and ultimately sell out. Large landowners cannot function efficiently without tractors, and many become overburdened with debt. Technological change alters social circumstances. Clearly, the pros and cons of these changes are in the eyes of the beholders.

WOMEN AND THE WHITE REVOLUTION

India is experiencing a **White Revolution** as its dairy industry undergoes transformation. Western planners introduced new techniques to increase milk yields as they had improved crop yields. The development plan, introduced in 1970, was called Operation Flood (OF). India has more than 240 million bovines including cattle and water buffalo. Here, 18 percent of the world's cattle stock produces only 6 percent of the world's milk. OF determined that India had too many low-milk-yielding zebu cattle and that new breeds and dairy processing plants should be introduced. The results are mixed.

Urban dwellers now consume more milk, but rural people often cannot afford higher prices and therefore drink less. New cattle breeds require grain, and this means less grain is available for human consumption. Moreover, women who frequently tend and milk zebu were deprived of this central role. Women traditionally collect milk and make *ghee* (clarified butter), curds, buttermilk, and cottage cheese, which they can sell. Dairy processing plants deny women these activities. Farmers take their milk to the factory, where it is reduced to products for purchase by the urban elite. Thus, nutritional sources are diverted from rural areas.

New breeds and cross-breeds are not always suited to tropical or even subtropical environmental conditions, and an investment can be easily lost to disease and death. Dairy cattle require high-quality feed and clean water—items not available to millions of people. They more easily succumb to parasites. Further, milk cows are not well suited as draft animals.

The sacred zebu is well suited to India's environmental conditions. It does give some milk, but more importantly, it provides dung for fertilizer and fuel, and it is India's main beast of burden. Further, it devours anything from straw to newspapers.

The goal of getting more milk from an animal fails to recognize the whole animal and its integral role in traditional cultural complexes. Many scholars attack this reductionist approach to development. While OF claims to have "emancipated" millions of women, in fact it has eliminated an important role and source of income for them. Very few women are directly involved in the modern dairy industry. Modern dairy enterprises provide better care for dairy animals than many farmers can for their children. Kanti George (1985) notes that in certain parts of India, "it must be more comfortable to be a large landholder's crossbred cow than to be a small farmer's child."

Urban Landscapes

Indian cities have their roots in Harappan civilization. Later, they centered on the Indian princely states. Such indigenous cities were nodes in functional regions, offering protection in return for food. With British intrusion, powerful economic systems shifted the emphasis to coastal cities. Interior cities became peripheral in importance to the larger international economic system in which spatially peripheral (coastal) cities became central to economic interchange. Furthermore, two separate administrative systems operated: that of the Indian princes and that of the British.

Similarly when British institutions and infrastructure were superimposed on traditional landscapes, two urban structures emerged. English cantonments and civil lines operated differently from the rest of the city, and straight avenues brought order to the perceived chaos of the Indian areas. Calcutta, Madras, Bombay, and later New Delhi and others emerged as dual cities (Figure 7–12).

In 1947, there were 582 princely states and estates of which 115 were **gun-salute states.** The number of gun salutes—21, 15, 10—Indicated the importance of the prince, not the economic system tied to the princely city. When states were reorganized in 1950 and 1956 along linguistic lines, princely states lost primacy, prestige, and power. Only five of these are still state capitals: Bhopal, Hyderabad, Jaipur, Srinagar, and Thiruvananthapuram (Trivandrum).

According to geographers Brian Berry and Howard Spodek (1982), in the residential geography of the traditional Indian city, socioeconomic status was the dominant theme. However, modernization is transforming communal caste status to communal class status. High-status areas are central by location; low-status areas are peripheral by location. With rapid growth and change after independence, by the

Figure 7–12

The Roman-style India Gate on one of the broad avenues of New Delhi. The arch holds the names of victims of the Afghan Wars and World War I. Such monuments were commonly built in the planned cities of Europe in the eighteenth and nineteenth centuries. Photograph courtesy of B.A. Weightman.

1960s, distinct familial areas and male migrant areas had emerged around factories and port facilities. And there was some suburbanization.

In the 1960s, planning direction shifted from controlling metropolitan growth to diffusing it. Greenbelts and satellite towns were installed but were largely ineffective in stemming urban growth (Figure 7–13).

Mumbai (Bombay) is located on a series of small joined islands and reclaimed land. Traffic congestion is exacerbated with north-south commuting. The government decided to develop a New Mumbai (Navi Mumbai) on the mainland to decentralize the ever-choking stream of commuters. Success was deemed likely because similar plans had been effective in Japan, China, South Korea, and Singapore. With jobs and housing available in Navi Mumbai, congestion would be relieved and more industry would be attracted away from the central city. Similar schemes were planned for all of India's major cities. Unfortunately, success at decentralization has been limited, at best.

India is a place where conflicting interests and powerful lobbies can distort any plan. Industry and business, real estate and construction, middle class and poor: all have their different goals. Besides, the Indian government has generally stood as champion of the poor, so how could it legitimately support projects benefiting the other groups? Maharashtra state government showed little interest in the plan. Moreover, it permitted further land reclamation and building development at the southern end of Bombay Island, adding to the existing congestion.

In Navi Mumbai, 50 percent of the housing is for the poor. Tiny apartments house extended families in

Figure 7–13

Located on the Arabian Sea, Marine Drive in Mumbai is lined with crowded apartment buildings. Middle- and upper-class people live here as well as squatters and the homeless. Overcrowding and the tropical climate combine to give the buildings a run-down appearance. Photograph courtesy of B.A. Weightman.

Figure 7–14

A Kolkata street scene. The sidewalks are crowded and the pavements broken. The man to the right is bathing at a public standpipe. These have been installed all over the city to accommodate the homeless and those with no running water. Note the confusion of wires at the top of the picture. Electricity was added after the buildings were constructed. Many people simply tap into the electrical system illegally, with little official reprimand. Photograph courtesy of B.A. Weightman.

large apartment blocks. Hurried building has resulted in poor-quality structures, which now appear no better than inner city *chawls* (slums). But middle-class housing shows sensitivity to both space and privacy needs. This reveals the class bias on the part of the designers. Today, half the residents of Navi Mumbai live in slums, and the city is ringed with squatter settlements. Nevertheless, in the 2000s it has become a center for computer-related industries. Some call it "Silicon City."

Indian cities are notoriously crowded, dilapidated, and polluted. Kolkata (Calcutta) is regarded by many as a lost cause (Figure 7–14). New Delhi is one of the world's most congested and polluted cities. With 16.5 million, Mumbai adds 300 migrant families every day. Many end up as ragpickers in Dharavi, Asia's largest slum; others join the city's throng of 120,000 prostitutes.

Government decentralization with developments in interior cities and rural areas have not stemmed this tide of migrants to larger urban centers. Thousands sleep in the streets, yet millions—the wealthy, the middle class, and the poverty-ridden —do manage to survive in often appalling conditions (Figure 7–15).

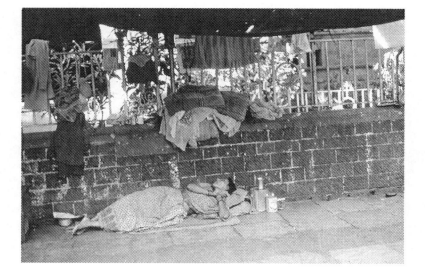

Figure 7–15

Homeless woman in Mumbai. Thousands of individuals and families live in the streets of India's major cities. Photograph courtesy of B.A. Weightman.

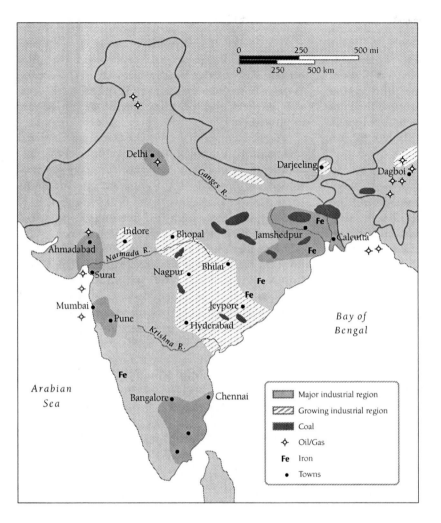

Figure 7–16
Raw material and manufacturing regions. Note the role of former colonial cities as centers of activity.

Hunting the Economic Tiger

Figure 7–16 shows India's raw material and manufacturing regions. The primary cores associated with Mumbai, Kolkata, and Chennai (Madras) are products of historic inertia. Mumbai is the largest port on the Indian sub-continent and handles about 25 percent of India's foreign trade. It also accounts for about 11 percent of India's industrial employment. But there are new, secondary regions of activity generated by decentralization policies and exploitation of new raw materials and other resources such as labor supply. Kolkata now anchors an industrial district, which focuses on jute and cotton manufactures, engineering, and chemical industries. Coal mining and iron and steel manufacturing center on Nagpur. The region incorporating Maharashtra and Gujarat specializes in textiles, food processing, chemicals, and engineering. Chennai also specializes in textiles and light engineering. This region incorporates what has become known as **India's Silicon Valley**, centered on Bangalore. Bangalore's story illustrates some of the problems and trends in modern economic development in India.

Bangalore, an army cantonment during the Raj, later became known as India's Garden City. In the 1980s, a spate of colleges and skilled workers, along with a pleasant climate, drew the attention of foreign corporations. Multinationals like Lipton Tea and Unilever; and American firms such as Texas Instruments, IBM, and Compaq opened plants in the area. Bangalore soon was called India's Silicon Valley.

The 1990s brought more companies, stores, malls, and restaurants, and then an array of woes ground the city to a halt. Population increased from 2.9 million in 1981 to more than 5 million in 1999. There are shortages in supplies of water and electricity. Cars and motor scooters vie with pedestrians in a dangerous rat race. Bangalore is now one of India's most polluted cities. Suburbs are built and occupied before service facilities are installed. Commuting times have doubled. Slums that sprung up to house migrant construction workers have become permanent. Some say that the city is collapsing under the weight of its own success. Nevertheless, liberalized government policies and continued availability of highly trained workers, a relatively high literacy rate (56 percent), and the use of English for business combine to make Bangalore's Karnataka State desirable for investment.

For years, government policies founded in socialist ideals served to block economic development. Government-operated institutions, state-run enterprises and entangled regulations made efficiency unachievable. As one Indian author noted, "If there is to be an Indian tiger, it must be freed from the shackles that keep it languishing in its small cage."

In 1991, the Congress Party (then in power) began a program of economic liberalization, including industry deregulation, privatization of state monopolies, and easing of foreign investment rules. The transition to a more market-oriented economy accelerated with the BJP which came into power in 1999. The notorious system of quotas and import licenses for machinery and consumer goods has been dismantled. Foreign ownership of Indian firms is now possible and transnationals such as Pepsi, Coca-Cola, IBM, Sony, and Phillips have entered the Indian marketplace.

Foreign trade is growing particularly in the service sector. India's software industry is spreading from its Bangalore and Hyderabad base and finding customers abroad, especially in the United States which outsources many services there. Software exports have been growing at an annual rate of 50 percent.

Fresh from its success in the global software and information technology markets, India is fast becoming a key outsourcing center for making everything from cars, to steel, to pharmaceuticals. India has a plethora of engineers and designers capable of creating low-cost, high-end products. China and Southeast Asia cannot compete in this arena.

Transnationals come to India to establish their presence in the huge domestic market. The South Korean car giant Hyundai provides an example. Hyundai built a car assembly plant in Chennai in 1996. In 2002, the company grabbed a 20 percent share of India's rapidly growing passenger car market. A confident Hyundai is investing millions to increase its production capacity to 250,000 cars a year. Hyundai is not alone in India. Ford and Suzuki are using their Indian operations to produce compact cars for markets in Asia, Latin America, and Africa.

India still has problems in its manufacturing sector. Its infrastructure is rickety and high tariffs on many imported products saddle manufacturers with bloated costs. Nevertheless, as of 2003, India's economy has been growing at a rate of 7.5 percent a year.

Bollywood

Mumbai has the world's largest movie industry. Each year, **Bollywood** (Bombay/Hollywood) produces more than 700 films.

Most of Bollywood's films are in Hindi and plots are filled with music, dance, and clichés. Many are romance stories but whatever the theme, good always overpowers evil. Movies typically exhibit nine cultural elements—love, hate, sorrow, disgust, joy, compassion, pity, pride, and courage—as instructed by the ancient text *Naya Shastra* (Science of Theater).

After showing in India, Bollywood films are exported to other parts of Asia and to Africa and the Middle East where they are very popular. Interestingly, only one in five films turns a profit. Audiences are frequently poor and pay only 25 cents to see one. Obviously, large audiences are requisite for profitability. The entire industry generates US $1.3 billion a year.

Financing is difficult and comes with exorbitant interest rates—24 to 36 percent. Mafia-type organizations have stepped in to fill the financial void. These organizations extort protection money from producers, directors, and stars. Like Mumbai's construction industry, its film industry is riddled with organized crime. Nevertheless, Bollywood's studios and stars add to the city's magnetism and bolster its ranking among the world's great cities.

Getting the Energy

THE PETROLEUM BACKBONE

The petroleum industry forms the backbone of India's national economy. One of the oldest oil fields in the world was established in 1889 at Dibrugrh in Assam. Exploration and discovery expanded from the folded mountainous regions to the alluvium-covered shelf in the Brahmaputra Valley. After independence, exploration was expanded to nearly all the sedimentary basins of the country. Several discoveries of oil and gas were made in the Bombay Basin (Gujarat), Assam and Arakan Basin, and the Bombay Offshore Basin. In the 1990s, additional fields were discovered offshore Mumbai (Bombay).

Another recent development has been to commercialize reserves of natural gas. Over the years there has been a significant shift in the domestic pattern of energy consumption. In 1947, more than two-thirds of India's total energy consumption was of non-commercial fuels such as wood, harvest

waste, animal dung, and commercial coal. Now, natural gas provides more than half of India's commercial energy, and the private sector has diversified into infrastructure construction, marketing, import facilities, pipeline manufacturing, and the production of an array of byproducts such as naptha, toluene, and lube oil.

India must still import fuel. Multinationals as well as private firms are contracted to seek imports in addition to finding new sources. One Australian-backed scheme involves a plan to build an undersea pipeline from Iran via Oman and the Arabian Sea coast to India's west coast. But problems are apparent.

The Arabian Sea is 9,840 feet (3,000 m) deep, and its seabed is a series of rugged mountains and ravines. This means that the pipeline would have to be laid in shallower coastal waters—territorial waters of Pakistan. Pakistan could bar passage of the pipeline through its territorial waters, or demand an inordinate share of the oil. Given the rocky nature of India-Pakistan relations, the future of this project is uncertain.

DAMS AND THE DAMNED

Inadequate water control—timing, amount, and distribution—has been a bane of Indian farmers for centuries, and the rapid expansion of irrigation systems has been one of the country's highest priorities since independence. Dam construction aims to alleviate these stresses, thereby increasing food production. Numerous projects have been installed, mainly large-scale operations involving foreign aid and modeled on America's Tennessee Valley Authority's (TVA) principles of regional hydroelectric planning. Tremendous investments have been made in several areas but the consequences have not always been positive.

The Krishna River, most important for irrigation on the Deccan, was developed with a series of dams, tanks, and canals dating back to the Raj. Unfortunately, planners did not take into account farmers' perceptions of their water requirements or desired crops. State boundaries divided stored water supplies from users' lands. Moreover, construction

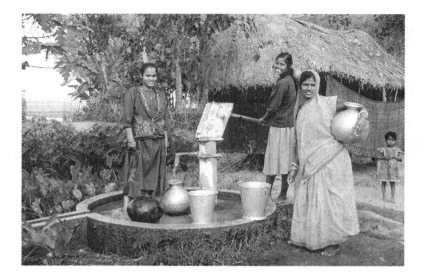

Figure 7–17

This poor village in Madya Pradesh has just received a well for potable water. Getting water is a twice-daily chore for women and children. Photograph courtesy of B.A. Weightman.

integrity was poor, resulting in silted dams and leaky canals. In fact, dams and canals are rarely able to provide water to even 50 percent of their designated service areas.

Another earlier dam project concerns the Damodar River, which drains Bihar and West Bengal, an old industrial core founded on iron and coal supplies. For years, excessive rainfall in the upper valley produced flooding in the lower valley, with devastating consequences. The Damodar Valley Corporation (DVC) was designed with the assistance of the TVA. Floods have been reduced, but the DVC's greatest success is in the area of power generation.

Cheap electricity attracted a number of industries into this once-stagnating region. Urbanization and industrialization increased with clusters of cities focusing on power plants. The area around Chandrapura became a primary growth pole at the expense of some of the other industrial centers. Devised initially to control floods, the DVC has stimulated industrial development in one of India's most heavily populated regions.

South Bihar is one of the richest areas of India in terms of raw materials such as coal, iron, bauxite, and mica. These minerals have been systematically exploited, first by the British, then by Indian industrialists such as Tata, as well as the Indian government

with money from the WB. The results have been lucrative for these outside developers but disastrous for the region's inhabitants and the environment.

Mining displaced thousands of poor people. For example, Tata Iron and Steel displaced nearly all the original inhabitants to make room for its massive complex at Jamshedpur. Other consequences are deforestation, soil erosion, and ground water pollution. In addition, the bulk of government expenditure is plowed into industries producing consumer goods for the urban wealthy and upper-middle classes.

The industrial region around Jamshedpur straddles the borders of four states: Bihar, West Bengal, Orissa, and Madhya Pradesh. Development has bypassed or negatively affected millions in this otherwise poor tribal region (Figure 7–17). In the 1980s people began to organize politically. Ultimately, a separatist movement emerged. Protesters advocated a new state to be called **Jarkhand.** Jarkhand was created in the year 2000.

Ganga Ma Is Sick

Ganga Ma—Mother Ganges, 1,560 miles (2,510 km) long, is the ancient symbol of creation, preservation, and destruction. Its sacred waters emanate the powers of Brahma, Vishnu, and Shiva. But it has become fouled with the profane evils of poison, death, and disease. More than 400 million people depend on its monsoonal waters to flood their fields with nutrient-rich silt. By 2030, there could be as many as a billion people living within its reach. But, an increasingly sick river could spell catastrophe.

Microorganisms that break down organic wastes require oxygen. **Biological oxygen demand** (BOD) is the measure of this need. Imagine how much BOD is needed to destroy the harmful components of the raw sewage of more than 30 cities, nearly a hundred towns, and thousands of villages. Although sewage comprises most of the waste material, industrial waste is far more dangerous.

Chromium and organic wastes from tanneries, bleaches and dyes from textile factories, thick goop from sugar processing, toxic farm chemicals (many of which are banned in the United States), sludge from oil refineries, human ashes, body parts, and whole bodies—all become part of the sacred river.

An estimated 35,000 bodies a year are brought to the sacred city of Varanasi alone for cremation. But the required sandalwood is expensive. Many bodies are only partly burned. Others are simply placed in the river by the devout.

Pollution worsens as the Ganges makes its way to the Bay of Bengal. The World Health Organization standards for drinking water call for no more than 10 coliform counts per 100 milliliters. In Varanasi, coliform counts are as high as 100,000. Typhoid, cholera, and viral hepatitis are common here. In this region, one person dies of diarrhea every minute. Amoebic dysentery, gastroenteritis, and tapeworm infections are part of the lives of millions. About 150 factories are lined up along the Hooghly River at Kolkata (Calcutta). These contribute 30 percent of the waste present in the mouths of the Ganges. An experiment showed that fish put in the water upstream from Kolkata survived only five hours.

Industrial discharge from the Ganges is growing at 8 percent a year. With increased population and industrialization, how can the river possibly possess enough BOD to clean itself and flush out the dangerous chemicals?

Consider these facts. Nearly 70 percent of India's available water is polluted, and waterborne diseases such as typhoid and cholera are responsible for 80 percent of all health problems and a third of all deaths. Only 7 percent of India's cities have any kind of sewage treatment services.

In 1985, there was a plan to clean up the river. However, corruption, mismanagement, and technological errors rendered the plan useless. "A fundamental reason for the failure was that most of those who have a stake in the river's health were never included in the planning" (Sampat 1996).

India has more than 1,500 dams, making it one of the world's largest dam-building nations. The Narmada River project, the largest ever undertaken, includes two super dams: the Sardar Sarovar in Gujarat and the Narmada Sagar in Madhya Pradesh. In addition, 30 large dams, 130 medium-sized dams, and 3,000 minor ones are planned. The scheme should help feed 20 million, foster an industrial boom, and protect 750,000 downstream residents from floods.

Critics claim that the Narmada scheme is a disaster in the making. Both forests and agricultural lands will be flooded and up to 1.5 million people will be displaced. The destruction of forests will be particularly devastating because they have housed a diverse community of tribal peoples such as the Bhils and the Gonds, as well as endangered wildlife, including the tiger. Tribal representatives say they have nowhere to go, and when the valley is flooded, they will join India's growing number of **environmental refugees.**

India's Oustees

The twentieth century was a century of uprooted people: oustees. These are people who flee from war and violence, political and/or ideological oppression, natural and other disasters, economic stagnation, and human-induced environmental problems. Dams produce the largest segment of environmental oustees, followed by mines.

About half of oustees are landless or sharecroppers with no recorded or "legitimate" claims to the land they may have worked for generations. No more than half of oustees are ever resettled. Some are compensated monetarily, but many squander their new-found "wealth" on consumer items and end up landless, jobless, and moneyless.

Many oustees become seasonal laborers in agriculture or construction. In northern India, husbands tend to leave their wives and children with a male relative, perhaps for years at a time. In southern India, husbands tend to take their family with them. These workers join the vast sprawls of squatter settlements on urban fringes. Many become indebted to ruthless bosses and spend the rest of their lives virtually enslaved.

An important point to note is that as India's population increases, every project displaces greater numbers of people. The Narmada River project is only one of many. The Sardar Sarovar in Gujarat will flood out 193 villages in Madhya Pradesh. All the associated works will displace an estimated 200,000 people. More than half are scheduled tribes. Thousands of people have marched in protest but the waters continue to rise as the project continues.

Estimates of those displaced by large dams in India in the last 50 years vary from 21 to 56 million people. Around 40 percent of these are *adivasis*—tribal people. The government claims that 30 percent of India's food production increases can be directly related to improved irrigation from dam projects.

HAVING THE BOMB

Gandhi called the atomic bomb the "most diabolical use of science" ever. Prime Minister Nehru proposed a "standstill agreement" on the nuclear arms race in 1954. These Indian ideas were essentially ignored by the world.

Geopolitical considerations are key to understanding why India "needs" military nuclear capability. In 1962, China invaded the disputed territory east of Kashmir. Two years later, China conducted its first nuclear test. India-Pakistan wars ensued from 1965 to 1971. China provided assistance to Pakistan. By 1974, India had exploded its first "peaceful" nuclear device. By 1988 it had tested a missile capable of delivering a nuclear warhead into Pakistan. In 1996 both countries refused to sign the International Test Ban Treaty. Two years later in 1998, Pakistan tested its first nuclear missile. India responded by conducting five underground tests.

Pakistan's missile is named the *Ghauri* after an Afghan, Muslim king who defeated a twelfth-century Hindu ruler named Prithviraj Chauhari. India's missile is called the *Prithvi*.

Kashmir has been a flashpoint in stimulating the nuclear competition between India and Pakistan. The Bharatiya Janata Party (BJP) has long advocated a more militant India and is committed to making the nation a nuclear power. Pakistan is just as committed to proving its nuclear strength.

Military expenditures on the part of both nations are exorbitant in light of their socioeconomic needs. The Indian government spends 3 percent of its GDP on defense—the same as that spent on education and twice that spent on health. In Pakistan, defense spending gobbles up almost 30 percent of the national budget, far more than is spent on education, health, and other social programs. Clearly, military costs are devastating to social and economic needs.

China's role is pivotal in this ongoing arms race. Not only has China supported Pakistan in its anti-India hostilities, but also it has sent it missiles capable

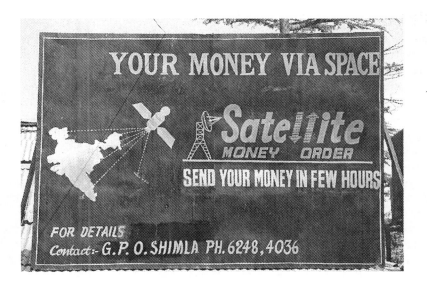

Figure 7–18

With its own satellite, India is now part of the global system of corporate finance. Instant transferral of funds takes just a few hits on a computer keyboard.
Photograph courtesy of B.A. Weightman.

of carrying nuclear warheads. Moreover, it has stepped up its military assistance to Myanmar (Burma). China is displeased that New Delhi views Tibet as an "autonomous" region within China. To make things worse, Tibet's Dalai Lama and his anti-communist, expatriate followers are headquartered in India at Dharamsala. You can read more about the "Tibet question" in Chapter 10.

PARADOX: REALITY BELIES DEVELOPED NATION IMAGE

India, with its nuclear capability and its average annual economic growth rate of 7.5 percent, is the leading economic power of South Asia. Scores of millions comprise a dynamic middle class and urban India, by itself, could be counted as the world's third largest country. Thirty million households are already in the market for various durable consumer products, and millions more are climbing into or aspiring to this category of economic behavior. Fired by the engine of its burgeoning middle class, India is on the move toward becoming a "developed" nation. This is the message delivered by an array of Indian government studies and politicos in the 1990s. These perspectives are at once correct and misleading. The truth is a paradox.

The Indian middle class certainly exists, but it is the middle class of an overall poor country. Only one in five Indians owns a wristwatch; only 48 percent of households own a bicycle; less than 4 percent of households have a telephone. In this country of over a billion, only 1.4 million are regarded as rich. The middle class is, in effect, a poor one unable to generate the kind of demand that can sustain rapid growth, even in elite consumption sectors.

Economic liberalization policies of the 1990s have granted the middle class a role it cannot assume. The credit card industry has mushroomed. Consumer giants such as Hindustan Lever are spending millions on advertising in the consumer sector (Figure 7–18). But as advertising expenditures have grown 70 percent, sales have risen only 30 percent. Much of the so-called middle class is not able or willing to march to the drum of consumerism.

Satellite TV has exploded, offering multiple channels, foreign programming, and images of a well-off, cosmopolitan, and urban-centric upper-middle class. This has reinforced the notion that a consumption-driven middle class will lead the masses down the road of development, progress, and prosperity. But the middle class watch television and confirm their lifestyles and ambitions. This does little for the masses. With respect to TV sets, India only has 76 sets per 1,000 people. Malaysia has 167. For those convinced that India is on the move, the fact that the country has the largest number of illiterates in the world is glossed over. Glitzy commercials mask harsh realities of urban slums, poor villages, and wasted countrysides.

Deprivation is sanitized by statistics and TV-screen "realities." Even the Indian government admits that 39 percent of the people live below its own definition of poverty. In effect, the small elite and the larger middle class live far removed from the worlds of the masses of destitute Indians. According to one critic, economic liberalization, "gave a flamboyant ideological justification for the creation of two Indias, one aspiring to be globalized, and the other hopelessly, despairingly marginalized" (Varma 1998).

India's rich-poor gap is being perpetuated. We know that education is an important vehicle for change because through education comes opportunity, empowerment, and self-realization. In Malaysia, the government spends US $128 per person on basic education. India invests US $9 per capita. Furthermore, the bulk of India's huge work force is far below the rest of the world in productivity and quality. For example, the value added per worker in India's manufacturing sector is a tenth that of Japan's.

Although multinationals and joint venture enterprises manufacture cars, electronics, machinery, chemicals, and war material, the sad fact is that poverty and ignorance prevent at least three-fifths of Indians from participating in any direct way in bettering their condition in any national context. Many of the middle class "have consigned the poor to being a fixture on a landscape they do not wish to see" (Varma 1998).

🌐 Main Point Summary

1. India is the world's largest democracy. Yet, reconciling cultural, economic, political, and social differences is its challenge in the twenty-first century.

2. With a population of more than one billion and a NIR of just under two percent, population numbers and distribution will continue to be at the forefront of many perplexing issues.

3. Regional imbalances in population result from gender and caste inequities as well as migration. Kerala exemplifies accomplishments made possible through the education and empowerment of women.

4. India is characterized by linguistic diversity. Fundamental differences in language families are at the core of divisiveness. Hindi, the most widely spoken language in central and northern India, belongs to the Indo-European language family. The southern states are dominated by Dravidian languages such as Tamil.

5. Although discrimination based on caste is illegal, the rules and regulations of caste are evident in all strata of society. Affirmative action programs to benefit unscheduled castes and tribes are resented by many members of regular castes. Tribal regions such as Assam are of particular concern because they are rich in natural resources. Caste affiliations are becoming increasingly significant in politics as parties jockey for group loyalties.

6. Religious conflicts, which underlay the history of the entire region, continue to stir up trouble on a regular basis. Religious differences have intensified with the emergence of the Bharatiya Janata Party (BJP).

7. With the accession to power by the BJP in 1999, the philosophy of *Hindutva* was put into place. This idea of Hindu nationalism has stirred opposition from Muslims and other non-Hindu people.

8. Many improvements have been made in agriculture: expanded irrigation, high-yield seeds, use of chemicals, biogas, land reclamation, and mechanization. Nevertheless, crop yields are below world averages and food imports will continue to rise in light of population growth and environmental limitations.

9. Changes in the structure and organization of agriculture have social impacts, as illustrated by ruptured caste relations in Punjab and curtailment of women's roles in the cattle industry.

10. India has a long history of urbanization. Those cities developed or strongly influenced by colonial powers emerged as dual cities with distinct Indian and European areas. Kolkata (Calcutta), Chennai (Madras), Mumbai (Bombay), and New Delhi are examples of this phenomenon.

11. In the traditional city, socioeconomic status was the dominant theme in residential patterns. Modernization is transforming communal caste status into communal class status. High status areas are central; low status areas are peripheral.

12. Indian cities are overcrowded, run-down, and polluted. Efforts to decentralize have mixed results, as in the case of New Mumbai. Rural to urban migration continues to overpower attempted urban improvements.

13. Patterns of industrialization are uneven with concentrations around former colonial and coastal centers such as Kolkata, Mumbai, and Chennai. Maharashtra is India's most important industrial state. Government efforts to decentralize have spawned new areas in the interior. One important example is Bangalore in the south, which has earned the sobriquet "India's Silicon Valley." Mumbai has the world's largest film industry: Bollywood.

14. Government regulations including limits on foreign investment have long controlled industrial development. The results of loosening these controls in the 1990s remain to be seen.

15. Modern developments have brought catastrophe to thousands of people. Dam construction, forestry, and mining operations have displaced countless individuals and groups powerless to claim their traditional territories. These are the oustees, many of whom end up in India's already overburdened cities. The Narmada River dam project is one of the world's most controversial.

16. The Ganges, Hindu India's most sacred river, is dangerously polluted.

17. India, like its rival Pakistan, is a nuclear power. Self-esteem and political disagreements with both Pakistan and China fuel India's participation in the nuclear race.

⊕ Bibliography

Adhikari, Sudeepta. 1986. "Some Aspects of Indian Federalism: A Study in Political Geography." *Singapore Journal of Tropical Geography*, 7: 1–11.

Berry, Brian, and Howard Spodek. 1982. "Comparative Ecologies of Large Indian Cities." In *Urban Patterns: Studies in Human Ecology*, pp. 444–53. University Park: Penn State.

Beteille, Andre. 1996. "Caste in Contemporary India." In *Caste Today*, ed. C. J. Fuller, pp. 150–79. Delhi: Oxford.

Catliff, Nick. 1990. "Killer in the Night." *World*, 47: 76–80.

Cohen, Stephen. 2000. "India Rising." *Wilson Quarterly*. Summer.

Critchfield, Richard. 1980. "The Greening of Ghungrali." *Asia* (May/June): 38–44.

De Blij, Harm, and Peter Muller. 2004. *Geography: Realms, Regions and Concepts*. New York: Wiley.

Duara, Prasenjit. 1991. "The New Politics of Hinduism." *Wilson Quarterly*, 15: 42–50.

Dutt, Ashok, K., and George M. Pomeroy. 2003. "Cities of South Asia." In *Cities of the World*, eds. S. Brunn, J. Williams, and D. Zeigler, pp. 331–371. New York: Rowman and Littlefield.

Dreze, Jean, and A. Sen. 1996. *India: Economic Development and Social Opportunity*. Delhi: Oxford.

Etienne, Gilbert. 1998. "Alarm Bells for South Asia." In *Feeding Asia in the Next Century*, eds. G. Etienne, C. Aubert, and J. L. Maurer, pp. 35–55. Delhi: MacMillan.

Farmer, B. J. 1981. 'The "Green Revolution" in South Asia.' *Geography*, 66: 202–07.

George, Kanti. 1985. *Operation Flood: An Appraisal of Current India Dairy Policy*. Delhi: Oxford University Press.

Gupta, Pranay. 1985. "Looking Ahead in India." *Harvard International Review*, 8: 29–32.

Halarnkar, Samar, and Uday Mahurkar. 1999. "Burning the Cross." *India Today*, 11 (January): 22–27.

Haub, Carl. 1996. "Spotlight: India." *Population Today*, 24: 7.

"India Bombs the Ban." 1998. *The Bulletin of the Atomic Scientists* (July/August).

Jacobson, Jodi. 1991. "India's Misconceived Family Plan." *World Watch*, 4: 18–25.

Jayaraman, N. et al. 1996. "GloomTown." *Far Eastern Economic Review*, 18 (January): 40–2.

Khan, Nizamuddin, and Almatar Ali. 1998. "Periodic Markets and Rural Transformation in Gonda District, Uttar Pradesh." *Focus*, 45: 34–73.

Kalam, A. P. J., and Y. S. Rajan. 1998. *India 2020: A Vision for the New Millennium*. New Delhi: Viking.

Karp, Jonathan. 1995. "To Have and Have Not." *Far Eastern Economic Review*, 17 (August): 62–5.

Kunakrishnan, K. 1991. "Handwriting on the Wall." *Far Eastern Economic Review*, 30 (May): 57.

Madeley, John. 1984. "Progress Has its Price." *Geographical Magazine*, 56: 226–27.

Maloney, Clarence. 1990/91. *Environmental and Project Displacement of Population in India Part I: Development and Deracination*. Universities Field Staff International: Field Staff Report No. 14.

Mandelbaum, Paul. 1999. "Dowry Deaths in India." *Commonweal*, October 8: 18–20.

Marwah, Bhushan. 1995. "India Striving for Self Reliance in Oil." *India Perspectives*, 7: 24–6.

Maslen, Geoffrey. 1995. "Asia's Aching Hearts." *Asia Inc.* (Sydney Morning Herald Magazine) October : 64–65.

McDonald, Hamish. 1993. "Fuel Injection." *Far Eastern Economic Review*, 19 (August): 42–43.

McDonald, Hamish. 1994. "Rebel Redoubt: New Delhi Fights Tribal Insurgencies in the Northeast." *Far Eastern Economic Review*, 9 (June): 31–35.

McKibben, Bill. 1996. "The Enigma of Kerala." *Utne Reader* (March/April): 102–12.

Mehta, Pratap, B. 1998. "India: The Nuclear Politics of Self-Esteem." *Current History*, 97: 403–406.

Mendelsohn, Oliver, and M. Vicziany. 1998. *The Untouchables: Subordination, Poverty and the State in Modern India*. London: Oxford.

Mohan, Rakesh. 1999. "What Ails the Indian Tiger?" *Far Eastern Economic Review*, 8 (April): 61.

Nelson, Toni. 1996. "Closing the Nutrient Loop." *World Watch* 9: 10–17.

Noble, Allen G., and Asok K. Dutt, eds. 1982. *India: Cultural Patterns and Processes*. Boulder: Westview.

O'Brien, Conor, C. 1988. "Holy War Against India." *Atlantic Monthly*, 262: 54–64.

Rain, Nick. 1994. "Living With the Landscape." *Geographical*, 66: 24–27.

Rao, Radhakrishna. 1992. "On Fertile Ground." *Far Eastern Economic Review*, 13 (August): 70.

Repetto, Robert. 1994. The *"Second India" Revisited: Population, Poverty, and Environmental Stress Over Two Decades*. New York: World Resources Institute.

Saha, Suranjit, K. 1979. "River-Basin Planning in the Damodar Valley of India." *The Geographical Review*, 69: 273–87.

Sampat, Payal. 1995. "India's Low-Tech Energy Success." *World Watch*, 8: 21–23.

Sampat, Payal. 1996. "The River Ganges' Long Decline." *World Watch*, 9: 24–32.

Sarangi, Satinath, and Carol Sherman. 1993. "Piparwar: White Industries' Black Hole." *The Ecologist*, 23: 52–56.

Sarkar, Aditi Nath. 1990. "Oh, Calcutta." *Natural History*, 5: 74–78.

Sharmer-Smith, Pamela. 1984. "The Sikh Identity." *The Geographical Magazine* (September): 442–43.

Solomon, Jay. 2003. "India Catches Up With East Asia." *Far Eastern Economic Review*, 9 (October): 50–52.

Tharoor, Shashi. 1997. *India: From Midnight to the Millenium*. New York: Arcade.

The Economist. 1997. "A Survey of India: Time to Let Go." *The Economist*, 22 (February): 1–26.

Varma, Pavan, K. 1998. *The Great Indian Middle Class*. New Delhi: Viking.

Visaria, Pravin, and Leela Visaria. 1981. "India's Population: Second and Growing." *Population Bulletin of the Population Reference Bureau*, 36: 1–56.

Wallach, Bret. 1984. "Irrigation Development in the Krishna Basin Since 1947." *The Geographical Review*, 74: 127–44.

Wallach, Bret. 1996. *Losing Asia*. Baltimore: Johns Hopkins.

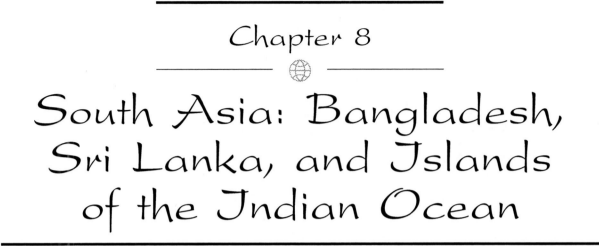

Chapter 8

South Asia: Bangladesh, Sri Lanka, and Islands of the Indian Ocean

"When unity is evolved out of diversity, then there is a real and abiding national progress."

MANHAR-UL-HAQUE (1866–1921)

Bangladesh: Nation on the Edge

Imagine 147 million people eking out an existence in a country the size of Wisconsin (population of about 5 million). Think about the facts that two-thirds are illiterate and two-thirds of children under five are malnourished. Add the fact that storms and floods wipe out people and livelihoods by the hundreds or even thousands yearly. Then, consider that this impoverished, primarily agricultural country is one of the largest exporters of ready-made garments in the world. What is the story of this relatively new nation of Bangladesh, and what are its future prospects?

Established in 1971, Bangladesh occupies a drainage basin where the Brahmaputra, Padma (Ganges), and Meghna rivers converge (Figure 8–1). Most of the country is a deltaic, alluvial plain although there are hill districts in the northeast and southeast. Its marshy coastline extends 370 miles (600 km) into the Bay of Bengal.

BENGAL NATION

Bangladesh means "Bengal Nation." This region, once called Bengal, is renowned for its tigers and as South Asia's "rice bowl." Aside from tribal areas in the Chittagong hills, Bangladesh is linguistically homogeneous. Bengali, known locally as *Bangla,* serves as a unifying force within this politically fractious nation.

Although not officially a Muslim state, most of its people are Islamic. Bengal, especially in the east, had a tradition of Tantric Buddhism and Hinduism. This was conducive to the introduction of new faiths such as Sufi Islam brought by Arab sea traders in the thirteenth century. The Hindu caste system made Islam and its notions of equality an attractive option. By 1947, more than half of all Bengalis were Muslim. Today, around 85 percent are Muslim, 12 percent are Hindu, and the remainder are Buddhist, animist, or Christian.

Christianity was introduced by Portuguese traders in the sixteenth century. Again, many low-caste Hindus were drawn to Christian ideas of equality

191

Figure 8–1

Bangladesh and city populations, 1991. Source: 1994 Demographic Yearbook. From H. J. de Blij and P. O. Muller, *Concepts and Regions in Geography,* Second Edition, 2005, p. 291. Originally rendered in color. © H. J. de Blij and P. O. Muller. Reprinted with permission of John Wiley & Sons, Inc.

and salvation. The Portuguese also introduced cashew nuts, papaya, pineapple, and guava to Bengal.

The region has a history of political subjugation and fragmentation dating back to the sixth century. British *divide and rule* policies only added to this legacy. Calcutta had served as the capital of British Bengal since 1772, but this changed in 1905 when eastern Muslim regions were linked with Assam and were designated as East Bengal and Assam, with Dacca (Dhaka) as the capital. At the same time,

majority Hindu districts of West Bengal, Bihar, and parts of Orissa were designated as Bengal.

In 1912, following years of mass resistance, East Bengal was divided into three provinces and West Bengal was reunited with Calcutta. Muslims insisted on having their own region and so boundaries were again redrawn with the creation of East Pakistan in 1947. Consequently, Bangladesh's borders are politically, not naturally, defined. Surrounded by India on three sides, Bangladesh also shares a 120-mile (193-km) frontier with Myanmar. Bangladesh has signed agreements with both countries, but territorial control remains controversial among some tribal groups. Marine boundaries are still undefined and problematic because of undersea oil discoveries in the Bay of Bengal.

This history of jurisdictional change has created regional dissension and political factionalism. For example, dissension continues to be problematic in the Chittagong Hill Tracts (CHT) situated in the southeastern corner of Bangladesh, bordering on both India and Myanmar. The indigenous Jumma, incorporating more than a dozen ethnic groups, are being overwhelmed by the Bangladeshis who have continued to move into this territory.

Requests for a degree of sovereignty were denied, and the nationalistic Jana Sanghati Samiti (JSS) was formed. Rebellion ensued, and the Bangladeshi army went on a rampage of killing and destruction in the northwest CHT. Thousands of refugees fled to India's Tripura state. At least 8,000 rebels have died in this two-decade struggle.

A peace accord was signed in 1997, and about 50,000 refugees returned. The Jumma were given some autonomy in the border areas, but the JSS continues to press for further changes. The future of this region remains uncertain.

Regional instability also engendered several core areas focusing on such centers as Calcutta and Dhaka. When East Pakistan was created, the key industrial center and port of Calcutta was lost to India, leaving East Pakistan without a viable industrial base or seaport. Cotton and jute mills were built in Dhaka. Ports were developed at Chittagong in the far south and Mangla in the southwest. Dhaka and Chittagong remain the focus of the modern economy. This economy is only in its initial stages relative to other countries in South Asia.

THE IMPACT OF FLOODS AND CYCLONES

"Bangladesh" is often perceived as synonymous with "disaster." It is true that millions of lives have been lost to natural forces. In August 1998, 10 million people were rendered homeless by floods! Countless others were affected in 2002, 2003, and 2004. As geographers, we want to know why this happens and what can be done about it.

The erratic nature of Bangladesh's climate produces uncertainty. Normally, the monsoon generates 60 to 80 inches (150 to 200 cm) of precipitation from April to October, with rainfall typically peaking from June to September. Excessive rains cause rivers to overflow their banks and inundate surrounding regions (Figure 8–2). Flooding frequently derives from heavy snowfall in the Himalayas. With deforestation and rapid runoff, rivers become deluges. In 1974, two-thirds of the country was under water; the subsequent famine killed 30,000 people. This degree

Figure 8–2
These buildings outside of Dhaka are built on stilts to avoid annual flooding. However, they have been washed out on several occasions when floods and storms are especially severe.
Photograph courtesy of B.A. Weightman.

of inundation occurred again in 1998, although with only a few thousand losing their lives.

It is understandable to associate Bangladesh and flood disaster. But Bangladesh and drought? In fact, the country does suffer from drought, especially in the northwest. Since governments are geared to dealing with floods, little if any aid reaches drought-stricken regions. Victims are left to their own devices and must rely on friends and relatives for support.

Bangladesh is also beset by **cyclones**. These tropical storms swirl in a northeasterly direction from the Bay of Bengal and wreak havoc across the nation. In 1970, a cyclone accompanied by a 20-foot (6.10 m) **storm surge** took 300,000 lives. A storm surge is water pushed toward shore by the force of the winds swirling around the storm. When combined with a tide in an increasingly narrow channel across low, flat terrain, it can cause an increase in water level of 15 feet or more. In the case of the Bay of Bengal, as the landform funnel narrows, the wave heightens and wind velocities increase to hurricane force.

The 1970 storm surge wiped out fishing communities, killed fishers, and destroyed boats, to the extent that 65 percent of the nation's fishing capacity was eliminated. This spelled human disaster in light of the fact that two-thirds of the country's protein is derived from seafood.

While rain and run-off floods can occur anywhere, the impact of cyclones is most evident in the south. In April 1991, Bangladesh faced the worst cyclone of any since independence. The death toll was 138,000 people, most of whom lived in the south.

The delta comprises 3,720 square miles (6,000 sq km) of ever-shifting mangrove, mudflats, islands, and distributary channels. This region is called the **Sunderbans**. Settlement, although discouraged by the government, is virtually immediate after a flood. Land and water configurations may be different, but any land is space to be seized at the first opportune moment. Even islands barely above sea level are occupied, at least until the next flood. Tigers are also a threat to those living in the Sunderbans.

WATER, LAND USE, AND SETTLEMENT

India affects the water supply in Bangladesh. In 1974, India installed the Farakka water diversion channel on the Ganges (Padma in Bangladesh) near the Bangladeshi border, allowing less water to flow to Bangladesh. Bangladesh claimed that reduction in water levels increases salinity in the Padma River basin, thereby preventing some 35 million people from getting irrigation water. After years of deadlock, the two countries signed a treaty in 1996 to share the water from the Farakka Barrage.

Instability of land and water patterns affect settlement and land use patterns. As rivers sway between valley walls, they assume a braided pattern separated by small, sandy islands called *chars.* As river waters shift laterally, erosion is exacerbated on the right banks and the whole system exhibits a westerly shift. *Chars* disappear and new channels are formed.

Populations move with *chars,* taking advantage of naturally fertilized land. If a *char* is developed, it can act as a magnet for more people who inhabit the adjacent riverbanks in makeshift shelters. Over half of these displaced people are sharecroppers or rent from large landowners. Others survive in the informal economy. Whatever system is established can be quickly erased in a subsequent flood.

In 1972, it became policy for newly emerged *chars* to revert to the government. The idea was to redistribute these lands to the landless and poor, small farmers. However, the power of landlords, inequities in land ownership, intergroup conflict, and political intrigue have hampered the implementation of resettlement. In fact, over half of rural householders are landless. With little chance of participating in the larger economy, they remain as semipermanent clusters in the countryside or become urban squatters.

What is being done to mitigate these yearly disasters? In 1967, tidal gauges were installed along the coast. A Flood Action Plan was developed in 1990. A year later, a Cyclone Warning Service was established. However, the system was overused and people became complacent. When warnings of "great danger" failed to materialize on several occasions, many people began to ignore the warnings.

Today, with help from the World Meteorological Organization, Bangladesh has its own Space Research and Remote Sensing Organization. With links to Indian systems, it can broadcast more accurate warnings ahead of time. In addition, concrete shelters, channels, and embankments have been built

Figure 8–3

New embankments have been constructed around Dhaka to minimize flooding. However, this one is being repaired from damage in a recent flood. Photograph courtesy of B.A. Weightman.

across the country (Figure 8–3). Unfortunately, many of these failed to hold up in the floods of 1998, and even parts of Dhaka were flooded. The nation's main road, connecting Dhaka with Chittagong, elevated more than 6 feet (1.7 m) with World Bank aid, remained above the floodwaters. Unfortunately thousands of miles of roads were washed out and need to be repaired and raised.

Embankments are also magnets for human habitation. Impromptu settlements cling to their sides. In fact, the right bank of the Brahmaputra has been described as "the world's longest linear housing development."

AGRICULTURAL RHYTHMS

As in all of South Asia, the rhythm of life in Bangladesh follows the seasons in which flooding is integral. Flooding also has a spatial pattern and temporal rhythm. For example, the Brahmaputra and Meghna rise in March and April, fed by Himalayan snowmelt. At this time, pre-monsoon rains occur in Assam and northeastern Bangladesh. The Ganges rises in May. By June or July, all rivers are swollen with monsoon rains. The Ganges reaches its peak flow anywhere from August to October.

In terms of planting crops, there are three main seasons (Figure 8–4). In addition, cropping practices are adapted to flood conditions. For example, virtually

	= sowing						= harvesting				
Crop:	JAN	FEB	MAR	APR	MAY	JUN	JUL	AUG	SEP	OCT	NOV	DEC
Aus Paddy												
Broadcast Aman Paddy												
Transplant Aman Paddy												
Boro Paddy												
White Jute												
Tossa Jute												
Masar Pulse												
Khesari Pulse												
Wheat												
Sugar Cane												
Tobacco												
Potato												
Bhadi Chillies												
Rabi Chillies												
Onion												
Tumeric												
Ginger												
Coriander												
	JAN	FEB	MAR	APR	MAY	JUN	JUL	AUG	SEP	OCT	NOV	DEC

Figure 8–4

Agricultural calendar. Note that June and July are the least busy months. Why do you think this is?

thousands of rice varieties have been planted over the centuries in context of variation in water levels. These include deep-water types that grow in 15 feet (5 m) of water. This rice is harvested from boats.

Bangladeshi farmers produce a wide array of crops, yet rice covers 80 percent of the cultivated area. Unfortunately, the high-yielding rice of the Green Revolution is not suited to flood conditions. Consequently, high-yielding rice is limited to *rabi* or dry season and other varieties are planted for the rain-fed and flooded *kharif*. Bangladesh must import food to meet the needs of its ever-growing population.

Given climatic factors and the limited availability of irrigation facilities and fertilizer, most of Bangladesh is limited to one crop a year. About 40 percent of the land can be double-cropped and 8 percent triple-cropped.

Recent efforts to diversify have seen regional specialization. Virginia tobacco is grown in the north, and the Sylhet region in the northeast produces 94 percent of the nation's tea. Harvested nine months a year, exported tea is an important earner of foreign exchange. Rubber and coffee are concentrated in the CHT.

With the installation of tube wells, high-yielding wheat has become prominent in the north, with yields exhibiting a tenfold increase between 1970 and 1983. However, in much of Bengal overuse of tube wells is drawing arsenic from pyrite bedrock into the water system. In Bangladesh, between 40 and 60 million people are developing symptoms of arsenic ingestion. This causes ever-worsening skin lesions and leads to death. The World Health Organization calls this the largest mass poisoning in human history.

Fiber crops are critical to the Bangladeshi economy. For instance, two-thirds of all the manufacturing jobs in the country are related to the cotton industry. Second to rice in terms of acreage is **jute**, a fiber crop used for sacking, twine, rope, carpets, car interiors, and lining for asphalt roads (Figure 8–5). Green jute is finding increasing use in fabrics. In fact, special jute blankets have been manufactured to protect poor, northern peoples from the winter cold. Moreover, jute stalks are used as fuel, the leaves can be eaten, and any remaining debris can be plowed back into the soil as natural fertilizer.

Sown from March to May and harvested in August, jute can be grown in uplands or lowlands. One variety can withstand 5 feet of flooding. In fact, Bangladesh produces two-thirds of the world's jute. Until recently, synthetic fibers threatened the jute industry and production declined. Fortunately, environmental concerns have made jute desirable once again.

Figure 8–5

Jute fiber is one of Bangladesh's most important products. These mats, however, are made of another fiber called coir. *This fiber comes from coconut husks.* Photograph courtesy of B.A. Weightman.

Livestock are of increasing significance in the economy. In fact, livestock rearing—employing about 20 percent of the labor force—is the most important activity for the rural poor. Cattle and water buffalo are not only draft animals, but also provide dung for fertilizer, milk, and meat for food, and leather, horn, bone, and oil for export. Bovine animals work as much as 80 percent of the agricultural land and are indispensable for threshing grain and crushing oilseeds and sugarcane.

Despite the large numbers of cattle, buffalo, goats, and chickens available, Bangladesh is deficient in meat, milk, and eggs. This can be attributed to genetic deficiencies of livestock, ignorance of breeding, feeding, and maintenance procedures, and lack of veterinary care. As a result, Bangladesh is a net importer of live cattle and milk products.

While Green Revolution technologies have increased yields, beneficiaries are primarily middle and rich peasants who dominate cooperatives and local power structures. It is important to note that more than half the land is owned by 10 percent of households and that 75 percent of rural households are either landless or possess tiny plots around 2 acres (0.8 ha). Geographer Abu M. S. Ali (1998) found that small farm size is the leading cause of unemployment and underemployment in agriculture (i.e., frequency of unemployment rises with decreasing farm size). Land reform or any other improvements in agriculture must be viewed in context of population requirements.

POPULATION AND GENDER ISSUES

Bangladesh experienced a population explosion after 1947 with the introduction of health care and improved sanitation. The death rate fell dramatically. Increasing numbers moved into marginal lands, where their needs could not be adequately met. This situation was compounded by migration streams from India.

In Chapter 3, I reported geographer Bimal Kanti Paul's findings that Bangladesh has made great strides in lowering its birth rate. However, a 2003 natural increase rate of 2.2 remains considerable in light of a large base population and a total fertility rate of 3.6. Only about half of Bangladeshi women use contraception and about two-thirds are illiterate.

In this Muslim society, women are traditionally subjugated by men, marriages are arranged, and dowry deaths still occur. The sex ratio is imbalanced with 105.6 males per 100 females. Although there is a tradition of *purdah,* most women must work outside. Consequently, strict *purdah* is practiced primarily among the middle classes.

Changes in Female Activity Space

Geographer Bimal Kanti Paul (1992) studied mobility and employment opportunities for rural women in Bangladesh and found that as poverty increased, more women were forced to seek work outside the home. This trend jeopardized traditional spatial contexts defined by village structure, *purdah,* and patriarchy.

The basic spatial unit is the *bari,* two to six houses built around a square or rectangular courtyard. Within the *bari* are patrilineally related, nuclear families. Clustered *bari* form a neighborhood, and several of these make up a village. Tradition holds that women's work is confined within or close to the *bari.* However, various factors such as religion and economic status influence the spatial extent of **activity space**. And, the status and position of women change at different times in their life cycle.

Rich women, young wives, and spouses of Islamic religious leaders have the smallest spaces and rarely venture from the home or *bari*. Other women are confined by more extensive neighborhood boundaries. Elderly, Hindu, and poor women have an even greater range of excursion; they can go farther away from the *bari.* Males and tribal women, with freedom of movement, have the largest activity spaces. Of course, this scheme is a model, and there is much variation in reality.

Figure 8–6

Bangladeshi women. Self-help programs are improving the lives of many women in Bangladesh. These women are middle class. Photograph courtesy of B.A. Weightman.

Patterns of women's activities and associated spatial realms are in flux. Certainly programs intended to empower women have encouraged and facilitated the expansion of female activity spaces (Figure 8–6). However, poverty is the basic cause of working outside the home. While paid labor within other households remains the major employer of women, increasing numbers are seeking wage employment beyond *bari* and village boundaries. As landlessness and destitution increase, traditional familial support structures fail and the rates of divorce, abandonment, and wife-beating rise. Consequently, more women are working alongside men in the fields, in road and building construction, in brick-making, and numerous manufacturing enterprises. Usually, increased activity space outside the home is accompanied by improvement in living standards, but as Dr. Paul observes, "Unfortunately, this relationship has not emerged in Bangladesh."

Recent governments, headed by women, have taken great strides in improving women's lives through education and employment. Widespread media propaganda promote the necessity of empowering women to fully participate in the country's development. Employment in labor-intensive industries and craft and other cooperatives have markedly advanced living standards for some Bangladeshi women.

Bangladesh's population dilemmas derive more from distribution than from sheer numbers. Rural-to-urban migration has overwhelmed cities, especially Dhaka and Chittagong. In 1961, a mere 5 percent of Bangladeshis lived in cities. In 1991 it was 18 percent. Currently, 23 percent live in urban places, and by 2015 that will increase to 37 percent. In other words, 80 million Bangladeshis will be urban residents by 2020. This is an amazing number in context of infrastructural requirements.

Take power availability as an example. Only 16 percent of Bangladeshis have access to electricity. The country often faces riots over blackouts. At the opening of the largest generating plant, 30 miles (50 km) northwest of Dhaka, Prime Minister Sheikh Hasina urged people to save electricity at home and at work by shutting off lights. Public and private sector projects were expected to boost the national grid's power to 4,700 megawatts by 2002. But there is another problem: People tap into power lines illegally, and nearly 30 percent of power production is lost to theft.

DILEMMAS OF GROWTH AND DEVELOPMENT

Cities and Industries

Rural push factors drive urban migration streams while growth in production is not expected to meet employment needs. As population experts Professor Abdul Barkat, U. R. Mati, and M. L. Bose (1997) point out, "urbanization in Bangladesh will remain poverty-driven." Poor migrants typically end up living in slums, squatter settlements, or as pavement dwellers. There are already about 25 million of these individuals, more males than females.

Dhaka, the country's capital and primate city, attracts the most migrants. In 1992, Dhaka had 7.4 million inhabitants. It now has 10 million residents. By 2010, it is expected to have 17.6 million, with at least 50 percent living in slums or worse. Dhaka's Old City can only be described as a hive of humanity (Figure 8–7). Every crumbling building, room, cubicle, alley, doorway, and indentation reveals people living in every imaginable condition. The narrow streets are crammed with people, carts, bicycle rickshaws, trucks, and buses, all carrying stupendous loads of oil drums, cotton bales, vegetables, sacks of ice, sheets of metal, thousands of rubber thongs, and baskets of poultry. The city is a limitless kaleidoscope of colors, sounds, and smells.

Child labor is common. This is reflected in the fact that only 20 percent of slum children attend school. In addition, at least one-third of slum dwellers are sick at any one time. Slum communities have the highest infant and maternal mortality rates in the nation, in part because health care programs have been geared to rural areas as well as more affluent urban areas.

Outside Dhaka's Old City are still relatively old areas that are becoming increasingly densified. Here, unchecked settlement outpaces government efforts to develop housing and other infrastructural needs (Figure 8–8). Incentives to decentralize and create communities outside the city have met with limited success.

In the 1960s, industries were given incentives to locate outside Dhaka. However, they located just beyond an established boundary to claim "less-developed area" status and still enjoy the advantages of urban location. These and other efforts at decentralization were disrupted by the 1971 war of independence.

In the 1980s, subdistricts called **upazilas** were created in Bangladesh. Each of these 500 subdistricts was to house an average of 186,000 people and have revenue-raising powers and measures of self-government. However, related employment development has not slowed out-migration.

Figure 8–7

Traffic crush in Dhaka. In all my travels, I have never seen anything to equal the traffic jams in Bangladesh.
Photograph courtesy of B.A. Weightman.

Figure 8-8

New multistory buildings arise out of the old skyline of Dhaka. Often the money for construction runs out and the buildings are never finished. Buildings that are completed often stand empty for want of occupants. Photograph courtesy of B.A. Weightman.

Although foreign investment and new construction are changing Dhaka's skyline, labor strikes called *hartals* are frequent, and infrastructural problems remain overwhelming. Power outages are daily events. Inadequate garbage and sewage removal results in clogged canals and drainage backups and overflow. Raw sewage gushes into the Bay of Bengal. Vehicles belch diesel fumes, and industries pour raw effluents into canals and poisonous chemicals into the air. Dhaka is now one of the most polluted cities in the world. Lead poisoning is more than double unsafe levels. Thousands of premature deaths and millions of cases of illness result from the city's pollution.

There are more than 175,000 motor vehicles in Dhaka, including more than 40,000 auto rickshaws that use a deadly combination of engine oil and leaded gas. Beginning in 1999, the government took steps to reduce lead pollution by banning the import of leaded fuels and calling for catalytic converters on new cars. Auto rickshaws were to be phased out. However, these are mostly still plans. Population and urban growth, natural calamities, and political or other instabilities or inefficiencies can easily slow or halt implementation.

Chittagong is about half the size of Dhaka. It has good rail, road, and water connections and is a designated export-processing zone. Its industries are based on primary products including oil from

offshore facilities and bamboo harvested from the surrounding hills. Yet, half of Bangladesh's formal employment sector is in Dhaka. Investors prefer Dhaka because of its larger pool of managerial and skilled workers and other facilities.

Jute and cotton mills and leather industries cluster in and around Dhaka, and these labor-intensive industries remain the country's leading employers. In recent years the government has promoted diversification into new fields such as food processing.

The frozen food industry is founded on fish, prawns, and similar products. More than 11 million individuals are connected either directly or indirectly with the fisheries sector for their livelihood. Oceans, rivers, lakes, canals, and more than a million man-made ponds and tanks are fishery sources. Shrimp exporting began in the early 1970s, earning US $3 million. By 1996–1997 this had increased to US $279 million. Even so, there are now more processing plants than fish. This calls for the development of fish farms and hatcheries. Chittagong and Mangla are the most important processing centers.

The garment industry is of great importance to the economy of Bangladesh and accounted for nearly 80 percent of the country's export earnings in 2003. Garment manufacturing earns the nation US $350 billion a year. Well-known brands such as Nike, Ralph Lauren, Tommy Hilfiger, and others are made in Bangladesh. The United States is the largest purchaser of Bangladeshi-made garments. Garment factories employ as many as 1.8 million workers—half the nation's industrial workforce.

A 30 year-old agreement that regulates world trade in textiles expires in 2005. This Multifibre Arrangement (MFA) guarantees poor countries like Bangladesh access to apparel racks in the United States and Europe. In turn, the MFA uses quotas to limit competing exports from more efficient producers such as China. Starting in 2005, however, there will be a more open market for textiles and garments overseen by the World Trade Organization (WTO). The WTO generally prohibits the use of trade quotas because they are protectionist. Free trade proponents hope the end of the MFA will mean cheaper clothes for consumers around the world.

This leveling of the playing field will hurt countries like Bangladesh which cannot compete with producer giants like China. China is expected to control nearly half of the world's clothing exports by 2010, up from 20 percent in 2003. An estimated one million jobs will be lost in Bangladesh. Also in danger are the jobs of many of the 15 million people who work in related industries. The prospect of massive job loss is worrisome in this country where lawlessness and Islamic fundamentalism are on the rise.

The garment industry in Bangladesh has been a significant employer of women—85 percent of the workers are female. Until recently, children were also employed. In 1993, the U.S. Congress passed a bill to ban Bangladeshi clothing made with child labor. Consequently 50,000 children were dismissed from the industry. Bangladesh has since agreed with international organizations to ban children from the garment sector.

Non-Government Organizations (NGOs)

Bangladesh is one of the world leaders in NGOs. More than 80 percent of Bangladeshi villages have them. According to most reports, NGOs function far more effectively than government organizations such as the Bangladesh Rural Advancement Committee (BRAC). However, BRAC has more than two million members, mostly female. It promotes craft and cottage industries and offers flexible schooling taught by locals.

The most well known NGO in Bangladesh is the **Grameen Bank**, which offers small loans to the poor—again mostly women—for shops, sewing machines and other labor-saving devices. They also loan money to purchase livestock or anything else to empower women. Women are regarded as better managers of household finances and more responsible in paying back loans. Loans are made to small groups. Both peer pressure and group responsibility are critical in ensuring repayment.

World Bank (WB) studies indicate that 50 percent of borrowers get out of poverty in 5 years and that both divorce and birth rates decline. With Grameen as a model, thousands of other organizations have provided micro-credit to the poor worldwide.

Critics point out that the bank is heavily supported by foreign funds and argue that it would collapse otherwise. Some say that Grameen is really controlled by the government. Some say it is anti-Islamic. Others argue that peer pressure drives out the weakest members, leaving them destitute and compelled to a life of bonded labor. Case studies indicate that perhaps two-thirds of Grameen loans are actually controlled by male family members. Stringent payback requirements drive some women to usurious moneylenders. Grameen authorities say that such stories are rumors circulated by a jealous BRAC or larger organizations like the WB, which are promoting their own schemes. Whatever the case, microloans apparently have helped millions of women in Bangladesh. The full story remains to be seen.

The Rise of Islamic Fundamentalism

Bangladesh's secular political tradition is in retreat. In 2001, the Muslim fundamentalist Jamaate-Islami (JI) Party captured 17 seats in the 300-seat parliament. The JI calls for an Islamic state. Senior JI members now control the powerful ministries of industry and social welfare. In both city and countryside, the popularity of religious schools is growing. There are 65,000 *madrassas* promoting fundamentalist Islam in the country. There have been attacks on women who use contraception, have an abortion, or opt for sterilization. Schools where large numbers of girls attend have been burned down. The houses of Christians and Hindus have been destroyed and non-Muslim groups subjected to extortion. Remote areas of the country have become ungovernable havens of Islamic militants who oppose the governments in neighboring India and Myanmar (Burma). As poverty worsens and unemployment rises, more militancy can be expected.

Sri Lanka: Teardrop Weeping

Arabs called it **Serendip**, the Portuguese named it **Cilao**, and the Dutch spoke of **Zeylan**. All were referring to *Simhala Dvipa*—Island of the Lion Tribe. Known as Ceylon until 1972 when it was renamed Sri Lanka, this tear-shaped island is divided from the Indian subcontinent by the 22-mile (35-km) wide but shallow Palk Strait. While separation fosters cultural distinction, propinquity favors spatial interaction. In the case of Sri Lanka and India, interrelationships have been filled with inherent antagonisms.

THE PHYSICAL LANDSCAPE

The island of Sri Lanka includes both up-country and low-country. About 30 percent is up-country, where mountains rise above 8,000 feet (2,500 m). From these heights, rivers flow in a radial pattern downward through lush forests to the coastal plains. The longest of these is the Mahaweli Ganga, running in a southeasterly direction for 206 miles (332 km). In the low-country, hills are separated by abrupt escarpments and cut by deep valleys. They descend in a step-like fashion to lowlands, mostly under 1,000 feet (300 m). In the north, these lowlands extend as an undulating plain to the Jaffna Peninsula (Figure 8–9).

The island experiences four seasons in a yearly cycle. From May to August, the southwest monsoon brings orographic rainfall to the southwestern hills. This season culminates with tropical cyclones from the Bay of Bengal, which can affect all areas from October through November. Next comes the northeastern monsoon, which supplies 40 to 80 inches (102–203 cm) of rainfall in the north and northeast

through February. From March to April, summer insolation and forested uplands generate **convectional rainfall**. Moisture-sodden warm air rises, cools, and condenses, resulting in late afternoon downpours.

Sri Lanka is tropical with little temperature variation. In fact, nocturnal-diurnal variations are greater than seasonal ones. Although temperatures rarely rise above the low 80°F (25°C), the sun's verticality induces high evaporation rates. This makes irrigation essential, especially in the east. Aridity is a traditional problem. Ancient civilizations fell not simply to invasion but to the destruction of their irrigation systems. Moreover, dry zones were malarial until the end of World War II and the introduction of the use of DDT. With lack of moisture and the presence of killer mosquitoes, colonization schemes of the dry zone were dismal failures before 1945. Karst topography in the Jaffna Peninsula allows the drilling of wells into limestone aquifers.

Water availability was pivotal to ancient settlement. Around 500 B.C., Sinhalese settlers developed elaborate water storage and distribution systems. Property ownership was based on the control of water, not land. Canals reached 50 miles (31 km) in length, and the Sinhalese devised water-control valves at dams to accommodate variation in water pressure. These preceded European devices by 1,500 years!

THE INDIAN CONNECTION

Sri Lanka's history is replete with invasions from India. The Sinhalese, descended from the Aryan invaders of Northern India, migrated via the islands and sandbars of the Palk Strait from the fifth to third century B.C. In time they became Buddhist. In the seventh century there were Hindu, Tamil invasions. In 1017, Ceylon became part of India's Chola Empire. Its history is marred by friction and fragmentation.

The Buddhist Sinhalese complex of Anuradhapura, with its 15-mile (24-km) walled perimeter, supported more than a million people based on an irrigated rice economy. In an elaborate interlocking system, rivers were channeled to storage tanks, from which diversion channels delivered water to villages. This was one of the most elaborate irrigation complexes ever conceived. However, by the time the

Figure 8–9
Sri Lanka and Tamil territorial claims. From H. J. de Blij and P. O. Muller, *Concepts and Regions in Geography,* Second Edition, 2005, p. 294. Originally rendered in color. © H. J. de Blij and P. O. Muller. Reprinted with permission of John Wiley & Sons, Inc.

Figure 8-10

Picking tea in Sri Lanka. Typically, supervisors are male and the pickers are female. Photograph courtesy of B.A. Weightman.

Europeans arrived, warfare had reduced the settlement to ruins and the area had become an overgrown wilderness.

EUROPEAN IMPACTS

The sixteenth century witnessed dramatic alterations in Sri Lanka's state of affairs. The Portuguese, striving to capture the Arab sea trade, stopped at Galle and set up spice-trading operations. They also demolished Buddhist and Hindu religious sites and forced conversions to Christianity. Further, they captured the mountain kingdom of Kandy and removed its most sacred relic—an alleged tooth of Buddha. The tooth eventually found its way back and is currently housed in Kandy's Temple of the Tooth.

A century later the Dutch arrived (1658) and ousted the Portuguese. They pursued commercial development, but as a result of the Napoleonic Wars in Europe, were forced to turn over the island to the British in the nineteenth century. The Dutch introduced the island's first commercial crop: coffee. But the coffee trade was under challenge from Brazil, and it was ruined by blight during the British period. The British then turned to **cinchona** and tea.

Cinchona is extracted from tree bark and was the only treatment for malaria, widespread at the time. Eventually numerous other plantation/estate crops were grown (e.g., rubber, coconuts, cacao, and pepper). Estates were nationalized in the 1970s but reverted to private management in the 1990s. Sri Lanka is still one of the world's top tea exporters. Tea accounts for about 75 percent of the country's income (Figure 8-10).

Tea: The Global Beverage

More people drink tea than any other beverage (aside from water) in the world. Ancient Chinese texts describe tea as an elixir. Today tea is touted for its antioxidant properties. Tea is widely employed as a social medium. I have enjoyed tea on many occasions around the world: "China" cups of black tea in England; small glasses of mint tea in Morocco, cups of steaming milk-tea in Pakistan; glasses of spicy *chai* in India; and cups of aromatic green tea in China. Socioeconomic interchange is eased via the ceremony of tea drinking.

Tea diffused from China to Japan, where at first it was used medicinally and in Buddhist monasteries to enhance meditation. Subsequently it spread to Korea. The British introduced tea to India and Sri Lanka in the nineteenth century and the tea trade contributed significantly to the fortunes of the British East India Company.

Tea flourishes best in the cooler air from 3,000 to 7,000 feet (900–2,150 m). Tea trees are kept small with hand pruning, which stimulates growth. Leaves are hand picked and graded according to size and position on a branch. The bud and top two leaves are the best and earn the label "Orange Pekoe." Tea bags contain small, broken leaves known as "fannings."

The flavor of tea is derived from local growing conditions and processing. For example, Darjeeling tea has a peach aroma and is called the champagne of tea. Earl Grey tea is flavored with oil of bergamot, extracted from an orange-like citrus. Jasmine tea is green tea mixed with jasmine flowers.

There are three basic categories of tea: "black" tea is fermented; "oolong" tea is partly fermented; and "green" tea is unfermented. Fermentation includes bruising, wilting, and drying, followed by rolling. As more surface is exposed to the air, tea juices oxidize. Hot air drying stops the process. In the case of green tea, the freshly picked leaves are dried immediately.

China, India, and Sri Lanka are the world's largest tea producers. However, it is also grown in Bangladesh, Kenya, Indonesia, Japan, Georgia, and Azerbaijan. Are you a tea drinker? Investigate the variety and origin of the types you enjoy most.

The British brought Hindu, Tamil laborers from Tamil Nadu in India into predominantly Buddhist Sri Lanka. These nineteenth-century plantation workers were distinguished from the Sri Lankan Tamil settlers of earlier times and are known as **estate Tamils.** In fact, because of pressure from the powerful Buddhist Sangha (a non-government group representing all forms of Buddhism), they were disenfranchised in 1948. Since India refused to recognize them as citizens, these Tamils were virtually stateless.

Other minorities are the aboriginal **Veddas** and the **Burghers,** descendants of the Portuguese and Dutch. The Burghers are Christian and speak English. There are also Muslims, a religious minority in this Buddhist country. In 1931, Britain granted Ceylon limited self-government and in 1948, full independence.

A controversial development project was launched with British and other aid in the 1950s, an era when dam building was at the forefront of foreign aid and regional development. Incorporated in the Mahaweli Development Program, the Victoria Dam was the world's largest aid project at the time. Although the dam was to provide electric power and irrigation, it required the displacement of 1.5 million people. Typical of many grandiose schemes, construction costs tripled and the government was forced to raise taxes. Displacees were settled in isolated areas with poor infrastructure and adverse climatic conditions. They also suffered psychologically from being uprooted from their traditional homeland. Eventually they became known as *Victoria's Victims.*

The Victoria Dam was part of a larger scheme for agricultural colonies in the eastern zone. Here, the climate was drier and the hilly land less fertile. More significantly, Sinhalese colonists clashed increasingly with Tamil inhabitants, presaging the problems to come.

CAUSES OF DISSSENSION

Religious, linguistic, political, and economic differences fomented dissention between Sinhalese and Tamils. Political assassinations and violence continue to revolve around these fundamental differences, although the cleavage is not simply two-sided. Anthropologist S. J. Tambiah (1986) makes a salient point: "Sinhalese and Tamil labels are porous sieves through which diverse groups and categories of Indians intermixed with non-Indians . . . have passed through." Moreover, this already complex situation is exacerbated by international involvement.

The Sinhalese are of Aryan extraction with ancient roots on the island. They are Buddhist and speak **Sinhala,** an Indo-Aryan language with its own script. Sinhalese make up 74 percent of the population and are predominent in administration.

The Tamils, who make up 20 percent of the population, are a Dravidian people. They are Hindu and speak **Tamil**, a Dravidian language with a script differ-ent to the Sinhala one. It was not until the 1970s that Tamil was given recognition as a national language.

Since the Sinhalese dominate government, they have tried to inculcate Sinhalese history and values into schools, textbooks, and the like. There have also been efforts to divorce Sinhalese and Tamils from any common history, such as their relationship with India. These practices have encouraged social distancing and hostility.

English is regarded as a **link-language**; it is sup-posed to serve as a *lingua franca,* or common lan-guage. But English is spoken and understood by an elite minority, and years of anti-English and pro-Sinhala policies mean that it is not a common lan-guage. Students have rioted and burned books at Colombo University because 90 percent of the books are in English. For many rural Sinhalese, English is known as Kaduwa—the sword that separated the privileged from the masses.

Violence erupted in the 1950s with a succession of Tamil uprisings against the Sinhalese. After her husband was assassinated, Sirimayo Bandaranaike became the first woman head of state in the world. She nationalized oil fields, companies, and planta-tions. Communal strife worsened in the following years as Tamil nationalism hardened. In 1972, Cey-lon was declared a republic and renamed Sri Lanka. Nevertheless, organized terrorism continued.

In the 1980s, transnational capital entered in the form of an International Monetary Fund (IMF) development program and U.S. aid "to stabilize the region." Subsequently, U.S. warships acquired refu-eling rights at Trincomalee on the east-west sea route via the Suez Canal. Sri Lanka was to be developed as an "export center" like Hong Kong, and this led to the creation of a free trade zone in Latunyabe. But, interethnic and other discriminatory practices chan-neled most benefits to the Sinhalese.

The killing of 13 Sinhalese soldiers by Tamils in Jaffna in 1983 sparked a spiral of violence that has only tentatively ended. Clashes between Sinhalese and Tamils resulted in 40,000 Tamils seeking refuge in India's state of Tamil Nadu. A series of political as-sassinations followed, and foreign investment faded.

More fighting led to the intervention of Indian troops. In 1987, the Tamils were granted a degree of autonomy in the north and northeastern provinces, and their language rights were guaranteed. Mean-while, Tamil militancy had intensified in the form of a strong guerrilla force with international ties. Known as the **Tamil Tigers,** the group called for the establishment of a Tamil state to be named **Tamil Eelam**. In 1991, India's Prime Minister Rajiv Gandhi was assassinated by a suicide bomber. The Tamil Tigers are blamed for this act.

December 1995 witnessed the largest military operation in the island's history. Government forces routed the Tamil Tigers and took control of Jaffna. Per-haps 2,000 died, and half a million civilians were dis-placed in this conflict. A cease fire was broken in 1996 when a bomb exploded in Colombo, killing 91 and in-juring 1,400 people. The explosives were determined to be Ukrainian RDX smuggled via Dhaka. Kandy's holy Temple of the Tooth was bombed in 1998.

Another cease fire was established in 2001. The killings that took around 60,000 lives have largely subsided. The government is working on rehabilitat-ing the north and east, removing land mines, and relocating displaced people. It is doubtful that the Tamils will get their own state but they may gain some degree of autonomy. Only time will tell.

More Islands of the Indian Ocean

Four island groups punctuate the Indian Ocean near the sub-continent: the Maldives and Laccadives to the west, and the Andaman and Nicobar to the east. Only the Maldives is independent. The latter three island chains belong to India. Virtually all of these islands were devastated in the 2004 tsunami. Thousands were killed and the damage to infrastructures is inestimable.

MALDIVES

Twenty-seven atolls including 1,200 islands known as Maldives are located atop a submarine ridge rising from the depths of the Indian Ocean.

Each atoll is a ring-shaped, coral reef supporting 20 to 60 islands with 5 to 10 inhabited. Dry during the northeast monsoon, the islands experience heavy precipitation with the southeast monsoon—from April to October. Although coral reefs serve as breakwaters, they could not protect the islands from death and destruction caused by tsunamis in 1991 and 2004.

Hills and rivers are nonexistent in Maldives. Breadfruit trees and coconut palms shade expanses of dense scrub, shrubs, and flowers. Soils are poor, and only about 10 percent of the land is cultivated.

Agriculture, mainly taro, a starchy root crop, plus bananas and coconuts, is concentrated on the highest island of Fua Mulaku. At an elevation of 9 feet (c. 3 m), its fresh groundwater is less prone to the intrusion of salt water. However, with global warming and rising sea levels, all the islands are experiencing a depletion of fresh groundwater. This situation threatens their existence.

Situated on historical maritime routes, Maldives became inhabited by diverse ethnic and linguistic groups including Dravidians from the subcontinent, Sinhalese speakers from Sri Lanka, Arabs, Australians, and Africans. Maldivians speak Dhivehi, derived from an archaic form of Sinhalese. The word *atoll* is Dhivehi.

Maldives was once the major source of cowrie shells, widely used as currency in Asia and East and West Africa. The renowned explorer Thor Heyerdahl discovered that Maldives was integral to the trade routes of the Egyptian, Mesopotamian, and Indus Valley civilizations as early as 2000 B.C. He postulated that the first settlers were the Redin, sun-worshipping seafarers. Even today, many mosques in Maldives are oriented to the east rather than to Mecca.

Islam of the Sunni school was introduced by Arab traders in the twelfth century and quickly replaced Buddhism, established by Sri Lankans in the fourth century. Both civil law and Islam-inspired law or *sharia* are used. Under sharia, the testimony of one man equals that of two women. However, women do not veil, nor are they rigidly secluded.

In 1558, the Portuguese became the first Europeans to occupy Maldives, administering it from their colony at Goa. The Dutch followed in the seventeenth century but were subsequently forced out by the British. Maldives was a British protectorate from 1887 until it achieved independence in 1965. It became a republic in 1968. In 1988, a coup by mercenaries, was squelched after military intervention by India. This move symbolized India's hegemony over Maldives.

More than a quarter of Maldives' 200,000 people are concentrated in Male, the capital. High birthrates and lack of an official family planning policy means intense population pressure. As rice, the dietary staple, is not grown in the islands, food shortages could be problematic.

Maldives has an active merchant shipping fleet, but fishing employs almost 25 percent of the islanders. Coconut wood ships have been mechanized, leading to depletion of fish stocks. Additional refrigeration and canning facilities are being built with aid from the World Bank and Japan. The Japanese have long participated in Maldives' fishing industry.

In the industrial sector, small-scale cottage industries are dominated by women who weave rope and mats from reeds and coconut and palm fiber. Women also work in Hong Kong-owned apparel factories at the former British air base on Gan.

The most important source of foreign exchange is tourism. As most Maldivians are uneducated, Sri Lankans, Indians, and other groups fill all but the menial jobs. Tourist resorts, deliberately isolated from settlements, offer gambling, alcohol, and Thai "massage girls." It is government policy to keep Maldivians away from Western influences in order to protect local Islamic values. The majority of tourists are from Germany, Italy, Britain, and Japan. Unfortunately, the seasonal employment pattern for the tourism industry coincides with that of the fishing industry: from April to October during the northeast monsoon.

While economic circumstances have improved for some, Maldives remains one of the world's poorest and most isolated nations.

ANDAMAN AND NICOBAR UNION TERRITORY

The Andaman and Nicobar islands, in southeastern Bay of Bengal, are considered an Indian Union Territory. The Andamans, consisting of more than 300 islands, sit on an ancient trade route between Burma and India. The islands served as penal colonies in the nineteenth century. Only a few indigenous hunters and gatherers remain, with the majority population being Indian. In 2001, the population of the Andaman and Nicobar islands was 356,265.

The Andamans comprise a series of highly dissected longitudinal valleys and are densely forested with mangroves to the north. There are few rivers, and fresh water is difficult to get. Only the south Andamans have roads. The main activity is agriculture, focusing on coconut, betel, and other palm nuts, cassava, and spices such as tumeric.

The Andamese language is interesting in that it has no numeric component. There is a word for "one" and another word for "more than one." Andamese is unrelated to any other language.

LAKSHADWEEP UNION TERRITORY

The Laccadive islands, a collection of coral reef islands in the Arabian Sea off the southwest coast of India, were renamed Lakshadweep in 1973. There are only 13 islands of which 8 are inhabited. The inhabitants are of Indian and Arab descent and practice Islam. The islands are covered with a sandy coral soil. Coconut groves are everywhere and the production of *coir* (coconut fiber) is the major industry. It is conducted by women. Men build boats and indulge in fishing.

⊕ *Main Point Summary*

1. Separated from Pakistan in 1971, Bangladesh occupies the largest deltaic region in the world where the Ganges, Brahmaputra, and Meghna rivers converge and flow into the Bay of Bengal.

2. Bangladesh is linguistically homogeneous. Most of its people are Muslim. Language and religion are unifying forces in a region historically beset by change in both political boundaries and jurisdiction. Regional dissension continues in the southeastern hill tracts.

3. Independence in 1971 left Bangladesh without a major port. Chittagong was developed for this purpose. Dacca (Dhaka) and Chittagong remain the focus of the economy.

4. Bangladesh suffers from frequent natural disasters. While floods and cyclones take the greatest toll of lives and property, drought also occurs in the northwest.

5. Decisions made in India regarding the Ganges (Padma) River impacts water supply in Bangladesh. A case in point is the Farakka Barrage.

6. Shifting water channels create permanent and impermanent land spaces. Sandy islands called *chars,* in braided river channels, are important for settlement, although patterns quickly shift with the next disaster. More than half of rural householders are landless.

7. A system of embankments and improved warning systems have lessened loss of life, but property damage from floods and storms is still severe.

8. Farming activities in Bangladesh correspond with the seasons, with field preparation, planting, and harvesting geared to monsoon patterns of wet and dry periods. Most of Bangladesh is limited to one crop a year. Rice is the most important food crop and jute is the most important commercial crop.

9. Livestock are of increasing importance, yet Bangladesh is deficient in meat and dairy products.

10. Green Revolution technologies have increased yields but beneficiaries are primarily middle and rich peasants who dominate local power structures. Even with improvements in production, Bangladesh must import food.

11. Bangladesh's population explosion has slowed with dramatic declines in NIRs. However, with

a large base population and a relatively high TFR, the population continues to grow. Education and empowerment of women are seen as means to slowing population growth. Problems with gender inequities are very apparent in the country's imbalanced sex ratio.

12. Female activity space is changing, with poorer women having greater access to larger areas outside the home. This is generated by economic necessity.

13. Cities such as Dacca (Dhaka) and Chittagong are overwhelmed with migrants from the countryside. Infrastructure is unable to handle the pressure of humanity. Areas ringing cities are more crowded than ever, and efforts to decentralize have met with limited success.

14. Bangladesh's urban (and rural) environments are seriously polluted. Increased industrial concentration adds to the problem. High rates of disease are related to poisonous air and water, especially in the cities.

15. Fiber crops are important components of industry. Jute products and textiles dominate industrial production. Food processing is increasing.

16. NGOs such as the Grameen Bank are significant factors in improving the lives of peasants, especially women, by providing monetary and other means to economic start-ups and maintenance.

17. Islamic fundamentalism is rising with increasing attacks on women and minorities.

18. Sri Lanka, only a few miles across the Palk Strait from India, is a tropical, mountainous island with rivers flowing to coastal plains. Convectional rainfall and the monsoon bring moisture to different regions throughout the year. Irrigation is essential, especially in the east.

19. Dealing with aridity was the bane of ancient civilizations. Even so, the Sinhalese became masters of water control long before Europeans.

20. The European presence began with the Portuguese and continued with the Dutch and the British. Plantation (estate) systems were established for the production of spices, coffee, and later tea and other tree products such as coconuts and cinchona.

21. The British imported Tamils from southern India to work on the estates. These "estate Tamils" are differentiated from Indians already living in Sri Lanka. Moreover, both these Hindu, Dravidian groups are differentiated from the Buddhist, Aryan Sinhalese. Language differences exacerbate the situation. English, supposed to be a link-language, is spoken mainly by an elite minority.

22. Tamils are concentrated in the north and northeast of the island. Since the 1950s, a revolutionary group called the Tamil Tigers has been fighting for an independent state to be named Tamil Eelam. Assassinations and bombings have taken a heavy toll. Government efforts to end this violence were not successful until 2001 when a cease-fire was established. Years of violence have hurt the economy of Sri Lanka, especially the tourist industry.

23. The Maldives, an independent group of atolls off the southwest coast of India, has little in the way of modern development. Anxious to protect its Muslim way of life, the government is promoting controlled tourism whereby tourists have contact with natural environments in resort settings, and minimal contact with Maldive citizens. The densely forested Andamans, historically used as penal colonies, remain sparsely populated.

⊕ Bibliography

Acker, Fabian. 1985. "Adam's Tears Turn to Electricity." *New Scientist*, 14 (February): 14–16.

Alam, Mahbubul, ed. 1998. *The Independent Yearbook: Bangladesh* 1998. Dhaka: Media.

Alam, Shamsul, ed. 1997. *Jahan Atlas*. Dhaka: Jahan.

Ali, Abu Muhammad Shajaat Ali. 1998. "Toward an Ecological Explanation of Agricultural Unemployment in Bangladesh." *The Professional Geographer,* 50: 176–91.

Amin, Sajeda, and Sara Hossein. 1995. "Women's Reproductive Rights and the Politics of Fundamentalism: A

View From Bangladesh." *The American University Law Review* 44: 1319–1343.

n.a. 1991. "Bangladesh in Troubled Waters." *The Economist*, 11 (May): 4–5.

Barkat, Abul, Mati Raham, and Mark Bose. 1997. "Family Planning Choice Behavior in Urban Slums of Bangladesh: An Econometric Approach." *Asia-Pacific Population Journal* 12: 17–32.

Bigelow, Elaine. 1995. *Bangladesh: The Guide*. Dhaka: ABP.

n.a. 1998. "Capital and Kindness: Micro-credit is No Miracle." *New Internationalist* 308: 4.

Brammer, H. 1990. "Floods in Bangladesh I: Geographical Background to the 1987 and 1988 Floods." *The Geographical Journal*, 156: 12–22.

Daneels, Jenny. 1998. "Defying Mother Nature." *Asian Business*, 34: 6–8.

Davis, Anthony. 1996. "Tiger International." *Far Eastern Economic Review*, 22: 30–38.

D'Costa, Jashinta. 1994. "Changes in Urban Structure in Bangladesh. Urban Geography of Hinduism." *Wilson Quarterly*, 15: 42–50.

Elahi, K. Maudood. 1989. "Population Displacement due to Riverbank Erosion of the Jamuna in Bangladesh." In *Population and Disaster*, eds. John I. Clarke et al., pp. 81–97. International Geographical Union. London: Blackwell.

Foster, Gerard. 1969. "The Concept of Regional Development in the Indigenous Irrigation Systems of Ceylon." *Association of Pacific Coast Geographer's Yearbook*, 31: 91–100.

Frank, Neil L. 1971. "The Deadliest Tropical Cyclone in History?" *Weatherwise*, 5: 438–44.

Fritsch, Peter. 2003. "Bangladesh Stares Into the Abyss." *Far Eastern Economic Review* 27 (November): 46–49.

Government of Bangladesh. 1998. *Statistical Pocketbook: Bangladesh 97*. Dhaka: Bureau of Statistics.

Haque, C. E., and MD. Z. Hossain. 1988. "Riverbank Erosion in Bangladesh." *The Geographical Review*, 78: 20–31.

Hye, H. A. 1992. *Case Studies on Rural Poverty Alleviation in the Commonwealth of Bangladesh*. London: Commonwealth Secretariat.

Islam, M. Aminul, and Nazrul Islam. 1997. "Urban Environmental Issues and Governance in Bangladesh." In *Urban Governance in Bangladesh and Pakistan*, eds. N. Islam, and M. M. Khan, pp. 69–80. Center for Urban Studies Dhaka: University of Dhaka.

Ives, Jack. 1991. "Floods in Bangladesh: Who Is to Blame?" *New Scientist*, 13 (April): 1–4.

Madeley, John. 1984. "Progress Has its Price." *Geographical Magazine*, 56: 226–27.

McDonald, Hamish. 1991. "Learning From Disaster." *Far Eastern Economic Review*, 30 (May): 23–4.

Murphey, Rhoads. 1957. "The Ruin of Ancient Ceylon." *The Journal of Asian Studies*, 16:181–200.

Paul, Bima Kanti. 1998. "Coping Mechanisms Practiced by Drought Victims (1994/95) in North Bengal, Bangladesh." *Applied Geography*, 18: 355–73.

———. 1992. "Female Activity Space in Rural Bangladesh." *The Geographical Review*, 82: 1–12.

Potten, David. 1994. "The Impact of Flood Control in Bangladesh." *Asian Affairs*, 25 Part II: 156–162.

Rain, Nick. 1994. "Living With the Landscape." *Geographical*, 66: 24–27.

Rao, Radhakrishna. 1992. "On Fertile Ground." *Far Eastern Economic Review*, 13 (August): 70.

Saenger, P. 1993. "Land From the Sea: The Mangrove Afforestation Program of Bangladesh." *Ocean & Coastal Management*, 20: 23–39.

Tambiah, S. 1986. *Sri Lanka: Ethnic Fratricide and the Dismantling of Democracy*. Chicago: University Press.

Taylor, Alice ed. 1968. "Ceylon." *Focus*, 18: 1–10.

Chapter 9

⊕

East Asia: Center of the World

"To understand is hard. Once one understands, action is easy."

SUN YAT-SEN (1866–1925)

For some 7,000 years, East Asia has been dominated by *Chung Kuo:* the Middle Kingdom. Not only is China the largest country in the region but also its cultural, economic, and political impact has been profound. Japan, Korea, and Taiwan—each owes significant aspects of its heritage to the wonder that was China. This chapter will tell you about the physical characteristics of the region and introduce you to the Chinese worldview relative to those within its sphere of influence. Then, China's history will be reviewed up to its inception as the People's Republic. Although I will mention Taiwan, Korea, and Japan where appropriate in this chapter, I will save the more salient details of their histories for Chapters 11 and 12. Be sure to examine the pronunciation guide before reading further (Figure 9–1). Pinyin, a system of romanization (phonetic notation and transliteration to Roman script), will be used in this chapter as the primary spelling. Pinyin pronunciation is based on the Mandarin dialect of Chinese. Following the Pinyin word, the Wade-Giles system of spelling is given in parentheses.

Topography

Mountains, hills, and plateaus dominate the East Asian landscape. For example, mainland China is 65 percent and Japan is 75 percent mountainous. These physiographic barriers, while permitting the evolution of unique cultures, were penetrable enough to allow outside influences to wend their way from one valley settlement to another. Mountain passes, river arteries, and ocean passages were instrumental in the diffusion of people and ideas across the East Asian landmass to the Korean peninsula and the islands of Japan and Taiwan.

Figure 9–2 shows the topography of East Asia. The mainland comprises rivers, plains, and basins separated by mountains. Geographers demarcate the rugged and arid environment of the west—**China Frontier**—from the gentler and more humid environment of the east—**China Proper.**

CHINA FRONTIER

The region demarcated as China Frontier is dominated by spectacular mountains and plateaus, as well as

211

PRONOUNCING AND SPELLING CHINESE PLACE NAMES

Since Chinese is written with characters, it is necessary to transliterate these phonetically into the Roman alphabet. The nineteenth century Wade-Giles System of romanization was commonly used before the 1970s. At that time a new system was introduced: Pinyin. Pinyin was made official in 1979.

While Pinyin is found in contemporary books and atlases, Wade-Giles remains in pre-1950 publications. Therefore it is important to be cognizant of both systems. If you learn to pronounce the names of all China's provinces and autonomous regions, you will have a head start on pronouncing other Chinese words.

PROVINCE IN PINYIN	PRONUNCIATION	WADE-GILES SPELLING
ANHUI	AHN-WAY	ANHWEI
ZHEJIANG	JUH-JEEONG	CHKIANG
FUJIAN	FOO-JEEEN	FUKIAN
HEILONGJIANG	HAY-LOONG-JEEONG	HEILUNKIANG
HENAN	HUH-NON`	HONAN
HEBEI	HUH-BAY	HUPEH
HUNAN	HOO-NAN	HUNAN
HUBEI	HOO-BAY	HUBEI
NEI MONGGOL	NAY-MUNG-GOO	INNER MONGOLIA
GANSU	GAHN-SOO	KANSU
JIANGXI	JEEONG SHE	KIANGSI
JIANGSU	JEEONG SU	KIANGSU
JILIN	JEE-LYNN	KIRIN
GUANGXI	GWONG-SHE	KWANGSI
GUANGDONG	GWONG-DOONG	KWANGTUNG
GUIZHOU	GWAY-JOE	KWEICHOW
KIAONING	LEEOW-NING	LIAONING
NINGXIA	NING-SHEAH	NINGSIA
SHANXI	SHAHN-SHE	SHANSI
SHANDONG	SHAHN-DOONG	SHANTUNG
SHAANXI	SHUN-SHE	SHENSI
XINJIANG	SHIN-JEEONG	SINKIANG
SICHUAN	SSU-CHWAN	SZECHWAN
XIZANG	SHE-DZONG	TIBET
QINGHAI	CHING-HI	TSINGHAI
YUNNAN	YUOON-NAN	YUNNAN
CITIES		
BEIJING	BAY-JING	PEKING
GUANGZHOU	GWONG-JOE	CANTON
CHONGQING	CHONG-CHING	CHUNGKING
SHANGHAI	SHONG-HI	SHANGHAI
TIANJIN	TEEN-JIN	TIENTSIN

Figure 9–1

Pronunciation and spelling guide for Chinese place names. Be sure to study this guide. Once you know how to pronounce and spell the province names, pronouncing and spelling city names will be easier.

Figure 9-2

Physical geography of East Asia. Study this map carefully and use it in reference to the text. Your atlas map will give more detail. Notice how mountainous China is and how distinct lowlands are. Look at a population distribution map and check the relationship between population distribution and China Frontier and China Proper.

forbidding deserts and swampy basins. The Altai Mountains and the Tien Shan (Heavenly Mountains) encase the dry Junggar (Dzungarian) Basin. Just to the east of the Tien Shan is the Turpan (Turfan) depression. At 505 feet below sea level (155 m), this is the second lowest place in Asia (after the Dead Sea). The southern slopes of the Tien Shan fade into the Tarim Basin, bordered on the west by the Karakoram and on the south by the Kunlun and Altun Shan. Within this basin lies the Taklamakan, one of the

world's largest sand deserts. Its name means: "You can go in but you can't come out."

Settlement is possible in desert basins because of the presence of water. The Junggar and Tarim basins and the Hexi Corridor are **artesian basins** with plentiful water trapped in underground **aquifers** (water-bearing rock), especially within alluvial fans at the base of mountains. With seasonal rainfall, rivers frequently disappear from the surface and flow underground. The use of *karez* (see page 138 in

Figure 9-3

Karez wells are vertical shafts connected to an underground horizontal canal that penetrates an aquifer. The water flows by force of gravity to a surface location, where it is diverted to homes and fields. Both vertical and lateral canals must be kept clear of material that could interrupt the water flow. Both digging and clearing wells are dangerous ventures.

Chapter 6) irrigation systems render mountain-base oases productive (Figures 9-3 and 9-4).

Directly south and southeast of the dry wastes of the Taklamakan lies the Qinghai-Xizang Plateau, with an average elevation of more than 12,000 feet (4,000 m). The Qilian Shan separate the swampy Tsaidam Basin from the steppes of Nei Mongol (Inner Mongolia). These short grasslands fade northward into the Gobi (desert) of Mongolia. Immediately north of the Qilian is a narrow strip of land known as the Hexi or Gansu Corridor, an integral part of China's famous trade route known as the Silk Route. To the northeast are the Da (Greater) and Xiao (Lesser) Hinggan (Khingan) mountains, sloping inland to the Manchurian Plain and the Liao River basin. Northward is the Amur River, forming the border between China's northeast and Russia. The Yalu and Tumen rivers separate China from North Korea.

CHINA PROPER

China Proper is dominated by rivers and plains. To the north is the Huang He, or Yellow River. Flowing 3,395 miles from the Qinghai Plateau, it sweeps northward, wrapping the Ordos Desert in a big loop, then cuts southward across the **loess** region of Shaanxi and Shanxi provinces. Loess, resulting from the last glacial period, is fine, wind-blown soil that is readily eroded. At the confluence of the Wei River, the Huang He turns northeast across the North China Plain and empties into the Bo Hai Gulf of the Yellow Sea.

The Huang He is literally yellow, due to the immense amounts of silt it carries in suspension through erosion of the loess plateau. At one time, the Shandong (Eastern Mountain) Peninsula was an island, but eons of alluvial deposition have joined it to the mainland. Its highest peak is the sacred Tai Shan at 5,056 feet (1,517 m).

Just south of the Wei River is the Qin Ling Range, an extension of the Kunlun. East of the Qin Ling is the Huai River. As you will discover, both the Qin Ling and the Huai are sentinels of change in both climatic and agricultural patterns in China.

Figure 9-4

Road and water channel in Dunhuang. Poplar trees line this dirt road. Straight poplar wood is important for construction. Note the irrigation canal alongside the road. The water comes from a karez and is fundamental to irrigation in this arid region. Photograph courtesy of B.A. Weightman.

The North China Plain is the largest alluvial land-form region in eastern China. It covers 125,000 square miles (323,750 sq km). It blends into the Central Chang Jiang (Yangzi River) Plain and its associated lake basins. The Chang Jiang, the world's third longest river, flows an incredible 3,900 miles (6,279 km) from Tibet to the East China Sea. To the west, surrounded by mountains, is the fertile Sichuan Basin and the Yunnan Plateau. Mountains dominate southern China until breaking at the Xi River Basin by the South China Sea. Southwestern China is dominated by the longitudinal extensions of the Tibetan-Qinghai ranges.

Climatic Variation

East Asia is characterized by continentality modified by the monsoon. Seasons are accentuated in the interior due to its remoteness from the moderating effects of water bodies. Mountain patterns add to these contrasts. Winter and summer temperatures are extreme in the interior of China, while coastal regions have more moderate climes. Latitude is another consideration. While rice is being harvested in the Xi Basin, ice sculptures are being created in the Liao Basin.

The summer monsoon (detailed in Chapter 2) brings ever-decreasing amounts of precipitation from south to north (Figure 9–5). Precipitation ranges from around 120 inches (305 cm) on Hainan Island to less than 10 inches (35 cm) in the Tarim Basin.

The Qin Ling, with peaks reaching 12,000 feet (3,600 m), form a topographic barrier to monsoon rainfall from the south and to dust-laden winds from the north. Significantly, the Qin Ling and Huai River approximate both the 20-in. (51-cm) isohyet (a line on a map joining points of equal rainfall) and the northern limit of double-cropping.

In winter, winds sweep outward from the Siberian high pressure cell near Lake Baikal, bringing relatively dry, cold air to the continent as well as to Japan. Sometimes the chilling air is filled with loess, carrying yellow dust as far as Japan's west coast.

Typhoons (the same as hurricanes in the United States) are important sources of moisture during the summer and fall. These develop in the South and East China seas, striking China's southern coast around May and reaching Taiwan and Japan in July or later. Typhoons bring a succession of rain squalls and winds that can reach velocities of 150 miles (242 km) per hour, causing severe damage.

Average Annual Rainfall	
120	3000
80	2000
60	1500
40	1000
30	750
20	500
10	250
0	0
INCHES	MILLIMETERS

Figure 9–5

Yearly precipitation in East Asia. Note how the amount of precipitation decreases as you move inland. How does this pattern compare with population distribution? Compare this map with Figure 10–10 and see the relationship between precipitation and agricultural patterns.

Seasons are highly differentiated between both hot and cold and wet and dry. Temperatures contrast greatly with latitude, averaging $-13°$ F ($-25°$ C) in northern Manchuria in January and 68° F (20° C) in Hainan in the south. July temperatures average 70° F (21° C) and 84° F (29° C), respectively. Summers are oppressively hot and humid, in Taiwan, Korea, and Japan as well, except in the dry west.

Peoples of East Asia

It is a common but erroneous perception that all East Asians are of Mongoloid stock. Although this is true of Chinese, Koreans, Mongols, and Tibetans, it is not true of all groups. The Japanese, for example, are thought to be a Mongol-Malay mixture. And, in China Frontier, there are several Turkic populations, such as the Uyghur. As you will learn in the next chapter, China has many minorities, and **Sinicizing** (making Chinese) these groups is a major policy issue for the Chinese government.

A brief look at language patterns reveals the dissimilarities of the peoples of East Asia. Mandarin and Cantonese are Chinese languages belonging to the Sino-Tibetan language family. Tibetan is included in this family but belongs to the Tibeto-Burmese group. Some of the minority people in southwestern China speak Mon-Khmer (of the Austro-Asiatic language family), which is related to languages in Southeast Asia. The Uyghur, Kazak, and other Turkic people speak Altaic languages of the Turkish group. Mongolian belongs to a different group of Altaic languages. Neither Korean nor Japanese are related to any of these!

The Chinese see themselves as Han Chinese, descendants of the glorious Han Dynasty (202 B.C.–220 A.D.). All others are non-Han people. China's historic perception of its own non-Han people mirrors its perception of other non-Chinese, Asian people. Furthermore, China's perception of itself relative to outsiders has been a major force in shaping its relationships with the world.

China's Worldview

The written characters for China translate as *Zhong Guo* (*Chung Kuo*), meaning central or Middle King-dom. For much of China's history China perceived itself as central to the known world. This notion has its roots in antiquity. To appreciate it we must venture back in time 4,000 years.

An ancient creation myth has the confusion of heaven and earth divided into light and bright, and heavy and dark segments, with a primeval being called **Pan-gu** (P'an-ku) growing between them. Pan-gu's body parts transmuted into various parts of the cosmos. Pan-gu's function prepares us for the important role of humans in the transformation of chaos into order and explains the Chinese view of apparently opposing forces as complementary in an organic harmony. These ideas are seeded into the philosophy of Daoism (Taoism) discussed later.

One seminal work regarding the origins of human society is the *Yi jing* (*I-ching*), a work in which hexagrams depict the subtle harmonies of complementary opposites (Figure 9–6). According to the *Yi jing*, humankind was ruled by a succession of

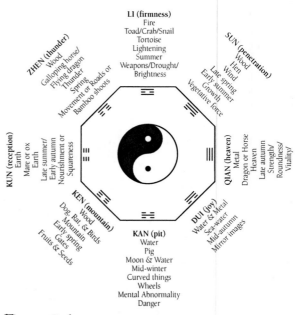

Figure 9–6

This diagram expresses harmony and interrelationships among all things. Understanding these relationships is important in interpreting East Asian landscapes. This diagram is part of the Korean flag attesting to its importance in East Asian culture.

sage kings such as Fu-xi, who invented nets and baskets for fishing. His wife invented feminine arts and patched a rip in the fabric of heaven. Shen-nong, the Divine Farmer, invented agriculture and herbal medicine, and taught people how to hold periodic markets.

Oracle Bones

The rulers of the ancient Shang Dynasty (originally thought to be 1766–1122 B.C., now 1600–1046) practiced divination by heating certain bones such as animal scapulae (shoulder blades) and interpreting the resultant cracks. Questions and interpretations were often inscribed on the bone. These **oracle bones** were first discovered by scholars in 1899 near the last Shang capital at Anyang in the Wei River valley.

The character Huang Di was of particular importance. Huang Di, the Yellow Lord (emperor), constructed roads and set up a bureaucracy to run his domain. He distributed grain at appropriate times and clarified knowledge of heaven and nature. Educated Chinese believe that humanity reached a peak of order about 2350–2200 B.C. during the reigns of subsequent emperors. This belief in human order is a fundamental theme in Chinese thinking and behavior.

Another fundamental belief is that an ordinary person can achieve great things and that the best government is government by virtuous individuals. Shun is held as an example of a man who rose from humble origins. The Confucian thinker Mencius (c. 390–305 B.C.) wrote that when Shun "heard a single good word, witnessed a single good deed, it was like water causing a breach in the dikes of the Yangzi or Yellow River. Nothing could withstand it."

Water control has been central to China's development since the earliest of times. Shun appointed a man named Yu, who assembled a massive labor force and spent 13 years supervising water control and irrigation works. Many of China's unruly rivers, including the Huang He, were rechanneled. Subsequently, agriculture flourished. The concept of water

management and agricultural productivity as responsibilities of the ruler is another salient theme in Chinese history.

Yu, the model of a competent minister, also surveyed the empire, compiling information on soils, local products, and land revenues. He reported these in a lengthy document called the *Yu Gong* (Tribute of Yu). The obligation of ministers to report to and advise the ruler is another important element of proper governance and social order. As Shun said to Yu, "The mind of a man is perilous; the mind of the True Way is hard to discern. Concentrate! Be single minded! Hold fast to its center." Years later, Yu's son came to power and his descendants ruled China by hereditary right for 400 years. Yu is regarded as the founder of China's first dynasty: the Xia (Hsia).

Control by diminishing degree, according to distance from the capital, has haunted Chinese history and, by association, the history of East Asia. An important component of the model of Chinese rule was the tributary zone. People of this zone acknowledged Chinese sovereignty through trade and giving gifts (paying tribute) to the Son of Heaven, the official title of the Chinese emperor.

Tribute payments were not always a one-way process. China often paid tribute to barbarians to keep them at bay or to employ their assistance in thwarting other barbarian groups. The northern frontier was of constant concern.

The northern frontier formed a zone where opposing ways of life—farming and herding—met and commingled. Sometimes pasturage would give way to sedentary farming. Historian Jacques Gernet (1983) observes, "The problem of defense against incursions from the steppe must be seen in a context which is as much cultural, political and economic as military, thanks to the phenomena of assimilation, diplomatic combinations, and commercial exchanges." Here were allied tribes who collaborated in the defense of the empire, outposts, forts and garrisons, military colonies, lands developed by deportees, horse breeding, and nomad encampments.

Two types of troops were located in the frontier: *farmer-soldiers,* known as "soldiers of the irrigation canals" or "soldiers of the granaries," and *soldiers on garrison duty* in forward posts. Each post could contact the other through a system of signals: red and blue flags, smoke by day, and fires by night. Transmission of information was swift, and all was recorded in writing. Precise information about troop movements is given in texts as early as the second century B.C.

The Ways of Writing Chinese

Chinese characters originated as *pictograms,* where a symbol stood for an object. Over time, pictograms evolved more complexly as *ideograms,* where symbols represent ideas. There are thousands of Chinese characters. To be reasonably literate, you need to know at least 10,000.

Chinese is a tonal language. Different dialects have different numbers of tones. Mandarin has four tones while Cantonese has nine although most of the time only six are used. The same sound, pronounced with different inflections, can mean different things or ideas. For instance *ma* might imply "mother" or "horse" depending on what tone you give it. However, these two words are written with their own unique characters.

Writing Chinese characters in the Roman alphabet requires "transliteration," a way of transforming one to the other. One system, employed until well after the establishment of communism in China (1949), is the Wade-Giles system. More recently, the Communists have introduced a more simplified character system. How Chinese words are transliterated into English is called **Pinyin.**

Shan-tung under Wade-Giles becomes *Shandong* in Pinyin. Ch'ing Dynasty becomes Qing Dynasty, and Mao Tse-tung becomes Mao Zedong. Pinyin more closely approximates the sound of the word in the Mandarin dialect of Chinese. While most books and atlases now use Pinyin, English publications from Taiwan continue to use Wade-Giles.

Perhaps the most obvious evidence of China's relationships with outsiders is the Great Wall (Figure 9–7). Dates for initial construction and subsequent connections of the wall(s) are more suggestive than accurate. It is said to have been built during the Qin (Ch'in) Dynasty (221–207 B.C.) when the Yellow Emperor consolidated existing sections with watchtowers along the northern frontier. More or less permeable throughout history, it was consolidated again under the Ming Dynasty (1368–1644 A.D.). It is the Ming wall, located north of China's capital Beijing (Peking), that most tourists visit today. The western sections have either vanished or are in a state of disrepair. The measured length of the Great Wall also varies, depending on which sections are measured. Estimates range from around 1,500 to 4,000 miles (2,415–6,440 km).

For most of Chinese history, the Great Wall was actually many walls—sections constructed as the need arose in time and space. Their positioning depended on the state of China's relations with the barbarian peoples of the time. China spent millions in terms of finances and manpower in warding off such nomadic hordes as the Khitan, Xiongnu (Hsuing-nu), Mongols, and Manchus, not always with success. The Mongols and Manchus, respectively, defeated Chinese forces and took control as the Yuan Dynasty (1279–1368) and the Qing (Ch'ing) Dynasty (1644–1912). China's influence in East Asia peaked under the Mongols, who sent expeditions against Japan, Burma, Vietnam, and Java using both Chinese and Korean troops and fleets.

The Terra-Cotta Army

In 1974 the discovery of an entire life-size army that had been buried for more than 2,000 years astonished the world. Thousands of terra-cotta soldiers, each with unique facial features, along with war-horses and chariots, are regarded by some as the "eighth wonder of the world." The army is believed to be guarding the tomb of Emperor Qin Shi Huang of the

Figure 9–7

The Great Wall of China was begun in the Qin (Ch'in) dynasty (221–206 B.C.) "China" comes from Ch'in. The wall actually linked several older walls, and in its full length reached nearly 4,000 miles (6,000 km). It averages 25 feet (7.8 m) in height and 19 feet (5.8 m) in width. Only parts of the wall survive today. This section of the wall near Beijing was refurbished under the Ming dynasty in the fourteenth century. Photograph courtesy of B.A. Weightman.

Qin Dynasty (221–206 B.C.). This site, near the city of Xian (Hsi-an), is one of China's top tourist attractions.

CONFUCIUS

K'ung-fu-tsu (Kong Fuzi)—or **Confucius**—was born around 551 B.C. in the ancient state of Lu in Shandong Province. After holding several political posts, he became a teacher of ethics, ritual, and philosophy. His ideas were written down by his students as the *Analects.* In addition to transmitting Confucius' teachings over time, the *Analects* state his belief in hierarchical structures of responsibility and duty: of ruler to Heaven (the source of all power and authority); of subject to ruler; of son to father; of younger brother to older brother; and of wife to husband. Each person within society should fulfill the responsibilities implied in their name be it ruler, son, or subject. Rulers should behave like rulers and take care of the people, sons should behave like sons and respect their father, and subjects should act like subjects and obey the ruler.

Another important aspect of Confucian philosophy is filial piety or filial devotion. This is one of the "right relationships" that govern family life and, ultimately, the entire social order. By filial piety Confucius meant the respect and care children owe their parents during their lives and the reverance they pay to their memories after death. Being an obedient and loving brother to one's older brother is also a kind of filial piety. All behaviors within the family and society must be accompanied by a loving consideration or kindness (called humaneness by Confucius) for the other person. If all individuals behaved according to their name, practiced filial piety, and treated others as they themselves would wish to be treated, the society would enjoy peace, harmony, and prosperity.

Self-cultivation and education as the way to achieve a virtuous character were also key elements in Confucian philosophy. The cultivated person enjoyed life in the form of music, painting, good food, the company of his family, and the attainment of a wise, old age. The greatest sadness was to have no descendants (i.e., male offspring) to perform funeral rites and rituals of remembrance. It was not necessary to be rich or noble to practice self-cultivation, according to Confucius. Each person could learn to be good. And, through dedication to learning, one could become a sage. No priests were necessary to guide one on this path. Self-development was the key to total fulfillment.

Confucius also affirmed the right of the people to rebel against an unjust ruler. China's rulers held power through the **Mandate of Heaven**. If they failed to fulfill their role as ruler and lapsed from virtue, or if a disaster occurred, they forfeited the Mandate of Heaven. Loss of Heaven's favor could lead to the downfall of one ruling family and its replacement by another.

Confucius was also a fervent believer in the importance of culture and learning, as well as hard work. A good and just ruler would surround himself with ministers who achieved their goodness through moral education. Regarding government, Confucius said: "If you set an example by being correct, who would dare to remain incorrect?" Confucius also recognized the need to exert control over distant peoples: "Ensure that those who are near are pleased and those who are far away are attracted." In all these matters, appropriate rituals were critical to facilitating and maintaining reciprocal relationships.

Confucius created the role of the scholar-minister, a calling of the highest value in ancient China. He believed that poverty was not a barrier to advancement and that any young man could obtain an education and achieve an administrative or even a court position. Confucius died in 479 B.C., but has always been known in China as the "First Teacher." His example underpins the extraordinary respect for education and teachers in China today. Under the Han Dynasty, Confucianism was adopted as the state philosophy and a system of rigorous examinations was established for admittance to the Imperial Civil Service.

In the twelfth century, a Confucian philosopher Zhuxi (Chu Hsi) (1130–1200) speculated about the existence of a Supreme Ultimate, a cosmic force. However, Confucius never discussed priests or religious rituals, leaving any religious aspect in question. Nevertheless, faithful Confucianists built temples for the purpose of worshipping the sage. It was this neo-Confucianism that diffused to Korea, Vietnam, and Japan, where it became the prevailing philosophy, especially for the educated.

In the twentieth century Confucian ideas were disparaged as contributing to China's backwardness relative to the industrialized world. Under communism, the sage was ridiculed and vilified as a symbol of all that was wrong with China's imperial past. In recent years the rigors of communism have relaxed a bit and Confucius has been extolled for his teachings of loyalty and obedience to government. "Confucian

principles" are frequently referred to in context of the economic successes of Singapore, Taiwan, South Korea, and other Asian countries.

DAOISM (TAOISM)

A contemporary of Confucius called **Laozi** (Lao-tze), meaning "The Old One," supposedly debated with Confucius on the right way of being. Daoist philosophy as recorded in the **Dao de Jing** (*Tao-te-ching* or *Classic of the Way*), is concerned with the Dao, or the way the universe works. The sayings contained in the *Dao de Jing* are quite cryptic and even mystical. For example: "The Way is void, yet inexhaustible when tapped;" or "Men should be bland, like melting ice, pure and peaceful like a block of uncarved wood."

Daoism evolved as a religion as it mingled with folk beliefs centered on the worship of nature and spirits. It stressed the desirable unity of people and nature and the futility of worldly strivings. A person's place in nature's enormous, eternally changing expanse was insignificant, "a drop of water in a flowing stream." Individuals should practice a kind of effortless action as water does when it flows around obstacles. This is the ideal of the Way. Daoist priests and monastic orders developed and temples were built. What started as an esoteric school of thought became a mass religion, eventually incorporating a pantheon of gods and immortals.

Later Daoists practiced alchemy—the mixing of metals to produce gold. They also searched for the elixirs of immortality, roaming hills and mountains in search of appropriate roots and plants. Their search for medicinal herbs contributed significantly to the evolution of Chinese medicine.

Even Confucianists found Daoism attractive. Daoism was particularly suitable when things were not going right, or in old age. Confucian activism and social reformism were counterbalanced with Daoist passivity and "go-with-the-flow" attitude. This dualism was very attractive in the context of even older ideas concerning the integral nature and harmonious balance of all things and the fundamental principles of **yin and yang** (Figure 9–8).

Yin and yang are two opposing, yet complementary, primordial forces that govern the universe and

FENG-SHUI RELATIONSHIPS

The concepts of the five elements or forces with their yin and yang relationships were introduced as early as the fourth century B.C. All of the elements are connected: Wood makes fire which burns it to make earth. Earth has metal (ores). Metal has condensation—water, and water nourishes plants and trees—wood.

The relationships of the five elements are in accordance with the laws of nature and heaven. It is important not to interfere with the order of nature or disrupt the natural conditions designed by heaven.

The ten stems and twelve branches were introduced prior to 2000 B.C. Each stem and branch is related to the five elements and yin-yang principles. All these forces are significant for proper *feng shui* location and a geomancer must be consulted.

FIVE ELEMENTS	YIN-YANG	TEN STEMS	TWELVE BRANCHES	DIRECTION	ANIMAL
WOOD	YIN	SPROUTING	BUD; BEGINNING	N	RAT
	YIN	SPREAD OF GROWTH	GROWTH	N/NE	OX
FIRE	YANG	BLOOMING	SPREAD OF GROWTH	E/NE	TIGER
	YIN	MATURITY	ABUNDANCE	E	RABBIT
EARTH	YANG	ABUNDANCE	PROGRESS	E/SE	DRAGON
	YANG	ORDER	RENEWAL	S/SE	SNAKE
METAL	YANG	FULLNESS	MATURITY	S	HORSE
	YANG	RESTORATION	SMELL OF MATURITY	S/SW	SHEEP
WATER	YIN	HEIGHT OF FUNCTION	EXPANDED MATURITY	W/SW	MONKEY
	YANG	PREPARATION FOR SPRING	RIPENESS	W	COCK
	YIN		DEATH	W/NW	DOG
	YIN		NUCLEUS	N/NW	PIG

Figure 9–8

Here the concept of yin and yang are outlined. Yin and yang balance opposite forces to ensure harmony between humans and nature. Yin and yang represent the interrelationship of all phenomena and are critical to feng shui *orientation.*

symbolize harmony. For instance, without dark there is no light. Without life there is no death. All phenomena contain elements of yin and yang that interact and create cyclical change. Night rolls into day, winter into summer, and so forth. This is the process of the Dao—the universal situation.

Did you know that East Asia has its own "geographic tradition?"

Feng Shui: The Chinese Art of Placement

The Chinese have always stressed relative position in the landscape, the world, and the cosmos. The forces of *feng shui* (wind and water) must be dealt with appropriately. For example, Chung-kuo infers the Middle Kingdom, a nation at the hub of the universe. Cities were usually oriented according to the cardinal directions. Numerology was also employed, with the number three and multiples thereof considered auspicious. The Heavenly Altar in Beijing, where the emperor conducted ceremonies for rain and good harvests, is reached by nine steps. The top surface comprises nine rings of stones. Gates to the Forbidden City are covered with nine rows and nine columns of brass studs. Four, which sounds like the word "death" in Cantonese, is not a lucky number. So you would not buy a house numbered 444. This belief is practiced mostly in southern China where Cantonese is spoken.

Decisions regarding location are made in the context of yin and yang forces in addition to the force of *Ch'i*. *Ch'i* is the most important element of *feng shui*. It is the vital force, the cosmic breath that gives life to all things. Without *ch'i*, there would be nothing. *Ch'i* is a pervasive concept in Chinese traditional medicine and acupuncture, and martial arts such as *Gung fu* (Kung fu).

The land most influenced by *ch'i* is the most habitable. *Ch'i* spirals around the earth, sometimes "exhaling" mountains or volcanoes at the earth's crust. When *ch'i* falls short of the surface, there is no water or flowers, but bad luck. Mountains are viewed as dragons exhaling cosmic breath. Mountains and hills are important as defensive sites and are typically kept to the north in orientation of settlement. "Back to the north; front to the south." The worst location would be on a featureless plain with no protection from anything. Atmospheric *ch'i* molds human *ch'i*. Balance and harmony of yin, yang, and *ch'i* are essential to repel negative forces.

Individuals called **geomancers** are experts in *feng shui*. They are consulted to determine the proper siting of buildings and even interior design and placement of furniture. Once a house is carefully and correctly designed and oriented, a Daoist priest is called to perform a moving-in or consecration ceremony. The Hong Kong and Shanghai Bank stages dragon dances when opening a new branch, even in overseas locations.

Even colors are subject to *feng shui* principles. The south is associated with red, representing joy and festivity. A south-facing house is auspicious, and the placement of red door guardians will ward off evil spirits. Green is related to wood and foliage and evokes youthful energy and growth. It represents the dragon and the east. Yellow was a royal color and is worn by Daoist priests for burying the dead and for geomantic blessings. White is the color of autumn. It stands for metal, represents purity, and denotes western orientation. Black is the north and winter. Its element is water. Black denotes the consequence of humans, death, mourning, and penance. It is the color of calamity, guilt, and evil.

Feng shui's goal is maximum harmony with natural forces to induce happiness, prosperity, and longevity. Widely practiced among Chinese communities worldwide, it was banned in

Communist China as harmful superstition. However, in the new, modernizing China, rules have relaxed, and *feng shui* and the practice of geomancy have reappeared.

When I visited rural and urban homes in several regions of China, I noticed that many new pieces of furniture were painted either red or green.

THE COMING OF BUDDHISM

Buddhism entered China from India in the first century A.D. This new religion spread quickly along the Silk Route (Figures 9–9 and 9–10). By the time of the Sui Dynasty (589–618), it was the dominant faith among the merchants and soldiers traveling the trail. Buddhist monks constructed monasteries at oases, and these became inns and even banks for travelers. Dunhuang became a significant staging point: this is where travelers would choose between the northern or southern route as the safest passage westward.

Dunhuang housed a large and elaborate Buddhist art treasury. Around 366, monks began to cut caves into cliffs and then carving and painting the interiors. This labyrinth of caves contains nearly 500,000 square feet of murals and 2,000 statues of various sizes. In the Tang Dynasty (618–906), there were more than 1,000 caves. Today there are 486. These Mogao Caves are one of China's prized tourist attractions.

Figure 9–9

Note how the Han Empire focused on major river plains. Compare this map to Figure 9–2 and a physical geography map in your atlas. Trace the arduous journey along the Silk Route where thousands died attempting to complete the trip. Note the preponderance of mountains and deserts.

Figure 9–10
This is called "White Horse Pagoda." It was built by a Buddhist monk in honor of his faithful horse who carried him along the Silk Route. Photograph courtesy of B.A. Weightman.

In the political chaos following the Han Dynasty (see the next section), Buddhism's nonsecular belief system appealed to many people and Buddhist monks were able to assert their authority. Because there were few texts and many questions, in 399 the pilgrim Fa Xian (Fa Hsien) journeyed to India for manuscripts and advice. His account of his trip, *A Record of the Buddhist Countries,* revealed developments in Mahayana Buddhism, as well as China's communications with the outside world.

Fa Xian traveled the Silk Route to Afghanistan and later went on to India. From the Ganges Delta, he visited Sri Lanka and Sumatra before returning to Shandong in 414. Then, with the support of various rulers who hosted thousands of visiting monks from India, Buddhism made a phenomenal rise in China.

Dynasties Rule the Middle Kingdom

For more than 3,000 years of its history, the vast region of China was held together—with varying degrees of success—by a series of **dynasties**. These were periods of rule by one family. Under the dynastic system an emperor ruled for life. Upon the emperor's death, his heir became the new emperor. If the dynasty kept the Mandate of Heaven, its future was safe. If it lost this all-important favor, it could cease to exist altogether.

Natural disasters and disturbances of order were seen as portents of Heaven's displeasure with the mistakes of rulers and as cause for rebellion. During the Han Dynasty (206 B.C.–220 A.D.) the emperor presided over special rituals at the imperial capital to intercede with Heaven for appropriate rain, bountiful harvests, and freedom from natural and human-made catastrophes. Even so, many dynasties, having become weak and corrupt, collapsed in coincidence with calamitous events. Floods were often precipitated by the emperor's failure to maintain water control works. Eventually, a just leader would come to power and bring order to the land and the **dynastic cycle** would begin again.

THE HAN DYNASTY

The gathering of territory and coping with barbarians are epitomized by the conquests of the Han Dynasty (206 B.C.–220 A.D.). At first, the Han made agreements with the barbarian Xiongnu, mollifying their leader with gifts, and later became powerful enough to wage war against them. Real expansion was made under Emperor Han Wu Di (Han Wu-ti), who ruled from 140 to 87 B.C. He established control over the regions through military colonies whose job was to clear and irrigate the land and defend the area.

Wu Di developed "forward policy" regarding the west, which was occupied by the Xiongnu. By 115 B.C., the Chinese had forced the Xiongnu north of the desert and had occupied the Tarim Basin. The Great Wall was extended and fortified with outposts.

Wu Di was particularly interested in the relatively large horses of the "Western Regions" because they could be used to carry heavily armed soldiers against the Xiongnu and their smaller Mongolian ponies. In 101 B.C., Wu Di's forces conquered the Ferghana Valley and the "Western Regions" were declared a Chinese Protectorate.

Wu Di also brought northeastern and southern regions under his control. In Korea, a large military colony was established near present-day Pyongyang. The southern kingdom of Nan Yueh around Canton was also captured by 111 B.C. Kingdoms in the border areas of Sichuan and Tibet also accepted Chinese supremacy. Yunnan, the terminus of an important trade route to India, was subjugated as well. In the years 42–43 A.D. southward expeditions captured Hainan Island, the Tonkin region (North Vietnam), and Annam (central Vietnam).

Assisted or forced population movement was another policy of the Han. Mass movement of populations to border regions was both a means of control and of evening out the distribution of people. More than 2 million settlers trekked to northern areas under Wu Di's reign alone. Even more migrated to the fertile lands of the Yangzi.

The original Han capital was at Ch'ang-an (near present-day Xian) in the Wei River Valley. The capital was later moved to Loyang. By 140 A.D., Loyang had a population of more than a million.

Throughout the Han and later dynasties, borders waxed and waned with the relative strength of Chinese and barbarian. Furthermore, China itself was unified or fragmented depending on internal strengths or weaknesses. Overspent and overburdened, the Han dynasty collapsed in 220 A.D. The Han dynasty was followed by several others. One of the most important of these dynasties was the Tang.

THE TANG (T'ANG) DYNASTY (618–906)

Under the Tang, China became the richest and most powerful empire in Asia. Its frontiers extended even further than those of the Han (Figure 9–11) and its population numbered more than 50 million. Two million people lived in and around their capital at Ch'ang-an, then the largest city in the world! This was a period unparalleled in poetry, painting, pottery, printing, and other art forms. The Son of Heaven— the title of the emperor—was perceived as the lord of "all under heaven" (i.e., the known world). This was acknowledged by all others who offered praise and paid tribute, and imitated everything Chinese.

Ch'ang-an was likely the largest planned city ever built. Laid out in a checkerboard pattern, it covered some 30 square miles (78 sq km). Its wide avenues ran north-south and east-west to great gates at the cardinal compass points. The walled city was further divided into quarters and lesser administrative units. The people lived in rectangular wards, each with gates that were closed at night. Markets and entertainment districts served residents and visitors alike. Ch'ang-an was notably cosmopolitan, housing an international group of merchants, traders, artists, pilgrims, students, scholars, adventurers, and others. Official tributary emissaries came from Korea and Japan and regions of the south on a regular basis, returning to their homelands with all they could carry.

Despite its accomplishments, the dynasty failed to endure. Tang forces were defeated by a coalition of Turks and Arabs at the Talas River near Samarkand in 751. The Silk Route was thereby closed and trade with the west cut off. This battle is significant because certain captured Chinese transmitted the recently developed art of wood-block printing and an older method of papermaking to the West. By 1030, the Chinese had developed movable-type printing, at least 400 years before it appeared in Europe. In fact, China hosted a wide spectrum of inventions long before the West (Figure 9–12).

A great drought in the north precipitated a series of rebellions against the Tang, exacerbating the dissatisfaction already apparent throughout the empire. Rival generals vied for control but the Tang empire collapsed in disarray in 907.

Figure 9–11

Note the expansion of the Tang Empire into Central Asia. The empire, although larger than the Han, was still basically confined to the lowlands and river valleys of the east. Note how the Grand Canal links the region of the Huang He and the Yangzi River basins.

Foot Binding—
The Golden Lotus

Golden Lotus is a Chinese euphemism for the **foot binding** of women. The first reliable reference to this atrocity is from the Southern Song (1127–1278). Supposedly, it began in court circles and spread among the upper classes of society. The idea was to keep women's feet very tiny by folding the toes under the foot and tying the bent foot tightly with cloth. In this manner, the feet would be deformed, but small. This very painful process was started between the ages of five and ten. Needless to say, the victims were incapable of walking normally. Judging by poetic references, the bound foot was an erotic sight to some men. Moreover, foot binding created a gait that would eventually restructure the pelvic bones to enhance men's pleasure during sexual intercourse.

The Mongols of the Yuan Dynasty (1279–1368) did not allow their own women to have bound feet but encouraged it among Chinese women. Foot binding was at its height during the Ming Dynasty (1368–1644). The Manchu or Qing Dynasty (1644–1911) tried to ban the practice but it continued into the early twentieth century. Very few women with bound feet are alive today.

50 AMAZING CHINESE INVENTIONS / DISCOVERIES

INVENTION	IN CHINA (century)	IN THE WEST (century)
Row cultivation of crops	6th BC	16th AD
Iron Plow	6th BC	16th AD
Multi-tube Seed Drill	2nd BC	16th AD
Horse Collar	3rd BC	7th AD
Quantitative Cartography	2nd AD	15th AD
Mercator Map Projection	10th AD	16th AD
Cast Iron	4th BC	13th AD
Crank Handle	2nd BC	9th AD
Steel from Cast Iron	2nd BC	18th AD
Water Power	1st AD	13th AD
Suspension Bridge	1st AD	19th AD
Deep Drilling for gas	1st BC	20th AD
Belt Drive	1st BC	13th AD
Chain Drive	10th AD	18th AD
Underwater Salvage	11th AD	19th AD
Lacquer (Plastic)	13th BC	19th AD
Petroleum and Gas as Fuel	4th BC	19th AD
Paper	2nd BC	12th AD
Wheelbarrow	1st BC	12th AD
Fishing Reel	3rd AD	17th AD
Horse Stirrup	3rd AD	6th AD
Porcelain	3rd AD	20th AD
Biological Pest Control	3rd AD	19th AD
Mechanical Clock	8th AD	14th AD
Printing Movable Type	11th AD	15th AD
Playing Cards	9th AD	14th AD
Paper Money	9th AD	17th AD
Spinning Wheel	11th AD	13th AD
Circulation of Blood	6th BC	12th AD
Deficiency Diseases	3rd AD	19th AD
Thyroid Hormone	7th AD	19th AD
Immunology for Smallpox	10th AD	18th AD
Compass	4th BC	11th AD
First Law of Motion	4th BC	16th AD
Seismograph	2nd AD	16th AD
Modern Geology	2nd AD	17th AD
Kite	5th BC	15th AD
Relief Map	3rd BC	13th AD
Parachute	2nd BC	18th AD
Ship's Rudder	1st AD	12th AD
Multiple Mast Ships	2nd AD	14th AD
Watertight Compartmetnts	2nd AD	19th AD
Canal Pound-Lock	10th AD	14th AD
Rotor and Propeller	4th AD	19th AD
Paddle-wheel Boat	5th AD	15th AD
Chemical Warfare	4th BC	19th AD
Crossbow	4th BC	2nd BC
Gunpowder	9th AD	12th AD
Fireworks	10th AD	12th AD
Land Mines	13th AD	14th AD

Figure 9-12

How many of these inventions are you familiar with?

THE YUAN DYNASTY (1279–1368)

Genghis (Chinghis or Ghenghiz) **Khan** (1155–1227) was a military genius who amassed his nomadic hordes into the most deadly cavalry armies the world had ever seen. In 1211, he crossed the Gobi into northern China. Carrying sparse rations of grain and mare's milk, his horsemen could cover 100 miles a day. They could fire arrows at full gallop—all in all, a terrifying force. A Chinese army of at least 150,000, stationed behind high walls and moats and armed with catapults and flame-throwers, was effectively massacred by the mobile Mongols who swarmed into Beijing in 1215 to enjoy "a most glorious slaughter." Genghis Khan died in 1227.

Genghis Khan's three grandsons ruled the North China Plain, Mongolia, and the Far East. One of these grandsons, **Kublai Khan,** mastered the North China Plain and moved his headquarters to Beijing. He began his invasion of southern China in 1271. Sweeping into Hangzhou, he crushed the last Chinese resistance at Canton in 1279. Kublai Khan thus became the Mongol Emperor of China.

Korea, northern Vietnam, and the previously non-Chinese southwest were also conquered, driving forced migrations southward into Southeast Asia. Southern Vietnam, Siam (Thailand), Burma, and Tibet were then compelled to accept tributary status.

Kublai Khan did have one important defeat. In 1274, he launched a force to occupy Japan. His main army was prevented from landing on the southern island of Kyushu by an impending storm which forced the Mongols to beat a hasty retreat. They returned to China where they continued to subdue their enemies. In 1281, they launched another invasion of Japan. More than 100,000 Mongols landed on Kyushu, where they battled for 53 days. Then, a typhoon struck the island and destroyed most of the Mongol transport ships. The Japanese commemorated the failed Mongol invasion by naming this typhoon *kamikaze:* the divine wind.

The Mongols were now in control of the largest empire in the world (Figure 9–13). While they filled some top administrative posts with their own people and abolished the civil service exams, they left much unchanged in the traditional methods of Chinese government. They reconstructed defenses, extended the Grand Canal to link Beijing and Hangzhou, and turned battle-scarred Beijing into a beautiful city. The period 1280 to 1300 was marked by peace and prosperity: *Pax Mongolica.*

Merchants, traders, diplomats and myriad others, including Marco Polo, journeyed the Silk Route to the cities of the vast Mongol Empire. Exchange of goods and ideas between East and West proliferated. But while commerce and the arts flourished once more, hatred of these foreign overlords did not wane. Kublai Khan was followed by increasingly inept leaders and by 1300 revolts against Mongol rule were widespread. Famine, earthquakes, and floods took the lives of millions. Populations were further decimated by the Black Death that was ravaging Europe. One peasant-rebel leader welded the Chinese forces together and drove the Mongols back to the steppes and, in 1368, announced the formation of the Ming Dynasty.

THE MING DYNASTY (1368–1644)

What was Chinese culture like at this time? With the Sinification begun under the Han and solidified under subsequent dynasties such as the Tang, Chinese culture had reached great heights. The scholarly bureaucracy was in place, as was a landed gentry. Government was administered through the multiple bureaus of the civil service and the military. The countryside was managed largely by the gentry and county magistrates. Strict Confucianism was enforced at all levels of society. Most important was the family that, in Confucian form, controlled members' actions and settled disputes.

The Yung-lo Emperor commissioned 3,000 scholars to produce an encyclopedia of all knowledge, a task which took five years. This was followed by a great medical encyclopedia that listed more than 10,000 drugs and prescriptions mostly unknown to the West. This text also recorded the use of inoculation to prevent smallpox, a medical advance not discovered in Europe until the eighteenth century. There was also a handbook of industrial technology describing techniques in canal lock construction, weaving, the manufacture of porcelain, shipbuilding,

Figure 9–13

China under the Mongols. At this time China was linked with other parts of Asia ruled by various allied khans. Some scholars believe that Marco Polo never made his journeys. He failed to mention chopsticks in his diaries which seems strange given the detail of his other observations. Read more about Marco Polo and see what you think.

papermaking, and many other items and pursuits. China during the Ming Dynasty led the world in science and technology.

In Ming Beijing the walls surrounding the city were 40 feet (12 m) high and nearly 15 miles (24 km) in circumference. There were nine double gates with watchtowers. Inside the city was the Imperial City with its own walls, 5 miles (8 km) around. Inside the Imperial City was the moated Forbidden City— 2 miles around—containing the emperor's palace, throne halls, audience halls, and other buildings situated along a north-south axis. Buildings with gold-colored tile roofs were set on terraces of white marble. The entire complex was designed to awe and impress (Figure 9–14).

Rice was the dominant element in the diet, supplemented by wheat noodles (brought back to Venice by Marco Polo as spaghetti). Food was eaten with chopsticks, a method quickly adopted in Korea and Japan. Bean curd (dofu or tofu) was a major source of protein, as were fish and chicken. Oxen and water buffalo were used as draft animals but eaten only if they died naturally. Whatever was put on the rice or noodles was sliced finely and cooked quickly in a wok. The shortage of fuel (as a result of deforestation) fostered the need for stir frying.

Humans altered the landscape with irrigated and terraced rice paddies and fish ponds. New towns and villages emerged as craft centers and periodic markets. From Tang times, the house became a common center for socializing and negotiating business deals and marriage contracts. It was not necessary to travel far—people rarely ventured beyond the nearest village. The wheelbarrow and the flexible bamboo carrying pole were the major means of transporting goods on land. These methods are still used in China and much of Southeast Asia. Peasants adhered to a folk proverb: "Work when the sun rises, rest when

Figure 9–14

Hall of Supreme Harmony viewed from the Gate of Supreme Harmony. This is the largest and most important building in the Forbidden City. Leading to the Hall is a three-tiered "dragon pavement" over which the Emperor was carried. Dragon motifs cover the pavement. Photograph courtesy of B.A. Weightman.

the sun sets. The emperor is far away." In essence, while dynasties rose and fell, the fundamental order of Chinese society persisted.

The first Ming emperor made valiant attempts to reopen the Silk Route, sending repeated expeditions against the Mongol rulers. Further, he repaired the northern frontier and made Korea a vassal state. More important, however, were Ming sea voyages.

Cheng-ho, a Muslim eunuch, sailed south to Malacca (Melaka) in 1403 in order to establish diplomatic relations and expand China's tributary system. Tributary states were required to trade with China and to pay homage to the Chinese emperor. Two years later, Cheng-ho's junks reached Ceylon, East Africa, and the Arabian Sea. But it was Malacca that would become the main regional entrepôt. Through it, thousands of Chinese merchants spread out to Malaysia, Sumatra, and Java. Then, in 1509, six Portuguese ships dropped anchor at Malacca. Soon they were seeking the land they called "Cathay." In 1514, the Portuguese were in Canton's harbor.

Between 1400 and 1600, China's population rose from 70 to 140 million. This growth spurt was accompanied by an agricultural revolution which made rice 70 percent of food production. The 100-day maturing Champa rice covered valleys and

hillsides. Then, in the 1550s came the "American" crops: potatoes, maize, and peanuts. The Chinese planted formerly undeveloped hills with these crops and established new farming communities. This agricultural revolution was not one of mechanical devices, it was one of crops. Unfortunately, clearing hillsides caused so much erosion that the rivers silted up and flooded more than ever before.

Meanwhile, the Portuguese pressed their case for trade. In 1557, the Chinese allowed them to build a colony on the isthmus of Macao. Spanish colonists in the Philippines were also trying to break into the lucrative China trade. They even contemplated attacking China in 1586.

The Japanese had been harassing China's coastal trade for years. Things were so bad that all settlements 30 miles (48 km) inland were removed, and maritime trade was officially forbidden. This edict was largely ignored however and in 1592 an invasion force of 200,000 Japanese attacked Korea and occupied Pusan, Seoul, and Pyongyang. Since China held sway over Korea, the Chinese in 1593, crossed the frozen Yalu River and began years of fighting on the Korean peninsula. The extraordinarily talented Korean Admiral Yi Sun-sin, who had developed iron-covered ships called turtle ships, managed to sink 70 Japanese ships in a single engagement. As the Ming armies contrived to defeat the Japanese on land, they were suddenly faced with an invasion from the northeast. The Chinese defeated the Japanese in Korea in 1593 but the Manchu tribes stood on the horizon. They would ultimately take over all of China.

THE QING (CH'ING) DYNASTY (1644–1912)

Internal chaos, peasant rebellions, and Manchu invasions combined to destroy the Ming, China's last native dynasty. The Manchus ruled as the Qing (Ch'ing) Dynasty. Qing means "pure." In many ways the Manchus became more Chinese than the Chinese themselves. However, there were looming problems facing Manchu rulers: the population was growing at an unprecedented rate, and food was in short supply. Foreign powers with their gun boats were eroding Chinese sovereignty and missionaries were pursuing converts. Western merchants were bargaining for silk, tea, and art treasures. Meanwhile, peasant uprisings threatened the entire fabric of imperial government.

The Manchus were cousins of the Mongols, and their appearance, clothing, customs, and writing system were different from those of the Chinese. They insisted that all males adopt the Manchu queue (pigtail), otherwise they left Chinese culture alone. In fact, the Manchus tried to inject more honesty and enthusiasm into a government which had become corrupt and complacent.

The Russians had been penetrating China's northeast borders, and Russian frontiersmen clashed with Manchu border patrols from 1650 to 1689. Then the Manchu emperor negotiated the Treaty of Nerchinsk, the first of many attempts to stabilize the Chinese-Russian frontier. Emperor Ch'ien Lung settled colonists in Xinjiang and sent troops to Tibet to counter Gurkha incursions from Nepal. He also forced Annam and Burma to recognize Manchu authority.

Meanwhile, China's population was increasing, probably doubling between 1700 and 1800. Ch'ien Lung authorized huge public works to provide peasants employment in times of crisis. He encouraged farmers to grow more wheat and, wherever possible, to plant cash crops such as cotton, sugar, and indigo. Related industries were also established. His foresight provided a long period of peace. Unfortunately, this was not to last. European powers and overseas Chinese were pressing at the gates of commerce.

A New Breed of Invaders

THE CHINA TRADE

The Chinese intelligentsia did not regard trade as an honorable or worthy profession. Nevertheless, thousands of Chinese were involved in trade within and without China. Portuguese, British, Dutch, Spaniards, and Danes were all anxious to gain trading footholds on the mainland. By the time the first American ship sailed into Canton in 1784, the emperor had already decreed that all trade must be channeled through Canton and that factors or traders must confine themselves to *factories* (storehouses) leased from local landlords. Foreigners were thus crowded into small compounds the Chinese called *hongs*.

In 1793, a British trade delegation arrived at the court of Ch'ien Lung. The emperor accepted the gifts of clocks, guns, and telescopes but, in a letter to England's King George III, pointed out, "There is nothing we lack. . . . We have never set much store on strange or ingenious objects, nor do we need any more of your country's manufactures." Ordinary Chinese saw things differently, and thousands went abroad to participate in the business of trade. The favorite destination was the island of Luzon in the Philippines, which became the first major center of the China trade. Silks, tea, porcelain, and rhubarb were exchanged for high-grade silver mined in Spanish America.

The emperor would not deign to have contact with the "foreign devils." Instead, civil servants carried on negotiations and reported their activities via horse relay to the Son of Heaven. Misleading reports and mixed interpretations of information soon caused trouble between the traders and port officials. The accidental deaths of three Chinese led to the execution by strangulation of a British and an American sailor according to Chinese law.

Meanwhile, the British discovered that the most lucrative import of all was opium, which was grown in abundance in Bengal (then under British rule) and could be sold in China. When the emperor tried to stem its import because of rapidly increasing addiction among Chinese, demand increased and the illegal trade boomed. Several efforts were made to end this pernicious trade, but to no avail. At one point, the British were forced to seek refuge on their ships anchored in the Hong Kong harbor. There, they used the ships as trading headquarters for opium. When Chinese junks tried to intercept the British transport ships, British gunboats fired upon the junks, sinking four.

Britain responded to China's obstruction of the opium trade by attacking the Chinese coast and threatening Nanjing (Nanking). This first Opium War (1839–1842) culminated in the Treaty of Nanking in 1842, whereby Britain acquired the island of Hong Kong, a cash indemnity, and permission to trade at five *treaty ports:* Canton, Xiamen (Amoy), Fuzhou (Fuchow), Ningpo, and Shanghai. Shortly thereafter, Britain was given "most favored nation" status, which meant that it would enjoy the provisions of treaties signed with any other nations. In 1844, the United States signed a trade treaty. China's long period of isolation was over.

The Doctrine of Extraterritoriality

Extraterritoriality refers to consular jurisdiction. This means that a foreigner is subject to the laws of his own country. For example, a British sailor who committed a crime in China would be punished under British law. This practice was introduced to China by the Russians in the Treaty of Nerchinsk in 1689. Under the terms of that treaty the privilege was mutual. Extraterritoriality was one of the provisions of the Treaty of Nanking of 1842. In that case, however, it was not a reciprocal arrangement. Europeans were subject to European laws and so were the Chinese. This was one reason why trade treaties with the West were called the Unequal Treaties.

The Nationalist Chinese began to negotiate the removal of extraterritoriality in 1928. Unsuccessful, they abolished it by decree in 1932. New treaties, minus the Doctrine of Extraterritoriality, were signed with Britain and the United States in 1943.

THE MISSIONARIES

Another intrusion involved Christian missionaries. By the end of the eighteenth century, the Jesuits had amassed around 300,000 converts among the peasantry. Although they failed to impress scholars steeped in Confucianism, they did introduce the Chinese to new ways of looking at the world. Initially, British and American missionaries built their missions, schools, and hospitals within the confines of the treaty ports. They and other missionary groups later spread to many regions such as Sichuan and Manchuria. In Sichuan, those who accepted food from the missionaries were called "rice Christians."

The French Sisters of Mercy set up an orphanage in Tianjin (Tientsin) but unwisely offered to pay the Chinese for "orphans" brought under their care. In Chinese eyes, these foreign women were really kidnappers. When the Chinese gathered to protest, the French opened fire. This sparked the 1870 Tientsin Massacre of Christians. Western gunboats steamed into port ending any semblance of good relations between China and Europe.

Heavenly Kingdom of the Great Peace

At the time of the Opium Wars, a Christian schoolmaster named Hung Chiu-chuan started a revolutionary movement in southern China. Hung taught that land should belong to the people, crop surpluses should go to areas of shortage, and women should be equal to men.

Resistance to the ineffective and corrupt Manchus began in 1850. After announcing the coming of the Taiping Tien Kuan: Heavenly Kingdom of the Great Peace, Hung and his followers attacked along the Yangzi valley and occupied Nanjing. By 1860, the "God Worshippers" had brought a third of China under their control. This was the **Taiping Rebellion.**

It seemed that Hung and his forces would soon rule all of China. However, they underestimated the reaction of the Western powers. The British decided to give the emperor just enough guns and ammunition to defeat Hung. Nanjing surrendered in 1864 and Hung eventually committed suicide. The Taiping Rebellion, which cost China 20 million lives, was over. However, its real significance, in terms of modern history, is the fact that it demonstrated the potential power of the peasantry. Properly equipped and led, China's peasant masses could achieve anything.

RESPONSE TO THE WEST

After the Opium Wars, many Chinese emigrated to Australia and the United States. Thousands of Chinese *coolies* (laborers) were hired to work in mines and build railroads. Australians resented their capacity for hard labor over long hours. Restrictive laws were passed, race riots took place, and widespread anti-Chinese feelings persuaded the government to adopt a "Keep Australia White" policy. In the United States, *Chinatowns* emerged and race riots were frequent. America's first zoning laws were designed to keep the Chinese out of "white" areas. People spoke of the "Yellow Peril," and in 1882, strict immigration laws were passed to stem the Chinese influx. Nevertheless, Chinese communities and business ventures prospered in the U.S.

Within China a *Self-Strengthening Movement* was taking place. Certain senior scholars recommended that China should learn from the West in order to get rid of the foreigners, including the corrupt Manchus. They pointed to Japan, a small nation that

was busy industrializing like the West. Many young Chinese went to Japan, Britain, and the United States to study. However, there were few lasting results in terms of reform.

Work began on China's first railroad out of Shanghai in 1875. Peasants believed that the steel road was disturbing the *feng shui* so they smashed it to bits. Others, acting out of fear of modernization, destroyed textile machinery. The government did, however, manage to buy guns and warships for the beginnings of a navy.

But the Self-Strengthening Movement failed to communicate any sense of urgency to the common people. Even the reclusive Empress Dowager Ci Xi (Tzu-Hsi) seemed unaware that China was disintegrating—losing territory to the "foreign devils."

DOWNFALL OF THE MANCHU

Figure 9–15 shows China's territorial losses in the nineteenth century. Vietnam was lost to France in 1885 and Burma fell into British hands the following

Figure 9–15

The Qing (Manchu) Empire in the nineteenth century. This map shows the extent of the Qing Empire including areas that paid tribute to China. It also shows treaty ports and other areas forcibly ceded to foreign powers in the era of Imperialism.

year. Russia, desperate for an ice-free port in the Pacific, gained access to Manchuria. It also occupied the chaotic Xinjiang border regions. In fact, the Russians stayed in the Ili Valley for ten years. To get rid of them, the Chinese gave the Russians navigation rights on Manchurian river systems. The Russians then began road and railway construction in Manchuria.

Between 1875 and 1880, Japan forced its way into two more tributary states: Korea and the Ryukyu Islands. This and related events precipitated the Sino-Japanese war of 1894–1895. Chinese troops were decimated on land and at sea while the Empress Dowager spent money repairing the Summer Palace.

The war ended with the Treaty of Shimonoseki (1895), whereby China recognized Korea as an independent kingdom. China also ceded the island of Taiwan, the Pescadores, and the Liaodong (Liao-tung) peninsula to Japan. Later, the Russians pressed the Japanese to leave the peninsula because of their interest in the ice-free port of Dalian (Dairen or Port Arthur). The Japanese obliged, and with that move, the Chinese saw the Russians as their friends. A secret treaty was signed giving the Russians rights to build railroads across Manchuria as far as Vladivostok.

After the Treaty of Shimonoseki concessions, there was a torrent of requests for land, mining rights, and other trade deals. The Germans took territories in Shandong in 1897, Russia took Dalian the following year, and France claimed Guangxi as a French sphere of influence. Britain wanted control of the Yangzi and a region north of Hong Kong known as the New Territories.

Most Chinese wanted the foreigners to disappear and were prone to support a secret anti-foreigner society called The League of Harmonious Fists—the Boxers. The Boxers were also anti-Christian and believed themselves to be invincible to bullets. Efforts at reform by the Manchu government were dismissed by the Boxers.

The country was in a state of disruption and distress worsened by Huang He floods. In 1900, the Boxers entered the capital and stormed the foreign legations. After 55 days of seige an international relief force arrived. The Empress Dowager fled in a cart to Xian, leaving the Imperial Palace open to looters.

Once again, the foreign powers humiliated China with war reparations. Scores of new treaty ports fell firmly into the hands of foreign bankers, shipping and railway companies, merchants, and missionaries. The vast interior of China was now open to foreign commercialization and exploitation. Western gunboats patrolled China's rivers and Western soldiers guarded settlements. Then, the Empress Dowager returned to Beijing, took up her role as empress, and began to negotiate with the foreigners. Many Chinese now began to seriously consider the overthrow of the Manchu Dynasty.

The Empress Dowager entertained Western diplomats, promised to outlaw foot binding, permitted Manchus to marry Chinese, and abolished the state examinations that had supplied the Imperial Civil Service for 2,000 years. She sent military officers to train with the Japanese who had just won the Russo-Japanese War (1904–1905). But it was too late for the "Old Buddha," as she was known in the north. Ci Xi died in 1908 as did her nephew, her heir and the new emperor. His successor was the two-year-old Pu-yi, whose father became Regent. The Manchu Dynasty was near its end. A modern revolution was at hand.

Asia's First Nationalist: Dr. Sun Yat-sen

Sun Yat-sen was born of a peasant family in 1866. He attended a mission school in Hawaii and became a Christian. He then studied at Hong Kong Medical School and graduated as a doctor in 1892. In 1895, he joined a revolutionary group, but this ended with the arrest and execution of most of his compatriots. Sun spent the next 16 years in exile trying to gain support for the liberation of China. He founded the Nationalist Party (known as the Kuomintang) to promote democracy, nationalism, and equalization of land rights. His efforts made him known around the world.

Between 1908 and 1911, the Manchus encouraged European firms to build new railroads. This policy infuriated the people of Sichuan and a peasant revolt ensued on October 10, 1910, later referred to as the *Double Tenth*. Revolution spread to central and southern China. Sun Yat-sen returned from abroad and was quickly elected as provisional president. He proclaimed a Chinese Republic on January 1, 1912. But Sun Yat-sen stepped down after six weeks in favor of a powerful warlord whom the Manchu regent had named as successor.

In 1910, Japan annexed Korea and in 1914 seized Shandong and extended its influence in Manchuria. Rebellions broke out in practically all regions of China as warlords fought one another for territorial gain. Interminable military campaigns ruined the harvests, causing widespread starvation and suffering.

Europe was locked in the battles of World War I (1914–1918), and China hoped that by supporting Britain and France against Germany, it would gain their support against the Japanese. The terms of the 1919 Treaty of Versailles therefore came as a shock to China. Japan was allowed to keep its territorial gains in China and the Europeans kept all their treaty ports and privileges both along the coast and along the Yangzi.

Students led a nationalistic demonstration that sparked a nationwide strike. One of the strike leaders was a young teacher named **Mao Zedong** (Mao Tse-tung). Communism attracted many young students with its promises of equality, and the logic of Marx and Engels appealed to many Chinese Confucian intellectuals. A Chinese Communist Party (CCP) was created and held its First National Congress in Shanghai on July 1, 1921. Warlordism continued, however, as the Kuomintang (KMT) tried to revitalize itself with Russian assistance. Unfortunately, Sun Yat-sen, often referred to as the first great Asian nationalist, was unable to complete his mission. Suffering from cancer, he died on March 12, 1925. KMT leadership was assumed by **Chiang Kai-shek**.

The Road to Communism

Outwardly allies, Mao Zedong's CCP and Chiang Kai-shek's KMT proceeded to fight the warlords. Strikes, riots, and rebellions wracked the countryside,

yet foreign powers continued to pour money into urban industrialization. Disruption in the rural areas drove millions of refugees to these cities which became breeding grounds for secret societies, strike units, and rebel organizations. Meanwhile, the CCP was working toward a Communist revolution and had formed peasant organizations in the countryside to carry it forward. These would eventually rise up and annihilate the KMT.

Following a northward sweep by the combined KMT and CCP, which together conquered most of southern China by 1927, Chiang Kai-shek decided to rid China of the CCP. Launching a reign of terror in newly captured cities such as Shanghai, he routed out and executed as many Communists as possible. The Communists fled the cities and vowed to garner the loyalty of the peasants and regroup. Chiang set up his government in Nanjing and proclaimed it capital of the Nationalist Republic of China in 1928.

Chiang Kai-shek and his troops followed the Communist forces, determined to exterminate them. But in 1931, the Japanese invaded Manchuria, with profound consequences for all factions. The following year they made a savage attack on Shanghai. Because Chiang was unwilling to give up his campaign against the Communists to fight the Japanese, the Japanese were able to set up the puppet state of **Manchukuo** in Manchuria—China's most important industrial region.

The Nationalists continued their extermination campaign against Mao and his Red Army. In eastern China, the Communists, who were not faring well in battle, decided to move elsewhere to reorganize and reconsider their options. In October 1934, 85,000 made the now famous **Long March** (Figure 9–16). Pursued relentlessly by the Nationalists, they trudged 6,000 miles (9,660 km) through jungles, swamps, hostile tribal areas, across torrential rivers, and through 16,000-foot (4,800 m) mountains. About 7,000 people survived the march and set up their capital at Yenan in the loess hills of Shensi Province in October 1935.

The Long March became a legend, an epic story of the heroism of soldiers and loyal peasants who placed national interests before individual

Figure 9-16

The Long March, from October 1934 to October 1935, was one of the most arduous journeys ever made by people. Mao and his fellow Communists traveled more than 6,000 miles (9,660 km) across snow-covered mountain ranges with passes at 16,000 feet (4,800 m), into lands inhabited by wild tribal people, and through deserts and swamps. Only 7,000 of Mao's forces completed the route to the relative safety of the loess caves of Yenan.

Guerrilla Behavior and Tactics

Mao Zedong advocated strict rules of behavior for his Red Army: 1. Speak politely to the people and help them when you can; 2. Return doors and straw matting to the owners; 3. Pay for any damage you cause; 4. Pay a fair price for any goods you buy; 5. Be sanitary—establish latrines well away from houses; 6. Do not take liberties with the women folk; 7. Do not ill-treat prisoners; and 8. Do not damage the crops. Unlike the relatively undisciplined Nationalists, Communist soldiers were much better able to win the support of the peasants.

Mao's guerrilla tactics can be summarized as follows: "The enemy attacks, we retreat; The enemy camps, we harass; The enemy tires, we attack; and, The enemy retreats, we pursue." The Nationalists were unable to cope with this type of warfare.

concerns. The Long March exemplifies struggle for an ideal.

In 1937 the Nationalists and Communists decided to form a temporary united front against Japan. As the Japanese advanced, Chiang moved his capital to Chongqing (Chungking), which he proceeded to industrialize with dismantled factories from the east.

The Japanese pressed on with "Kill all, burn all" tactics. In the now-infamous *Rape of Nanjing,* they butchered an estimated 300,000 people. Nevertheless, Chiang Kai-shek continued to expend much of his energies rooting out Communists. After a failed attack against the Japanese, Mao and his troops fell back to Yenan. In 1940 the war in China was a stalemate. The British and French were occupied fighting Germany, and the Japanese pressed southward to French Indochina.

The Japanese attack on Pearl Harbor on December 7, 1941, changed the conduct of the war. China became allied with Britain and the United States. With U.S. aid, Chiang prosecuted the war against the Japanese with the aid of a motley group of warlords from Chongqing. Much U.S. support was wasted in battle losses and mass defections to the Communists ensued. However, Chiang's propaganda machine declared nothing but success to his Western supporters. Meanwhile, Mao was consolidating his position in Yenan. While masterminding guerrilla warfare against the Japanese, he spent most of his energy wooing the peasants of North China with Communist propaganda. With the Japanese war's end at hand in 1945, American efforts to bring Communists and Nationalists together failed. Their war continued.

Industrial Manchuria was regarded as an essential prize of war by both Mao and Chiang. In 1946, Chiang moved north. It was a fatal decision because at that point, the Americans—fed up with bogus

Figure 9–17

This picture represents both old and new China. The tile art shows a Silk Route caravan while the bicycles represent the major form of transportation in China today. I shot this photo in Dunhuang, a major gateway to the Silk Route.
Photograph courtesy of B.A. Weightman.

stories of military successes and not wanting to take sides in what was now a civil war—decided to cut off aid to the Nationalists. Meanwhile, inflation and constant threat of conscription ruined morale among civilians. Their despair spread to the troops, who defected to Mao's Red Army by the millions. Chiang Kai-shek, conceding defeat on the mainland, nevertheless transferred his "government" to Taiwan, where he remained until his death in 1975.

By September 1949, Mao was ready to declare a new Republic of China (Figure 9–17). Speaking to the Chinese people from the Gate of Heavenly Peace in Beijing, he said, "Our work shall be written down in the history of mankind, and it will clearly demonstrate the fact that the Chinese, who comprise one quarter of mankind, have from now on stood up . . . we announce the establishment of the People's Republic of China."

Main Point Summary

1. East Asia has been dominated by China for around 7,000 years. China has impacted Japan, Korea, and Taiwan, culturally, politically, and economically.
2. Each country of East Asia is at least two-thirds mountainous. Mountains have acted as barriers to diffusion but have also formed cultural sanctuaries. Mountain passes, river valleys, and ocean channels have encouraged the diffusion of people and ideas throughout the region.
3. China is commonly divided into two parts: the sparsely populated, rugged regions of China Frontier and the densely populated lowlands of China Proper. Water availability is critical to settlement in both regions.
4. The North China Plain and the Chang Jiang (Yangzi) lowlands are the most important regions of settlement, agriculture, and urbanization in China.
5. East Asia's monsoon is modified by the presence of both continental land mass and ocean expanse. As precipitation decreases northward, seasonal temperature extremes increase. In China, double-cropping is only possible south of the 20-inch isohyet approximated by the Qinling Shan and the Huai He.

6. Populations in East Asia are diverse. Although most people are Mongoloid, many groups are of non-Mongoloid origin. This fact is reflected in linguistic and other cultural differences.

7. Historically, China perceived itself as the *Middle Kingdom,* at the center of the world. The Chinese "People of the Han" were regarded as superior to non-Han peoples. These perceptions fostered the idea of the civilized Han versus the non-civilized or barbaric "other."

8. Ancient philosophies, such as Daoism, instilled the ideas of harmony between people and nature and the proper balance among all forces. These ideas are expressed in the concepts of *feng shui,* the art of placement. Confucian philosophy promoted government by men of virtue. It also stressed the maintenance of proper relationships between different levels of a hierarchical society, ranked from ruler to children and female spouses.

9. Relationships with "barbarians" along the northern frontier and elsewhere beyond the core of China have pervaded the history of the Chinese and other groups in East Asia. The Chinese came to regard tribute-paying regions as legitimate parts of their empire. Landscape evidence of the ever-changing nature of Chinese-barbarian relationships is provided by the Great Wall.

10. In China, time frames of rule were called *dynasties.* Rulers were said to rule by the *Mandate of Heaven.* When conditions were positive, the ruler had Heaven's mandate; when conditions were negative, the mandate was lost. Thus, each dynasty witnessed a cycle of rise and fall. At the same time, borders expanded and shrank, depending on the strength of the dynastic rule.

11. Barbarians ruled all of China on two occasions: the Mongols (Yuan Dynasty) 1279–1368 and the Manchus (Qing Dynasty) 1644–1911.

12. Buddhism entered China from India in the first century A.D. and diffused along the Silk Route from western Xinjiang to Xian.

13. Under the Tang (618–907), China became the richest and most powerful empire in Asia and Ch'ang-an became the world's largest city. But the Tang forces were defeated by the Turks and Arabs in the far west, thereby closing the Silk Route and redirecting China's attentions southward.

14. The Mongols created the largest empire in the world. Merchants and traders traversed the Silk Route to exchange goods and ideas between Europe and China. Kublai Khan attempted, unsuccessfully, to invade Japan. The glories of Kublai Khan's *Pax Mongolica* did not last. Under succeeding incompetent rulers and natural disasters, the Yuan Dynasty collapsed to the Ming in 1368.

15. The Ming (1369–1644) led the world in science and technology. They repaired the Great Wall and redeveloped their capital at Beijing. Population growth and technological improvements transformed the natural landscape into an increasingly cultural one.

16. The Ming were known for sea exploration. Malacca (Melaka) became the main regional entrepôt, and Chinese merchants flocked to Southeast Asia. Between 1400 and 1600, many new crops were introduced and the country experienced an agricultural revolution.

17. In 1514, the Portuguese arrived at Canton. Contacts with the West increased from that point on. In the latter part of the sixteenth century, the Chinese helped the Koreans fight Japanese invaders. As Ming troops dealt blows to the Japanese on land, they were faced with both internal rebellion and Manchu invasions. At the same time, foreigners were carving out spheres of influence and Christian missionaries were pursuing converts.

18. By the eighteenth century, the Europeans had established several factories at Canton. Although the Chinese elite were not interested in trade, millions of Chinese were anxious to become part of this growing industry. The British forced China to import opium from India, and this ultimately led to war and Britain's acquisition of Hong Kong.

19. One of several attempts to oust the Europeans was the Taiping Rebellion, which resulted in the loss of 20 million Chinese lives. This rebellion demonstrated the potential power of the peasantry.

20. Particularly galling to the Chinese was the Doctrine of Extraterritoriality, which allowed foreigners to be subject to their own laws in Chinese territory.
21. Many Chinese left to study or emigrated abroad for work. Chinatowns emerged in Australia and North America, where Chinese often faced discriminatory behavior and policies.
22. In the latter part of the nineteenth century, the by-then corrupt Manchus were forced to yield increasing territorial privileges to both the Europeans and the Japanese, who were on their own imperialistic drive. The Boxer Rebellion in 1900 failed, and China was further humiliated as more treaty ports were opened to foreign capitalists, missionaries, and the like.
23. In the fray of a failing Manchu Dynasty, three individuals emerged who would shape the future of China and East Asia. An early leader of the Kuomintang, Sun Yat-sen became the first Asian nationalist. Another nationalist, General Chiang Kai-shek, inherited the Kuomintang Nationalist leadership after the death of Dr. Sun. A young revolutionary, schooled in the ways of communism, became the leader of the Chinese Communist Party. His name was Mao Zedong.
24. Internal war broke out among various war lords, the Nationalists, and the Communists. Meanwhile, Japan was forging its own empire in East Asia. Communists and Nationalists presented a united front against the Japanese, but upon Japan's defeat in 1945, returned to battling each other.
25. With strong peasant support and expertise in guerrilla warfare, Mao and his Communists were able to thwart the inept and corrupt Kuomintang forces. Chiang Kai-shek and his supporters sought refuge in Taiwan. Mao Zedong proclaimed the People's Republic of China in September 1949.

Bibliography

Ballas, Donald. 1968. "An Introduction to the Historical Geography of Han China." *The Professional Geographer,* 22: 155–62.

Blunden, Caroline, and Mark Elvin. 1983. *Cultural Atlas of China.* New York: Equinox.

Cotterell, Arthur. 1988. *China: A Cultural History.* New York: Penguin.

de Bary, William Theodore 1999. *Sources of Chinese Tradition,* Volume 1. New York: Columbia University Press.

Fung Yu-Lan. 1948. *A Short History of Chinese Philosophy.* Edited by Derk Bodde. New York: The Free Press.

Gernet, Jacques. 1983. *A History of Chinese Civilization,* 2nd edition. Trans. J. R. Foster and C. Hartman. New York: Cambridge.

Levathes, Louise. 1994. *When China Ruled the Seas.* New York: Oxford.

Levy, Howard. 1966. *Chinese Footbinding: A History of a Curious Erotic Custom.* New York: W. Rawls.

Lip, Evelyn. 1989. *Feng Shui: A Layman's Guide to Chinese Geomancy.* Union City, CA: Heian.

Murphey, Rhoads, 1996. *A History of Asia,* 2nd edition. New York: HarperCollins.

Nemeth, David. 1995. "Feng Shui." *Asian American Encyclopedia,* 2: 414–16.

Ronan, Colin, and Joseph Needham. 1981. *The Shorter Science and Civilization in China: 2.* (Abridged). New York: Cambridge.

Spence, Johnathan D. 1993. "Confucius." *Wilson Quarterly,* 17: 30–38.

———. 1999. "Paradise Lost." *Far Eastern Economic Review,* 15 (April): 40–45.

Temple, Robert. 1986. *The Genius of China: 3,000 Years of Science, Discovery, and Invention.* New York: Simon and Schuster.

Waldron, Arthur. 1991. *The Great Wall of China: From History to Myth.* Cambridge: University Press.

Wills, John, E. Jr. 1994. *Mountain of Fame: Portraits in Chinese History.* Princeton: University Press.

Xinzhong Yao. 2000. *An Introduction to Confucianism.* Cambridge, UK: Cambridge University Press.

Chapter 10

⊕

China:
Great Dragon Rising

WITH CINDY FAN

*"The comrades must be helped to
preserve the style of plain living
and hard struggle."*

MAO ZEDONG (1949)

"To get rich is glorious."

DENG XIAOPING (1980)

These contradictory quotes illustrate the political and economic transition that China has made in the latter half of the twentieth century and into the twenty-first century. A grim situation faced the new government in 1949. The economy was severely disrupted; 40 percent of the arable land was flooded; and industrial and food production were well below pre-war levels. This chapter will tell you about China's rollercoaster ride from the strains of rural-based communism to the relative prosperity experienced under the current policy of market socialism.

Agrarian and Industrial Revolution

In 1950, the government abolished the "land ownership of feudal exploitation" and redistributed land to landless peasants. Two years later, 300 million peasants owned a sixth of an acre apiece and thousands of landlords had been eliminated. The following year witnessed **collectivization**, which put an end to individual land ownership.

Collectivization involved several stages. Farmers first were organized into *mutual-aid teams,* pooling their tools and helping one another during planting and harvest seasons. The second stage involved *producers' cooperatives,* which pooled tools, labor, and land, although theoretically retaining individual land ownership. The third stage was the fully *socialized cooperative,* modeled after the Soviet collective farm in which all members collectively owned the land. By 1957, there were close to 800,000 cooperative farms, each averaging 160 families, or 600 to 700 persons.

Large industry was seen as the foundation of socialist society. Plans were made for the building

Figure 10–1

This is Hohhot, the "Green City." Hohhot has been the capital of Inner Mongolia since 1952. Part of Mao's decentralization and industrialization plans, the city now has more than a million residents. Hohhot is one of China's most polluted cities. These are workers' apartments. Chimneys attest to the presence of industry, mostly iron and steel and related products. Hohhot is dominated by ethnic Chinese. Why do you think this is the case? Photograph courtesy of B.A. Weightman.

of hundreds of industrial projects, many with Soviet assistance. Spectacular advances were made in both industrial and agricultural production, and by 1962, the national income had increased by fifty percent.

THE GREAT LEAP FORWARD

The **Great Leap Forward** was announced in 1958. Cooperatives were reorganized into **People's communes.** There were 26,000 of these, embracing 98 percent of the farm population. Each comprised about 30 cooperatives, totaling approximately 5,000 households or 25,000 people. Communes took on the responsibilities of villages, collecting taxes, operating schools, banks, and health clinics. The commune was seen as "the morning sun above the broad horizon of East Asia." However, unrealistic goals, lack of expertise, ill-founded programs, and general mismanagement caused agricultural production to actually decline.

Communization was expanded to include urban and industrial regions (Figure 10–1). Collective living, public eating halls, child care centers, and the like served workers who were organized into brigades, battalions, and platoons to produce every conceivable kind of output.

Everyone was to participate in industrialization. Thousands of backyard steel furnaces sprang up across the country. Production quotas were raised sky-high, and outputs burgeoned. However, quality was sacrificed for quantity—in 1959, the government admitted that almost a third of the steel made was unfit for industrial use.

THE SINO-SOVIET SPLIT

The problematic circumstances of the Great Leap Forward were compounded by the Sino-Soviet split. A formal Moscow-Beijing axis had been established in 1950, and thousands of Soviet scientists, technicians, and advisors were sent to the PRC. In 1957, the Soviets agreed to help China in its nuclear development. It sent them a heavy water reactor the following year. But by the end of the decade, strains were evident in the relationship.

The Sino-Soviet pact, once described as "lasting, unbreakable, and invincible," degenerated into a bitter ideological and territorial dispute. While Mao counted on the rural masses as the basis for communist transformation, Soviet Marxists scorned their potential, stressing the importance of the urban proletariat in organizing the revolution. Disagreement over the path of international communism meant that Mao was challenging Soviet hegemony over the movement.

This conflict led to the withdrawal of Soviet technicians—along with tools, supplies, and blueprints—in 1959. China, determined to go it alone, successfully completed the major construction projects. In 1964, China exploded its first atomic device in Xinjiang. However, all was not well. Bad weather, poor harvests, and the departure of the Soviets created severe economic dislocations.

Overall, the Great Leap Forward was a disaster for China. Normal market mechanisms were disrupted, distribution systems became dysfunctional, and products were shoddy and useless. Most serious was the fact that famine and disease swept the countryside, resulting in anywhere from 14 million to 26 million deaths between 1959 and 1961.

The Great Proletarian Cultural Revolution

The Great Leap Forward, communes, and the Sino-Soviet rift fomented dissension in the Chinese Communist Party (CCP). Some members questioned the wisdom of campaigns stressing "redness" over technological expertise. Others held that Soviet aid was essential for continued economic, military, and scientific development. In 1966, ideologies collided in the **Great Proletarian Cultural Revolution**—a giant social, political, and cultural upheaval.

Pure communization in agriculture was altered, with a policy of "Three Privates and One Guarantee." This allowed peasants to cultivate their own plots (about 5 percent of the arable land), operate private handicraft enterprises, and sell their products at rural free markets. The *One Guarantee* called for the fulfillment of agricultural quotas set by the government.

Worried about the lack of revolutionary zeal within the party apparatus, among young people, and within the educational system, Mao ordered a Socialist Education Movement in 1962. Suspected anti-Maoists were "sent downward" to the countryside to experience peasant living and working conditions. This was known as the *hsia-fang* (*xiafang* in Pinyin) (**downward transfer**) **movement**. Thousands of cadres and intellectuals were "rehabilitated" in this manner. In addition, millions of urban-based middle-school students were dispersed to "strike roots firmly" in the countryside. By May 1975, nearly 10 million youths had been resettled in rural areas.

Cultural revolutionary groups were established at all levels to promote Mao's mass line and class struggle which promoted the equality of all members of society. The youthful Red Guards saw themselves as true revolutionaries, dedicated to the obliteration of old thought, customs, and habits. Knowledge was deemed the source of reactionary and bourgeois thought and action. Eliminating their perceived enemies, the Red Guards rampaged cities, ransacked private property, and attacked anything remotely modern. Ultimately, Mao employed the army to restore order.

Although the Maoists proclaimed the Cultural Revolution a great victory, it actually ushered in a decade of turmoil and civil strife that drove the country to the edge of bankruptcy. The CCP was decimated through purges. Members were killed or imprisoned. Industrial and agricultural production suffered major setbacks, and disruptions in education reduced the availability of trained workers. It is estimated that nearly 100 million people were targeted during the Cultural Revolution. Tens of thousands were killed. Added to this was a series of natural disasters. The Huang He flooded seven times, and a major earthquake wiped out Tangshan and killed some 470,000 people.

Mao Zedong's death in 1976 set off a scramble for succession. Purges, including that of Mao's wife Jiang Qing and her associates, culminated in 1977 with the installment of **Deng Xiaoping** as party chairman. Deng, a colleague of Mao and a survivor of the Long March (Chapter 9), had been purged twice during the Cultural Revolution. His "rehabilitation" ushered in a new era for the PRC. But the China inherited by Deng exhibited widespread changes in light of agricultural and industrial policies pursued since 1949.

Altered Spatial Patterns

Spatial change was evident in both rural and urban landscapes. Communes, although reduced in size, had transformed the face of agriculture. Significantly, private plots were producing the bulk of vegetables and pork. Rice acreage had been dramatically increased, acknowledging rice's significance as China's most important crop of production

Figure 10–2

Threshing rice. Rice and other grain quotas are set by the government. Farmers work in groups to fulfill these quotas before spending time on their private plots. This threshing machine is operated by a foot pedal. The harvesting operation involves cutting the rice stalks; threshing to remove the grain; bundling the stalks for animal feed and other uses; sacking the rice kernels; and carrying the sacked rice to the road for truck pickup to a rice mill. Photograph courtesy of B.A. Weightman.

and consumption (Figure 10–2). But, as peasants devoted more energy to private plots, rice had begun to decline as a preferred crop in favor of more profitable cash crops (Figure 10–3). The most important cash crops are cotton and tobacco (for both of which China is the world's leading producer), as well as silk, tea, ramie, jute, hemp, sugarcane, and sugar beets.

Industrial expansion resulted in changing spatial patterns in the distribution of both industry and infrastructure. From 1949, the government sought to decentralize industry to the interior. At that time, egalitarianism in regional development was deemed more important than efficiency. In 1949, most industrial centers were concentrated in the northeast from southern Manchuria through Shanghai. Subsequently, other centers were developed in line with discoveries of new resources such as coal and oil. As a result, the share of industrial output increased in the interior provinces. China's single oil well at Yumen became only one of hundreds

Figure 10–3

In this privately owned market garden outside the city of Guilin, women are pouring "night soil" on their vegetables. Night soil is a mixture of human waste and water. Although the use of chemicals has increased dramatically, the use of night soil remains common in many places. Never eat unwashed or uncooked vegetables in China. Photograph courtesy of B.A. Weightman.

Figure 10–4

Note how surface communications are concentrated in China Proper. Why is China Frontier so devoid of road and rail transportation?

across the nation. Xinjiang became a cornerstone of natural resource exploitation (Figure 10–4).

Railroads were extended into the interior. No longer was the network largely concentrated in northern and northeastern China. New links were established between north and south and extensions were built to the northwest and southwest. Railroads, although built for strategic reasons, were critical facilitators of population redistribution in concert with the *hsia-fang* program and industrialization strategies. They were also important channels of rural-urban migration streams deriving

from the dislocations of the Great Leap Forward. Lanzhou, a city of 190,000 people in 1949, mushroomed to nearly 2 million by 1960. Urumqi, 80,000 in 1949, burgeoned to 200,000. Towns became cities, and new cities appeared on the landscape.

Although Chinese definitions of *city* have been inconsistent, the number of designated cities increased from 132 in 1949 to 188 in 1976. Within this framework of flexible definitions, the percentage of urban population changed from 5 percent in 1949 to 8 percent in 1976. Between 1953 and 1961, 39 new cities were established in context of industrialization

Figure 10–5

Compare this map with Figure 10–4. How many "over a million cities" does China have?

and decentralization. This trend toward urbanization continues—1990 saw 300 more Chinese cities than in 1953; by 2003, 666 cities (Figure 10–5). Fifteen percent of these were rootless, in that they had no historic foundation and were built at or near sites of rich natural resources. China is expected to have more than 1,000 cities by 2015.

The 2000 census indicated that China is 36 percent urbanized and although enormous urban growth has occurred in the past 50 years, the broad spatial pattern has not changed dramatically. Despite government efforts, more than 90 percent of the population continues to reside in the eastern half of the country.

Under Deng Xiaoping's platform of reform and economic liberalization, China has become an economic and political force to be reckoned with on a global scale. Whereas Mao Zedong decried foreign interference and stressed "self-reliance," Deng Xiaoping pursued more pragmatic, flexible policies opening the nation to the outside world. Deng said, "I don't care whether the cat is black or white as long as it catches mice." Foreign investment and even capitalism propelled China into the twenty-first century.

Emerging Modern China

ADMINISTRATIVE FRAMEWORK

Figure 10–6 shows that the PRC has four categories of administrative structure. There are 22 provinces; 5 autonomous regions (ARs); 4 central government controlled municipalities known as *zhixiashis* (shortened to *shi's*); and 2 special administrative regions (SARs)—Xianggang (Hong Kong) and Macau.

Autonomous regions were created to recognize the existence of non-Han peoples. There are 55 officially recognized minority groups in China (Figure 10–7). Although minorities constitute only a small percentage of China's overall population, they occupy large regions such as Xinjiang and Nei Mongol ARs. China's population is 94 percent Han Chinese. Minorities make up a mere 6 percent, yet they are spread over 60 percent of China's territory, mostly concentrated in the international border regions (Figure 10–8).

Autonomous regions have a legacy of population transfers in context of Sinicization policies. Han Chinese actually now outnumber minorities in some places such as Nei Mongol AR (Inner Mongolia).

Figure 10–6

Note how China's autonomous regions are on the periphery. Why do you think this is the case? Which cities are designated as shi? *From H. J. de Blij and P. O. Muller,* Concepts and Regions in Geography, *Second Edition, 2005, p. 309. Originally rendered in color. H. J. de Blij and P. O. Muller. Reprinted with permission of John Wiley & Sons, Inc.*

Figure 10–7

This Kazakh woman and daughter were spending the summer in the Tien Shan where they were tending animal herds. Their way of life is being altered as the Chinese government is promoting tourism in this region. I asked the daughter what she was going to do with her life. Her eyes filled with tears as she sadly proclaimed, "I don't know what will happen to me." Photograph courtesy of B.A. Weightman.

Only the larger minority groups such as the Uyghurs and Chuang have their own ARs. Numerous other non-Mandarin speakers such as the Hakka, the Min, and the Yue (Cantonese) live in the provinces, especially in southern China.

The four *shi's* are the capital, Beijing; its port, Tianjin; the largest city, Shanghai; and the Chang/Yangzi river port of Chongqing. When Chongqing was designated a *zhixiashi* in 1996, its municipal boundaries were extended to incorporate a huge hinterland totaling 30 million people! In reality, the central urban area has around 4 million inhabitants. However, because Chongqing *zhixiashi* is separate from the provincial administrative area, its creation actually reduced the population of Sichuan Province.

Figure 10–8

This man is clearly not Chinese. In fact, he is a Uyghur and speaks a Turkic language. Uyghurs are one of China's largest minority groups and dominate the Xinjiang Uyghur Autonomous Region. I took this photo at the Sunday market in Kashgar. Photograph courtesy of B.A. Weightman.

Xinjiang: New Frontier

Occupying one-sixth of China, the Xinjiang Uyghur Autonomous Region is the country's largest administrative area. Divided into a series of basins and ranges (Chapter 9), it is home to more than 40 different ethnic groups. The Uyghurs are a Muslim, Turkic-speaking people who constitute almost half of Xinjiang's population. Other groups include the Mongols, Hui (Chinese Muslims), Kazakhs, Tajiks, Tatars, Russians, and Xibo.

Due to Sinicization efforts, Han Chinese now make up one-third of its population. In 1949, the Han numbered 200,000. In 1993, there were 6 million Han in a population of 16 million. About 250,000 to 300,000 Han enter the region yearly. In 2000, Xinjiang had 19.25 million people 37.5 percent of which are Han. Sinicization is important to keep any notions of separatism at bay. Such movements on the part of the Uyghurs in the 1990s have been ruthlessly crushed.

In 1992, the capital of Urumqi was decreed a port and given tax incentives like other port cities such as Shanghai. This is interesting because Urumqi is one of the most landlocked cities in the world. However, joint-venture economic development makes sense in light of the fact that Xinjiang gained three new (and nearby) trading partners with the collapse of the Soviet Union: Kazakhstan, Kyrgyzstan, and Tajikistan. Joint markets and factories have emerged along these borders.

Because of its dry climate, Xinjiang relies almost entirely on irrigation for cultivation. Both spring and summer wheat are grown, along with corn, rice, cotton, millet, and *kaoliang* (sorghum). Xinjiang is also noted for its Hami melons, seedless Turfan grapes, and Ili apples. Sheep and horses are the principle livestock. While Han Chinese tend to dominate business, minority peoples dominate agriculture.

Unfortunately, the Chinese government has deemed nomadic ways of life politically impractical and economically inefficient. Consequently, it has made every effort to sedentarize groups such as the nomadic Kazakhs, many of whom have resisted (Figure 10–9). Provision of electricity is one of the tactics employed, as people are more likely to stay settled in an electrified, permanent dwelling. Only by settling down can the nomads "shake off poverty." However, since 1996, the government is focusing more on tourism to engage the Kazakhs in the regular economy. Increasing levels of cultivated land, encroachment by Han farmers, tourism, and other entrepreneurial activities all ensure the decrease in opportunity to pursue a nomadic lifestyle.

Similar to schemes of the eighteenth century, in the 1950s, soldiers organized the Production and Construction Corps. Continuing to the present day, this structure of state farms runs an empire of 2.2 million mostly Han Chinese with direct loyalties to Beijing. State farms are complete entities with their own factories, oil refineries, processing plants, and marketing networks. Many of the workers are under contract.

Xinjiang is rich in natural resources, especially in the Junggar and Tarim basins with their wealth of oil. The Junggar Basin is of particular significance because of its rail links to Kazakhstan and Central Asia's oil and gas deposits. Pipelines between the two regions are already under construction.

Xinjiang is a potentially strategic region especially in light of its new connections with the Central Asian Republics. Pan-Islamic movements threaten stability, and historic Russian interests in this region always weigh in the background. Xinjiang is also the heartland of China's space program and nuclear weapons development. This is one of many regions of China worth keeping an eye on.

Figure 10–9

The Kazakhs practice transhumance, dividing their time between the alpine slopes of the Tien Shan in summer and the lowlands near Urumqi in winter. Their summer homes are yurts made of wood, felt, and canvas. Yurts are divided into male and female sections with a common area in the middle. Photograph courtesy of B.A. Weightman.

AGRICULTURAL CHINA

Overall crop patterns have remained essentially the same (Figure 10–10) but within this larger framework, landscapes are being notably transformed. Now, food needs are not only aligned with an increasing population but also with heightened consumption, including more meat, eggs, and poultry.

Large-scale commodity production has become the essence of Chinese agriculture. Double-cropping (two crops a year) is practiced almost everywhere south of Beijing, where rice is the dominant crop. While two rice crops are garnered in the south, in the Yangzi valley rice is alternated with wheat. Wheat dominates the north. On the North China Plain, where soils do not retain water well, wheat is grown through the winter. It is replaced by cotton, maize, millet, tobacco, or sweet potatoes in the summer. In the north and northeast, it is too cold for winter wheat so maize and millet are single-cropped. Vegetables are largely concentrated in market gardens around settlements.

Crop yields are higher in the south than in the north but in both areas diminish westward. Nevertheless, some eastern provinces such as Shandong, Hebei, and Henan, have made spectacular gains under the new policies.

Chinese Pigs

China has the world's largest agricultural economy and one of the most varied. One of the most significant elements is the pig. In 1955, China had an inventory of 88 million. This number had increased to 331 million by 1985. Nearly 71 million pigs live in Sichuan Province alone, almost three times as many as in the United States hog belt of Illinois, Indiana, Iowa, and Missouri. Today, China has 126 of the 300 breeds of pigs in the world and 51 percent of the world's pig population. Of the world's total meat increase in recent years, China's pork accounts for two-thirds.

China has had a long-standing relationship with the pig's ancestor, the wild boar. Farmers hated the boar for destroying their vegetable crops but admired it for its strength, speed, and the perceived therapeutic value of its body parts. Once the boar was domesticated as the pig, it provided meat, bristles, fertilizer, and countless other by-products. It required no large space and ate anything. Its fast growth and reproductive ability made it a symbol of wealth and prosperity. Even today, you can

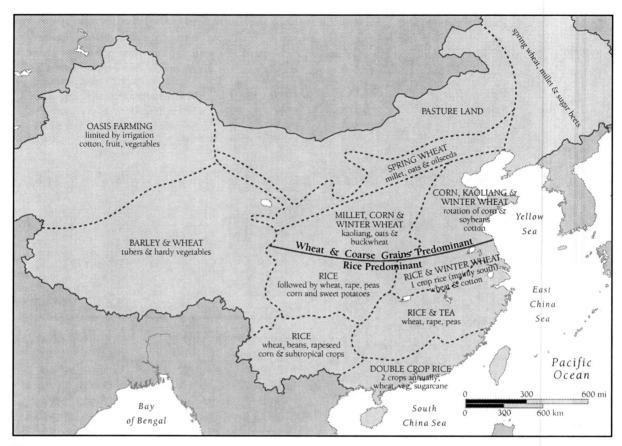

Figure 10–10

This agricultural map of China shows the division between north and south in terms of crops grown and the ability to double-crop. The north grows more wheat while the south grows more rice. The division coincides with the 20 inch isohyet and the Qinling Shan and the Huai He.

purchase gilded golden pigs to bring prosperity to your home or office.

Before 200 B.C., social stratification defined what one could eat. Pork could be eaten by the emperor and lower ministers. Others, such as feudal lords, state ministers, generals, and commoners were required to eat other meats. Pigs also played a role in life and death rituals, and pig effigies and bones were often included in burials. Pigs were also thought to be efficient guardians of children.

Although the most salient product in China's livestock sector, the family farm pig could decline in importance because of constraints in available feed grain. Furthermore, increasing numbers of commercial feed-lot operations are appearing in urban hinterlands.

Agriculture is rooted in the **household responsibility system**, representing a return to family farming. Under these 1978 reforms, families are able to lease land for up to 30 years. This land may also be inherited. In return, farmers must fulfill state quotas in grain or industrial crops. Surpluses can be used or sold to the state or free markets for even higher prices. Such incentives have increased crop yields and rendered labor more efficient and productive. Many farmers have become relatively rich and are constructing new and larger homes (Figure 10–11). Countless farmers have gone into rural business enterprises such as pig farming, fisheries, or small factories.

By 1991, the production of major crops had increased dramatically. Most increases came about through the use of fertilizers on increased grain acreage. However, overall cropland decreased in area. Generally, the responsibility system has boosted agricultural production and brought prosperity to many rural areas.

There are several drawbacks to the system. Peasants tend to underproduce most cereal grains and cotton in favor of more specialized and higher value crops such as fruit and vegetables. They do this in part because government price controls are less stringent on these crops, thereby allowing for greater profits. Moreover, the number of free markets has increased dramatically providing more opportunities for peasants to sell their products. And, more affluent consumers have raised their demand for specialty crops as components of their increasingly varied diet. Another drawback derives from the fact that agriculture has become more efficient calling for less labor. By 2001, the surplus of rural labor was estimated at 200 million!

A significant amount of agricultural land has been lost to urbanization and environmental degradation. These factors plus the fact that many farmers are not growing grain have resulted in a substantial decline in grain production. Current production figures place the output at 1990 levels. China is now a net importer of food.

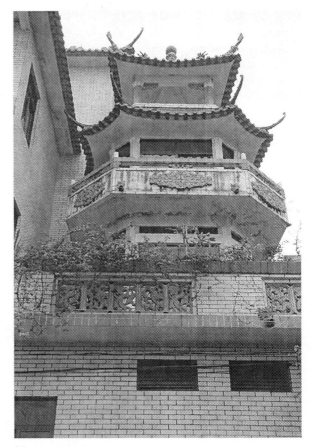

Figure 10–11

In the era of market socialism, numerous people have accumulated new wealth and have put it on display. This is the home of a farmer-businessman on the fringes of Guangzhou. It shows a lot of traditional symbolism, things that would have been forbidden a decade ago. The dragon stands for power and change; the phoenix for virtue and grace; the clouds for fortune and happiness; and the pearl for purity. Photograph courtesy of B.A. Weightman.

THE *HUKOU* SYSTEM

Socioeconomic position and opportunities for mobility are controlled by a household registration system known as *hukou*. Every Chinese citizen is registered as nonagricultural (mostly urban) or agricultural (mostly rural). *Nonagricultural* persons are heavily subsidized by the state in grain rations, housing, medical care, and so forth. Registration includes one's particular geographic location, and it is very difficult to alter one's classification. *Hukou* denotes one's identity and place in society.

Until the 1980s, the system rooted people to their birthplace and allowed little movement elsewhere. A migrant was excluded from most desirable jobs and subsidized housing and benefits. It is extremely difficult to change one's *hukou*, especially from agricultural to nonagricultural and to thereby shift location from rural to urban. Nonagricultural *hukou* are considered superior to agricultural *hukou*.

Since market reforms have produced a sizable labor surplus, the state has allowed temporary migration to towns and cities, although permanent change of status remains strictly controlled.

Geographers Cindy Fan and Youqin Huang (1998) have discovered a trend of marriage migration. Many women in poorer areas employ kinship and friend networks as well as marriage brokers to marry someone located in a more desirable area. Women gain access to better economic opportunities, and socially or otherwise disadvantaged men are able to get brides. Marriage migration streams have been demonstrated from poorer southwestern provinces such as Yunnan, Gweichou, Sichuan, and Guangxi to the more affluent coastal provinces of Guangdong, Hebei, Jiangsu, Zhejiang, Fujien, and Anhui. Brides then take on the *hukou* of their spouse and are able to participate in the local economy.

TOWNSHIP AND VILLAGE ENTERPRISES

Rural industry is springing up around cities and towns, and it is estimated that 60 to 100 million have left the land for better pay in small industries around towns and along transportation lines. Others work in industry on a part-time basis, as in Korea and Japan. Farm types, based on income sources, can be categorized as traditional agricultural, mixed activity, and industrially focused.

These trends illustrate the formal policy of "Leave the land but not the township or village." This results in a pattern of partially urbanized rural people who engage in non-farm sector employment without migrating to cities. In the 1990s, rural industry contributed more than 40 percent of rural total output value. Agriculture contributed 45 percent. Moreover, the growth in non-farm activities was significantly greater than farm ones. Some 23 million township and village enterprises employed 135 million people in 1996. These enterprises are important in absorbing surplus rural labor.

Millions migrate to urban areas in search of employment opportunities. Most of these moves are

planned as rural individuals hear about jobs from friends and relatives already in the city. Furthermore, there is a sizable **floating population** in urban places. These are people who no longer reside in their *hukou* location. These people find work in construction, factories, hotels, and other services. This population is now estimated at 120 million.

Free-flying industrial development without the building or maintenance of rural infrastructure has caused serious problems. Accessible lowland is becoming scarce in the east because of demands of state and city occupance, peasant house-construction, soil erosion, and consumption by local industry. Farmland in the east has been lost at the rate of, 16,556 acres (6,700 hectares) annually since 1979. Another problem is the return to fragmentation of land with family farming. Pollution, from increased use of chemical fertilizers and pesticides, as well as the accumulation of refuse, especially plastics, has become a serious problem in small towns and villages. New economic enterprises are also polluters. Increasing infringement on farm space is a critical issue with regard to the feeding of China's still burgeoning population.

Urban and Regional Development

Since 1978, China's development policy has stressed efficiency over equity and an open door instead of self-reliance. However, in order to defray the impact of radical change on the political system, Deng Xiaoping focused his new economic policies on the country's Pacific Rim. The plan incorporated a system of **special economic zones (SEZs)** and **economic development zones (EDZs)** designed to attract technologies and foreign investment from abroad.

SPECIAL ECONOMIC ZONES AND ECONOMIC DEVELOPMENT ZONES

SEZs and EDZs are founded on a variety of incentives such as relaxed import and export regulations and low taxes. Land can be leased and labor can be contracted. Goods produced can be sold abroad and even within China. Investors can reap profits at home.

There are many open economic zones and development zones in China but formally, there were only five SEZs: Shenzhen, Zhuhai, and Shantou in Guangdong province, Xiamen in Fujian province, and the entire province of Hainan Island. Initially, there were four SEZs along the coast of China (see Figure 10–12). Each of these possesses particular locational advantages. Shenzhen is adjacent to Hong Kong (Xianggang) on the Pearl River estuary in southern Guangdong Province. In the same region is Zhuhai, adjacent to the former Portuguese colony of Macau. Shantou and Xiamen are former treaty

Figure 10–12

Economic zones of China. From H. J. de Blij and P. O. Muller, Concepts and Regions in Geography, *Second Edition, 2005, p. 322. Originally rendered in color. © H. J. de Blij and P. O. Muller. Reprinted with permission of John Wiley & Sons, Inc.*

ports across from Taiwan. Hainan Island, approaching Southeast Asia, was added in 1988.

Fourteen additional cities were opened to foreign enterprise in 1984. Most of these were once treaty ports with a history of foreign relations. They also had established industry and transport infrastructures as well as local labor pools. Moreover, all the open cities had connections to overseas Chinese communities.

In 1988, three *open economic regions* were designated as special zones for foreign investment: the Yangzi (Chang Jiang) Delta Economic Region around Shanghai; the Pearl River (Zhujiang) Delta Economic Region around Guangzhou; and the Minnan Delta Economic Region around Xiamen. These represent efforts to diffuse the benefits of an open policy from SEZs to other parts of the country.

More freedom and opportunity for development and profit have generated unrest. University students and other dissidents decried the greed and corruption they saw among CCP officials and affiliates taking advantage of the new economic situation. In 1989, a huge protest demonstration took place in Beijing's Tiananmen Square and in several other cities. Deng Xiaoping called in the military and the demonstrations were ruthlessly crushed with a significant loss of life. Whether this incident is interpreted as a "cry for democracy" or a "counterrevolutionary rebellion," it is a black spot on China's modern history.

In 1990, the Chinese government opened the Pudong New Area in Shanghai as an SEZ, transforming what had been farmland and countryside into a financial hub. Pudong, a district of Shanghai, was created as an SEZ following the opening of other coastal areas to further expansion as open economic zones. Pudong has succeeded in attracting foreign capital and developing into China's financial capital. The special preferential policies granted to Pudong include reducing or eliminating customs duties and income tax, the right of foreign business people to open financial institutions, and the right for Shanghai to set up a stock exchange. Pudong's Gross Domestic Product

(GDP) for 2002 was 25.13 billion dollars (see Figure 10–13).

Because of these economic policies, China has become the world's fifth largest exporter, trading extensively with the European Union, Japan, and the United States. China, now consuming more steel, coal, meat, and grain than any other nation, has one of the ten largest economies in the world and is a prime force for economic growth across the globe.

FOREIGN INVESTMENT IN MODERN CHINA

Deng Xiaoping's economic policies opened China to the global economy and spurred sharp increases in tourism, imports and exports, and foreign investment. In 2002, there were 98 million tourists, 54 times the number in 1978. The leading countries of origin are Japan, Korea, Russia, and the United States. Imports and exports have risen dramatically since 1978. Since 1987, there has been a trade surplus because of China's growing exports, especially to the United States, Hong Kong, and Japan.

Exports make up 80 percent of China's foreign currency earnings, and there has been a shift in the structure of these as light industrial products have overtaken agricultural products. China's top exports are electrical machinery and equipment, power-generating equipment, apparel, and toys.

Initially, foreign investment was concentrated in the south but ultimately spread northward when the 14 cities were opened in 1984. A second spatial pattern emerged with the infusion of investment in interior cities, especially along the Yangzi and Huang He. Developments in Wuhan and Chongqing have given a new name to the Yangzi valley: China's Soaring Dragon. Most of the foreign capital has come from Asian sources, directed into places with Chinese family ties. Most Western investment pours into large cities such as Shanghai, Beijing, and Tianjin.

Figure 10–13

Pudong SEZ in Shanghai. Pudong's development began in 1990. Landmark buildings such as the Oriental Pearl TV Tower (center, tower continues for 1,535 feet) mark Pudong's new status as one of the most productive economic zones in China. Tibor Bognor/Corbis Images.

INDUSTRIAL REFORM

In 1949, industry accounted for a mere one percent of the workforce and was concentrated in the northeast and along the coast. The communist government made a valiant effort to build new industry and decentralize. All industry was state-owned. The Cold War of the 1960s and 1970s encouraged the development of inland provinces. The focus was on heavy industry particularly iron and steel. However, not all regions with iron and steel factories had sufficient iron ore. Plants were inefficient and, in many cases, managers were bureaucrats with little if any experience with industrial production. Economic

efficiency was sacrificed to the goals of regional self-sufficiency and equality.

The 1970s ushered in a period of change. State-owned enterprises (SOEs) were forced to answer to market forces and become capitalistic ventures. In an effort to reduce losses, the central government ordered a reduction of workers. As a result, tens of millions of factory employees were thrown out of work. When workers lose their jobs, they also lose their pensions and health care benefits. In 1998, there were approximately 20 million laid-off workers.

The government cannot afford to close all inefficient plants. Consequently, it supports them with monetary loans while they restructure. However, restructuring has frequently failed to make factories competitive. SOEs account for 80 percent of loans from government banks. This leaves little for the energetic private sector.

Shanghai: Head of the Dragon

The "Paris of the East." Asia's most sophisticated city. The world's most crime-ridden city. This was Shanghai before 1949. Today, Shanghai is China's largest city, its foremost economic center, and a rival to Beijing as the nation's cultural capital. In fact, Shanghai rivals Hong Kong in economic importance.

Shanghai thrived long before the arrival of Europeans. It had a population in excess of a million in the Ming Dynasty (1368–1644) and three million by 1816. From 1895 to 1945, the city's population increased tenfold. Migrants accounted for a significant portion of this growth because the city was safely distant from most battlefronts and had economic opportunities associated with the multiple foreign concessions. While migrants came from the surrounding provinces, they also came from Guangdong to pursue business enterprises. This connection remains significant in China's new economic landscape.

In 1949, Shanghai accounted for one-quarter of China's industrial production. The city housed a highly educated population and continued to be a migration target in Mao's era. However, when large numbers of people were forcibly moved out of Shanghai to populate China Frontier as part of the *xiafang* (*hsia-fang*) movement (from 1958–1976), the city experienced a net out-migration of more than 1.2 million. With its educated population, Shanghai had a tremendous impact on development in the northeast and western regions. With the end of the Cultural Revolution, the city once more experienced net in-migration.

Since 1992, when liberal policies were "given" to Shanghai, the city has undergone substantial reform and is in the midst of an economic boom. One advantage Shanghai has is that it is of sufficient size to become the "Dragon Head" of the Huangpu and Yangzi River commercial ribbon. It is already a key commercial center with proximity to both the north and south coasts. It has the largest pool of experts in science, technology, and management, as well as skilled workers.

These advantages led to the 1990 decisions to promote both Shanghai on the west bank of the Huangpu River and Pudong on the east side of the river. Pudong was to be developed and opened to foreign banks, finance, and insurance companies, department stores, and supermarkets. Since 1992, Shanghai's growth rate has exceeded that of all China!

While the Shanghai economy is mainly industrial, the service sector is becoming increasingly important. The tertiary sector is designed to serve the entire nation. Within the industrial sector, there are six pillars of industry: automobiles, iron and steel, petrochemicals, power equipment, telecommunications, and electrical equipment. The Jinqiao Export Processing Zone in Pudong has played an essential role in this development.

The high cost of land and labor in Shanghai has forced it into high-tech and high-value industry. Other kinds of industry are gravitating to the Nanjing and Hangzhou areas, where a new freeway is expected to enhance their competitiveness. A new Ring Road, six-lane elevated highways, and subway lines now facilitate spatial interaction in the web of industry and commerce. Pudong International Airport opened in 1999, and a maglev train connects the new airport with a downtown metro stop.

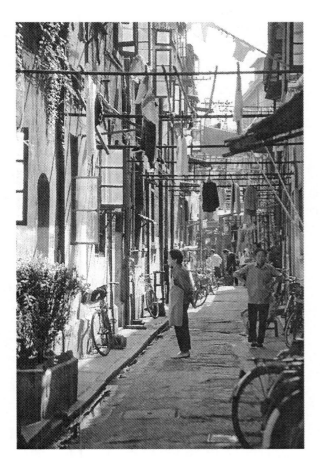

Figure 10–14

This is a traditional housing area in Shanghai known as a lilong. These apartments do not have modern facilities. Bathrooms are in a building down the street. Laundry is hung outside for lack of space inside. Photograph courtesy of B.A. Weightman.

Housing and services are of great concern. Shanghai has five times the density of Beijing with 12,000 persons per sq km (Figure 10–14). A new housing plan is based on the Singapore Model, whereby the residents own their own housing. An owner must occupy the dwelling for five years before renting or selling it. New houses and apartments are rising everywhere, especially in the former market gardening suburbs (Figure 10–15).

Shanghai is one of the greatest urban renewal stories in history. Nearly 17 billion dollars in foreign investment has transformed the once drab skyline to one of glittering skyscrapers and neon. But all is not perfect. There is a vast influx of "floating" and "temporary" populations. There is a growing income disparity between rich and poor. State-run industrial

Figure 10–15

These new apartments in Shanghai illustrate the impact of market socialism on the Chinese landscape. These blocks are guarded by gates and have a small kiosk for the purchase of basic essentials, the delivery of mail, and use of a public telephone. This building is located on former farmland. Photograph courtesy of B.A. Weightman.

enterprises are still money-losers generating massive layoffs. Government regulation still hinders essential quick decision-making. Therefore, market strategy shifts are slow and competitiveness sluggish. Some are looking to Hong Kong as a model for new policy.

Observers use Pudong as an analogy. *Hardware* in Pudong means infrastructure; *software* means services. Both must exist in tandem for Pudong to work. Change is in the offing and as Shanghai becomes more affordable for foreign investors, even greater progress is expected.

Globalizing China

A quarter century of open-door policy has paved the way for China to embrace globalization and become a powerful trading economy (Figure 10–16). China is now the sixth largest trading nation in the world, with total trade (the sum of imports and exports) greater than US $500 billion and nearly 15 times larger than in 1980. Moreover, China has surpassed Japan as the country with the largest trade surplus with the United States. More than 40 percent of China's exports go to the United States, which buys not only garments, electronics, and toys, but also a wide range of other products from fine furniture to swimming pool supplies.

China's exports are competitive primarily because its large population keeps labor costs low. The large population, likewise, makes China an attractive market. After years of debates and negotiations and after China agreed to further open its economy, the country was admitted to the World Trade Organization (WTO) in 2001. This symbolizes and formalizes China's prominent status in world trade.

Entering the WTO means that China's market is more accessible to the world. Multinational corporations searching for populous and unsaturated markets were among the first to advertise their products to the Chinese. At the same time, the Chinese economy became increasingly dependent upon the world market, and its industries must improve in order to survive. China's unemployment rate, officially listed at 4 percent, is thought to be much higher by Western analysts. This is one of the most significant social problems facing China today.

WORLD EVENTS

On December 3, 2002, the sky of Shanghai was sparkling with fireworks and the streets were filled with festive banners and Chinese national flags. The Shanghainese were celebrating their winning the bid to host the 2010 World Expo. Seventeen months before, billboards in Beijing publicized messages that welcomed the world to come to the city as it was selected to be the site of the 2008 Olympics. Although Beijing previously lost to Sydney in the bid for the 2000 Olympics, two years in a row in the new millennium China was successful in being selected to host a major world event.

Beijing and Shanghai are global cities attaining world-class status. Most importantly, their selections depict a resounding international recognition of China as a powerful nation, which is in no small part a product of its increasingly central role in the global economy. To many Chinese, their successful bids to host world events are evidence that once again "the

Figure 10–16

Teens in Maoming, a small city in southern China, smoke cigarettes and connect to the world in an Internet cafe.
Cindy Fan.

Middle Kingdom" is occupying the center stage of the world. Against the backdrop of one century of failed resistance against Western imperialism and an even longer period of domestic turmoil, the perception and image of global-scale successes are hotly sought after by the Chinese and contribute further to a heightened sense of nationalism.

NEW CITIES AND RURAL URBANIZATION

One of the most spectacular achievements in the 1980s and 1990s is the increase in the number of cities in China. Policy was relaxed to permit the formation of more cities, and city status became the mechanism to attract foreign investment. The total number of cities rose from 194 in 1978 to 467 in 1990, to 666 in 2003 with a concomitant increase in city population. A relatively large number of small- and medium-size cities were added to the system.

The *household responsibility system* effectively raised agricultural productivity, stimulating growth in rural commercial and manufacturing activities and the revival of market towns. Rural savings are reinvested in rural enterprises, fostering town growth. In some regions, such as the Pearl River Delta, small towns are growing faster than cities. However, it must be noted that rural urbanization is highly concentrated in the eastern coastal region.

NON-PERMANENT POPULATIONS

Temporary residents obtain permission to stay in a city for a fixed amount of time to work in factories or on construction sites. These individuals make up a quarter of some cities' populations. In Guangzhou they comprise 38 percent, and they outnumber permanent residents in Shenzhen. Most of the temporary people work in the building industry, followed by retailing, especially individual street enterprises (Figure 10–17). Temporary populations place great stress on social facilities, transport, and housing.

REGIONAL DISPARITIES

Geographer Cindy Fan (1995) notes that, "The paradox of a post-Mao regional policy is that a policy

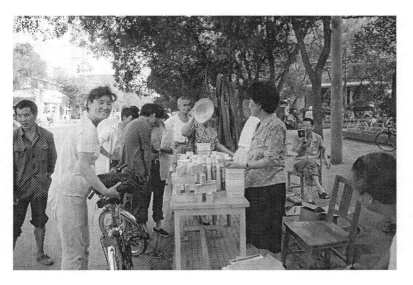

Figure 10–17

Selling soap and shampoo as a private enterprise in Shanghai—another example of market socialism on the landscape.
Photograph courtesy of B.A. Weightman.

designed to favor China's most developed eastern region has reduced inequality among all provinces in the nation." This happened because policies favored a number of regions not already highly developed and these, from Shandong to Guangdong, constitute China's growth corridor. Excluded from this corridor is the older, relatively stagnant region of the northeast.

China's Rust Belt

Home of the Manchus, founders of China's last dynasty; a region of historic contention among China, Japan, and Russia; a northeastern extension of the agriculturally productive North China Plain. This is former Manchuria, resource-rich and responsible for 25 percent of China's industrial output in the 1970s. Today, the northeastern provinces of Heilungjiang, Jilin, and Liaoning are trying to overcome their rust-belt status.

In the early years of communism, planners concentrated state-run industries around the coal and iron fields of Shenyang, Fushun, and Anshan. Shenyang became the nation's leading steel making complex. Discovery of the Daqing oil fields west of Harbin rendered China self-sufficient in oil. Directed on giant state farms, agriculture boomed and further illustrated what communist planning could accomplish.

With the new market reforms of the post-Mao era, state-run industries stood as monoliths of obsolescence, unable to cope with privatization and restructuring. Now the northeast contributes a mere 10 percent of China's industrial output.

Nevertheless, there is light on the horizon for former Manchuria. Its resource wealth and its proximity to Russian, Korean, and Japanese ports make it a potential core of development in East Asia. The launching of an economic development zone at Dalian, as part of the Liaoning Economic Development Area, has drawn the attention of foreign investors, especially the Japanese.

In 1994, one-third of Japan's investment in China was pumped into Dalian to manufacture everything from motors to chopsticks. Spatial association, economic incentives, a huge harbor, and a cheap labor pool have drawn more than 800 firms to Dalian.

Although much of the northeast still struggles with its rust belt heritage, some economic geographers see economic potential in light of future possibilities. A reunified Korea or a prosperous Russian Far East could alter northeastern fortunes.

Professor Fan (1995) further observes that neoclassical development theories, while influential in shaping policy, do not satisfactorily explain recent spatial trends in regional development because they downplay the influence of the state. State decisions have shaped development policies and resultant spatial patterns. Improvement of inter-provincial equality has occurred and regional imbalances between coastal and interior regions are being redressed in context of new policies encouraging interior investment. Furthermore, studies in Guangdong Province exemplify this significant spatial restructuring at the intra-provincial level.

GUANGDONG

Guangdong fits the definition of a **regional state**. It is a natural economic zone shaped by the global economy in which it participates. Guangdong Province—and especially the Pearl River Delta's agglomeration of Guangdong, Shenzen, and Zhuhai—has become a place apart from the rest of China. Guangdong's southern location has particular political-economic and socio-cultural implications. Proximity to Hong Kong and the incentive of low production costs have transformed the Zhujiang Delta into an outward processing zone. Since many Hong Kong residents once resided in Guangdong, business ties are facilitated. This explains the flows of labor-intensive industries and investment into Guangdong from Hong Kong. Economic structural change has seen a decline in agricultural participation in favor of industrial participation.

Shenzhen has become Guangdong's wealthiest area and the most important channel for foreign investment. It has established trade relations with 120 countries and has attracted 29,000 foreign investment projects. At least 250 of the world's top 500 corporations have invested here. Shenzhen was destined to align with Hong Kong in its transition from a British colony to a socialist state. It would be a showcase for exhibiting the concept of "one country, two systems."

The state has continued to expand open zones in Guangdong. More rural and peripheral counties are being drawn into the picture as investors continue to seek lower wages and subcontracting opportunities in context of established social relations and linkages. The province now has 75 economic function zones, many for the processing of high-tech equipment.

These development dynamics have generated new migration streams in Guangdong and elsewhere. Although migration was once strictly controlled by the government, recent economic reforms have created new push and pull forces fostering population movement. The household responsibility system with its boom in agricultural productivity has released a surplus labor supply. Foreign investment has instituted a plethora of labor-intensive industries looking to recruit workers. In 1984, a State Council directive allowed rural residents to move to market towns and even larger urban areas, provided they did not require the state-rationed grains to which the non-agricultural population was entitled. The eastern region, including Guangdong, is now the recipient of migration streams seeking economic opportunity in the industrial production system.

If this growth continues in Guangdong, the Pearl River region will have three cities with more than 5 million inhabitants, five with more than a million, and six with more than half a million. Per capita incomes in the province will be 10 times the national average. Guangdong now has the largest GDP of any province and has the highest industrial output. It is the largest recipient of Foreign Direct Investment (FDI), has the largest tourist income, and is the largest exporter in terms of value.

HONG KONG (XIANGGANG)

Here on 400 square miles (1,000 sq km) of fragmented, rugged land live 6.8 million people, many of them refugees from the mainland. During British colonial rule, Hong Kong was divided into three segments: Hong Kong Island with the capital of Victoria; Kowloon Peninsula and an array of other islands including the largest, Lantau; and the mainland section known as the New Territories. While Hong Kong Island and Kowloon were ceded in perpetuity to Britain in 1842 and 1860, the New Territories were leased by Britain in 1898 for 100 years.

The expiration of that lease in 1997 impelled Britain to negotiate the return of the entire colony to China. Then, Hong Kong, British Crown Colony, became the Xianggang Special Administrative Region, although the name Hong Kong remains in general use. The return of Hong Kong in 1997 (and Macau in 1999) was celebrated throughout China. On July 1, the handover date, Beijing's hand-picked Chief Executive of Hong Kong Special Administrative Region announced, "For the first time in history, we, the people of Hong Kong will shape our own destiny. We will finally be master of our own house."

Hong Kong is governed by a mini-constitution known as the Basic Law. The Basic Law gives Hong Kong a high degree of autonomy except in matters of foreign policy and defense, which are under the control of Beijing. Beijing is committed to preserving Hong Kong's capitalist system and lifestyle for 50 years. Capital is free flowing in and out of the region; Hong Kong is a member of the World Trade Organization, the World Bank, and a variety of other

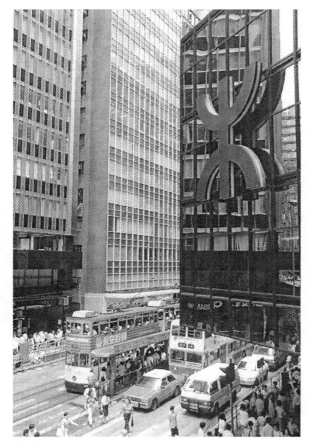

Figure 10-18

This scene is in the central business district (CBD) of Hong Kong. It shows the density of land use, orderly traffic, and a giant symbol for the city's spotless metro. Hong Kong is a very crowded city, but several forms of public transportation make getting around quite easy. Photograph courtesy of B.A. Weightman.

harbor, Hong Kong soon became a hub of economic activity in East Asia. It has become one of the world's largest container ports, a giant in banking and finance with its own stock exchange, a major processing and re-exporting center, and the world's tenth largest trading economy (Figure 10–18). However, China's economic policies in adjacent Guangdong Province have changed the structure of Hong Kong's economy.

When Shenzhen SEZ was established, thousands of factories moved there from Hong Kong to take advantage of cheap land and labor. Workers poured in, especially from Guangdong and Guangxi-Zhuang Autonomous Region. The former fishing and duck-farming community grew to nearly 4 million, and Shenzhen took on the appearance of high-rise Hong Kong with corporate offices, hotels, department stores, and gambling casinos.

Geographer Chi Kin Leung (1993) found that the subcontracting activities of Hong Kong were widely distributed in the Zhujiang Delta with Shenzhen and Guangzhou two of the major centers. Leung discovered that existing kinship and established business ties were even more important than proximity in forging these links. Another factor was the efforts of Chinese authorities to attract foreign investment into the Delta. An interdependency has arisen, with Hong Kong relying on low-cost labor to maintain its export competitiveness, and the Delta depending on Hong Kong's technological and marketing expertise to procure overseas markets.

In the 1990s, a quarter of China's foreign trade was funneled through Hong Kong. Today, more than 75 percent of the Hong Kong work force is in the service sector with only 13 percent remaining in manufacturing. More than 5 million Chinese laborers work in Hong Kong-owned factories on the mainland. However, costs have risen in Shenzhen, so many Hong Kong factories are moving elsewhere in Guangdong stimulating the growth of other cities and triggering more intra-regional migration.

An economic downturn in East and Southeast Asia in the late 1990s reduced this rapid regional development. Hong Kong's inflated real estate market collapsed, and service industries such as tourism

economic and trade structures. Many of the people of Hong Kong resent the fact that Beijing appoints the Chief Executive and other government members. They have demonstrated in significant numbers for true democracy with an elected government. The Beijing government continues to deny these rights to the people of Hong Kong.

As a free port, unencumbered by tariffs or other restrictions, with a magnificent, deep-water endured significant losses. For now, Hong Kong's fortunes are unstable.

"OPEN UP THE WEST"

Since the founding of the PRC, the Chinese government has been active in designing and implementing regional policy that has far-reaching impacts on the geography of development. In 1999, a new policy called "Open Up the West" (*xibu da kaifa*) was formally adopted. In general terms, "the West" refers to the western interior portion of China and comprises ten provincial-level units defined as the "western region" in the Seventh Five-Year Plan (1986–1990): Xinjiang, Tibet, Qinghai, Gansu, Ningxia, Shaanxi, Sichuan, Chongqing (which became a centrally-administered municipality and provincial-level unit in 1996), Yunnan, and Guizhou. More recently, two other provincial-level units—Inner Mongolia and Guangxi—were added to the grouping. Most of these provincial-level units share relatively low levels of development and adverse environments such as high relief and dry climate. Among them are the five autonomous regions and other areas that have high proportions of ethnic minorities.

Announcement of the campaign invited much attention from the public in part because of business potential in the West. In 2001, under the arrangement of the Hong Kong government, a business delegation of more than 200 toured the West and initiated a number of investment projects.

The campaign is also noteworthy because it signals an apparent departure from the reform era's regional policy that focused heavily on the eastern coastal region. Since the early 1990s, the Chinese government has been under severe criticism for allowing and even enabling regional inequalities to worsen. Observers warn of political and social instability if this trend continues. These criticisms echo those voiced against the persistent inequality between peasants and urbanites that is institutionalized by the *hukou* system and those highlighting the widening gap between the haves and have-nots. Moreover, unrest in the inland region, where ethnic minorities do not identify with the Communist

regime and do not share the culture and history of Han Chinese, can seriously threaten the integrity of the PRC. It is believed that development of the West can not only alleviate political and social tension but also accomplish the goal of Sinicization of minority regions.

Despite the fanfare and the government's large investment accompanying the campaign, its likelihood for success is questionable. In addition to environmental challenges, the West suffers from poor human resources, as its population has the lowest level of education in China. Moreover, migrants leaving the West have led to a loss of skilled labor. Though the resettlement of Han Chinese into minority regions can alleviate labor loss, it also induces conflict due to cultural and social difficulties of assimilation.

Population Issues

From the beginning, Mao Zedong believed that China's people were the country's greatest resource, and that to reduce population growth was a capitalist, anti-communist plot. Therefore, initial population efforts were to reduce mortality. A program of sending minimally trained "barefoot doctors" into the countryside, coupled with anti-sterilization and anti-abortion policies, forged this goal. Only in the 1970s, after disastrous famines associated with the Great Leap Forward, did consensus shift from the idealistic view of people as producers to the realistic understanding of people as consumers.

In the early 1970s, China's yearly natural increase rate was about 3 percent. Beginning in 1971, couples were encouraged (often by coercive means) to marry later, have fewer births, and space children farther apart. A more vigorous population control program was pursued after Mao's death in 1976. Families were ordered to have only one child, with severe economic and social consequences for failure to follow this rule. By the mid-1980s, China's natural increase rate was reduced to 1.2 percent. That rate has been further reduced to an officially claimed 0.6 percent. Nevertheless, China's population exceeds 1.3 billion—the largest

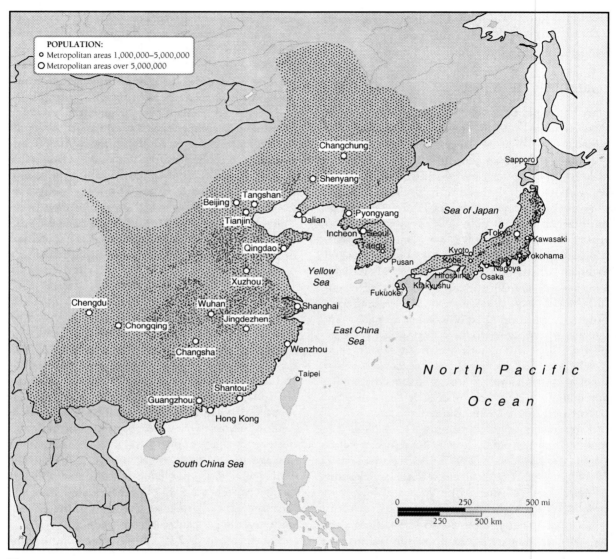

Figure 10–19

China's population distribution. Notice the difference between China Proper and China Frontier. What roles have history and current policies played in shaping this distribution?
ROWNTREE, LESTER; LEWIS, MARTIN; PRICE, MARIE; WYCKOFF, WILLIAM, GLOBALIZATION AND DIVERSITY: *GEOGRAPHY OF A CHANGING WORLD,* 1st edition, © 2005, pp. 289, 322. Reprinted by permission of Pearson Education, Inc., Upper Saddle River, NJ.

population in the world! Figure 10–19 shows population concentrations in relatively humid lowlands. Figure 10–20 charts the male/female population.

The one-child policy has been most successful in the cities, where it is rigorously enforced

(Figure 10–21). Policies vary in rural areas, depending on local needs. Minorities are permitted to have several children. Consequently, both rural populations and minority populations are growing faster than urban and Han populations.

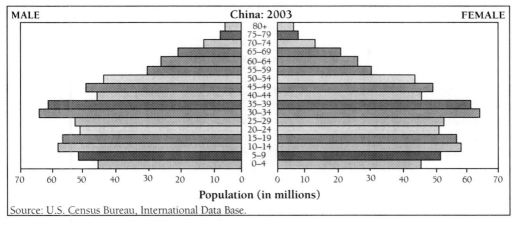

Figure 10–20

This population pyramid reflects the impact of major events. See if you can select the age cohorts for events such as the Great Leap Forward, the Cultural Revolution, and the one-child policy. How have these events affected population numbers and sex ratios?

The biggest census in the world was China's 2000 census, which documented a total of 1.27 billion people in November of that year. Census-taking in China is extremely labor intensive. Census workers visit every individual household to fill out or assist in the filling out of census forms. This is necessary because a large portion of the population, especially those living in rural areas, have low levels of education. In November of 2000, six million census workers went from door to door to complete the census.

Between 1990 and 2000, the average annual rate of population growth in China was 1.1 percent, the lowest since the founding of the PRC. No doubt,

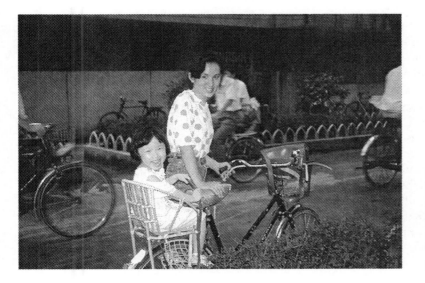

Figure 10–21

I met this woman and her daughter in Guangzhou. When I asked to take the photograph, the woman proudly pointed to her child and announced in English: "One child—girl." Photograph courtesy of B.A. Weightman.

the decline was in large part due to the one child policy. Currently, the average number of children per woman is 1.8, which is among the lowest in the world. Fertility decline more than offset the effect of a large childbearing cohort, namely those born during the 1960s and early 1970s when no birth control policy was enforced. Demographers predict that by the year 2020 China will have achieved zero population growth but will still have a population of 1.5 billion.

Despite the success of birth control, its ramifications are far-reaching and can be problematic. Fertility decline, in conjunction with increased life expectancy, accelerates the aging of the population. Given shrinking numbers of youngsters and poor social security, the Chinese are increasingly susceptible to lack of care in their old age. In addition, census results confirm the persistence of male preference among Chinese. In 1982, for every 100 girls born, 108.5 boys were born. In 2000, that ratio jumped to 100:116.86. Official surveys report that the sex ratios of the second and third born are considerably higher. Imbalance of the sexes has already had adverse effects on those in matrimonial age groups, as the number of men seeking wives significantly outnumbers women of similar ages—a phenomenon referred to as *marriage squeeze* by demographers. Thus, both the age structure and sex ratio of the population are expected to induce new social problems in China.

China is primarily an agrarian society but the 2000 census documents a steady increase in the level of urbanization to 36.09 percent. Two processes contributed to urbanization. First, a rural area, because of changes such as rural industrialization and intensified interaction with cities, may be redefined as an urban area, thus boosting the size of the urban population. Second, rural-urban migrants directly add to the size of the urban population. Both have happened in China and will continue to elevate the level of urbanization.

Population mobility in China is significantly lower than that in most Western economies. However, the 2000 census documents that mobility has increased significantly. Inter-provincial migration surged in volume from 11.53 million for the 1985–1990 period to 33.24 million in the 1995–2000 period. More than 60 percent of inter-provincial migrants in the latter period originated from inland provinces and moved to coastal provinces, thus exacerbating the unevenness in regional population distribution. Areas with rapid economic growth, such as Beijing, Shanghai, and Guangdong, are especially attractive destinations of migration. The 12 coastal provinces account for less than 14 percent of the country's land area and more than 42 percent of its population. Their relatively rapid economic growth and population gain through migration suggest that the development gap between coastal and inland China will continue to widen.

Women's Work

"Women hold up half the sky." So said Mao Zedong in the 1950s. Women achieved many gains under communism, especially during the Cultural Revolution (1966–1976) when people tried to be gender neutral and participation of women in political and economic activities reached nearly 100 percent. Now, as communism gives way to capitalism, women are falling back into old situations of inequality in line with traditional Confucian and Daoist thinking, which defined women as culturally inferior to men.

In China, didactic thinking is common and is reflected in the fact that experience is classified in paired concepts that are frequently polar opposites. Furthermore, contrasting concepts often have one that is more highly valued. Gender differences reflect this system.

One fundamental dichotomy is that women are associated with "inside" domains of work and men are associated with "outside" domains of work. Women do "light" work while men do "heavy" work. "Skilled" work is done by men, and "unskilled" work is done by women. The introduction of new technology has exacerbated these divisions. Men dominate in work involving labor-saving machinery while women are relegated to performing more menial, labor-intensive tasks. Today, women work in the fields while many men leave the village to work in a new factory enterprise. Agriculture has been reclassified as lesser "inside" work.

Factory work is more important "outside" work although millions of women do work in factories. Moreover, women are responsible for family members and interactions within the household but external representations of family business to the outside world falls to males.

China expert Tamara Jacka (1997) found that although male-female work patterns have changed with recent reforms, "expectations embodied in the traditional saying 'Men rule outside, women rule inside' . . . continued to exert an important influence on work patterns." It is true that modernization has given women opportunities to work outside the home. Still, a woman is entirely responsible for household affairs and so carries a double burden of responsibility. Women are still expected to be mothers and keepers of social stability in the home. Although the content of male and female work is changing, the conceptual divisions remain.

Extra assembly or piece-work done in the household courtyard is also included in the inferior "inside" domain. Commodity production in the household realm has become important, yet women are poorly paid for this work. When "courtyard economy" expands into a larger enterprise, management is typically seized by men.

The government mandates equal opportunity, but things are different in practice. At present, large numbers of women work in township and village enterprises yet they are poorly paid, do the most menial tasks, and are the first fired. Very few are involved in management over male workers. As many as 70 percent of laid-off workers are women, and women are forced to retire at age 50, five years earlier than men.

In the SEZs, female workers are desirable because they will work longer hours for lower pay. Many women are working in deplorable conditions, locked in their unsafe factories at night. Hundreds have died in fires, unable to escape the barricaded structures. Managers find that women, who make up nearly half of the floating population streaming into the coastal regions, especially Guangdong, are subservient and docile, and easy to replace.

Newfound affluence has also revived the notion of women as property. This has led to an upsurge in prostitution, concubinage, and the abduction of women into forced marriages.

Professor Jacka notes that gendered dichotomies have been bolstered over the years by state-sponsored shifts in social and political values. Mao's notion of "What men can do, women can do too," has been rejected in the wake of promoting gender divisions of labor as essential to the success of development and modernization. The critical role of women as mothers and promoters of spiritual civilization is also a central element in modern thinking.

Environmental Dragons

China has a spate of environmental problems, of which it is possible to touch on only a few here. In industrialized China Proper, the air is black with soot and smoke. Only about five of China's hundreds of cities are deemed to have clean air. People in the iron and steel city of Baotou have brittle bones and decayed teeth due to decades of hydrogen fluoride being pumped into the air. Up to 30 percent of acid rain in Japan is linked to sulphur dioxide from coal burning in China, where such emissions are expected to exceed those of the entire world by 2035.

Nearly half of China's major river systems are seriously polluted—in part because 80 percent of industrial waste is dumped untreated into them. Shanghai's drinking water is contaminated with oil, ammonia, and nitrogen among other chemicals. Huge quantities of silt and chemicals from intensive farming pour into the Yellow Sea. This silt is laced with heavy metals from industry. In 1989, the Huang He dumped 751 tons of such minerals as cadmium, zinc, arsenic, and chromium, along with 21,000 tons of oil into the Bohai.

In China's coastal waters, metals in fish have risen dramatically. Oysters have ten times acceptable chromium content. In many areas, fish have disappeared altogether. The more coastal development and migration occurs, the more marine habitat is lost or destroyed.

Increased urbanization has diverted water needed for irrigation to non-farm use. Aquifers are being depleted. Now more than 300 Chinese cities are short of water. For example, the annual drop in the North China Plain's water table has increased from an average of five feet (1.5 m) in 1993 to 9.8 feet (3 m) in 2003. Deep wells drilled around Beijing now have to reach below the surface more than half a mile (1,000 m) to get fresh water, adding dramatically to the cost of supply.

As aquifers are depleted, the ground subsides. Shanghai is sinking half an inch (1.5 cm) every year. Planners are considering putting limits on the heights of buildings because the weight of skyscrapers contributes to the sinking process.

Deforestation and desertification are also serious problems. Throughout history, China's forests have been decimated through human occupance and farming, the requirements for home construction, and the need for domestic fuel. Forests were also ravaged to make industrial charcoal and to manufacture ships. During the last three decades alone, a quarter of China's forests have vanished. Soil erosion and desertification are direct consequences of deforestation.

One of the most severely eroded regions is the Loess Plateau of north central China, which is the most extensive loess region on the face of the earth. Wind and water erosion have taken their toll over the centuries but human activities aggravate the situation. Grazing of sheep and goats on semiarid steppe lands removes the already sparse vegetative cover. Wind and water then erode and dissect the region, and thousands of tons of loess blow or flow away.

Unwise use of water for poorly planned or inadequate irrigation schemes also contributes to land degradation. This will probably get worse as most global warming models predict increasing aridity for countries in China's latitudes.

Efforts have been made to reforest areas, erect wind breaks, and lock the sand in place with layers of straw or clay basins. If successful in soil retention and irrigation, these areas can be reclaimed for such crops as apples and walnuts. Aerial seeding of grasses such as alfalfa (lucerne) has been employed to vegetate bare hills. But salinity is an ongoing hindrance as high potential evaporation rates cause salts to rise to the surface through capillary action. Then,

flushing out the salts and special drainage are required, making the whole process very costly.

Two weak links in China's economy are transportation and energy. Around 30 percent of the country's industrial capacity is idle at any one time due to lack of energy. Many factories operate on a three- or four-day workweek. When I visited the last steam-locomotive factory in the world in Datong, I was told how very lucky I was to find it open. Interestingly, luxury hotels in tourist and business centers are rarely short of electricity. Furthermore, China's energy supplies are not necessarily where they are needed, and transportation is woefully inadequate to link source and destination. Most of China's industry is concentrated in the coastal provinces but they possess only 10 percent of the energy stock.

China has the world's largest hydroelectric power potential yet it is developed to less than 10 percent. One massive project has come to world attention for its controversial impacts. This is the **Three Gorges Dam** on the Yangzi.

The Chang Jiang (Long River) is the longest river in Asia and third longest in the world (Figure 10–22). Known as the Yangzi to foreigners, the Chinese use that name only for the last 300 or 400 miles (480 or 645 km) of its 3,937-mile (6,300 km) course. Yichang, 1,000 miles (1,600 km) from the sea, is the head of navigation for river steamers. Ocean vessels may navigate the river to Hankou, about 600 miles (1,000 km) from the sea. Between Yichang and Chongqing, at an altitude of 130 feet (40 m), are the spectacular Three Gorges, noted for their geological magnificence and cultural significance.

The river and its tributaries drain 650,000 square miles (1,683,500 sq km) of territory. The entire region is subject to floods. From the Han to the Qing Dynasties (206 B.C. to 1911), devastating floods occurred once every 10 years. During the last 300 years, dams have been breached more than 60 times and flooding continues to be a problem all along the river. About one-third of China's population lives along its banks.

The Three Gorges Dam project is the world's largest water conservation project, including the world's most powerful hydroelectric dam. The normal water level of 574 feet (175 m) will be reached in 2009, with a reservoir covering 408 square miles (1,045 sq km). Its generating capacity will be 18.2 kilowatts

Figure 10–22

The Yangzi River and the Three Gorges Dam. Ocean-going vessels can now reach Wuhan. The new dam permits larger ships to reach Chongqing. What impact do you think this will have on Chongqing? How does the Three Gorges Dam impact this whole region?

with a yearly output capacity of 84.7 billion kilowatt hours. This is equivalent to burning 50 million tons of coal or 25 million tons of crude oil. Should water power actually replace mineral power, massive amounts of carbon dioxide, sulphur dioxide, and dust would be eliminated from the air. The hydroelectric power is intended for use in the energy-deficient eastern and central regions as well as in Sichuan Province.

Shipping is expected to improve markedly and affect Chongqing's status as a *zhixiashi*. In effect, the city would be tied to the Pacific Rim economy. However, its greater fortunes would appear to lie westward with the rapidly developing Sichuan Basin, a densely populated, resource rich, agricultural, and industrial region virtually isolated by topography from the rest of China. Chongqing might evolve as the most important growth pole in China's interior.

Between 1.2 and 1.9 million people will be forced to leave their homes, and some 620,000 acres (248,000 ha) of farmland will be lost. Approximately 8,000 cultural sites and some 1,000 industrial enterprises will be inundated as well contributing their pollution to the reservoir.

The Three Gorges Dam is extremely controversial. Forced removal of people is seen by many as a human rights issue. Although China claims that people are treated fairly with appropriate compensation and special privileges, many Chinese and others claim that resettlement is racked with corruption and unfairness. Compensation is inadequate, promised farmland is unproductive, and guaranteed jobs do not exist.

Scientists are also concerned. Some fear that the dam will ultimately collapse. In 1975, 62 dams in Henan Province gave way, killing anywhere from 86,000 to 230,000 individuals. Disease and famine affected over 10 million in the aftermath. A Three Gorges flood would be 40 times as great! Environmentalists claim that the Yangzi River ecosystem will be irreversibly upset. The Chinese alligator, the finless porpoise, the river dolphin, and the Chinese sturgeon will be severely endangered or even wiped out. Hydrologists postulate that sedimentation will cause sewage backups and floods upstream around Chongqing. Downstream, people worry about a lack of silt for agriculture.

Another aspect worth consideration is the fact that control of the dam by the northern power structure means control over water supply to the south. Water could be used as a weapon against any separatist movements in the south.

The debate over the potential benefits and detriments of the Three Gorges Dam project continues around the world. Both the United States and the World Bank have elected to withdraw their support of the project.

Severe Acute Respiratory Syndrome (SARS)

SARS is a highly contagious pneumonia-like disease that infected more than 8,400 people and killed more than 800 worldwide between November 2002 and July 2003. China was identified as the source of the disease and topped the list of SARS hotspots with more than 5,300 cases and approximately 350 deaths during the nine-month period. After the World Health Organization (WHO) issued travel advisories and made concerted efforts to halt the spread, the global outbreak was finally contained in July 2003, though health experts caution that it may be a seasonal disease and can return in cold weather.

It is believed that SARS originated in Guangdong in southern China. Travelers quickly spread the disease to Hong Kong, Taiwan, Vietnam, Singapore, and as far as Canada, South Africa, Australia, and Brazil. Asia was hit the hardest and Toronto—a city with a large Asian population and frequent travelers from and to Asia—had the most cases outside Asia. The scope and speed of the outbreak, unfortunately, is in large part explained by globalization. As China is increasingly connected to the rest of the world, its influence is no longer limited to global economic affairs but also includes health issues. Densely populated cities such as Hong Kong and Singapore are most vulnerable to the spread of infectious diseases. In Hong Kong, within just a matter of several days, more than 300 residents in one single apartment block were infected, probably via faulty sewage pipes.

The SARS outbreak has been both a challenge and an opportunity for the Chinese Communist Party (CCP). China was severely criticized for covering up the epidemic and underreporting its magnitude. International travel to China dropped sharply which adversely affected tourism. It became clear that the public health care system had major flaws.

But the CCP quickly admitted that mistakes had been made and vowed to implement measures to control the spread of the virus. SARS patients were isolated through mandatory quarantines, and schools and businesses were shut. The one-week May Day holiday was slashed to three days in order to discourage travel. The country's health minister and the mayor of Beijing were removed from their posts, which signaled that the Chinese government was serious about enforcing accountability and honest reporting. The CCP's quick response to criticisms suggests that we may well be observing the beginning of a more open society.

Xizang/Tibet

Tibet, according to geographer Antonia Hussey (1992), "is indeed a special place, at the top of the world, a landscape of light and shadow dancing over towering mountains, gravel plains and lush green valleys." A mysterious place, isolated by massive topographic barriers and profoundly influenced by Tibetan Buddhism, the region is centered on a previously forbidden city—Lhasa, the soul of the nation. But Professor Hussey declares that Tibet is no longer mysterious. "The distinctive landscape of this ancient theocratic state with its Buddhist traditions and institutions has been shattered and replaced with the utilitarian socialist landscape of China and renamed Xizang."

Back in the seventh century, Tibet was a great military power with borders extending far beyond its current ones. In fact, the Chinese paid a tribute of 100,000 rolls of silk to Tibet. By the tenth century Tibet had foresworn the arts of war and had withdrawn from its Chinese and Indian territories.

In 1207, Tibet fell under Mongol domination but maintained an amenable priest-patron relationship with the Mongol court, many of whom had adopted Tibetan Buddhism. During this period, the Dalai Lama system emerged. A specially chosen child would become the Dalai Lama, spiritual head of the Tibetan people. Great monasteries such as Drepung and Potala were built, eventually housing thousands of monks.

By the mid-seventeenth century, Mongol power weakened. In 1720, the Qing Dynasty sent a garrison of Chinese soldiers to Lhasa. The next year they proclaimed that Tibet had always been a vassal of China.

By the late nineteenth century, Tibet was caught up in an imperialist power struggle between the British and the Russians. British troops were placed in Tibet, and Tibet agreed to keep other powers out and trade with British India. When Chinese forces entered Tibet in 1910, the Tibetans asked the British for help and Britain refused. Meanwhile in 1911, corrupt China had fallen to the Nationalists. In 1912, the Dalai Lama proclaimed Tibet's independence.

The Tibet of the early twentieth century was dominated by powerful monasteries and feudal landlords. There was no secular education, and the masses lived in dire poverty. The Dalai Lama's attempts at reform were crushed by the traditionalists. The victory of the Communists in China in 1949 had little meaning to most ignorant and isolated Tibetans.

The story for China was different: in 1950, Radio Beijing announced, "The task of the People's Liberation Army for 1950 is to liberate Tibet." On October 7, 84,000 Chinese soldiers entered Tibet. India and Nepal expressed sympathy but did nothing, and Britain again refused to get involved. The 15-year-old Dalai Lama was forced to agree to return Tibet to the "Motherland." The resistance movement headed to the mountains to pursue guerrilla warfare.

In 1959, Tibetan resistance exploded and the Dalai Lama went into exile in India. Chinese retaliation caused anywhere from 87,000 to 430,000 deaths. Torture and humiliation were ubiquitous. Peasants and nomads, forced into communes, were made to grow wheat instead of their traditional barley. Tens of thousands starved to death in the famines following the Great Leap Forward. Tibet was declared a Xizang Autonomous Region of the People's Republic of China in 1965.

The Cultural Revolution, led by the Red Guards, decimated Tibet. Thousands were killed or sent to labor camps. Monasteries and religious artifacts were destroyed. By 1978, the number of monasteries had plummeted from 2,700 to 8. Holy texts were turned into toilet paper. Only Mao's death in 1976 ended this hideous onslaught.

In the 1980s, the Chinese apologized and proceeded to rebuild monasteries and restore the cultural landscape. However, about 130,000 Tibetans lived in exile and the Dalai Lama, headquartered in Dharamsala, India, began to seek understanding and aid from abroad. Support for Tibet's independence has grown worldwide. But, in the absence of the Dalai Lama, socialist transformation had begun in earnest.

China now sees Tibet as a storehouse of natural resources and is determined to Sinicize it. Coal, lithium, uranium, gold, chromium, and tin are only some of the minerals sought by the Chinese. The Qaidam Basin (once part of Tibet) is of special interest for oil.

Chinese settlers have streamed into Tibet, a region of barely more than 2 million inhabitants. Lhasa and other towns are becoming dominated by Han Chinese. Out of some 13,000 shopkeepers in Lhasa, for instance, only 300 are Tibetans. Tourism is being promoted, and huge hotels and related tourist attractions are being built. Much of traditional Lhasa has disappeared. The caretaker monks of the Potala Palace have been replaced by Chinese-trained "guides." The Drepang monastery on the outskirts of Lhasa, once home to 10,000 monks, now has around 700. Most of the monastery lies in ruins.

Crackdowns on religion continue. As late as 1997, the Chinese authorities announced that religion would have to bow to communism. "Development" has meant that half of Tibet's forests have been felled and more than a quarter of its mineral resources have been extracted since 1959. A fiber-optic cable has been laid between Lhasa and Shigatse, and plans are underway for a rail link with Sichuan.

It is clear that the Chinese are attempting to marginalize the Tibetans and make Tibet more livable for the Han. Tibet's new name is a metaphor for its relationship with China: Xizang—Our Western Storehouse.

Reform in the Twenty-First Century

Before **Hu Jintao** turned 60 in November 2002, he was named General Secretary of the CCP. This is a top post in China that also cemented Hu's presidency, formalized in March 2003. Hu was not well known to the Chinese even though he had been a member of the Politburo Standing Committee—the most powerful political entity in China—for ten years and had been vice president since 1999. Analysts believe that Hu's low profile was in fact his strength as it eased transfer of power, which in the past was almost always accompanied by bloody struggle. The stepping down of past President Jiang Zemin was orderly, peaceful, and backed by negotiations that allow him to continue to exert influence over the inner circle.

Together with Hu in his rise to power came eight other men, who form the expanded nine-person Politburo. Wu Bangguo succeeded Li Peng as Chairman of the Standing Committee of the National People's Congress; Wen Jiabao replaced Zhu Rongji as premier; and Zeng Qinghong took the post of vice president. The nine men constitute the so-called fourth generation of leaders, after those led by Mao Zedong, Deng Xiaoping, and Jiang Zemin. They are also by far the youngest leaders, ranging in age from 58 to 67 in 2002, replacing predecessors who were well into their 70s. All nine men were trained in engineering or related fields; four graduated from the prestigious Qinghua (Tsinghua) University—often referred to as China's MIT. What is also remarkable is that for the first time in Communist China's history no peasants or workers are represented in the Politburo, thus sending a clear message that the days of proletariat revolution are over.

The Politburo members' relative youth and technocratic backgrounds reflect the development path that the country is dedicated to taking, one that emphasizes economic development and reform over socialist ideology. Shared outlooks among the leaders suggest that they will continue to pursue economic liberalization. Indeed, the CCP has begun to appeal to private entrepreneurs, once considered the party's enemy, to join. In this light, unlike the eras under Mao and Deng, who is the head of China does not matter as much any more.

China's Conundrums and Contradictions

China is a land of conundrums and contradictions. It is the world's most populous nation. Yet, it has to feed more than 10 million new mouths every year. It also has to create more than this number of jobs

every year. It has risen from a poor to a middle-income country with a per capita income of more than US $1000. However, income disparities between rural and urban populations are reaching a dangerous level. Average per capita urban incomes are now triple rural incomes. Half of the country's bank deposits are owned by the richest five percent of the population.

The foundation of the revolution was the peasantry, yet the government predicts that millions of farmers will become redundant in the next 5 years. China is a rural country but it has more cities of over a million than the rest of the world combined. Thirty more are likely to appear in the next decade. China has many provinces but just one of these—Sichuan—would rank as the world's seventh largest country in population. Nine additional provinces rank among the globe's top 20 populous nations. Clearly, scale is a significant issue, making change far more difficult and complex than it might be otherwise.

China has a highly limited resource base. Constraints derive from the fact that the country must support 20 percent of humanity with less than one-fifteenth of the world's farmland. Now, China is a food importer. In 1949, China had a mere dribble of oil. Thirty years later it was self-sufficient in oil. Today, it is again a net importer of oil. Self-sufficiency is dead; China is inexorably tied to the world economy. Moreover, the shortage of arable land, water, rising population, and overbearing pollution push the notion of sustainable development into the frontline.

Even if China could slow its economic growth considerably, annual pollution increases would still equal that of the entire United States. China has seven of the ten most polluted cities in the world. Half the rain in Guangdong is acidic. China's water table is dropping dramatically, and its waters are fouled with industrial effluents. The fact that more and more Chinese are acquiring electricity, refrigerators, and cars guarantees rampant increases in carbon dioxide emissions.

China is changing rapidly within the constraints of a culture notable for its longevity. Autocratic leadership, saving face, distrust of merchants, the subservience of women, and nationalistic and xenophobic leanings are all as much a part of modern China as they were in the past. Change is sweeping the landscapes of China Proper while emerging as mere oases in China Frontier's sea of tradition. Politics are liberalizing, yet change is taking place within the existing system.

China's character is centered in continuity. So much has changed, yet so much remains the same. While capitalistic ventures are making some people very rich, this happens under the umbrella of socialism. It is the vision of socialism that is changing. As one Chinese economic planner stated: "Socialism used to mean government planning. Now it means common prosperity." While the past carries into the future, policies will continue to evolve. As the former Vice Mayor of Beijing said: "It's like a fried egg. Once you've tried it, it tastes good, and if you wanted to make it raw again, you can't."

Mongolia: In Range of the Dragon's Breath

Mongolia, a landlocked country of 2.5 million and one of the world's most sparsely populated countries, is three times the size of France or four times the size of California. Here in this remote and arid land, livestock outnumber people 12 to 1. Mongolia, lacking in infrastructure, modern industrial skills, and managerial knowledge, is struggling to develop its economy in light of its 1990 freedom from Soviet control. It was the first country in Asia to experience a bloodless transition from communism to democracy.

THE PHYSICAL LANDSCAPE

Mongolia's physical environment is characterized by extremes as well as fragility. Well-watered mountains in the west and northwest descend to steppe on the eastern plateau. Steppe lands graduate into the *Gobi*—semidesert and desert (*Gobi* means "waterless place" in Mongolian, see map on p. 213). Rivers are concentrated in northern Mongolia but these flow to the Arctic, the Pacific, or into landlocked basins. The *Gobi* has no perennial rivers, therefore underground wells are important. But, most of these freeze in winter. Soviet-installed pumps have fallen apart, and

spare parts are not available. Getting water is problematic in the arid regions. At other times, dry river beds are subject to flash floods.

During the Mesozoic Period, much of the *Gobi* was swamp and favored by dinosaurs. In the early 1920s, the world's first known dinosaur egg was discovered here. In 1993, a 75-million-year-old fossilized embryo of a meat-eating dinosaur was found in the western *Gobi*, a region rich in vertebrate fossils.

Soil erosion is common throughout Mongolia. In many areas, soils are only a few inches thick and overlie granite. Dessicating winds easily whip up the earth, thereby filling the air with dust and leaving the ground barren. Overgrazing contributes to the problem.

Mongolia has an extreme continental climate with sharp fluctuations in temperatures both seasonally and diurnally. Temperatures average 15°F (−9°C) in January and 64°F (18°C) in July. Winter is long, lasting from October to May. Ulaanbaator (Ulan Bator) is the coldest capital in the world, with winter temperatures sinking to as low as −50°F (−45°C). Precipitation is limited, varying between 10 and 20 inches (25-51 cm). The Chinese used to refer to Mongolia as *Han-hai:* "dry sea."

HISTORICAL BACKGROUND

Cultural Mongolia is much larger than indicated by current political boundaries. There are many Mongol tribes speaking various dialects, and several of these live outside Mongolia and China's Inner Mongolia. These various warring groups were unified under Genghis Khan in 1206.

Inner and Outer Mongolia are political designations dating from the Manchu Dynasty (1644–1911). In 1622, the tribes of South and East Mongolia allied with China. This is why China claims legitimacy over Nei Mongol. In 1689, the Treaty of Nerchinsk laid down the frontiers between Russia and China. At that time Mongolia was part of China. Two years later, the Mongols agreed to pay a yearly tribute to the emperor known as the "Nine Whites"—eight white horses and one white camel. Mongolia then became known as Outer Mongolia. These historic connections provide rationale for China's notions of hegemony over this region.

Outer Mongolia persisted until 1911 when it became an autonomous monarchy ruled by a religious leader, a Lama-King. But, it soon fell under the influence of Soviet communism and in 1921, was taken over by the Mongolian People's Party. With the help of the Russians, the People's Republic of Mongolia was proclaimed in 1924, and Mongolia became the world's second Communist nation. Mongolia remained a Soviet satellite until the collapse of the U.S.S.R. in 1990.

SOVIET IMPACTS

The Soviet's chief aim was to transform pastoral Mongolia into a source of raw materials for Siberian industrial efforts. Concerted attempts were also made to settle both rural dwellers and urban workers out of their traditional *gers* (round felt or canvas dwellings) and into multistoried buildings associated with industrialization and regional development. By 1981, more than half of Mongolia's population was classified as urban as compared to less than 22 percent in 1956.

A major contribution to this rapid urbanization was the expansion of Ulaanbaatar with multistoried suburbs encroaching on *ger* communities. Second, administrative centers of Mongolia's 18 provinces acted as magnets for regional growth. Third, the establishment of new towns mediated the dispersal of industry.

Darkhan is an example of a new town that burgeoned from a railway station of 2,000 people to an industrial center of 25,000 by 1970. Located close to the Soviet border, Darkhan's thermal power (based on local coal), as well as its brick works, meat-packing plants, and other industries easily served Soviet needs. Other centers were developed to exploit and process new resources such as molybdenum and copper. Some administrative centers added manufacturing functions.

Mongolia got its first motor road in 1937 from Ulaanbaatar to Altanbulag on the Soviet Frontier. The Ulan Bator Industrial Combine was also developed in the 1930s, producing carpets, wool garments, felt, and leather goods.

To further effect control, the Communists established state farms and required the permanent settlement of the nomadic herders. Needs were met and subsidized by the state. When the subsidies ended after 1990, the herders returned to their nomadic lifestyle, moving frequently across the steppe in summer and clustering in more sheltered valleys in winter.

Another Soviet goal was to transform Buddhist Mongolia into a secular state. Most of the country's 700 temples were destroyed in the 1930s, and more than 100,000 lamas dispersed or killed. Five hundred truckloads of religious art treasures disappeared into the Soviet Union. Since 1990, monasteries are being rebuilt and the country's Tibetan-style Buddhism is experiencing a revival.

MODERN MONGOLIA

Mongolia is undergoing its own version of Perestroika—*shinechiel,* meaning a "renewal." A free-market economy has replaced state-controlled enterprises. People are studying English as well as their traditional Uyghur-Turkic script, which had been replaced by the Russian Cyrillic alphabet. Trade, once as high as 90 percent with the Soviet Union, has dropped to 50 percent and has increased with China, South Korea, and Japan. Foreign aid is pouring into the country, with Japan as the largest aid donor.

Not all aid projects are successful. One crossbreeding program resulted in Mongol-European cattle that died in the winter cold. A prefabricated mini-mill for steel production was brought from Japan. It was to manufacture steel from scrap metal. However, no one realized that Mongolia, with few cars or machines, possessed virtually no metal. Japan has also introduced windmill, battery, and TV assemblies. People once isolated in their round, felt *gers* on the steppe now can choose from 52 channels, including the Cartoon Network and MTV. What impact do you think this will have on Mongolia's development?

Even with the exploitation of mineral resources such as coal, oil, molybdenum, copper, and uranium, animals and animal products account for half of industrial output and 90 percent of exports. Material infrastructure is either lacking or obsolete, and few Mongols are educated in modern business and industrial management.

Three-quarters of Mongols are livestock herders practicing transhumance and living either in their large, permanent winter *gers* or in their smaller, portable summer *gers*. Because animal pastures are sparsely covered with low-yield grasses, families must move around 30 times a year, covering distances of 125–185 miles (200–300 km).

One of the most important animal products is cashmere from cashmere goats. Until recently, China was buying over half of Mongolia's raw wool, to the detriment of Mongolia's processing industry. Consequently, the Mongolian government has banned the sale of raw wool. It is now smuggled into China.

Development, in context of transition from rigid control under communism to freedoms presented by democracy, has its downside. Mongolia is short of food, and there are thousands of destitute people, mainly women and children. In 1996, the World Bank estimated that only 15 percent entitled to government benefits were receiving any. In addition, literacy has fallen dramatically and school dropouts

have increased. Maternal mortality has doubled. Although Mongolia's medium- to long-term economic prospects are good, social reforms are sorely needed.

Relationships with China are difficult, as Mongols do not wish to be dominated by China, as is the case in Nei Mongol where millions of Chinese have settled. China is anxious to intensify relations however, as it needs lumber, minerals, and animal products as well as markets for its farm produce and low-end manufactures. Now that Mongolia is out from under Russian control, China is exerting its authority based on historic relationships. For example, China has loudly protested visits of the Dalai Lama to Mongolia and has discouraged connections with the Mongols of Nei Mongol. For decades, Mongolia has served as a buffer state between China and the Soviet Union with its strongest links to the U.S.S.R. Today, Mongolia is a buffer state between China and Russia with increasing ties—wanted and unwanted—to China.

Main Point Summary

1. In 1949, in order to prevent famine and chaos, the Communist government took drastic action to rectify a severely damaged environment and disrupted economy.

2. An agrarian and industrial revolution that included collectivization reached its zenith with the Great Leap Forward in 1958. The failures of the Great Leap were compounded by the Sino-Soviet split, and millions perished in subsequent famines.

3. Policy conflicts led to the Great Proletarian Cultural Revolution in 1966, a giant social, economic, and political upheaval. Millions were forced to move to frontier regions to encourage development, and millions more were "sent down to the countryside" to learn the ways of the peasants and become rehabilitated. Revolutionary groups such as the Red Guards ransacked the land and killed thousands.

4. After the death of Mao Zedong in 1976, a "rehabilitated" Deng Xiaoping took the helm and promoted the Four Modernizations stressing

agriculture, industry, education, science and defense in a context of market socialism. Deng moved away from Mao's "self-reliance" edict and promoted an "open-door" policy along with economic liberalization.

5. By the late 1970s, agriculture had improved and private plots were the most productive parts of the farmland. Industry had been diversified and decentralized to some degree. Railroads diffused westward and southward, further decentralizing population and economic activity into China Frontier.

6. Rural to urban migration streams and decentralization led to the growth of towns to cities and the formation of completely new urban places. New cities tended to be associated with natural resource extraction. Nevertheless, most of China's population continued to live in the eastern provinces.

7. There are four categories of administrative structure: provinces such as Sichuan, autonomous regions such as Xinjiang, municipalities known as

zhixiashis such as Tianjin, and special administrative regions such as Hong Kong.

8. There are 55 officially recognized minority groups in China. Most live in border regions such as Guangxi Zhuang A.R. or Xinjiang Uyghur A.R. Xinjiang is strategically significant because of its wealth of natural resources, its proximity to the new Central Asian states, and its historic relationship with Russia.

9. China's overall cropping patterns have not changed drastically but modernization is improving crop yields. There is now greater emphasis on the production of grain and meat. Land is used intensively, with double-cropping the rule wherever climate or water supply allows. Efforts are being made to open new crop lands in China Frontier.

10. Agriculture functions by the household responsibility system whereby farmland can be leased and even inherited. Families must fulfill quotas for grain or industrial crop production, but then can use the land as they wish. Many farmers are becoming quite wealthy under this system. However, greater efficiency has released millions of workers from the land who must seek employment elsewhere.

11. All Chinese citizens are registered in a particular location and classified as non-agricultural or agricultural. Movement of location and change of classification are very difficult. Since the 1980s, the government has relaxed this *hukou* system to some degree and there is a large rural floating population seeking work in cities. There are also marriage migrations of brides from poor locations seeking economic opportunities in richer areas through arranged marriages.

12. Many seek work in town and village enterprises. Others become part of a floating population, moving around to find work. Some become temporary residents of towns and cities while holding a permanent job.

13. Special economic zones (SEZs) were created in the 1980s to open up regions to foreign investment and technology. Open zones and cities have diffused into the interior, especially along river arteries. Many of these regions have strong connections to overseas Chinese communities. Most of the foreign capital has come from Asian sources.

14. Industrial reforms have been pursued since the 1970s. The Chinese government is trying to make state-owned enterprises more efficient, correct regional imbalances, and encourage private and foreign investment in industry.

15. Shanghai is China's largest and most sophisticated city, the economic head of the Yangzi Dragon. Shanghai has been both a source and destination for migrants, and has played an active role in facilitating development of the interior. Pudong development is intended to draw significant foreign investment and technology into tertiary activities. Shanghai is becoming strongly integrated with Nanjing, Hangzhou, and other centers linked by new transportation corridors. Agricultural hinterlands are giving way to housing and industrial developments.

16. A quarter-century open-door policy has paved the way for China to embrace globalization and become a powerful trading economy. China's impact on the global scene is reflected in the fact that it has been awarded the hosting of the 2008 Olympics as well as the World Expo in 2010.

17. The number of cities in China has more than tripled since 1978. Small- and medium-sized cities are experiencing rapid growth. Urbanization remains concentrated in the eastern coastal region.

18. Temporary residents make up significant proportions of city work forces. Most work in the building industry. Others operate as street entrepreneurs. All place strains on urban infrastructures.

19. Regional disparities exist in China. For example, the northeast is suffering from industrial obsolescence and interior provinces have not taken the same economic strides as the coastal provinces. Classical economic theories do not satisfactorily explain these spatial trends as they fail to account for the intrusive role of the state in China's economic development.

20. Guandong fits the definition of a regional state—pursuing its own lines of development

with global enterprises. Proximity to Hong Kong and long-standing business and family connections have facilitated economic enterprise. Shanzhen has become the province's wealthiest area.

21. Hong Kong, no longer a British colony, has strengthened its ties with the mainland. Hundreds of industries now locate in Guangdong and Guangxi to take advantage of tax incentives and cheap land and labor. Shenzhen, Guangzhou, and Zhujiang are becoming increasingly integrated in the Hong Kong S.A.R. and Guangdong regional state phenomenon.

22. China has ventured on an "Open Up the West" campaign. This focuses on the economic development of frontier regions via private and foreign investment. Since the West is home to many ethnic minorities, China encourages the in-migration of Han Chinese to further its ongoing process of Sinicization.

23. Major efforts are being made to slow China's population growth. Vigorous population control programs have been pursued in recent years. One child per family is held up as the ideal. However, this has been successful mostly among the better-educated urban populations. Rural people want more children, and their preference for male offspring has led to imbalanced sex ratios.

24. The 2000 census indicates that China's NIR has fallen to 0.6. Population growth is expected to decline to zero by 2020. China's population is aging. Given the population structure, there could be a shortage of labor and not enough young people to care for a burgeoning elderly population.

25. Women are not equal in China. Didactic thinking polarizes duties and responsibilities into "inside" and "outside" categories, with "outside" being superior. Even as more women work outside in agriculture and courtyard enterprises, these roles are reclassified as inferior "inside" work. Even though the nature of women's work is changing under economic restructuring, inequitable conceptual divisions remain.

26. China has a spate of environmental problems, exacerbated by continued development. Air, water, vegetation, and soil are all under severe stress with little relief in sight. Many factories operate only a few days a week for lack of power.

27. The Three Gorges Dam on the Chang Jiang (Yangzi) is intended to ameliorate the power shortages in China's southeast and improve navigation on the river to Chongqing. However, the environmental and human consequences of this world's largest water conservation project and most powerful dam have been be devastating. Certain organizations such as the World Bank have withdrawn their support for the project.

28. SARS is a highly infectious pneumonia-like disease that infected more than 8,400 people and killed more than 800 worldwide in 2002–2003. It originated in China and diffused via Hong Kong and Singapore to several other countries. Criticism of the Chinese government's lack of immediate response has led to greater openness in its reporting of events.

29. Xizang A.R. (Tibet) is experiencing drastic change as traditional landscapes are being replaced with Han Chinese, socialist ones. Even though Tibet was once a vast empire to which the Chinese paid tribute, China claims that Tibet is inexorably a part of China. China is determined to Sinicize the region and harvest its wealth of natural resources by force if necessary. Chinese invasions have subdued the Tibetans, and their culture is constrained and practiced only under the watchful eye of the Chinese military. The Dalai Lama and other Tibetans living in exile have brought world attention to the plight of Tibet.

30. In November, 2002, Hu Jintao was named General Secretary of the CCP. He and the other members of the Politburo are relatively young and well educated. This represents a change in direction from the past when high-level administrators were relatively old and drawn from the working class.

31. China is a land of conundrums and contradictions. While so much has changed in recent years, much has remained the same, particularly in the realm of cultural values. While China's landscape is being irreversibly altered, China's character is centered on continuity.

32. Mongolia is struggling to develop its economy in light of its recent freedom from Soviet control in 1990. A sparsely populated land, it was exploited by the Soviets for natural resources. Nomadic culture and Buddhist religion, then squashed, are being revived in a situation of newfound freedom. Three-quarters of Mongolians are livestock herders practicing transhumance. Mongolia remains a very poor country with social reforms sorely needed.

33. Proximity to Nei Mongol with its influx of Han migrants puts Mongolia in a potentially precarious position in relation to China. China wants access to its timber, minerals, and other resources but does not want to facilitate any pan-Mongol movement. Mongolia remains a buffer state between Russia and China.

Bibliography

Allen, Thomas B. 1996. "Xinjiang." *National Geographic* 189: 2–43.

Baird, Vanessa, ed. 1995. "Tibet A Cause For Courage." *New Internationalist,* No. 274.

Benson, Linda, and Ingvar Svanberg. 1998. *China's Last Nomads: The History and Culture of China's Kazaks.* Armonk: M. E. Sharpe.

Brown, Lester R. 1994. "Who Will Feed China?" *World Watch,* 7: 10–19.

Brunn, Ole, and Ole Odgaard, eds. 1996. *Mongolia in Transition: Old Patterns, New Challenges.* Richmond, Surrey: Curzon.

Burstein, Daniel, and Arne deKeijzer. 1998. *Big Dragon.* New York: Touchstone.

Cannon, Terry, ed. 2000. *China's Economic Growth: The Impact on Regions, Migration and Environment.* New York: St. Martin's Press.

Cannon, Terry, and Alan Jenkins, eds. 1990. *The Geography of Contemporary China: The Impact of Deng Xiaoping's Decade.* New York: Routledge.

Chang, Kuei-sheng. 1961. "The Changing Railroad Pattern in Mainland China." *The Geographical Review,* 51: 534–48.

Fan, C. Cindy. 1996. "Economic Opportunities and Internal Migration: A Case Study of Guangdong Province, China." *Professional Geographer,* 48: 28–45.

Fan, C. Cindy. 1995. "Of Belts and Ladders: State Policy and Uneven Regional Development in Post-Mao China." *Annals of the Association of American Geographers,* 85: 421–49.

Fan, C. Cindy, and Youqin Huang. 1998. "Waves of Rural Brides: Female Marriage Migration in China." *Annals of the Association of American Geographers,* 88: 227–51.

Farley, Maggie. 1995. "China's Women Pay Cost of Change." *Los Angeles Times* (August) 31: A4.

Fullen, Michael, and David Mitchell. 1991. "Taming the Shamo Dragon." *Geographical Magazine,* 63: 26–29.

Gately, David. 1997. "Growth in China's Grain Imports an Opportunity for Exporting Countries, Not a Threat to World Food Supply." *International Food Policy Research Institute News Release.* Washington, D.C.: IFPRI.

Gong, Hongmian. 1995. "Spatial Patterns of Foreign Investment in China's Cities, 1980–1989." *Urban Geography,* 16: 198–209.

Goodman, David, S. G. ed. 1989. *China's Regional Development.* New York: Routledge.

Hall, Derek. 1987. "Economic and Urban Development in Mongolia." *Geography,* 72: 73–76.

He, Bochuan. 1991. *China on the Edge: The Crisis of Ecology and Development.* San Francisco: China Books.

Hsieh, Chao-min and Max Lu. eds. 2004. *Changing China: A Geographical Appraisal.* Boulder: Westview.

Hsu, Immanuel, C. 1995. *The Rise of Modern China,* 5th edition. New York: Oxford.

Hsu, Mei-ling. 1994. "The Expansion of the Chinese Urban System, 1953–1990." *Urban Geography,* 15: 514–36.

Hussey, Antonia. 1992. "The Landscape of Lhasa, Tibet." *Focus,* 42: 8–12.

Jacka, Tamara. 1997. *Women's Work in Rural China: Change and Continuity in an Era of Reform.* New York: Cambridge.

Jacobs, J. Bruce. 1997. "Shanghai: An Alternative Centre?" In *China's Provinces in Reform,* ed. David S. G. Goodman, pp. 163–93. New York: Routledge.

Judd, Ellen R. 1996. "Men Are More Able: Rural Chinese Women's Conceptions of Gender and Agency." *Pacific Affairs,* 63: 40–61.

Kaye, Lincoln. 1992. "Faltering Steppes: China Looms Large in Ethnic Upsurge Among Mongols." *Far Eastern Economic Review*, 9 (April): 16–18.

Kuhn, Anthony. 2004. "The Death of Growth at Any Cost." *Far Eastern Economic Review* 1 (April): 28–30.

Lai, David Chuenyan. 1995. *Land of Genghis Khan: The Rise and Fall of Nation States in China's Northern Frontiers*, 30, Western Geographical Series. Victoria, British Columbia: University of Victoria.

Lattimore, Owen. 1955. *Nationalism and Revolution in Mongolia*. New York: Oxford.

Leeming, Frank. 1993. *The Changing Geography of China*. Oxford: Blackwell.

Leung, Chi Kin. 1993. "Personal Contacts, Subcontracting Linkages, and Development in the Hong Kong-Zhujiang Delta Region." *Annals of the Association of American Geographers*, 83: 272–302.

Li, Dun J. 1978. *Modern China: From Mandarin to Commissar*. New York: Scribner's.

Lu, Jonathan J., and Ma Shengtang. 1990. "Changes in China's Rice Area: 1966–1985. *Asian Profile*, 18: 391–414.

Ma, Laurence J. C. 1976. "Anti-Urbanism in China." *Proceedings of the Association of American Geographers*, 8: 114–18.

McColl, R. W. "In the Shadow of Chinggis Khan." *Focus*, 44: 1–9.

Nemeth, David J. 1998. "Privy-Pigs in Prehistory? A Korean Analog for Neolithic Chinese Subsistence Practices." *MASCA Research Papers in Science and Archaeology*, 15: 11–25.

Pannell, Clifton, W. 1988. "Regional Shifts in China's Industrial Output." *The Professional Geographer*, 40: 19–32.

Pannell, Clifton W., and Jeffrey S. Torguson. 1991. "Interpreting Spatial Patterns From the 1990 Census." *The Geographical Review*, 81: 304–17.

Pearce, Fred. 1995. "The Biggest Dam in the World." *New Scientist*, 28 (February): 25–9.

Platt, Anne E. 1995. "Dying Seas." *World Watch*, 8: 10–19.

Sattaur, Omar. 1990. "Thoroughly Modern Mongolia." *New Scientist*, 1: 45–50.

Smil, Vaclav. 1993. *China's Environmental Crisis: An Inquiry into the Limits of National Development*. Armonk: M. E. Sharpe.

Smith, Christopher J. 2000. *China in the Post-Utopian Age*. Boulder: Westview.

Thayer, Nate. 1997. "Forward Steppes." *Far Eastern Economic Review*, 27 (March): 18–23.

Tien, H. Yuan, et al. 1992. "China's Demographic Dilemmas." *Population Bulletin* 47: 1–44.

Topping, Audrey, R. 1995. "Ecological Roulette: Damming the Yangtze." *Foreign Affairs*, 74: 132–46.

Veeck, Gregory, and Clifton W. Pannell. 1989. "Rural Economic Restructuring and Farm Household Income in Jiangsu, People's Republic of China." *Annals of the Association of American Geographers*, 79: 275–92.

Vriens, Hans. 1997. "Lhasa Lost". *Far Eastern Economic Review*, 29 (May): 44–5.

N. A. 1996. *White Paper—The Grain Issue in China*. Information Office of the State Council of the People's Republic of China. October: 1–12.

Wong, Shue Tuck, and Sun Sheng Han. 1998. "Whither China's Market Economy? The Case of Lijin Zhen." *The Geographical Review*, 88: 29–46.

Yabuki, Susuma. 1995. Trans. S. M. Harner. *China's New Political Economy: The Giant Awakes*. Boulder: Westview.

Yatsko, Pamela. 1997. "Work In Progress (Shanghai)." *Far Eastern Economic Review*, 7 (August): 66–69.

Yeh, Anthony Gar-on, and Xueqiang Xu. 1996. "Globalization and the Urban System in China." In *Emerging World Cities in Pacific Asia*, eds. Fu-chen Lo and Yue-man Yeung, pp. 219–65. New York: United Nations.

Zich, Arthur. 1997. "China's Three Gorges." *National Geographic*, 192: 2–33.

Zimmerman, Tim et al. 1996. "China Takes a Deep Breath." *U.S. News and World Report*, 9 (September): 36–39.

Chapter 11

Japan: A Tiger Risen

WITH ROMAN CYBRIWSKY

"Our greatest glory is not in never falling but in rising every time we fall."

CONFUCIUS (C. 551–479 B.C.)

Rich, highly urbanized, densely populated, and one of the world's largest users of natural resources, Japan is Asia's leading tiger. The Tokyo–Yokohama–Kawasaki urban area is the largest metropolitan region on earth. Japan consumes the products of more tropical forests than any other nation, a reminder that Japan is virtually devoid of the natural resources to exploit.

Having achieved full status as an economic tiger, Japan is reckoning with problems facing all developed economies in the twenty-first century. How did Japan reach this pinnacle of economic development? What has gone awry, and how is this affecting economic structure and spatial organization? And, what social changes are occurring in context of economic transformation?

Dynamic Archipelago

Japan, with its 127.5 million people, is the tenth most populous country in the world. However, most Japanese live on 3 percent of the country's land area. Japan's physical landscape comprises mountains thrust up from the Pacific Ocean. Only 25 percent of the land area has level to moderate slopes while 75 percent consists of hills and mountains too steep to be easily cultivated or settled. Consequently, the Japanese are densely crowded on very limited lowland.

Landforms show evidence of tectonic uplift with coastal terraces, and glaciation with ice-carved features in the Japanese Alps. In many places, the mountains descend directly into the sea. Plains are restricted to narrow river floodplains and deltas. The most extensive plains on the main island of Honshu are the Kanto Plain around Tokyo and Yokohama, the Nobi Plain around Nagoya, and the Yamato Plain around Osaka, Kyoto, and Kobe. Needless to say, these areas are very densely populated (Figure 11–1).

As a result of its mid-latitude location and the fact that it is surrounded by water, Japan is very humid with a humid continental climate in the north and a humid subtropical climate in the south. Japan is also affected by the monsoon. In summer, warm monsoonal winds from the south bring heavy rainfall and frequently, typhoons. In winter, cold winds from the northwest carry heavy snows to the Sea of Japan coast and the north. An important dividing line is the 37° line of latitude which marks the northern limit of double-cropping of rice.

Four main islands dominate the Japanese archipelago: Hokkaido, Honshu, Kyushu, and Shikoku. These and hundreds of smaller islands lie at the junction

Figure 11–1

The major islands and cities of Japan. From H. J. de Blij and P. O. Muller, *Concepts and Regions in Geography*, Second Edition, 2005, p. 330. Originally rendered in color. © H. J. de Blij and P. O. Muller. Reprinted with permission of John Wiley & Sons, Inc.

Figure 11–2

The earth in action. Note how the Pacific and Philippine plates push against the Eurasian Plate. This geological movement makes Japan one of the most unstable places on earth.

of three tectonic plates: the Pacific, Philippine, and Eurasian (Figure 11–2). The Pacific Plate dips between the Eurasian and Philippine plates in the Japan trench north of Tokyo. The Philippine Plate is being consumed by the Eurasian Plate in the Nankai trench south of Kyushu and Shikoku. The Pacific Plate drops at a rate of 2.9 inches (7.5 cm) a year.

Japan is the most seismically active country in the world. Such instability results in many earthquakes. In Tokyo, there is an average of three noticeable quakes every month. In 1923 an earthquake registering 8.3 on the Richter scale devastated Tokyo and the Kanto Plain around it. Cooking stoves set fire to the mostly wooden and paper structures. The quake, the fires, and a 36-foot (10.8 m) tsunami claimed 140,000 lives. This was known as the *Great Kanto Quake.*

Another severe tremor shook the Kanto region in 1983, resulting in much property damage. All Japanese know that major quakes strike roughly every 70 years. With acres of landfill to liquify, thousands of skyscrapers to crumble, miles of freeways to collapse, and miles of underground gas lines to rupture, Tokyo is a scary place.

In 1995, another ruinous earthquake struck: this time the epicenter was the city of Kobe (Figure 11–3). This quake was caused by a "strike-slip" (sideways) motion along a minor fault. Older, concrete buildings collapsed, while those of flexible construction or fitted with shock absorbers sustained less damage. Some buildings still had severe damage, however. Fires raged out of control for days; 5,000 people were killed, and 250,000 were made homeless

Figure 11-3

The Hanshin Expressway fell over as part of the disastrous Kobe Quake. Fortunately, this was early morning and there was little traffic. Even in the midst of chaos, the orderly Japanese waited for lights to change and used crosswalks. Risa Palm.

Volcanic activity is another hazard in Japan, which boasts more than 200 volcanoes. Sixty of these have been active in historic time. Mount Fuji, a dormant volcano towering to 12,385 feet (3,776 m), is regarded as sacred by the Japanese (Figure 11–4). Japan is also the world's foremost hot springs country, with close to 20,000 springs and some 2,000 resort spas, many in Kyushu.

Historical Development

ORIGINS

Popular myth says the Japanese and the first emperor are direct descendants of the Sun Goddess Amaterasu. The first emperor took the throne in 660 B.C., but recent archeological evidence tells a different story.

Figure 11-4

Mount Fuji is a dormant volcano towering to 12,385 feet (3,776 m) and is one of Japan's more than 200 volcanoes. This mountain is sacred to the Japanese and is indicative of the country's mountainous terrain. © Royalty-Free/Corbis Images.

About 400 B.C., people from the Korean peninsula migrated to the island of Kyushu. Bringing with them rice culture from China, their numbers rose about two percent a year. In a few hundred years, their culture overwhelmed that of the indigenous Jomon hunters and gatherers who had inhabited the region for at least 10,000 years. Using metal tools and the techniques of large-scale rice cultivation, the newcomers established a regional core on the Yamato Plain. Japanese culture diffused from the Kansai Plain northward throughout the islands. Even though this story is based on scientific evidence, it is difficult for many Japanese to accept the fact that they are related to Koreans, whom they long have considered inferior.

The language that originally diffused to Japan was not Korean as we know it today. It is thought that the Japanese language diverged from the Korean language at least 4,000 years ago, as they have few words in common. However, Japanese and Korean both employ common Chinese words.

Japanese Writing

Initially, the Japanese simply borrowed the Chinese character system and called it *kanji*. However, profound differences between Chinese and Japanese grammar meant that sentences were constructed awkwardly. So the Japanese developed a syllabary known as *hiragama* that permitted the expression of words and parts of speech not easily represented by Chinese characters. While there are thousands of Chinese characters, there are only 51 *hiragama* symbols. Each represents a syllable or a combination of a consonant and vowel. A different system of sounds called *katakama* is used for words of foreign origin.

Unlike Chinese, Japanese is easily written in the Roman alphabet. The resultant *romanji* style of writing is very important in advertising and computer use. Today's technology allows the translation of *romanji* into the traditional Japanese meld of *kanji*, *hiragama*, and *katakama*.

OPENING CLOSED DOORS

Japan continued to be influenced by its Asian neighbors for centuries to come. However, as an island nation it was isolated enough to also develop its own distinctive cultural patterns, including the unique Japanese language and many customs, art forms, and other traditions. Japanese people are justifiably proud of their distinctive culture, and continue to enjoy traditional foods, festivals, theater, music and other activities, even as they lead modern lifestyles and follow fashions and current trends from countries around the world.

For much of Japanese history, the capital was Kyoto, the seat of imperial rule and repository for many aspects of traditional Japanese culture. It is a beautiful city with spectacular temples and gardens, as well as Nijo Castle and the historic Imperial Palace, all outstanding examples of traditional Japanese architecture and landscaping (Figure 11–5). After a time, the emperor lost power to Yoritomo, the leader of one of Japan's many clans. Yoritomo was the first **shogun**, an all-powerful military leader who rewarded his followers with lands and castle towns from which they ruled the various provinces and often waged war on one another. Eventually, Japan was unified under the rule of the Tokugawa line of shoguns who governed Japan from 1603 to 1867 from a capital city of their own choosing, **Edo**. Edo grew quickly under the shoguns' influence to become one of the biggest cities in the world. It was renamed Tokyo when the Tokugawa shogunate ended and imperial rule was restored.

From 1603 to 1854, Japan was essentially closed to foreigners and Japanese themselves were not permitted to leave the country. This was the shoguns' response to the intrusion by Western powers that was changing other parts of Asia and that threatened Japan. However, some trading was permitted under close scrutiny in the southern port of Nagasaki, far from Japan's heartland. Then, in 1853, an event took place that would change Japan forever. This was the arrival in Edo (Tokyo) Bay of a squadron of armed American ships under the leadership of Commodore Matthew C. Perry. Because of superior military technology, Perry was able to engineer an unequal treaty with Japan that favored the intruders and opened Japan for trade to outsiders.

Figure 11–5
This is the Nijo castle in Kyoto. It is a landscape symbol of Kyoto's role as a castle town.

It became readily clear to the Japanese that they had fallen behind Western nations in technology during their time of isolation. The ensuing crisis hastened the demise of rule by the shoguns and brought back to power the nation's emperor, who in turn moved his seat of authority to then newly named Tokyo. Britain, an island nation in the Atlantic that had acquired the largest colonial empire in the world, became a model for Japanese modernizers who, under the forward-thinking Meiji emperor and his advisors, led the transformation of Japan. Confucian values were called upon to educate the Japanese and move the nation into the twentieth century. Foreign teachers of everything from languages to technology were imported in great numbers. Japanese also began to study abroad. Japanese industry developed especially quickly, with huge military-industrial conglomerates called *zaibatsu* rising to great power. This time of fast-paced, catch-up modernization is referred to as the Meiji Period (1868–1912).

Shintoism

While most Japanese are Buddhist, many follow the ancient religion of **Shinto** for certain aspects of their lives. However, most Japanese combine elements of Shinto with Buddhist practices. A person might have a Shinto wedding and a Buddhist funeral for example. Shinto began as the worship of spirits in nature and evolved into a belief system focused on the relationship between natural harmony and human existence. In the late 1800s, the Japanese government elevated Shinto into a nationalistic cult which worshipped the emperor as divine. The more salient nationalistic elements of the religion were expunged after World War II.

Shinto is centered on places and nature. Certain mountains, such as Mount Fuji, are considered sacred and are climbed by thousands of people. Major Shinto shrines attract large numbers of pilgrims. The most notable of these is the Ise Shrine south of Nagoya. It represents the cult of the emperor.

EXPANSIONISM AND EMPIRE

Japan's industrialization was accompanied by an expansion of its military might and acquisition of territory (Chapter 1). Flexing its new muscles, Japan launched wars with China in 1894–1895 and with

Russia in 1904–1905. In 1895, it annexed Taiwan. In 1920, Japan annexed the whole of Korea, which it ruled until 1945. Manchuria became another Japanese colony in 1931. Such expansion was Japan's response to the colonial ambitions of Europeans and Americans in Asia, and reflected Japan's vision of a Greater East Asia Co-Prosperity Sphere (GEACPS) in which "Asia would be for Asians." Yet, Japanese rule was often harsh and oppressive, and was opposed by rebellious patriots in the occupied lands. Many of Japan's Asian neighbors still harbor resentments against Japan for its time of colonialism, citing specific atrocities that the country has yet to officially acknowledge. The most famous of these is the December, 1937 Nanking (Nanjing) Massacre, named for a Chinese city in which up to 300,000 Chinese civilians were killed and 20,000 women raped by Japanese troops. For Koreans, a key unresolved issue is the conscription of thousands of women as sex slaves for Japanese soldiers during the occupation of their country.

Japan's expansionism was brought to a halt in 1941 after its attack on the American naval base of Pearl Harbor in Hawaii. Japan justified the attack by claiming that Americans had interfered with its oil supply lines in Asia. The United States entered World War II and began to push back against Japanese control in Asia. Although Japanese troops managed to advance across a wide area of China and Southeast Asia and the South Pacific, they eventually lost ground in fierce fighting against the Americans and their allies. Japan came within range of allied bombers (based in China and on Pacific islands), which destroyed Tokyo and almost all other Japanese cities. The war in the Pacific was brought to an abrupt end after the Americans dropped atomic bombs on the cities of Hiroshima and Nagasaki in 1945.

PHOENIX FROM THE ASHES

Defeat in the war meant the end of GEACPS but a new beginning for Japan. The country was almost completely devastated by the bombing, and its people were literally starving in the immediate postwar period. Yet, Japan's recovery in the ensuing years was the most rapid of any nation in history. Relying on

the Confucian values of hard work, loyalty, and cooperative efforts and with American assistance, infrastructure was rebuilt. A democratic form of government was put into place, the *zaibatsu* were broken up and divorced from a reduced military, and land reform gave land to the war-ravaged peasants.

Grounded in a fresh technological base, government, financial institutions, and industry banded together to build the new Japan. Another form of industrial conglomerate emerged: *keiretsu*.

Did you know that Toyota Motor Group is a keiretsu *with ties to Mitsui and that Honda has very close relations with Mitsubishi?*

Keiretsu

Keiretsu means something close to "link" or "affiliated with." A *keiretsu* is a hierarchical organization of companies. There are "horizontal *keiretsu*" and "vertical *keiretsu*."

Horizontal *keiretsu* revolve around a financial core including a major bank, a giant trading company, and a huge manufacturer. For instance, there is Mitsubishi Bank, Mitsubishi Corporation, and Mitsubishi Heavy Industries. There are also about 30 companies and more than 200 affiliates (10 percent equity ownership) that are part of this horizontal *keiretsu*.

Many vertical *keiretsu* lie within the borders of the major horizontal *keiretsu* such as Mitsubishi, Sumitomo, or Mitsui. Vertical organizations are either in production or distribution. A collection of firms assembles parts for a manufacturer who directs its end product through a distribution network of wholesalers and retailers. Most production *keiretsu* concern automobiles and electronics. Most distribution *keiretsu* are involved in consumer electronics.

A member, but very independent, of the Mitsui Group is the Toyota Motor Group.

Toyota was founded in 1894 by the Toyoda clan that had its start in the manufacture of wooden spinning machines. Toyota, with its automobiles, real estate firms, insurance companies, and other interests is becoming a combination vertical-horizontal *keiretsu*.

A LEGACY OF TENSION

More than half a century after Japan's defeat in World War II, there are still tensions between Japan and the neighboring countries that once comprised Japan's empire. Survivors and their descendants in China and Korea have little inclination to forgive the Japanese for their harsh and oppressive rule. Moreover, the Japanese have yet to atone for their atrocities. Lingering resentments damage international relations to this day.

One example of the bad relations concerns a recent episode of sex tourism by Japanese men to the southern Chinese city of Zhuhai, one of that country's fast-growing SEZs. Such cities attract poor migrants from the Chinese countryside in search of a livelihood and have a flourishing underside of prostitution and other illicit activities. A package tour group of some 400 Japanese men, all employees of an Osaka-based construction company, was entertained there for three days by 500 local prostitutes. Even though it is not unusual for Japanese males to travel in groups to China and other Asian countries for packaged sex, this particular orgy happened to coincide with an anniversary of Japan's pre-World War II invasion of China. The incident raised particularly strong indignation in the Chinese media, which described the event as an affront to Chinese dignity. It then became the center of a diplomatic flap between the two nations, causing Japanese officials to issue a public apology.

Japan's relations with neighboring Russia are also strained by historical tensions. The two countries could be enormous trading partners, given the proximity and richness of Siberia's natural resources. However, full exchange is held back by a territorial dispute that dates back to the end of World War II. At that time Russia snapped up a chain of small islands northeast of Hokkaido that had been Japanese territory. Japan claims that these "Northern Territories" are illegally occupied and insists on repatriation as a condition for full economic and diplomatic relations. This issue is a source of frequent protest by Japanese right-wing political groups who hold noisy demonstrations almost daily near the Russian embassy in Tokyo. Their favorite tactic is the use of sound trucks to blast ear-splitting demands that Russia give back the disputed islands.

Population and Society

AGING SOCIETY

Japan's population is expected to decline (Figure 11–6). At 127.5 million in 2004, this number is expected to peak in 2007 and plunge dramatically at the end of the twenty-first century. The average life expectancy for women is 81 years, for men 78 years. For the first time in history, Japan has more people over 65 than it does children under age 15. By 2035, the government estimates that more than 30 percent of Japanese will be 65 or older. Japan is preparing to launch a public nursing insurance program, and several companies, such as Sanyo and Mitsubishi, are jumping on the bandwagon to produce senior-oriented products and services.

Young Japanese are putting off marriage and having fewer children than in the past. The average age of marriage has risen to 26.3 for women and 28.5 for men. Furthermore, many women are seeking careers and are less likely to have or even want children. In 2003, there were 9.0 births per 1,000 people, compared with 28.1 births per 1,000 people in 1950. The average Japanese mother has about 1.3 children well below the 2.1 replacement level. The working age group began to decline in 1997. Now, numerous local governments are actually giving financial incentives for a first, second, or third child, and day-care centers are becoming commonplace.

Changing population dynamics have serious ramifications throughout both economy and society.

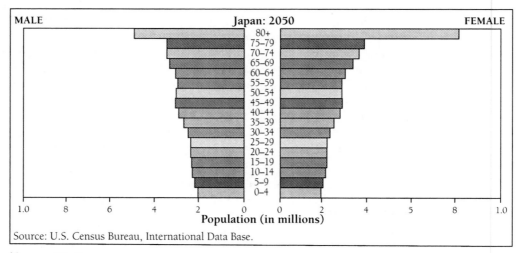

Figure 11–6

These population pyramids clearly indicate that Japan's population is declining and aging rapidly.

The Japanese government is concerned about its tax base and available work force. It must also face the fact that family structure is changing from extended to nuclear. In 1960, 87 percent of Japanese over 65 lived with adult children. By 1990 this number had dropped to 60 percent, and it has been falling ever since. The government must cater to older people because they are the most likely to vote. Aside from this, Japan is still a country where the elderly are revered and taken care of.

THE DARK SIDE

A myth of Japan is that it is an ethnically and culturally homogeneous society. In fact, Japan has three classes of minorities: native ethnic groups such as the **Ainu**; foreign ethnic groups such as the Koreans and Chinese; and social outcasts known as *Burakumin*. Ethnic and social minorities make up about four percent of the population, roughly five million people.

Native to Hokkaido, the Ainu are racially unrelated to the Japanese. Currently, about 30,000 identify themselves as Ainu. Their culture has been undermined by assimilative pressure; they suffer discrimination in all walks of life, and they live in relative poverty. Recent public awareness of the Ainu plight has not reduced discrimination, an issue most Japanese do not face. Media portray the adorable and folksy Ainu living close to nature. According to historian David Howell (1996), recognizing some aspects of Ainu identity may generate pride but this "also shields Japanese society at large from the need to reexamine its basic premises."

There are about 630,000 ethnic Koreans living in Japan, descended from the 2.5 million drafted as soldiers or forced laborers during Japan's 35 years of colonial rule in Korea. Most are classed as resident aliens; until the 1990s, even those Koreans born in Japan were routinely fingerprinted and denied citizenship. About 20,000 Koreans died in the bombing of Hiroshima, but the needs of bomb-diseased survivors are virtually ignored. Discrimination in housing, employment, and education is rampant. Consequently, the majority of Japan's Koreans are trapped in poorly paid, unskilled jobs.

The Chinese number more than 270,000. About 20 percent of these people are descendants of residents who were in Japan before World War II. The rest are temporary residents, guest workers and migrant laborers. Chinese are most visible in "Chinatowns" the largest of which is in Yokohama. While discriminated against, they fare better than the Koreans.

Another minority is the Okinawans from the Ryukyu Islands south of Kyushu. The Ryukyus were not incorporated into Japan until the seventeenth century. Okinawans are similar to Japanese and speak a variant of the language. They are treated as second-class citizens.

There are also foreign residents in Japan. These hail from places such as the Philippines, Thailand, Indonesia, and Vietnam. Coming to Japan for economic opportunities, they are forced into low-end jobs that no Japanese will take. There are also foreigners of Japanese ancestry—the *Nikkei*. Numbering about half a million, these people come from Brazil, Peru, and other parts of South America where they emigrated in the twentieth century. Now returning to their homeland for work, they look Japanese but cannot speak the language. Moreover, their cultural bent is Latin American. Consequently, they are treated as outsiders.

The largest minority group is the *Burakumin*, who are Japanese descended from underclasses defined in medieval times. They number 2 to 3 million. The *Hinin* (nonpersons) were beggars, prostitutes, entertainers, and the like. The *Eta* (filth) were a subclass of untouchables regarded as spiritually unclean because of their occupations burying the dead or slaughtering animals. Since myths of spiritual and racial purity still abound in Japan, *Burakumin* are shunned and remain segregated in urban ghettoes. Some attempt to quietly integrate into society but they are usually found during background checks for employment and marriage. Then they are rejected.

Many companies, including the major ones, will not hire minorities. Students from the best schools will not work for companies that do. To marry a minority is unthinkable.

Several minority action groups deal with government agencies to obtain improved conditions, but not all minorities are equally represented. Moreover, some minorities (e.g., the Koreans) eschew full participation in Japanese society. This results in a paradox: sustaining group identity in the face of discrimination and relying on organizational intermediaries to deal with major issues shields Japanese society and individuals from confronting their prejudicial behaviors and practices.

IMMIGRATION DEBATE

One of the most hotly debated topics in Japan is whether or not to relax the nation's strict controls on foreign immigration and allow more foreigners to enter. In comparison to other countries, Japan's population is almost entirely ethnically homogeneous, with non-Japanese accounting for less than five percent of the population. Foreign immigration would diversify the country ethnically, as well as help change the age structure of the population by bringing in younger people of childbearing age. Moreover, immigrants are said to be helpful to the economy by providing candidates for difficult, dirty, and dangerous (three words beginning with a "k" sound in Japanese) work that most Japanese avoid. The result would be more taxpayers to provide financial support to Japan's aging society by paying in to the increasingly stressed social security system.

However, Japan is not used to hosting immigrants, and anti-foreigner feelings are common. Immigrants from developing countries in Southeast Asia and Africa face an extra amount of discrimination although it is they that do the unpopular "3K" jobs in the country. "Japan for the Japanese" is a popular attitude. In many cities one sees signs at the doors of bars, bathhouses, and pachinko (Japanese pinball) parlors that read "Japanese only." Many Japanese unfairly associate increasing social problems in their country such as crime and sexually transmitted diseases with the presence of foreigners. When foreign visitors came to Japan in bigger numbers than usual in the summer of 2002 for the World Cup soccer tournament, the Japanese media was filled with warnings from police and political leaders that more prisons would have to be built to house the "hooligans" and that hospitals should prepare for an overflow in births nine months later. As it turned out, the "hooligan" problem never materialized and there was no spike in births.

Tiger's Transformation

AGRICULTURAL JAPAN

In 1920, 51 percent of Japan's labor force was in agriculture. Now less than 5 percent is farming. About 90 percent of these individuals are part-time farmers who earn most of their income from non-farm activities. In fact, much of the farm work is done by women and older people; men work other jobs, and young people move to the cities.

Japanese farms are very small, averaging 4 acres (1.6 ha). Since agriculture competes with other activities for level land, farmland is very expensive. Many farmers are independently rich after selling their land to industry. However, resistance to parting with land imbued with centuries of family tradition has made acquisition of farmland for urban and other development difficult and expensive.

Climatic factors and limited space call for intensive land use. The growing season varies from 260 days a year in Kyushu to 150 days in Hokkaido. The northern limit for double-cropping is 37° latitude.

The primary crop is rice, planted in summer. Rice is rotated with wheat or barley in the double-cropping zone. Tea and fruit such as apples grow on hill slopes. Much of Japan's agriculture is **market gardening**: producing vegetables and poultry products for the vast urban market. For this purpose, greenhouse agriculture is very important and organic farming is on the increase. Nevertheless, agriculture is highly mechanized and chemical fertilizers and pesticides remain widely used (Figure 11–7). Consequently, crop yields are very high with rice yields three times the average Asian yield.

Animal herds are rare other than in Hokkaido, where extensive plains make dairying feasible. An ordinance of 1896 established nonmilitary settlements on Hokkaido, and these were modeled on the township system in the United States. Here, dairy farms look like American farmsteads. Some beef cattle are kept near Kobe. These animals are beer fed

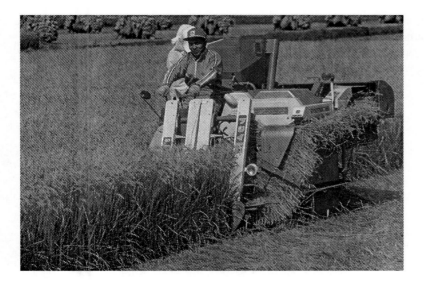

Figure 11-7
This rice harvester symbolizes Japan's high degree of mechanization in agriculture despite the small size of its farms.

and hand massaged daily. A steak from Kobe beef cattle can cost hundreds of dollars.

Japan's major source of protein is fish—the country has the world's largest fishing fleet. In addition, **aquaculture** is common, especially around the Inland Sea where shrimp, oysters, and seaweed are cultivated. Fish are also raised in ponds and in ocean-based fish farms.

Agriculture faces an array of problems in Japan. Government subsidies pay for surplus crops. Rice is six times more expensive to grow in Japan than elsewhere, yet it is a tradition supported by the government, which also prevents imports of cheaper Japonica rice from places such as California. Yet, food preferences are also changing. Many Japanese are turning to imported meat, vegetables, and dairy products. Traditional meals are being replaced by packaged instant noodles and McDonald's, and other fast foods. Changing diets have resulted in Japanese who are taller and heavier than before. In addition, heart disease and cancer are on the rise.

Farming has long had a special place in the Japanese psyche. In this nation, dominated by factories and urban sprawl, the countryside is viewed as a repository of pristine national values such as harmony and community spirit. These emotional ties have helped make agriculture sacred in Japanese politics. Moreover, rural votes have helped guarantee the conservative Liberal Democratic Party a half-century of rule. In return, the government has given crop subsidies and public works projects to farm communities. Trade barriers have protected farmers from cheaper imports. But, the farm lobby is losing its grip on government as demographic change shrinks the base of rural voters.

Rural areas are losing population as young people move to urban bright lights. Over the past decade, the number of Japanese living on farms dropped by 25 percent to 13.5 million. Forecasters predict that this figure will be halved by 2025. Furthermore, 53 percent of these farmers are age 65 or older.

As rural populations lose their influence in government, their products are no longer shielded by tariffs, which in the past have been as high as 1,000 percent. Pressure from trading partners and global treaties has forced the Japanese government to negotiate trade agreements with a variety of nations such as Mexico, Thailand, and China. Consequently, a growing number of farm imports are being allowed into the country.

In the early 2000s, Japanese farmers were shocked to see cheap Chinese vegetables stocking supermarket shelves. Tomatoes, onions, eggplant, and garlic grown in China can be sold for far less than the same Japanese produce. The Chinese have already garnered a 40 percent share of the shiitake mushroom market.

Nationwide, agricultural prices have fallen 16.4 percent over the last 10 years. Over the same period, farm incomes have dropped 29 percent. Japan now produces just 40 percent of the food it consumes.

As family farms close, a new phenomenon is appearing in the countryside—the factory farm. For example, a subsidiary of Daiei Inc., one of Japan's largest supermarkets, has opened a factory farm to produce 20,000 pigs a year to be sold in Daiei markets. The government is encouraging farms like this to make the country more competitive in an increasingly globalized market place. In 2005 it is cutting subsidies to smaller farms in order to encourage mergers into larger corporate entities. Still, there were only a few more than 1,200 factory farms in 2000—a mere 0.34 percent of all farms. The fate of the family farm is yet to be decided.

INDUSTRIAL JAPAN

The foundation of modern Japan's economic growth is manufacturing. From the mid-1800s, the economy responded to external trading pressures as well as a desire to become a military power. Factories were built in the coastal zone from Tokyo to Osaka. These specialized in iron and steel production, shipbuilding, and textile manufacturing. In the period prior to World War II, industrial expansion was based on links between business and government, cheap labor, large numbers of small businesses, and an improved infrastructure. Japan became known for its cheap but poor quality goods. This industrial base was devastated in World War II.

Rebuilding and restructuring of Japanese industry and infrastructure began with the infusion of American capital. Following the reconstruction of heavy industries such as steel and chemicals in the 1950s, Japan diversified into shipbuilding, automaking, and light manufacturing.

Japan's industry has always relied on imported raw materials. In fact, Japan has the most limited natural resource base of any of the world's major industrial powers. In post-war Japan, raw materials were cheap and, with its own low-cost labor force, this meant that Japan could easily compete in world markets.

Conditions changed in the 1970s with labor shortages, increased oil prices, and a slowdown of global shipping. As Japan now depended on more costly oil for 70 percent of its energy needs, the government decided to reduce the cost of hydroelectric and nuclear power and to subsidize unprofitable coal mines.

By the 1980s, older heavy industries were reaping the effects of overcapacity and competition from emerging tigers such as South Korea. Factory closures led to rising unemployment. Japan reacted to these changes by diversifying into light industries such as electronics, cameras, and appliances. Another venture was robotics—the building of mechanical devices to do the work of humans. Robots became widely used in the auto industry. However, their use has been cut back in recent years because they have proven to be very expensive.

At the end of the 1980s, Japanese corporations began investing more overseas. First, they bought out foreign mineral suppliers and then they built factories in other countries. After major investments in the United States and Britain to establish auto plants, they turned to Southeast Asia where economies were stimulated by the infusion of Japanese capital. In Japan, industries that lost jobs to cheap labor overseas were said to be **hollowed out**.

By the mid-1980s, Japan expanded its investment in cheap labor and assembly-line factories by moving research and design facilities overseas as well. It also built on its relationship with China especially in China's northeast. By the early twentieth century, most Japanese multinationals such as Sony, Hitachi and Toshiba and automakers such as Toyota, Honda, Mazda and Nissan were operating in China. But the 1990s proved that all was not well in Japan Inc.

AILING TIGER

Despite Japan's generally high standard of living and great wealth in comparison to other parts of Asia, all is not well in the country. After the unprecedented prosperity of the 1970s and 1980s, the economy went into a painful slump in the 1990s, so much so that the decade is now being called the "Lost Decade." In 2004 Japan still showed little sign of recovery. Countless companies, including several major banks, well-known supermarket and department store chains, and once prosperous manufacturers, have gone bankrupt. They have been forced to terminate long-term workers just to stay afloat. The factory sector has been particularly hard hit, as more and more employment has shifted to overseas locations where labor is cheaper. Companies in traditional industries such as steel production and shipbuilding have gone bankrupt altogether because of competition from South Korea, China, and Taiwan.

For Japanese employees in general, the promise of a guaranteed job for life has all but disappeared. Many workers face unemployment or moves downward to inferior occupations at lower wages. Likewise, recent college graduates, once accustomed to having a choice of jobs, struggle to find any employment at all and are often forced to settle for part-time work without benefits. Moreover, a new class of young workers called *freetah* has emerged. They eschew the "rat race" of full-time employment and instead go from one part-time job to another in accordance with short-term financial needs and personal inclinations. To lower the cost of living, many *freetah* delay or avoid marriage and live at home with parents long after the usual age of striking out on one's own.

There are many explanations for the malaise, including the troubling assessment that Japan's economy is chronically inefficient with redundant employees and outmoded business practices. A look at virtually any construction site or inside any bank shows the excess of workers, many of whom seem to do nothing but drive up the cost of business. Every road repair site, for instance, seems to have a bevy of traffic control assistants with flashing light batons to wave vehicles and pedestrians around the obstruction, whether they are needed or not, as well as what seems like an inordinately large number of workers and supervisors. The bad economy is also blamed on Japan's financial institutions because of the many poorly secured loans they made to cronies during the bubble. Many loans were for dubious investments such as speculative real estate. The national government has recently had to bail out several large companies with special loans and tax breaks. This has caused critics to point to Japan's cozy relationships between politicians and businessmen as still another flaw in the nation's economy.

Another major criticism of the economic situation in Japan is the disproportionate use of the construction sector. In comparison to the United States, which has about 7.1 percent of the workforce in construction and is fairly representative in this regard of advanced nations in the world, Japan has more than 10 percent of its workforce in construction. Likewise, construction accounts for a disproportionate percentage of Japan's GDP (7.3 percent versus 4.7 percent for the US), as well as a disproportionate percentage of Japan's government spending. These differences are attributed largely to the overly friendly relationships in Japan between politicians and construction companies, particularly in terms of politicians' reliance on campaign contributions from construction companies that want contracts to build public works. Campaign practices that would be outrageous in most other countries are common in Japan, as is bribery. The result is that Japan is sometimes called a "construction state," just as critics of the United States might refer to the US as being dominated by oil interests and the "military-industrial complex."

The influence of the construction industry is seen everywhere in Japan. Not only does the country have some of the longest tunnels and bridges in the world, such as the tunnel to Hokkaido and the bridge to Shikoku (Figure 11–8), but it also has countless other more ordinary tunnels, bridges, and highways built at great cost across the nation's difficult terrain. Similarly, Japan has the world's highest density of dams, which it continues to build even as other countries dismantle some of theirs for sound environmental reasons. Moreover, there is hardly a free-flowing river in the country as almost all have been reconfigured and realigned in concrete by various kinds of flood control projects. In small towns in the countryside, where the nation's economic woes are most keenly felt, it seems that the proposed fix is to apply government funds for wider roads and short-cut bridges, museums, and cultural centers. While such projects create construction jobs in the short run, they often have little or no long-run benefits for a community. This causes politicians and construction

Figure 11–8
This bridge links Shikoku with Honshu and is part of Japan's core region.

companies to "solve" the local unemployment problem, in turn, with still another expensive highway or bridge to nowhere, oversized public buildings in marble, or a concrete wall for a stream or hillside.

A town that is representative of Japan's economic woes is Ashio, a once prosperous copper mining center in the mountains of Honshu north of Tokyo. As mining declined over much of the twentieth century because of cheaper copper from other sources and depletion of the richest ores, the town faced a problem of unemployment and uncertainty about its future. Population dropped from a peak of nearly 40,000 around 1915 to less than 5,000 today, as residents relocated to Tokyo and other cities for jobs, and young people moved away soon after finishing high school because of a lack of opportunities. The proposed solution for the stagnation was to open Ashio for tourism. Government economic development funds helped with the construction of a museum about copper mining, as well as with construction of new roads and bridges to make the town less isolated. Unfortunately, the planners overestimated the number of people who would travel out of their way, even by improved roads, to learn details about copper. The museum and souvenir stands have few visitors and Ashio remains another depressed town in Japan's countryside.

Despite the current economic problems, it is important to remember that, on the whole, Japan still has the second largest economy in the world and is a country with a very high standard of living overall. The malaise is noteworthy because it is so unlike the Japan of recent times, when fast economic growth, expanding exports, and a leading role in technological development in the world were the overarching characteristics. Japan may well solve its internal ailments and become Asia's dominant tiger once again but will have to face severe competition from China.

The Big Five

Vehicles are Japan's largest export product, with more than 7 million being exported in 2004. The "Big Five" (Toyota, Honda, Mitsubishi, Nissan, and Mazda) account for about 80 percent of the 9 to 10 million vehicles produced each year.

The industry had a slow start with a small domestic market. Moreover, there was competition with U.S. companies producing cars in Japan in the 1920s and 1930s. In the 1950s, the government targeted the car industry as one with great potential. As incomes rose, the domestic market expanded. At the same time, there was a growing market in Europe where people were demanding small, well-developed, fuel-efficient cars. The industry took off and has reached the point where one-tenth of Japan's workforce is now in car manufacturing.

There is a trend in the worldwide auto industry to integrate national production thereby making autos international entities. For example, Toyota, Nissan, and Honda manufacture cars and parts in China. As of 2000, Toyota had invested more than US $10 million in 18 facilities in the United States that employ more than 27,000 workers. The Camry, for instance, is produced in Kentucky. In fact, more than 63 percent of Toyotas sold in the U.S. are made somewhere in North America. The "Big Five" are making big profits on overseas sales. It will be interesting to see how the vehicle manufacturers in Japan will fare in light of the nation's deindustrialization.

Urban Japan

City development in Japan can be traced back to the eighth century when the first permanent capitals of Heijokyo (now Nara) and Heiankyo (now Kyoto) were established. Soon other cities were built, primarily for political and military purposes. Some, such as Naniwa (now Osaka), were built to meet the needs of travelers. As the central authority of Japan's military government waned in the fifteenth and sixteenth centuries, towns increasingly evolved around castles constructed by regional warlords.

CASTLE TOWNS

The catalyst for urban growth was the fortified **castle town**. One of the most important of these was Osaka where a grand castle was built in 1583. This served as the nucleus for urban growth. Moreover, government policies stimulated the rise of Osaka and other towns. After the mid-1630s, the prohibition of foreign trade stimulated internal interchange of goods and subsequent town growth. Minor feudal castles were destroyed so that there was only one castle town per province furthering focus of activity in major centers. These centers became consolidated as civilians were encouraged to move to the more important castle communities.

Castle towns, with their political and economic functions, were typically located on level land near important landscape features. Osaka emerged as the principal business, financial, and manufacturing center in Tokugawa Japan. Cities were also linked by highways that stimulated growth and trade. The most famous was the Tokaido Road running from Osaka, eastward through Nagoya, to Edo (Tokyo) the capital. By the middle of the eighteenth century, Tokyo had one million inhabitants. With rapid post-war development in Tokaido Road cities, an urban-industrial region emerged to be known as the **Tokaido megalopolis**.

TOKYO RISING

Edo was founded in the fifteenth century when a minor feudal lord built a castle on a bluff near the sea. With a natural harbor, hills that could be fortified, and the Kanto Plain behind it for expansion, Edo had an excellent geographic situation.

Tokyo got its real start when the Tokugawa shogun Ieyasu decided to make Edo his capital in the sixteenth century. He and his descendants designed a vast area of palaces, parks, and moats in the heart of the city. This grand plan incorporated an imperial enclosure for the emperor. Reclamation of land from the sea set the pattern of development for many Japanese cities that faced a shortage of level land for expansion and needed good port facilities. By the mid-eighteenth century, Tokyo was one of the largest cities in the world.

Edo's initial growth was based on its function as a political center tied to other places by a network of roads. The rivalry between Tokyo as a cultural and political center and Osaka as a business center lingers to this day. With the restoration of the Meiji emperor in 1868, the royal court was transferred from Kyoto to Edo which was renamed "Tokyo" meaning "Eastern Capital." With its new political functions and the rapid industrialization and modernization programs undertaken from the 1870s onward, the city experienced profound growth into the twenty-first century.

RAPID URBANIZATION

Despite more than a century of industrialization, Japan did not reach the 50 percent urban mark until after World War II. Between 1950 and 1970, urban population increased to 72 percent, a figure that took the United States many decades to achieve. Since 1970, the proportion of people living in cities has risen more slowly reaching 78 percent in 2003 and projected to reach 85 percent by 2025. As the urban population grew dramatically, so did the number and size of cities. While small towns and villages (fewer than 10,000 people) declined sharply, medium and large cities expanded rapidly both in population numbers and spatial extent. This process reflected the country's phenomenal economic growth after World War II.

THE CORE REGION

A distinctive feature of Japan's urban pattern is the concentration of cities in a relatively small area. Almost all of the major cities are found in the core region extending from Nagasaki and Fukuoka in northern Kyushu, along both flanks of the Inland Sea on Shikoku and Honshu, through Nagoya to the Tokyo region (Figure 11–9). The core region, especially the southern coast of Honshu, contains strings of industrial cities such as Hiroshima which has recovered from the devastation created by the atomic bomb dropped on it at the end of World War II. The Tokaido megalopolis, with 45 percent of the Japanese population, is contained within the core area.

Japan has a classic core-periphery imbalance— the "developed" capital region focused on Tokyo, and the "less-developed" (in a relative sense) regions elsewhere. Since the 1970s, migration and growth have been increasingly toward Tokyo at the expense of other places, a phenomenon called **unipolar concentration.** As the city continues to expand, it drains both capital and people from other areas, many of which are in a state of stagnation. The Osaka region,

for example, has not seen the growth of many industries to replace the smokestack ones of steel and shipbuilding, In addition, Osaka businesses continue to relocate to Tokyo. This process stimulates out-migration and lowers personal consumption. To be in the mainstream of modern Japan is to live and work in or near Tokyo.

TECHNO-ARCHIPELAGO

Today, the Capital Tokyo Metropolitan Area (CTMA) houses 31 million people. CTMA's phenomenal growth has paralleled Japan's integration in the global economy. Its geographical location between New York and London is perfectly suited to the online, global, 24-hour, international telecommunication and financial transaction system. As a result, Tokyo was expected to provide an array of businesses and services not directly related to the domestic economy. Other sectors of the economy increasingly demanded specialized services in order to maintain their competitive edge. The crucial role of research and development was soon recognized. Kansai Science City began to take shape in the 1980s. Kansai Techno-Research Complex includes five cities and three towns in the Kyoto–Osaka–Nara prefectures and is an integrated, multinodal development. This structural format models plans for a new urban Japan.

The new Japan will possess a multi-centered regional and urban system founded on advanced information technology. The system is reorganizing from an assemblage of hierarchical structure to a uni-centered, functional network with CTMA as the central node. The **technopolis** is integral to the master plan that conceives a vast network of science cities across the nation. Fourteen industries such as biotechnology, robotics, software, and space have been targeted for development in tandem with research and development and other service functions. These are to fuel additional industry in the region. Each technopolis centers on a "mother city" and is to be linked by rapid transit to Tokyo, Nagoya, or Osaka by no longer than a one-day return trip. The new undersea tunnel to Hokkaido and bridge to Shikoku along with more expressways and high-speed bullet trains are intended to facilitate this goal.

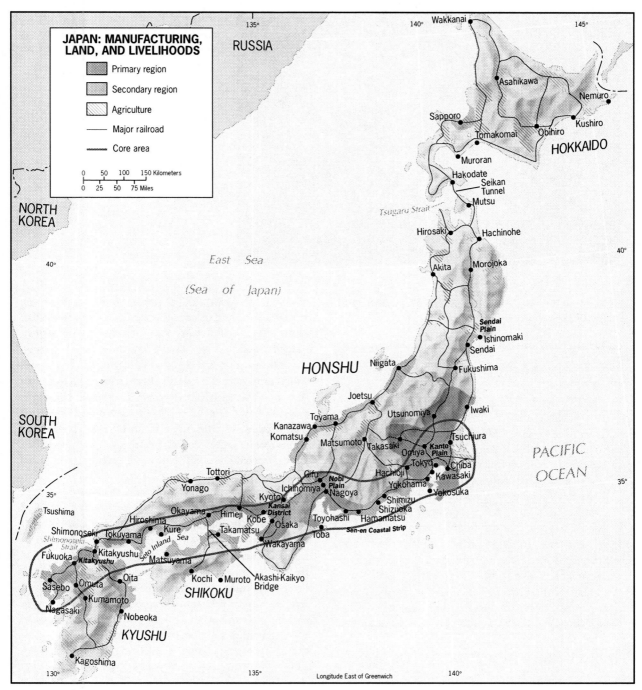

Figure 11–9

Japan's core region and major manufacturing areas. From H. J. de Blij and P. O. Muller, *Concepts and Regions in Geography*, Second Edition, 2005, p. 331. Originally rendered in color. © H. J. de Blij and P. O. Muller. Reprinted with permission of John Wiley & Sons, Inc.

Figure 11–10
Trains in Tokyo are so crowded that people are employed to push commuters into the cars.

This evolving Techno-Archipelago is Japan's landscape of the twenty-first century.

Tsukuba: Science City

Tsukuba Science City, 40 miles northeast of Tokyo, was conceived in the 1960s as a government-sponsored scheme. It is an extensive development covering 112 square miles (28,559 ha) and has at least 48 research and educational institutions. Until very recently, the city was regarded as spatially and psychologically isolated from Tokyo. Another major complaint was that too many institutions were working independently with insufficient communication among disciplines. With the expansion of the CTMA and greater attention paid to organizational linkages, Tsukuba is now enmeshed in Japan's larger urban system.

TOKYO: CROWDED BUT COMFORTABLE

Tokyo is famous for its crowding. Perhaps one of the most enduring images of the city is of "commuters' hell" when rush hour trains and subways are jammed with homeward-bound passengers who, like sardines in a can, are hardly able to move. In the morning the scene is repeated as inbound trains are packed with commuters, with many stations employing white-gloved "pushers" to cram in still more passengers as the train car doors slide closed (Figure 11–10). Some riders commute very long distances and have to endure the crowding for two hours or more. Tokyo highways and side streets are also crowded and standstill traffic jams are common. At times, even pedestrian zones can be jammed with people complaining about slow sidewalks and difficulties in walking even short distances.

Housing is another key issue. Because of the excessively high cost of land, most people in Tokyo feel that their housing conditions are inadequate: too distant, too small, and too expensive. It is not uncommon for parents and their children to share a single bedroom, or for families to share a housing unit with in-laws, grandparents, or other relatives. Most young couples can only dream about owning a single home in Tokyo, as increasingly overcrowded apartments in high-rise buildings and small condominium units that are ironically called *manshon* (mansions) are the modal dwelling types. However, recent data show that Tokyo rents are dropping because of the poor economy, and that "empty nesters" and two-income households in particular are increasingly able to afford housing in Tokyo's central wards. A new boom

Figure 11–11

One of the main problems in Tokyo and other large Japanese cities is the long commute, often under very crowded conditions, between home and workplace. As a result, planners are advocating constructing high-quality, high-rise housing, with ample amenities and community facilities in central locations. This photograph is from the upscale Roppongi Hills development near downtown Tokyo.

in inner-city apartment construction is a Tokyo variant of the gentrification process (Figure 11–11).

Planners and government officials in Tokyo have tried to address the housing shortage by building large residential towns, such as Tama New Town in Tokyo's western suburbs, and by requiring developers of large commercial properties in central wards to include residential buildings as part of their projects. The latter reduces commuting times for residents and helps to repopulate the city core, which had been hollowed out over the years by commercial expansion.

Yet, while living in Tokyo means living with crowding, the city can also be surprisingly comfortable and enjoyable. It is quite clean despite the number of people, and crime rates are among the lowest in the world. While bad things do sometimes happen, for the most part one can go just about anywhere at any time of day or night in Tokyo, without fear of being robbed or assaulted. This is true for both men and women. The city is a paradise for lovers of restaurants, cafes, music pubs, and dance clubs, and a magnet for the best entertainment from around the world, be it classical music or the hottest stars of rock and hip-hop (Figure 11–12). It is also an outstanding place for shopping, keeping pace with all the latest fashions in Europe and North America, as well as

Figure 11–12

The sport of Sumo is one of countless activities available in Tokyo. Sumo, along with baseball, is a national passion in Japan. Wrestlers have to be not only big, but also quick and wily.

Figure 11–13

Tokyo shoppers. Tokyo and other large Japanese cities are very crowded, especially so at major commuter rail stations. This is a view of a shopping district near Shinjuku Station in Tokyo.

setting fashion trends of its own (Figure 11–13). Furthermore, Tokyo has an enormous array of museums showing some of the finest art in the world as well as the art treasures of Japan, and dozens of small, special purpose museums such as one about beer, another called the Tobacco and Salt Museum, and one devoted to displaying sewing needles.

It is not surprising that Tokyoites are often on the move. Since dwellings are too small for comfort or for entertaining friends, a very common Tokyo lifestyle involves going out after work or school and going home just to sleep. Friends come together at designated meeting places at train or subway stations and then go out for dinner, shopping, a movie, a concert, or whatever other activity interests them. For those with intimacy on their minds, Tokyo has thousands of inviting "love hotels" that rent fantasy theme rooms by the hour. As much as any city in the world, Tokyo has it all.

Main Point Summary

1. Rich and highly urbanized, Japan is one of the world's largest users of natural resources. The Tokyo urban area is the largest metropolitan region on earth.

2. Most Japanese are densely crowded on the less than 25 percent lowland such as the Kanto Plain around Tokyo.

3. Japan has a humid continental climate in the north and a humid subtropical climate in the south.

4. Four islands dominate the archipelago: Hokkaido, Honshu, Kyushu, and Shikoku.

5. Japan is the most seismically active country in the world. Its most recent devastating earthquake was in Kobe in 1995.

6. Japan had its origins with people migrating to Kyushu from Korea around 400 B.C.

7. For much of Japanese history, Kyoto was the capital.

8. From 1603 to 1854, Japan was essentially closed to foreigners. In 1853, Commodore Matthew Perry of the United States sailed into Edo (Tokyo) Bay and soon engineered an unequal treaty with Japan.

9. Fast-paced modernization ensued under the leadership of the Meiji emperor (r. 1868–1912).

10. Beginning in 1894, Japan launched a series of wars to gain natural resources, land for expansion, and its own colonial empire. This instigated the Pacific War of World War II. The Pacific War ended when the United States dropped atomic bombs on Hiroshima and Nagasaki in 1945.

11. Japan's recovery—the fastest of any nation in history—was based on Confucian values of hard work, loyalty, and cooperative efforts, together with American aid.

12. An important institution that was part of Japan's recovery after the war was the *keiretsu* or industrial conglomerate.

13. More than half a century after Japan's defeat in World War II, there are still tensions between Japan and the neighboring countries that were once part of Japan's colonial empire.

14. Japan's relationship with Russia is strained by tension over a disputed string of islands northwest of Hokkaido.

15. Japan's population is declining and aging rapidly.

16. Japan has several minority groups such as the native Ainu, Koreans, Chinese, Okinawans, and social outcasts known as *Burakumin*. Minorities are discriminated against, making it difficult for them to participate fully in Japanese society.

17. Less than five percent of Japanese are farmers and most of these farm only part-time. Farms are very small and land use is intensive. The primary crop is rice.

18. Japanese diets are changing rapidly with more emphasis now placed on imported meat, vegetable, and dairy products. Cheaper foreign imports are undercutting the domestic market. Many people are giving up farming.

19. The foundation of modern Japan's economic growth is manufacturing and services.

20. Japan's industry has always relied on imported raw materials.

21. By the 1980s, Japanese industry was reaping the effects of overcapacity and competition from emerging tigers such as South Korea. Japanese corporations began buying up raw material sources and building factories overseas. Many domestic industries became hollowed out as they lost jobs to cheap overseas labor.

22. The 1990s was the "Lost Decade" in terms of economic growth. Foreign competition, a decline in domestic purchasing, and insolvent loans drove many industries and banks into bankruptcy.

23. Japan has too many workers in the construction industry. Many projects are unnecessary yet they are sponsored by politicians who rely on contributions from construction companies.

24. Motor vehicles are Japan's largest export. The industry is dominated by the "Big Five" of Toyota, Honda, Mitsubishi, Nissan, and Mazda.

25. The catalyst for urban growth was the fortified castle town. Two important castle towns were Osaka and Edo (Tokyo) that were linked by the Tokaido Road. Today, these and other cities along the former Tokaido Road have emerged as an urban-industrial region known as the Tokaido megalopolis.

26. By the mid-eighteenth century, Tokyo was one of the largest cities in the world. The royal capital was moved from Kyoto to Edo in 1868 and Edo was renamed "Tokyo" meaning "Eastern Capital."

27. Japan experienced rapid urbanization after World War II. Now 78 percent of the population is urban.

28. A distinctive feature of Japan's urban pattern is the concentration of cities in a relatively small area. Almost all of Japan's major cities are located in a core region stretching from northern Kyushu to Tokyo.

29. Expansion and growth continues to be focused on the Tokyo metropolitan region at the expense of other places. Thus, Tokyo is characterized by unipolar concentration.

30. In the twenty-first century, Japan possesses a multi-centered regional and urban system founded on advanced information technology. The technopolis is an integral part of this system.

31. Tokyo is crowded but comfortable.

🌐 Bibliography

Castells, Manuel, and Peter Hall. 1994. *Technopoles of the World*. New York: Routledge.

Chang, Iris. 1997. *The Rape of Nanking: The Forgotten Holocaust of World War II*. New York: Basic Books.

Cybriwsky, Roman. 1988. "Shibuya Center, Tokyo." *The Geographical Review*, 78: 48–61.

———. 1998. *Tokyo: The Shogun's City at the Twenty-First Century*. New York: Wiley.

Cybriwsky, Roman, and Aoi Shimuzu. 1993. "Rocks, Rust, and Tourism on Japan's Copper Mountain." *Focus* (American Geographical Society) 43: 22–28.

Diamond, Jared. 1998. "Japanese Roots." *Discover*, 19: 86–94.

1990. "The Sun Sets on Rice." *The Economist*, 21 (April): 44.

Edgington, David, W. 1997. "The Rise of the Yen, "Hollowing Out" and Japan's Troubled Industries." In

Asia Pacific: New Geographies of the Pacific Rim, eds. R. F. Watters and T. G. McGee, pp. 170–189. Vancouver: UBC.

Evans, Richard. 1990. "Japan's Untouchables." *Geographical Magazine* (November): 46–47.

Facklet, Martin. 2004. "The Decline of Japan's Farmers." *Far Eastern Economic Review* 26 (February): 12–15.

Fujita, Kuniko, and Richard Hill, eds. 1993. *Japanese Cities in the World Economy*. Philadelphia: Temple.

Harris, Chauncy, D. 1982. "The Urban and Industrial Transformation of Japan." *The Geographical Review* 72: 50–89.

Hoon, Shim Jae et al. 1993. "Rural Exodus: Seeds of Despair." *Far Eastern Economic Review*, 4 (March): 20–21.

Howell, David, L. 1996. "Ethnicity and Culture in Contemporary Japan." *Journal of Contemporary History*, 31: 171–90.

Kerr, Alex. 2001. *Dogs and Demons: Tales From the Dark Side of Japan*. New York: Hill and Wang.

Kingston, Jeffrey. 2001. *Japan in Transformation 1952–2000*. London: Longman.

Lamont-Brown, Raymond. 1995. "Earthquake—The Ultimate Japanese Nightmare." *Contemporary Review*, 266: 117–19.

Landers, Peter. 1998. "Arthritic Nation (Japan)." *Far Eastern Economic Review* 16 (July): 10–12.

Lie, John. 2001. *Multiethnic Japan*. Cambridge: Harvard University Press.

McCarthy, Shawn. 1997. "The Aging of Japan." *World Press Review* (August): 34–35.

Miyashita, Kenichi, and David Russell. 1996. *Keiretsu: Inside the Hidden Japanese Conglomerates*. New York: McGraw-Hill.

Rimmer, Peter, J. 1997. "Japan's Foreign Direct Investment in the Pacific Rim." In *Asia Pacific: New Geographies of the Pacific Rim*, eds. R. F. Watters and T. G. McGee, pp. 113–32. Vancouver: UBC.

Rothery, David A. 1995. "Kobe—Anatomy of an Earthquake." *Geographical Magazine* 67: 54–55.

Smith, Anne, and Tony Smith. 1986. "Tokyo On a Plate." *Geographical Magazine*, 58: 106–107.

Takahashi, Junjiro, and Noriyuki Sugiura. 1996. "The Japanese Urban System and the Growing Centrality of Tokyo in the Global Economy." In *Emerging World Cities in Pacific Asia*, eds. Fu-chen Lo and Yue-man Yeung, pp. 101–43. New York: United Nations.

Weiner, Michael, ed. 1997. *Japan's Minorities: The Illusion of Homogeneity*. New York: Routledge.

Williams, Jack F., and Kam Wing Chan. 2003. "Cities of East Asia." In *Cities of the World*, eds. Stanley D. Brunn, Jack F. Williams, and Donald J. Zeigler, pp. 413–454. New York: Rowman and Littlefield.

Yeung, Yue-man, and Fu-chen Lo. 1996. "Global Restructuring and Emerging Urban Corridors in Pacific Asia." In *Emerging World Cities in Pacific Asia*, eds. Fu-chen Lo and Yue-ma Yeung, pp. 17–47. New York: United Nations.

Chapter 12

❋

Korea and Taiwan: Tigers Rising

"An oppressive government is more to be feared than a tiger"

CONFUCIUS (C. 551–479 B.C.)

In this chapter I will tell you about two tigers of East Asia: South Korea and Taiwan. I will also talk about North Korea, which is definitely not an economic tiger. As you read on, it will become increasingly evident that the historical evolution of each of these regions is related to developments in China.

Korea: Two Countries, One Nation

PHYSICAL SETTING

Korea is one of the world's oldest known land areas, dating back 1,600 to 2,700 million years. Made of granite and limestone, the peninsula is tilted westward to the Yellow Sea where hundreds of islands dot the coast. Only 20 percent of Korea is level land. The rest consists of low-rising, sharp, and often bare mountain ridges. These ridges never reach above 9,200 feet (2,800 meters). The mountains are more concentrated in the north and east. Along the east coast, the Taebaek Range runs like a spine with southwestern spurs that cut the peninsula into narrow valleys and alluvial plains (Figure 12–1).

Korea is bordered by China on the north by the navigable Yalu and Tumen rivers, both of which flow from Paektusan, the highest point in Korea at 9,000 feet (2,744 m). A range of extinct volcanoes leads from Paektusan to the Sea of Japan.

Like the rest of East Asia, Korea has a monsoon climate with warm, humid summers and cold, dry winters. While North Korea is influenced by continental extremes, the south is warmed by the Japan Current. Southwestern Korea and Cheju Island are, in fact, subtropical and have considerably more rainfall than the rest of the peninsula. About half the rainfall comes in June, July, and August, amounting to about 30 to 40 inches (76 to 102 cm) in the south and less in the colder and drier north.

Figure 12–1

Geography of North and South Korea. From H. J. de Blij and P. O. Muller, *Concepts and Regions in Geography,* Second Edition, 2005, p. 334. Originally rendered in color.
© H. J. de Blij and P. O. Muller. Reprinted with permission of John Wiley & Sons, Inc.

Figure 12–2

These are fields of ginseng outside Kumsan, the main marketing center for ginseng in Korea. Here the root is known as insam. *The soil is a rich mulch of chestnut or oak leaves, and the plants are shaded with thatch or fiber mats. Once used, the soil is not cultivated again for another 10 to 15 years. Ginseng, especially the red variety, has been part of East Asian pharmacopoeia for centuries.* Photograph courtesy of B.A. Weightman.

North-South differences are also apparent in resource availability. Although the peninsula is only moderately endowed with mineral resources, the north is favored with substantial deposits of coal, iron, copper, zinc, and other ingredients needed for heavy industry. The north also has extensive hydroelectric power resources, particularly along the Yalu, Tumen, and Taedong rivers. In the case of the Yalu and Tumen, the power is shared with China. In South Korea, minerals are similar to those in the north but, with the exception of graphite and tungsten, exist in smaller quantities and are of lower quality. Consequently, the South relies on imported raw materials for its heavy industrial base.

Only about a sixth of mountainous North Korea is suitable for cropping. Coastal lowlands produce rice, corn, wheat, millet, and soybeans but food sufficiency is a problem. The hills have rich timber reserves, extensive orchards, and livestock grazing.

South Korea, with its population of 48.6 million, is the peninsula's traditional rice bowl, with about 20 percent level land. However, greater productivity is offset by a considerably larger population, more than double that of North Korea's 22.7 million in 2003. South Korea's agriculture is intensive, focused on rice, vegetables, fruit, and other types of market gardening. Kumsan is notable for its ginseng production (Figure 12–2). Cattle are raised on Cheju Island, where it is just warm enough to grow oranges and pineapples. Greenhouse agriculture is ubiquitous in South Korea.

South Korea is more urbanized than North Korea, which has only one city over a million people. This is the capital of Pyongyang with 2.6 million. The twin cities of Hamhung-Hungnam contain fewer than a million people. In the South, six cities have more than a million people. The conurbation of Seoul and its port Inchon is the largest, with a combined population of 18 million people. Seoul, which means "capital," has been Korea's (now South Korea's) capital for 500 years.

The quintessential difference between North and South is political. North Korea's strict communist system has produced a harsh landscape of rigidity and poverty, isolated from all except China and Russia. Variation in South Korea's government—from authoritarian to democracy—has produced a landscape of free enterprise, modernization, and economic development which have raised the nation to the status of Tiger in East Asia.

HISTORIC EVOLUTION

Archaeological evidence suggests that the Korean Peninsula was first settled 5,000 years ago when Tungusic tribes migrated to the area from Manchuria

Figure 12–3

This corner eave of the great Buddhist temple at Pulguksa illustrates the detailed craftsmanship of Korean temple builders. The central dragon is meant to ward off evil. Built in the sixth century, this is one of the oldest surviving Buddhist monasteries in Korea. Photograph courtesy of B.A. Weightman.

and northern China. Koguryo, the first state, developed around 100 A.D. The kingdoms of Paekche in the southwest and Silla in the southeast emerged around 250 and 350 A.D., respectively. These three kingdoms vied for control of the peninsula until the seventh century.

Aided by the Chinese, Silla was able to conquer Paekche and Koguryo by 668 A.D. Trading centers developed in Kyongju, the capital, as well as in provincial towns. Korean ships engaged in the region's sea trade. Buddhism (from China) was established as the state religion, and lavish temples and tombs were constructed (Figure 12–3). Woodblock printing was developed for the production of religious texts. From this time forward, Korea had an important cultural influence on Japan, including the introduction of Buddhism.

By 918, the kingdom of **Koryo** emerged in the central region of the peninsula and overthrew Silla in 935. Korea is the westernized version of Koryo. Buddhism remained the state religion but Confucianism emerged as a major force in government ethics and political structure. Even today, Korea is considered the most Confucian of all Asian nations. Check out its flag. Where do these symbols originate?

In 1231, Mongol armies moved into the peninsula. After 30 years of war, Koryo became a vassal state of China's Mongol Yuan Dynasty. Following the fall of the Yuan, Korea's Yi Dynasty continued the tributary relationship with China's Ming Dynasty (Chapter 9).

The Yi moved their capital to Seoul on the Han River and renamed their kingdom Chosun in 1392. Under Yi rule, Confucianism replaced Buddhism as the state belief system. The world's first movable, metal type was developed. More importantly, a Korean lettering system was developed in the fifteenth century. Known as **Hangul**, it reduced dependency on Chinese characters and made literacy achievable for the general public.

Hangul

In the fifteenth century, King Sejong (ruled from 1419–1450) commissioned scholars from the "Hall of Worthies" to develop a Korean alphabet system for those who had not mastered Chinese characters. After analyzing the phonetic structure of the language, they distinguished three separate elements of words: initial, medial, and final sounds. Of these, the medial element is a vowel. The new alphabet consisted of 29 letters—22 consonants and

7 vowels. Consonants and vowels are grouped into a syllabic block form to create a word. This new system, called *Hangul*, differed radically from the Chinese ideographic system in which a concept is conveyed by each character. King Sejong presented his "correct sounds for the instruction of the people" in 1443 and literacy became widespread.

Tangling with a Japanese Tiger

Japanese forces invaded Korea in 1592 and again in 1597. With the help of China, the Koreans fended off the Japanese but were forced to send artisans and scholars to teach Chosun's advanced technology in Japan. The Japanese ceramic and laquerware industries owe their origins to the Koreans. Later, the Japanese were given residence rights in a small area of Pusan.

The seventeenth century brought Manchu control and subsequent decline. After much struggle, Korea fell to Manchu power in 1637. Significant change was on the horizon. Factional strife within Korea weakened the political structure. Christian ideas, introduced via official tribute missions to China, clashed with Confucianism. New agricultural techniques fostered a commercial economy, despite aristocratic disdain for such activity. Corruption and intrigue hampered orderly administration. By the mid-1800s, Korean politics and the economy were in shambles.

In 1864, the Korean government closed the country to all except the Chinese and the Japanese at Pusan. It was during this period that Korea became known as the *Hermit Kingdom*. All other foreign interests were spurned until Japan forced Korea to enter into its first foreign commercial treaty in 1876. This period also marked a great divide in the nation's history. Prior history was China-centered. The future would be determined by Japan and the West.

Korea fell prey to Japanese imperialism in the late nineteenth century. The Japanese engineered the murder of the anti-Japanese Korean Queen Min in 1895. The Russo-Japanese War of 1904–1905 saw Japanese troops operating on the Korean peninsula

as they moved the war front into Manchuria. Korea became a Japanese protectorate in 1905, and despite anti-Japanese efforts, Japan was able to annex the country in 1910. Japanese rule was highly resented by the Koreans, who viewed themselves as culturally superior. Japan relied on the use of force to transform its colony.

While the traditional, Confucian agrarian order was retained, a modern administration was established, with Japanese functioning in key positions. Both agriculture and industry were structured to meet the needs of Japan and its plans for aggrandizement. Transportation, communication, mining, and other infrastructure were modernized and expanded. Thousands of Koreans were sent to Japan to work in factories. In Japan, the Korean language was banned and Koreans had to take Japanese names. In Korea, resistance movements were brutally crushed, deepening Korean hatred of their Japanese overlords. Japanese rule of Korea did not end until Tokyo surrendered to the Allied Forces in 1945.

North-South Division

In 1945, the Soviets and the Americans divided the peninsula into two spheres of influence and military occupation at the **38th parallel**. Free elections were supposed to follow, but disagreement on this issue ensued in the next two years. In the face of the adamant refusal of the Soviets to subscribe to United Nations-supervised elections, the Republic of Korea (ROK) was established south of the 38th parallel on August 15, 1948. The Soviet response came in September, with the formation of the Democratic People's Republic (DPRK) in the north. The United States withdrew its military forces the following year.

On June 25, 1950, North Korean armed forces attacked South Korea across the 38th parallel. The United Nations responded with the support of sixteen nations, led by the United States with 90 percent of the troops, and the Korean War ensued. The conflict was prolonged by the entry of Chinese troops in support of North Korea in 1950. After three years of fighting, an armistice was signed, with an agreement to find a political solution to the problem of division. The 1953 agreement led to the creation of a Demilitarized Zone (DMZ), which has become heavily militarized on both sides.

The DMZ is an interesting site. You can visit the southern flank and look across a "no-man's-land" to a phony "ideal" North Korean village where propaganda billboards tout the superior lifestyle of North Koreans. The route to the DMZ, just 25 miles (40 km) north of Seoul, is marked by military bunkers and bomb-filled overpasses ready to crush any invasion from the North. The North Koreans have made several tunneling attempts into the South, but these have been detected and destroyed.

The Korean War devastated the country, especially the north, where systematic bombing eradicated the industrial plants. Some 4 million perished, and millions of refugees remained trapped on the wrong side of the divide, most never to see their families again. Around 1 million refugees fled to the South with the retreating U.N. forces. Subsequently, the Russians and Chinese aided the North and the Americans supported the South in recovery efforts. Korea was now two countries, each pursuing its own path of development.

North Korea: A Closed Society

SOCIETY AND ECONOMY

The Democratic People's Republic of Korea (DPRK) is one of the, if not *the* world's most closed and rigid societies. Under the absolute power of "Great Leader" Kim Il Song, the north was brought from the brink of disaster and turned into an industrial economy based on central planning and state ownership of the means of production. Individual needs are subjected to the will of the state and ownership is limited to individual and household possessions.

The Great Leader espoused the *juche* philosophy, meaning self-reliance or self-sufficiency. *Juche* is based on the proposition that humans are masters of all things and that they, among all creatures, are endowed with the capability of self-reliance and independence. However, the masses cannot succeed in their struggle for these goals without a supreme leader to give form to their causes and direction to

their actions. This leader was Kim Il Song who was followed by his son, "Dear Leader" Kim Jong Il in 1994.

North Korean society is stratified according to Communist Party loyalty; level of commitment determines job prospects, and access to housing, education, and health care. Citizens can be detained arbitrarily; there are an estimated 150,000 political prisoners. Listening to foreign radio broadcasts is punishable by death. Subjugation to a tightly structured administration is a familiar mode of rule in Korea, and this reinforces the absolute authoritarianism that North Koreans now experience.

North Korea has a highly socialized command economy, in which farm land is collectivized and agriculture accounts for 36 percent of the work force. State-owned industry produces 95 percent of manufactured goods, mainly heavy industrial products and armaments. Light industry lags far behind. In 1992, output dropped sharply because of disruption in relations with Russia and China and reduction in their economic and technical support. There is a shortage of arable land and the country has not achieved self-sufficiency in food production.

North Korea needs at least 6 million tons of feed grain a year to get by. Its last crop in 1994 produced 4.1 million tons. Torrential rains in a badly deforested environment drastically reduced the harvests of 1995 and 1996. Aid workers and others who managed to sneak across the Chinese border to the Korean community there reported widespread starvation. Hillsides were stripped of vegetation. There were no cats, dogs, or chickens. It is said that the military and the Communist Party elite remain well fed, due to food aid from China, Russia, Japan, and the United States.

In 1995, peasants were allowed to cultivate private plots, but this resulted in reduction of yields at the collective level. The DPRK estimates that famine claimed at least 220,000 lives between 1995 and 1998. International organizations estimate that 1.5 million North Koreans have perished due to famine-related causes. Although climatic conditions have improved in recent years, the nation is still significantly short of food and lacks the minimal calories required each day to stave off malnutrition. The government has broadcast recipes for boiling grasses.

Data on urbanization in North Korea is notably unreliable. However, according to DPRK information, urbanization has increased from less than 20 percent in 1953 to nearly 60 percent in 2003. The definition of *urban* is unclear. Most urbanites live in 23 cities of more than 89,000. Pyongyang has over two million. Hamhung, the second largest city, is about a third the size of Pyongyang. The port of Chongjin is another large, industrial city.

Figure 12–4

Pyongyang. Tony Waltham/Robert Harding Picture Library.

Pyongyang

Pyongyang, North Korea's largest city, is also its national capital. Situated along the Taedong River near the west coast, it controls all commerce moving north and south along North Korea's western side of the peninsula. Leveled to the ground in the Korean War, Pyongyang was rebuilt with wide boulevards and massive, grim government buildings typical of those built in the Soviet Union in the late 1940s and 1950s (Figure 12–4). Drab apartment blocks have few amenities. It is the only city in North Korea to have an underground subway system. Although there are only two lines, the subway is considered a national accomplishment and ranks with political and cultural monuments as a point of pride.

In 2002, North Korea allowed wages and prices to rise, and created incentives whereby factories could turn a profit. Consequently, consumerism has increased, at least in Pyongyang. Billboards advertise a North Korean-made car—the *Huiparam* or "The Whistle." The Whistle is assembled in the port city of Nampo and is made with Fiat parts imported via Italy and South Korea. The enterprise is a joint venture with a South Korean automobile company. However, most North Koreans cannot afford to purchase a car.

Along major streets, sidewalk kiosks are open for business. Locally grown apples and peaches are sold alongside imported pineapples. Clothing, televisions, and other cheap goods from China, dishwashing soap from Thailand, and beer from Singapore are among the goods that are available. Prices are determined by customer demand and people pay with cash, not with coupons as in the past.

Since the changes were introduced, increasing numbers of cars and bicycles are seen in the streets. People are wearing brighter colored clothes from China instead of the grey and brown Mao suits and peaked caps that they wore before. Some young people even wear baseball caps and the sight of Mickey Mouse T-shirts is not uncommon.

Economic reforms have at least affected Pyongyang. Their impact in other North Korean cities is unclear. Some experts say that the country will achieve widespread market socialism like China. Others say that these reforms are not enough to generate much change overall. The United States views the changes as insignificant and considers North Korea to be a "rogue state."

Beijing has cast its relationship with North Korea as being "close as lips and teeth." It is in China's interest to prop up the North Korean regime while at the same time effecting its redirection along a more liberal path according to the China model. China wants regional stability. It does not want a floundering country on its northeastern border next to its own Korean population, nor does it want a unified Korea, friendly to the West and housing American troops.

North Korea, with its population of a mere 22.7 million, has the fifth largest standing army in the world after China, the United States, Russia, and India. More than 1 million troops hover over the half-million South Korean forces.

Today, many young North Koreans eschew higher education in favor of joining the armed forces. Here they will get good pay and sufficient food. It is estimated that military expenses consume one-quarter to one-third of the entire national budget despite near-famine conditions in many parts of the country.

NUCLEAR CRISIS

In 1994, the DPRK suggested that it could build a nuclear bomb by extracting plutonium from one of its test reactors. In response, the United States, South Korea, and other countries agreed to provide the North with oil until the South Koreans completed two nuclear power reactors in the North. The North Koreans were to dismantle their bomb-making program.

In 2002, North Korea admitted that it had not dismantled its bomb-making program. Interested parties, including South Korea, Japan, Russia, and the United States have made several attempts to reason with the North Koreans, but to no avail. U.S. President George W. Bush cut off shipments of oil and other supplies to the DPRK, which responded by evicting the United Nations inspectors who had been monitoring the nuclear program. The situation remains unresolved.

The DPRK also has an arsenal of missiles capable of striking Japan and beyond. It sells these to Middle Eastern countries such as Iran for badly needed foreign exchange. In 1998, while engaged in talks with the United States regarding normalization of relations and food aid, North Korea proposed that Washington pay US $500 million to stop missile sales. North Korea's actions have provoked concerns in the region and have revived support for "Star Wars" type missile defense systems in Japan and Taiwan. China is not pleased with this turn of events. Clearly, North Korea's actions are detrimental to any consideration of reunification of North and South.

South Korea: Korea, Inc.

South Korea has transformed itself from a poor, agrarian society to one of the world's most highly industrialized nations. With huge support from the United States following the Korean War, South Korea has become one of the newly-emergent, urban-industrial countries linked to the global economic system. Land reform, education, and economic expansion have eliminated the *yangban,* the landlord elite. The growth of a middle class has been accompanied by dramatic population growth and urbanization.

South Korea's governments have ranged from paternalistic democracies to dictatorships. Absolute power was used to promote personal savings and export industries. By the 1980s and 1990s, there was rising prosperity and large trade balances. Freer elections in the mid-1980s led to an expansion of democratic processes in the 1990s. Because of the influential role played by the government, South Korea is said to be a product of **state capitalism**.

STATE CAPITALISM

South Korea's transition from an agricultural to an urban and industrial society followed several phases. At the end of the Korean War, three-quarters of the

South Korean people were engaged in agriculture. In post-war reforms, the government took over poorly run large estates and encouraged a new group of small landowners to farm with the aid of imported chemical fertilizer. Productivity rose rapidly and the country became largely self-sufficient in food.

Government and business targeted specific industries for development. Initially, emphasis was placed on light manufactures and textiles. In the 1970s, heavy industries such as iron and steel and chemicals were emphasized. Then, the focus was shifted to making automobiles, ships, and electronics.

The lynchpin of industrialization is the *chaebol*. A *chaebol* is a giant corporate conglomerate. Hyundai, Daewoo, Samsung, and LG (Lucky-Goldstar) are leading *chaebols*. In the early years of development, preferential investment in exports discouraged the production of consumer goods. This policy was reversed in the 1980s. Then labor unions demanded wage increases and domestic consumption rose. Higher wages saw a decline in South Korea's international competitiveness in labor-intensive manufacturing, and overseas investment for cheap labor sources became essential.

Since 1990, South Korea has become a major investor in China where it has access to cheap labor. Much of this investment is in the relatively nearby coastal province of Shandong. Initially, the investors were small businesses but now it is the *chaebols* that are establishing small companies in China to produce their components and products.

Chaebols

Chaebols are the equivalent of the Japanese *keiretsu*. Both terms are written with the same Chinese characters meaning "fortune cluster." Owned and managed by family groups, they are enormous conglomerates of diversified businesses, each having hundreds of ancillary firms. Of about 30 such concerns, Samsung is the largest. Samsung manufactures products ranging from airplanes to semiconductors, to

flour and paper. It also operates an array of service firms such as insurance, hotels, and department stores.

The rise of *chaebols* to economic dominance can be attributed, in part, to Confucian ideals and loyalties. The *chaebol* is like a family with its structured loyalties intensifying toward the family head: the boss. Unlike Japan, where consensus rules, in Korea consensus is what the boss says it is.

In the Hyundai *chaebol*, Confucian spirit became the "Hyundai Spirit." A manual given to new recruits observes: "The indomitable driving force, a religious belief in attaining a goal, and a moral diligence tempered with the frugality of the Hyundai group have materialized as a major part in the heavy industry of Korea."

Another important *chaebol* anchor is the support of state-owned banks and guaranteed loans. Government-infused capital has fostered expansion and diversification. In 2000, the four top *chaebols* accounted for 40 to 45 percent of South Korea's economic output.

More recently, with rapid wage increases under democratic reforms, *chaebols* have become less competitive. Disbanding or selling off numerous unprofitable "family members" (ancillary firms), venturing more into research and development, and procuring cheaper labor overseas are part of recent restructuring efforts.

South Korean enterprises are found worldwide. For example, Daewoo, which means "Great Universe," has poured billions of dollars in planet-circling investments including electronics and components, hotels, autos, construction, finance, textiles, and heavy industries. Daewoo's globalization plans include hiring 250,000 foreign workers at 1,000 overseas subsidiaries. Daewoo Motors has plants in such countries as Poland, India, Iran, Uzbekistan, and the Philippines. Daewoo Electronics hopes to capture 10 percent of the world market in TVs, VCRs, PC monitors, refrigerators, washing machines, and microwave ovens.

ECONOMIC CRISIS AND CHANGE

The Asian financial crisis of 1997 was especially severe in South Korea. Having unlimited access to bank loans, many *chaebols* incurred serious debt. That year, the Korean currency was devalued 50 percent and many subsidiaries went bankrupt. Per capita income declined by 35 percent and 1.6 million people became unemployed. The International Monetary Fund (IMF) came to the rescue with the largest loan ever awarded a single country. Reforms were introduced to reduce the power of the *chaebols* and more loans were given to small and medium sized companies. State-owned industries are being privatized.

Unemployment continues to be a problem. About five million South Koreans between the ages of 20 and 34 are unemployed or underemployed. Graduates fare the worst in terms of having difficulty finding jobs. In 2002, six out of ten higher-education graduates failed to find employment.

There are two reasons for this situation. The first is demographic. Children of the country's second baby boom—the progeny of the first baby boomers born after the Korean War—are reaching adulthood and pouring into the job market. The second reason is economic. Since the Asian financial crisis of the late 1990s, companies have become more focused on profits and are reducing workforces. Many rely on temporary employees and others with more experience and skills. They have also turned to technology for labor-intensive tasks and are outsourcing jobs to China and Southeast Asia.

Improvements in the economy will not necessarily translate into positions for younger workers. Take the giant Lotte Department Store as an example. Lotte has developed a checkout counter system that replaces workers with personal digital assistants and is planning to install these in its 20 stores nationwide. The company expects to replace at least 70 percent of its workers, mostly in their 20s.

The government is trying to help. In 2004, it spent nearly 500 million dollars in subsidies to companies that hire jobless youth. Job creation programs last between six months and a year and are designed to give young people experience and a chance to improve their job skills. Government ministries have opened up 142,000 positions and hope to stabilize the youth unemployment rate at five percent over the next four years.

PLANS FOR CHEJU ISLAND

In 2002, the government announced plans to develop the island of Cheju off the southern coast. The island is very attractive with its mild climate, volcanic mountains, waterfalls, and sandy beaches. It is a traditional honeymoon destination and possesses fruit plantations, fish farms, and abundant seafood.

Officials who are developing the project boast of South Korea's and Cheju's geographic position between China and Japan. Cheju is two hours flying time from Hong Kong, Beijing, and Tokyo. The island is expected to become an international business and tourism hub as well as a free-trade area where English will be the second language. There will be shopping malls, a science and technology park, and several golf courses. A new airport and many new hotels will be built.

The project is supposed to be completed in 2010 but this depends largely on private investors. Moreover, the local population of half a million is complaining about lack of consultation, the potential disruption of their way of life, and ensuing environmental damage.

URBANIZATION

South Korea's population more than doubled from 20 million in 1949 to 48.6 million by 2004. A major contributor was the influx of 3.3 million Koreans who had been living in Japan, China, and the Soviet Union. An additional 2 million refugees migrated from North Korea. A population control program launched in 1965 curbed the birth rate, but improved living standards, lowered the death rate, and increased life expectancy. Rural-urban migration depleted the countryside and provoked a major urban housing crisis. Millions of apartment units (flats) have been constructed to house this 75 percent urban society.

Figure 12–5

This is a southern expansion of Seoul looking past the Han River and the Olympic Expressway to the suburb of Tongjakgu. More than half of Seoul's population lives south of the Han River. Photograph courtesy of B.A. Weightman.

Urbanization and apartment living have altered Korean traditional family structure. Most Koreans now live in nuclear family settings rather than in extended family systems. However, Confucian values prevail. Although women now have more control over their lives, they are far from equal to men in decision-making and economic opportunity. Women are found mostly in service or "nimble-finger" jobs. Corporations are overwhelmingly male enterprises at management levels.

Two-thirds of South Koreans live in cities of 50,000 or more with 47 percent living in and around Seoul. Greater Seoul has sprawled eastward and southward with an agglomeration of new satellite cities (Figure 12–5). Formerly focused on a single central business district (CBD) dating from Japanese times, Seoul has evolved from a mono-centric to a multi-centric city. Despite rapid construction of apartments, the average occupation rate in 1990 was 193 percent. As of 2004, in older, more densely populated areas, occupation rates are as high as 600 percent! This means that most flats house two families—one owner and one renter.

Much of downtown Seoul is underground. Miles of shopping arcades lie under the streets. All are connected to a network of subway lines and linked to major department stores and hotels above ground.

Another modern development is the rejuvenation of the Han River. Formerly a trash-filled, dry ditch in summer, its water level is now kept constant by a series of dams. The river is criss-crossed by bridges and lined with parks and high-rise housing developments. Whereas 80 percent of raw sewage used to flow into the Han channel, new sewage systems and treatment plants now treat all sewage. Today, the Han River carries sailing and passenger boats.

Inchon and Taegu are cities of more than two million, while Taejon and Kwangju have more than a million residents apiece. The port of Pusan has a population of close to four million. With their concentrations of industry, these large cities continue to grow, while medium size and smaller places tend to stagnate. However, even their rate of growth is decreasing as traditional industries have begun to decline.

With the bulk of industry still concentrated in the capital region of Seoul, the Korean government has devised a **technobelt** system whereby research and industry are to be linked via communication and telecommunication. The key nodes of this system are Seoul and Taedok, a new research and development center near Taejon. Each belt has designated specializations. For example, the Seoul to Kwangju technobelt will focus on energy and food production. The Seoul–Taedok–Pusan technobelt will concentrate on electronics and textiles. Other areas designated for growth (e.g., Masan and Chang-won on the south coast) have become engineering centers linked with the Pusan-Ulsan heavy industrial region. Ulsan is dominated by Hyundai car manufacturing and shipbuilding.

Overseas Koreans

Nearly 5 million Koreans live overseas. The largest group resides in China. As many as 2 million live in China's Manchurian region. The strongest concentration is in Yanbian Korean Autonomous Prefecture just north of the Tumen River. Here, Koreans are able to exercise a degree of autonomy and speak Korean in accordance with the rights given to all minority nationalities in China.

Koreans were pioneers in opening up the cold Manchurian region to paddy rice farming. Rice was first grown in Yanbian in 1877, and Koreans are still responsible for producing much of the rice crop in the most northerly province of Heilungjiang.

Large contingents of Koreans also live in the Central Asian republics of Kazakhstan and Uzbekistan and on Sakhalin off the east coast of Siberia. Unofficial estimates range up to 750,000. These populations are important links for South Korea's recent Central Asian investments.

As discussed earlier, the Korean community in Japan makes up 85 percent of its foreign population. Regarded as citizens of the Japanese empire during colonial times, the majority of Koreans continue to be classified as aliens and suffer from various forms of discrimination.

Emigration to the United States began in the early twentieth century. From 1903 to 1905 a group of 7,226 Koreans was allowed to live in Hawaii and work on sugar plantations. Under Japanese pressure, this migration was stopped and many Korean settlers abandoned the plantations and became shopkeepers. Some 2,000 moved to California. This was the seed of a subsequent migration stream to the United States. The majority of these Koreans are Christian.

About a third of America's Koreans live in California, mainly in the Los Angeles area. Other significant concentrations are in New York and Chicago.

THE U.S. AND SOUTH KOREA

For the past 50 years, South Korea's foreign policy has been in virtual lock-step with that of the United States. However, in 2002 when President Roh Moo-hyun was elected, he said that his country would no longer "kowtow" to the United States.

There has been a wave of nationalism in South Korea triggered by the acquittal of two American soldiers charged in the deaths of two girls killed during a training exercise. "Wrongs will be righted," declared President Roh as he referred to grievances associated with the bilateral treaty that governs the 37,000 American troops stationed in his country. Even North Korea has criticized the treaty because, in essence, it gives extraterritorial rights to the soldiers. In other words, American troops are subject to American and not South Korean laws.

As the relationship between the two countries has changed, the United States has taken steps to downplay its profile. It plans to close a large military base in Seoul and pull back its troops from the DMZ. South Korean conservatives are not happy with these moves as they feel vulnerable to attack from North Korea. Moreover, they do not want to pay for the moves. Only time will tell what the ramifications of this situation will be.

PROSPECTS FOR REUNIFICATION

North and South Korea have entered the twenty-first century as two diametrically opposed systems with divergent paths of development. Both Koreas remain pledged to reunification and occasionally present proposals. A strong emotional commitment to reunification exists among many Koreans, especially those with cross-border family ties.

Extant differences make reunification problematic. North Koreans are not even an eighth as rich as South Koreans. South Koreans are twice as numerous. North Koreans have virtually no idea of how to function in a modern, technologically advanced society. There are only 24 telephone lines between Seoul and Pyongyang, and these are rarely used. In the 1980s, a visiting DPRK delegation accused the South of rounding up every car in Korea to create traffic jams in Seoul. Even the language has evolved differently, with the South's technological vocabulary unknown in the North.

Raising the North's productivity level to stem a southward flow of migrants would cost billions. Although the South has traded with the North since 1990—food for raw materials—investment has been severely limited by the DPRK government.

South Korea wants to reunify in stages: diplomatic contacts to build trust; investment in special economic zones in the North; and, down the line when the economic gap has been narrowed, full unification. In 2000, an historic event took place. President Kim Dae Jung of South Korea met with President Kim Jong Il of North Korea to reduce tension between the two countries and promote peace and reconciliation. This decision, made by President Kim, became known as the "Sunshine Policy." As a result of the meeting, both sides have stopped propaganda broadcasts across the border and allowed a limited number of cross-border visits for family reunification. Agreement was made for the construction of a railway from Seoul to Pyongyang. Symbolically, North and South Korean athletes marched together under one flag at the Olympic Games in Sydney, Australia in 2000 and in Athens, Greece in 2004. More talks are scheduled in the future. Kim Dae Jung has been awarded the Nobel Peace Prize for his efforts towards peace in the region. However, North Korea's recent nuclear program threatens reunification considerations.

Taiwan: Little Tiger and Big Dragon

Small but rich, economically powerful but politically isolated: this is Taiwan, an island just 100 miles (165 km) from the big dragon, Mainland China (Figure 12–6). The history of Taiwan is inextricably tied to that of the People's Republic, and its current development remains couched in that relationship. It is one of East Asia's newly industrializing "countries" and has achieved tiger status in terms of economic development.

THE PHYSICAL LANDSCAPE

Taiwan incorporates the main island and 64 smaller islands called the Penghu Islands. The Penghu were colonized by the Portuguese, who named them the Pescadores (Fishermen). Here, fringing reefs provide good shelter for ships and there is an important naval base at Makung. Because of their position in the Taiwan Strait, the Penghu Islands are of great strategic importance to Taiwan.

The island of Taiwan can be divided both vertically and horizontally. The vertical division is created by the Chungyang Shan (Central Mountains) ranging from 10 to 35 miles (16–56 km) in width. The Chungyang are extremely rugged with more than 30 peaks reaching over 10,000 feet (3,000 m). The highest is Yu Shan at 13,113 feet (3,997 m). River valleys and canyons cut the mountainsides, which are covered with coniferous and mixed forests. With the exception of the Taitung rift valley, cool temperatures, thin soils, and severe erosion hinder cultivation. Still, the mountains are important for forest products and hydroelectric power.

Most people dwell on the western coastal plain, which can be divided horizontally by the Choshui River. To the north are hills and basins, including those of Taipei and Taichung. To the south is a low, flat, alluvial plain. The river also marks the climatic boundary between north and south. In the north, the rainy season is from October to March; in the south it is from May to September. Typhoons are

Figure 12–6

Taiwan. From H. J. de Blij and P. O. Muller, *Concepts and Regions in Geography,* Second Edition, 2005, p. 336. Originally rendered in color. © H. J. de Blij and P. O. Muller. Reprinted with permission of John Wiley & Sons, Inc.

common in July and August. Agricultural patterns differ as well with such crops as rice and tea in the north and soybeans, pineapples, sugarcane, and other tropical crops in the south.

Agricultural output has steadily declined over the years as farmland and employment shifts to urbanization and industry. Taiwan is 78 percent urban with major population clusters in the capital of Taipei, and in the ports of Chilung, Tainan, and Kaohsiung.

FROM *BAO DAO* TO FORMOSA TO TAIWAN

Ancient Chinese maps portray the island of Taiwan as the island of **Bao Dao.** Written accounts referred to it as "a mudball across the sea, not worthy of China." Bao Dao was known to Chinese fishermen and pirates but was inhabited by non-Chinese people. These aboriginal people are Malay-Polynesian and are related to the indigenous people of the mountains of northern Luzon, the closest island of the Philippines. There are nine distinct ethnic groups among the aborigines, each with its own language and culture. However, immigration from China pushed these people into the mountains and subsequent discriminatory policies of the Taiwanese government have rendered their cultures almost extinct.

Immigrants from China began arriving in the fourteenth century, a migration stream that continued until World War II. These were Han people from the Fujian coast who spoke the dialect known variously as Amoy, Minnan, or Hoklo. This dialect is called *Taiwanese*, and these people are known as **native Taiwanese**. Native Taiwanese are 70 percent of Taiwan's population. In recent years the term **Taiwanese** has come to be used to mean everyone who lives on the island.

In the sixteenth century, the Portuguese were struck by the beauty of the forested mountains and pristine beaches of the island and called it *Ilha Formosa*, "the beautiful island." Taiwan was known to the world as Formosa for much of the twentieth century.

Japanese traders also visited the island and even established a settlement near present-day Tainan. They called the island *Tai Wan*, meaning "Big Bay."

In 1624, the Dutch established Forts Zeelandia and Providentia near the by-then defunct Japanese settlement at Tainan. Spain established colonies in 1626 but was later ousted by the Dutch. The Dutch established sugar plantations and encouraged increased immigration from China as fieldworkers. The Dutch were thwarted by a Ming general in 1662.

In the seventeenth century, Hakka speakers from Guangdong arrived and settled in the foothills. There, they developed Taiwan's forest products industry including camphor wood and oil. Hakka account for 10 to 15 percent of the population and make up a disproportionate share of Taiwan's industrial workers.

Emigration from China had been illegal since the Ming Dynasty. Even so, people left to avoid taxes and seize new opportunities elsewhere. The Chinese regime regarded them as people who renounced civilization. To China, Taiwan was uncivilized.

In 1684, the Qing Dynasty achieved control of Taiwan for the first time. But the island's population rebelled against them repeatedly over the next 200 years. One official Qing document complained: "Every three years an uprising! Every five years a rebellion!" Chinese immigration continued, and by the late nineteenth century, the island had a population of 2.5 million.

Western traders reappeared after the Opium Wars of the 1840s. Britain and Canada expanded agricultural exports, sent Presbyterian missionaries, and established hospitals and schools.

The safety of foreign traders was a major issue. China denied responsibility for incidents of violence because these were taking place "beyond the boundaries of China." Nevertheless, China made Taiwan a province of China in 1887. These unclear relationships between China and Taiwan were a prelude to decades of uncertainty and difficulties which continue to this day.

JAPANESE RULE

At the end of the Sino-Japanese War (1894–1895), the Treaty of Shimonoseki gave Formosa and the Pescadores (along with Korea) to Japan. On June 1, 1895, a resistance movement proclaimed Asia's first republic: the Republic of Formosa. This republican group was quickly crushed, and the Japanese began their era of harsh rule.

Japan accelerated Taiwan's economic development. This was meant to provide Japan with agricultural and later, industrial products. The colonial authorities expanded electric power, and improved port facilities, internal transportation, and communication. A health system was established, and an educational system promoted Japanese language and culture. Many Taiwanese attended Japanese universities. The groundwork for future progress was laid at this time.

When the Qing Dynasty was overthrown in 1911, both the Chinese Nationalists and the Communists wanted Taiwan liberated from the Japanese. Both groups voiced support for an "independent and democratic" Taiwan.

Chiang Kai-shek became increasingly interested in Taiwan as World War II ensued. Fearing that Chiang's corrupt and incompetent troops would surrender to Japan, the Allied powers issued the Cairo Declaration in 1943. This was one of several communiqués promising to restore Taiwan and the Pescadores to China. Ever since, despite their earlier support of Taiwan's independence, the Nationalists and Communists have cited this declaration to justify their view that "Taiwan is a sacred and inseparable province of China." But the Cairo Declaration was, in fact, a statement and not a legally binding treaty. The island's uncertain status was furthered by the Allied-Japanese peace treaties that left the fate of the island to its inhabitants.

Another element of the island's population consists of immigrants who arrived from the mainland between 1945 and 1949. These are referred to as **mainlanders,** although their "mainlander" descendants are now born in Taiwan. In fact, more than 90 percent of Taiwanese were born on the island.

THE "ONE CHINA" QUESTION

In 1949, Chiang Kai-shek and his two million KMT loyalists retreated to Taiwan where they set up what they claimed to be the true government of the Republic of China (ROC) in exile. In that same year, Mao Zedong renamed China the People's Republic of China (PRC). Both Mao and Chiang claimed to head the government of "one China," of which Taiwan is a province.

In Taiwan, Mandarin was declared the official language by the KMT government. However, Taiwanese is widely spoken in informal settings, and current demand has resulted in the production of Taiwanese language media.

The United States was entrenched in the Cold War against communism and consequently supported Chiang Kai-shek and the KMT. However, this support met with world realities in 1971 when the ROC lost its seat at the United Nations in favor of the PRC. The seating of the PRC in the world body quickly eroded Taiwan's status as the sole representative of China. In 1972, U.S. President Nixon signed the Shanghai Communiqué by which the United States acknowledged the PRC view that "all Chinese on either side of the Taiwan Strait maintain there is but one China and that Taiwan is a part of China." It gave no credence to the Taiwanese desire for self-determination.

In 1978, President Carter established formal diplomatic relations with the PRC and reiterated the Shanghai Communiqué. The United States also severed formal relations with Taiwan, although economic and cultural offices maintain informal ties. The United States wants the issue resolved peacefully. The PRC says that the issue of reunification is an internal affair.

Regional considerations are also critical. If the PRC controlled both flanks of the Taiwan Strait, it would have a chokehold on the main sea lanes to South Korea and Japan.

As PRC-international relations increased, Taiwan became represented only as "Taipei" in international organizations. Taiwan has no official embassies abroad but allows its trade offices to handle similar functions. In the 1990s, the island began promoting itself as "China Taipei."

Both Taiwan and the mainland cling to their notions of one China. The PRC hopes that the examples of Hong Kong and Macau and their "one country, two systems" policy will stimulate the ROC to formally rejoin the PRC. But many Taiwanese, aware of conditions on the mainland, have little interest in this goal. There is also talk of changing the name PRC to simply "China." However, while everyone agrees that they belong to "China," political, economic and other practical matters remain at issue. The China-Taiwan problem remains unresolved.

The tenuous political situation has not precluded ongoing and intensifying economic relationships between China and Taiwan. In fact, big dragon China is of significant economic consequence to little tiger Taiwan.

RISE OF THE TIGER

Taiwan's achievements as an economic tiger are underpinned by its excellent educational system, in-place infrastructure, and American connections. Initially, KMT land reform gave land to tenant farmers and labor-intensive industries employed thousands. The state confined itself to key sectors such as finance and energy, leaving most industrial opportunities to private entrepreneurs. Farmers now focus on specialty crops such as mushrooms and aquaculture products such as eels. While agriculture is very efficient, the country must import increasing amounts of food. Industry has moved steadily up the value-added ladder from apparel, to footwear, to electronics.

Like Japan, Taiwan must import most of its raw materials to produce finished exports. Consequently, most of the industries are located on the alluvial plain along the west coast where the major ports are located and where raw materials are unloaded. This industrial corridor is anchored by Taipei and its port of Chilung in the north and the port of Kaohsiung in the south. A high-speed rail line between the two cities is under construction.

Taiwan has long met the criteria of a *kotadesasi* (discussed in Chapter 4) through its economic transformation and evolving spatial structure. Growth of population in the four metropolitan regions of Taipei, Taichung, Tainan, and Kaohsiung and the construction of a north-south freeway from Taipei to Kaohsiung have created a metropolitan corridor similar to that between Tokyo and Osaka. With its multiplex of service, industrial, and agricultural activities, linked by transportation and communication networks, this part of Taiwan exhibits all the features of a *kotadesasi* zone.

Hsinchu Science Park, established in 1980, has become the world's third largest high-tech industrial center with strong links to California's Silicon Valley. Hundreds of U.S.-schooled Taiwanese are now employed at Hsinchu. A second science park is being built near Tainan. Taiwan now leads the world in the production of notebook computers and is among the top three makers of desktop computers and peripherals. But a small domestic market, increasing labor costs, shortage of land for expansion, and lack of raw materials have compelled structural and spatial changes.

Taiwan has taken advantage of **commoditization** of the personal computer (PC) industry: the division of the PC market into two sectors—big name-brand marketing and sales and hundreds of no-name component suppliers. Taiwan has garnered the latter—the Original Equipment Manufacture market. Three-quarters of the country's electronic products is sold under another brand name. For example, more than 80 percent of flat-screen displays sold under such names as Acer and Dell, are manufactured by Taiwan and Korea.

In this export-driven economy, small and medium size firms are dominant. Unlike Japan and South Korea where large corporations were beholden to large banks in the financial crisis of 1997, Taiwanese companies borrowed from savings clubs and mutual aid associations. Bad loans were minimal and banks did not collapse.

A penchant for expansion, combined with cost efficiency, has led Taiwan to extend its operations overseas, where both space and cheap labor are available. More than 80,000 firms have moved offshore—half to Southeast Asia and half to the PRC. For example, Paochen footwear could employ 10,000 workers in Taiwan but employs 120,000 in the PRC. On the mainland, it is the world's largest producer of Nike, Reebok, and other name-brand shoes.

The China Connection

Taiwan is the second largest investor in China after Hong Kong and has invested up to $100 million in the mainland since 1980. Along with Hong Kong, it shares half the investment projects in Shanghai's Pudong development project and China's Xiamen SEZ. Some estimates put Taiwan's involvement in China's IT production at 70 percent. About one million Taiwanese—mostly business executives—now live in China. Half of these live in Shanghai. Moreover, China absorbs more than one-quarter of Taiwan's exports. The bulk of these exports are parts that will be assembled into products destined for the United States, Japan, and Europe.

China is Taiwan's third largest trading partner after the United States and Japan. Until recently, because of the political situation, most of this trade was indirect via Hong Kong. But Hong Kong, even with its SAR status, is now part of the PRC. In 1997, transshipment services commenced between Kaohsiung and Xiamen and Fuzhou in Fujian Province—the first authorized cross-strait shipping links. Beginning in 2003, direct air links were put in place for the Chinese New Year celebrations.

To further ease restrictions, vessels from China or Taiwan are no longer required to transfer cargo at a third port. Instead, they only have to pass through the waters of a third port on their way across the

Taiwan Strait. Taiwan is also creating an SEZ to handle the China trade. Foreign ships are not allowed to participate in this trade because the PRC considers cross-strait traffic to be domestic. Foreign ships can ply the Hong Kong–Shanghai route but not the Shanghai–Taiwan route.

As part of its efforts to lure Taiwan into the PRC fold, China is deepening and widening harbors to accommodate Taiwanese ships. It also gives preferential treatment to Taiwanese investors on the mainland who have built factories from Suzhou to Urumchi. Some fear that Taiwan's increased foreign investment and plant location in the PRC could ultimately hold the country hostage to the whims of the Communist government in Beijing.

Southward Policy

Taiwan's **southward policy** involves increasing investment in Southeast Asian countries to balance investments in China. By 1994, the ASEAN nations, including Vietnam, comprised Taiwan's third most important export destination after Japan and the United States. Subic Bay in the Philippines, Batam, Indonesia (just south of Singapore), and Vietnam are key targets for investment, construction, and development loans. Southeast Asia supplies Taiwan with raw materials including minerals and forest products.

This investment strategy is intended to upgrade Taiwan's industrial capacity, internationalize its business operations, and maintain its competitive advantage and economic growth in context of regional and global dynamics. Moreover, strong economic impacts can only strengthen Taiwan's *de facto* political influence in the region.

THE DOWNSIDE OF DEVELOPMENT

Development has brought a lot of benefits, but many Taiwanese are not happy with its negative side effects. For instance, there is a growing rich-poor gap as money becomes increasingly concentrated in the upper echelons of society. By 2004, the richest 20 percent controlled more than half the wealth in Taiwan. Business is notorious for corruption and nepotism. Gambling, prostitution, and other criminal activities are serious issues. Individualism, material-

ism, and hedonism are striking blows at traditional Confucian values, to the dismay of the older generation. Americanization is often blamed for these trends.

Environmental quality has sadly deteriorated with unchecked air, ground, and water pollution. Traffic congestion is out of control, and the infrastructure cannot keep up with demands for electricity and power. One good sign is that people are forming issue-oriented public interest groups and that

Figure 12–7

The Taipei 101 building—the tallest in the world. © Reuters/ Corbis Images.

environmental organizations have multiplied. However, environmental and social campaigns have not kept pace with rapid change and the uncontrolled growth of Taiwan's material culture.

One controversial project is the Taipei 101 building, the tallest structure in the world after the Petronas Towers in Kuala Lumpur, Malaysia (Figure 12–7). It houses shops, restaurants, and offices. Taiwan is hardly the place to build tall because of the many earthquakes that strike the island each year. A 7.3 quake centered in the middle of the island killed 2,415 people in 1999. Furthermore, typhoons with high velocity winds impact the island several times a year. Many believe that Taipei 101 is a dangerous building despite the designers claim that it is perfectly safe and able to withstand the forces of nature.

In spite of its difficulties, Taiwan will continue to be an economic tiger in Asia. It will increase its economic ties with the mainland but the future of its political relationship with the PRC remains in question.

⊕ *Main Point Summary*

1. Korea is one of the world's oldest land areas. Only 20 percent of the land is level.
2. North-south differences are apparent in climate, resources, and farmland availability. The north has a shorter growing season, less land suitable for agriculture, but has greater mineral wealth. South Korea is the peninsula's rice bowl.
3. South Korea is highly urbanized with several cities of more than one million people. Only Pyongyang, the capital of the North, has more than a million inhabitants.
4. The regions also vary politically, with the North being communist and the South being democratic.
5. The Korean peninsula was settled more than 5,000 years ago. Subsequent kingdoms were influenced particularly by the Chinese. Today, South Korea is considered the most Confucian country in Asia.
6. Hangul, an alphabet for spoken Korean, was devised in the fifteenth century. This enabled the Koreans to become highly literate.
7. In the seventeenth century Korea fell under Chinese control. Internal dissension led to the closing of the country to all but the Chinese and the Japanese. By the nineteenth century, Korea was known to the West as the "Hermit Kingdom."
8. Korea was occupied and annexed by Japan in 1910, and an elaborate infrastructure was put into place. Thousands of Koreans moved to Japan to work where they were discriminated against. Japanese rule of Korea did not end until 1945.
9. After World War II, Korea was divided at the 38th parallel. The intended reunification never happened, and different cultures emerged in the northern and southern parts of the peninsula. In 1953, a demilitarized zone was created, further separating the two Koreas. As a result of the Korean War and subsequent division, millions of families were broken and divided between the new countries of North Korea and South Korea.
10. North Korea is essentially a closed society. It adheres to the *juche* philosophy, which means self-reliance. Society is stratified according to Communist Party loyalty. The population lives under absolute authoritarianism. There are food shortages, and the 1990s have witnessed severe famine. Even so, North Korea has managed to build missiles capable of striking Japan and beyond.
11. China props up North Korea because it does not want instability along its border.
12. North Korea has the fifth largest standing army in the world with more than a million troops.
13. North Korea has a nuclear bomb-making program. Efforts by South Korea, China, Russia, Japan, and the United States to get the DPRK to dismantle this program have failed.
14. South Korea has transformed itself from an agrarian society to one of the world's most highly industrialized nations. It has been designated as an Asian Tiger.
15. South Korea is a product of state capitalism. The economy is founded on giant corporate

conglomerates known as *chaebols*. These are similar to the Japanese *keiretsu*. Samsung is the largest. *Chaebols* are run as families with strong Confucian regulations and loyalties. Industries and banks are closely related within *chaebol* complexes.

16. South Korea was severely impacted by the Asian financial crisis of 1997. Many *chaebols* were in serious debt and many subsidiaries went bankrupt. Restructuring of large companies and government support of medium and small companies are helping the economy to recover.

17. Unemployment, especially for young workers, is a current problem. There are too many youth entering the job market and technology is replacing humans in the workplace.

18. The island of Cheju is being developed as an international business and tourism hub.

19. Two-thirds of South Koreans live in cities of 50,000 or greater. South Korea is meeting its high demand for housing by constructing thousands of new apartment units. Many flats house two families. Apartment living has altered the traditional extended family structure.

20. South Korea has established a technobelt system linking research and industry via telecommunication. The key nodes of this system are Seoul and Taedok.

21. Nearly 5 million Koreans live overseas, the largest group in mainland China. Other large contingents are in Central Asia, Japan, and North America.

22. Many South Koreans resent having American soldiers in their country. The United States is planning to close its main base in Seoul and pull back troops from the DMZ.

23. There is potential for the reunification of the two Koreas but the probability is small given the extreme differences in the two countries' development. Most North Koreans would be unprepared to live in a modern, technologically advanced society.

24. In 2000, President Kim Dae Jung met with Kim Jong Il in Pyongyang to promote better relations between the two Koreas. This became known as President Kim's "Sunshine Policy."

25. Rich and economically powerful Taiwan is another designated Tiger of Asia. Its history is closely tied to that of the PRC.

26. The island of Taiwan is very mountainous. Most development has taken place on the western coastal plain. Agricultural output is steadily declining as farmland gives way to urbanization and industrialization.

27. Taiwan has a long history of relationships with the PRC. Known as *Bao Dao* on ancient Chinese maps, it was a migration destination for opportunists and refugees from southeastern China. These native Taiwanese make up 70 percent of Taiwan's population. The island was known as Formosa for many years until after the Japanese occupation.

28. The Japanese accelerated Taiwan's development by building an infrastructure geared to the needs of Japan.

29. Various proclamations were made by both the Chinese and the Japanese regarding the independence of Taiwan. However, the end of WWII saw the occupation of the island by Chiang Kai-shek and his Nationalist troops from the mainland. Many "mainlander" refugees migrated there as well.

30. The distinction between Taiwanese and mainlanders is becoming blurred. More than 90 percent of inhabitants were born on the island.

31. Both Taiwan and the PRC claim to be the legitimate government of all of China. Realities reveal sharp differences between the two entities—politically, economically, and socially. Nevertheless, the PRC insists that Taiwan is a province of China and Taiwan is hinting at its being an independent entity.

32. The United States does not want to see Taiwan as a part of the PRC because that would give the PRC complete control over the shipping lanes through the Taiwan Strait. The issue of unification or separation remains unresolved.

33. Taiwan's achievements as an economic tiger are underpinned by its excellent education system, infrastructure, and American connections. The

state is heavily involved in finance and energy but most businesses are privately run. Electronics are of increasing significance. Farmers now focus on specialty crops.

34. Taiwan meets the criteria of a *kotadesasi* through its economic transformation and evolving spatial structure. There is an evolving metropolitan corridor stretching from Taipei to Kaohsiung. This includes science parks such as those associated with Hsinchu-Taiwan's Silicon Valley.

35. Taiwan has "commoditized" its computer industry —making components for big-name brands.

36. Taiwan has extended its operations overseas to take advantage of space and cheap labor. Many of these new companies are in the PRC.

37. Taiwan is increasingly doing business with and investing in the PRC. China has become Taiwan's third largest trading partner after Japan and the United States. Taiwan also has a "southward policy" whereby it is focusing investment in Southeast Asia.

38. Rapid development has its downside with ever more noticeable pollution, congestion, hedonism, and a rich-poor gap.

Bibliography

Baum, Julian. 1998. "A Long Wait Ahead: Commercial Concerns Hold Back Cross-Strait Shipping Links." *Far Eastern Economic Review*, 30 (April): 68.

Chan, Gerald. 1997. "Taiwan's Economic Growth and its Southward Policy in Asia." In *Asia Pacific: New Geographies of the Pacific Rim*, eds. R. F. Waters, and T. F. McGee, pp. 206–222. Vancouver: UBC.

Chanda, Nayan et al. 1997. "On Borrowed Time (North Korea)." *Far Eastern Economic Review*, 26 (June): 22–28.

Cheng, Allen, T. 1997. "And Now For Taiwan." *Asia, Inc.* (June): 38–45.

Dean, Jason. 2003. "On Top of the World For Now (Taiwan)." *Far Eastern Economic Review*. June 12: 30–35.

Eberstadt, Nicholas. 1995. *Korea Approaches Reunification*. Armonk, N.Y.: M. E. Sharpe.

1996. "In the Shadow of the *Chaebol*." *The Economist*, 6 (July): 57–8.

1994. "Kim Jong Il's Inheritance." *The Economist*, 16 (July): 19–21.

1990. "The Sun Sets on Rice." *The Economist*, 21 (April): 44.

1998. "Taiwan Survey." *The Economist*, 7 (November).

Garrett, Banning, and Bonnie Glaser. 1995. "Looking Across the Yalu: Chinese Assessments of North Korea." *Asian Survey*, 35: 528–45.

Harris, Chauncy, D. 1982. "The Urban and Industrial Transformation of Japan." *The Geographical Review* 72: 50–89.

Hiebert, Murray. 2003. "Decision Time Looms Over North Korea." *Far Eastern Economic Review*, May 15: 12–13.

Hoon, Shim Jae et al. 1993. "Rural Exodus: Seeds of Despair." *Far Eastern Economic Review*, 4 (March): 20–21.

Howard, Keith, ed. 1996. *Korea: People, Country and Culture*. London: School of Oriental and African Studies.

Hsieh, Chiao-min. 1964. *Taiwan—Ilha Formosa*. Washington D.C.: Butterworth.

Kirk, Donald. 1994. *Korean Dynasty: Hyundai and Chung Ju Yung*. Armonk, N.Y.: M. E. Sharpe.

Landers, Peter et al. 1998. "Hard Target (North Korea)." *Far Eastern Economic Review*. September 24: 20–21.

Lee, Charles. 1997. "Tomorrow the World (South Korea)." *Far Eastern Economic Review*. May 1: 43–45.

Lintner, Bertil. 2004. "Shop Till You Drop (North Korea)." *Far Eastern Economic Review*. May 13: 14–19.

Macdonald, Donald S. 1990. *The Koreans: Contemporary Politics and Society*, 2nd edition. Boulder: Westview.

Palka, Eugene, and Francis Galgano. 2004. *North Korea: Geographic Perspectives*. Guilford, Connecticut: McGraw Hill/Dushkin.

Shambaug, David. 1998. *Contemporary Taiwan*. Oxford: Clarendon.

Song, Kimberly. 2004. "Daewoo's Long Road to Recovery." *Far Eastern Economic Review*. May 13: 40–45.

Tsai, H. H. 1996. "Globalization and the Urban System in Taiwan." In *Emerging World Cities in Pacific Asia*, eds. Fu-chen Lo and Yue-ma Yeung, pp. 179–218. New York: United Nations.

Yeung, Yue-ma, and Fu-chen Lo. 1996. "Global Restructuring and Emerging Urban Corridors in Pacific Asia." In *Emerging World Cities in Pacific Asia*, eds. Fu-chen Lo and Yue-ma Yeung, pp. 17–47. New York: United Nations.

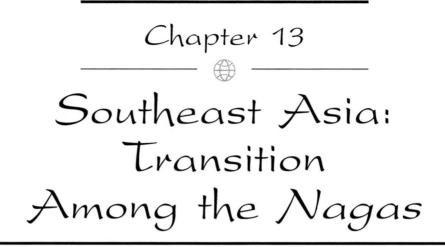

Chapter 13

Southeast Asia: Transition Among the Nagas

"To understand the present and
anticipate the future, one must
know enough of the past, enough
to have a sense of the history of
a people."

LEE KUAN YEW (1980)

"A more highly fragmented part of the globe could scarcely be imagined," said geographer Norton Ginsburg (1972). Southeast Asia, land of the *nagas,* is a realm of contrasts in both physical and cultural geography. Fragmented landscapes have formed the backdrop for the establishment of glorious kingdoms and trading states, while hills and valleys have complicated the imposition of boundaries and the demarcation of people and territory. Separation has enhanced cultural diversity and thwarted regional unity. Nevertheless, several Southeast Asian states have become prominent on the global economic map. Malaysia, Thailand, and Singapore are designated as economic tigers. Indonesia has tiger potential even though weighted with political troubles. Vietnam boasts of its new label: *baby tiger.* In contrast, countries such as Myanmar, Laos, and Cambodia lag far behind in both the human condition and regional stability.

Many geographers stress the region's diversity over unity. Geographer Jonathan Rigg (1991) notes, "Southeast Asia represents a residual region: it owes its regional identity not to internal coherence, but to external incoherence." In this chapter, you will learn about the differences, but also the similarities, among the peoples, landscapes, and cultures of this complex region.

Figure 13-1

Physical features of Southeast Asia. From H. J. de Blij and P. O. Muller, *Concepts and Regions in Geography,* Second Edition, 2005, p. 340. Originally rendered in color. © H. J. de Blij and P. O. Muller. Reprinted with permission of John Wiley & Sons, Inc.

Physical Landscapes

The difference between land and water is clearly evident. In fact, there is a mainland Southeast Asia and an island, or archipelagic, Southeast Asia (Figure 13–1). The mainland includes Burma (Myanmar), Thailand, Laos, Cambodia, and Vietnam. I will include Malaysia with the insular region as it not only shares the island of Borneo with Indonesia but also is connected to the mainland by the narrow Kra Peninsula.

Consequently, Malaysia, Singapore, Indonesia, Brunei, East Timor, and the Philippines form the insular or archipelagic region.

All told, there is four times more water area than there is land area. While civilizations arose in the mainland interior, marine landscapes and coastal locations dominated the fortunes of the island realm. Control of ocean passageways marked the rise and fall of many a power group and later fixed Singapore as the dominant port of the entire region.

Figure 13–2

Most temples in Laos exhibit nagas *as guardian figures.*
Photograph courtesy of B.A. Weightman.

The Naga

Naga, in Sanskrit, means serpent. The *naga* is an aquatic symbol that permeates the daily life of people everywhere in the watery realm of Southeast Asia. The *naga* appears in art form as a dragon and snakelike creature, and frequently as a mutation of a different being. For instance, Thailand has a *sang*, a cross between a *naga* and a *singh* (lion). The *naga* form is ubiquitous in Thai and other cosmological representations. The Khmer and numerous other Southeast Asian civilizations are portrayed as emerging from water. Sometimes whole populations have been identified as the *naga* people. In almost everything of significance, the *naga* symbol asserts itself (Figure 13–2).

Mountain formation remains active and crustal instability is a feature affecting the lives of millions (Chapter 2). The 2004 earthquake off the northwest coast of Sumatra in Indonesia and the subsequent tsunami, had a devastating effect on Thailand. More than 5,000 people, including more than 2,000 foreigners, were killed and thousands are still missing.

Throughout this unstable region, mountains dominate over plains. The hills of Myanmar are cut by the Ayeyarwady River and its major tributary, the Chindwin, which meet southwest of the city of Mandalay. The Sittang Valley lies to the east. Both the Sittang and Ayeyarwady deltas are used extensively for rice cultivation. The highly dissected Shan Plateau of eastern Myanmar rises over 4,000 feet (1,200 m) in many places.

To the south, in Thailand, is the delta of the Chao Praya. The Chao Praya flows from the Thai highlands to Bangkok. Since this river is confined in a geological trough, its delta is elongated rather than fan-shaped. Alluvial lowlands extend nearly 310 miles (500 km) upstream from the Gulf of Thailand.

Southeast of the Shan Plateau is the Khorat Plateau of eastern Thailand. The Khorat, at an elevation of just under 1,000 feet (300 m), is edged by higher escarpments. The Khorat is thus separated from the rest of the country. South of the Khorat Plateau lies the lowland of the Cambodian Basin with Lake Tonle Sap in the lowest, middle portion. In the wet season, the Mekong River drainage rolls inward to fill the lake up to six times its average depth. Tonle Sap extends from 1,600 square miles (2,580 sq km) in the dry season to 12,400 square miles (20,000 sq km) in the wet.

The Mekong River, called the *Mégõngk* or Tonle Thom ("great river") in Cambodian, flows to the east of the Cambodian Basin. The longest river of Southeast Asia, the Mekong has constructed an expansive delta starting at Phnom Penh in Cambodia and extending 175 miles (282 km) into the sea. This vast, alluvial delta has developed as the heartland of southern Vietnam.

The Mekong

The Mekong is the twelfth longest river in the world. Originating in the Tibetan Highlands of China's Qinghai Province, the river flows 2,600 miles (4,184 km) through southwestern China, along the borders of Myanmar, Laos, and Thailand, across Cambodia, and down through Vietnam where it sprawls across the 26,000-square-mile (67,340-sq-km) Mekong Delta to the South China Sea. One-third of the population of the latter four countries lives in the lower Mekong Basin.

The Mekong cultural realm is overwhelmingly rural. Most of the region is poor, with people engaging primarily in rice cultivation and fishing. However, as foreign aid and investment increase and modern development ensues, both landscapes and lifestyles are being transformed rapidly. Increased population pressure and resource demands, coupled with uncoordinated policies and regulations, are threatening both human and physical environments.

Although the river is an artery of human activity, upstream rapids and waterfalls have prevented any large-scale development for international shipping. Smaller seagoing vessels can travel upstream to Phnom Penh, but travel beyond is limited to local traffic. A bridge connecting Laos to Thailand was constructed in 1994.

In 1992, China initiated plans to open the upper reaches of the river to year-round navigation by large cargo ships. China's primary objective is to facilitate the export of raw materials from its landlocked Yunnan Province to ports in Thailand, Laos, and the rest of Southeast Asia. If fully implemented, the dredging and channeling would more than double the annual shipping capacity of the Mekong from 4 million tons in 2001 to 10 million tons by 2007 and involve the passage of at least a dozen 500-ton ships daily.

Cooperative efforts to develop the Mekong River basin have been discussed since 1957. Several hydroelectric power, navigation, flood control, and irrigation projects were planned but war and political upheaval have allowed only some to be implemented. For example, dam construction has made Laos a key exporter of power to Thailand.

In 2000, Thailand, Laos, Burma (Myanmar) and China signed a navigation agreement. Each promised to develop ports and facilitate the passage of ships. However, many plans have been scaled down because of both cost factors and environmental concerns. The first three countries are worried that their markets will be inundated with cheap Chinese goods. Farmers, who plant crops along the riverbank and rely on natural flood waters for irrigation, are concerned that the seasonal flow of water will be disrupted and that large ships will generate waves that will drown their crops.

Blasting rapids to create shipping channels has already damaged the Mekong's environment. Fishers already complain that waves from large ships cause their small boats to capsize and that the fish supply is not as abundant as it used to be. Rapids may impede shipping but the rocky outcrops underneath harbor more than 1,000 fish species.

Various groups have protested the development plans and projects in progress. In June, 2002, China announced that it would scale down the blasting and require environmental impact reports. Nevertheless, it has already built two dams on the upper Mekong and a third dam is due for completion in 2012.

> Villagers downstream are already complaining about the fluctuation in water levels. China appears driven to develop the river regardless of the consequences elsewhere.

The Annamite Ranges run through Vietnam between the Mekong and the South China Sea. In northern Vietnam, the Song Koi (Red) River flows in a trough-like valley to the Gulf of Tonkin, where it forms a 5,460-square-mile (14,000-sq-km) delta. The southern Annamites give way to the Mekong Delta.

The shallow sea enclosed by Cambodia, Malaysia, and the island of Borneo is less than 656 feet (200 m) deep, as it is part of the Sunda Shelf (Chapter 2). Borneo is the chief emergent part of the shelf, having several mountains above 6,560 feet (2,000 m). Mount Kinabalu in northern Sabah is the highest peak at 13,451 feet (4,101 m).

Southeast Asia's relief offers little opportunity for human settlement. The region's populations are found primarily along rivers. Most of the Sunda Shelf, which could have offered livable land, is overlain by the sea. Extensive coastal areas are mangrove swamps. The most densely populated areas in the island realm are Java in Indonesia and Luzon in the Philippines, both of which possess fertile volcanic soils.

Even within this tropical realm, straddling the equator, there are considerable temperature variations. Pinang, Malaysia, at 5° 21′ N averages 80°F (27°C) for twelve months, Hanoi, Vietnam at 21°04′ N might drop to 60°F (16°C) in December or January. Altitude also plays a role. With temperatures falling 3 degrees per 1,000 feet, Southeast Asia had its hill stations comparable to those of the Raj. Java's Bandung and the Malay peninsula's Cameron Highlands are examples of places where temperatures might be 10 to 20 degrees cooler than in the sweltering lands below.

Precipitation varies both in amount received and its seasonality (Figure 13–3). Sixty inches (152 cm) will support tropical rain forests or monsoon forests. Areas with less than this amount are moisture deficient. Most moisture-deficient areas are on the mainland and on the islands east of Java. Distinct wet and dry seasons increase as one moves away from the equator. In equatorial regions, the wet and dry monsoons are affected by convectional rainfall. Even so, topographic barriers exhibit seasonally humid and dry sides, depending on the monsoon and other wind patterns.

Length of dry season varies as well, lengthening both north and south of the equator. And, some "dry seasons" are merely less wet ones. For instance, there is no real dry season in Borneo, and in peninsular Malaysia it lasts only a month. But on the mainland the dry season lasts four to five months. Most of the rainfall is in the summer months.

THE ROLE OF WATER

According to Rigg (1991) "Of the environmental elements, water has possibly been the most instrumental in forging Southeast Asia's distinctive personality." Fairly predictable winds, wood supplies located close to shore for shipbuilding, and endless waterways made this region most favorable to maritime activity. Extensive trade existed well before the fourteenth century.

In prehistory, waves of migrants filtered southward from the mainland to the Indonesian Islands and on to the Pacific Islands and Australia. At the beginning of the first century A.D., there was an infusion of Hindu and Buddhist culture when Indian sea traders ventured for gold and spices. This Indianization of Southeast Asia peaked with the establishment of empires centered on great complexes such as Srivijaya on Sumatra and Angkor in Cambodia.

In the second century, both Arab and Indian ocean traders brought Islam to the island realm. Meanwhile, monks from Sri Lanka disseminated Theravada Buddhism to the mainland.

In the modern era, colonial countries introduced new technology and power into the region and created conditions conducive to the immigration of large numbers of Indians and Chinese. Commercial interaction ensured diffusion of products and ideas and linked the peoples of the region with each other. Interestingly, with the possible exceptions of Vietnam and Burma, no Southeast Asian country established strong ties with either India or China.

The amount and availability of water are crucial to life in Southeast Asia. Virtually all outdoor economic

Figure 13–3

Length of dry seasons and precipitation. Note that dry seasons increase in length toward the interior of the mainland.

activities are regulated by precipitation. Water is central to rice cultivation. In fact, there is no other staple crop so dependent on its presence. Many regions have rituals surrounding the "gift of water." In Thailand, female cats (the personification of dryness) are carried through the streets and soaked with water while people chant "Hail *Nang Maew* (Lady Cat) give us rain." In parts of northern Thailand, Bali, Java, Sumatra, and other islands of both Indonesia and the Philippines, elaborate irrigation systems have been constructed to ensure a stable water supply.

Bali's Subak Irrigation

In Bali, rice is a gift of the gods and the traditional rice growing cycle commences on a day set according to a religious calendar. Water regulation is an essential aspect of this highly regimented pattern of activities.

Rivers cut through Bali's volcanic mountains to the sea. The Balinese have settled the mountain

Figure 13-4

Irrigated rice terraces (sawah) *in Indonesia. Water is moved by gravity flow.* Photograph courtesy of B.A. Weightman.

ridges and covered the slopes with *sawah:* terraced rice paddies (Figure 13–4). Water runs by gravity but must be fairly distributed and regulated according to the needs of farmers and the system. For example, irrigated *sawah* are drained up to three times per year to encourage root development and allow weeding. Fields are drained for harvesting and then reflooded to allow other plants to grow. Water hyacinth is popular because it can be worked back into the soil or used as pig feed.

Balinese farmers belong to cooperatives called *subak.* Every farmer who owns or rents more than a quarter of an acre (one-tenth of a hectare) must join. Bali has more than 1,000 *subak.* Each *subak* elects a leader who runs the operation and liaisons with agricultural and other agencies. *Subak* law (*awig-awig*) meetings must be held on one auspicious day each month of the religious calendar. Monies are derived from fees, fines, and land sales.

Cooperation is the key to success. *Subak* members repair and maintain the irrigation system after the harvest in accordance with the amount of water they have received. Those not working pay a fee. The average farmer gives the communal system about 15 days a year. An important aspect of the *subak* is its religious connection. Each keeps a temple, which must be cooperatively maintained by the members.

The *subak* system diffused from Java a thousand years ago. It has survived in Bali because of its isolation as an island and because the *adat* system of local regulations, stressing personal responsibility and communal consensus, remains strongly entrenched.

It now seems clear that rice was first domesticated in Southeast Asia and diffused from there to China and India, replacing millet as the staple crop. Around 4000 to 3500 B.C., a new red-on-tan pottery form appeared in northern Thailand. This is known as Ban Chiang pottery. Shortly thereafter, bronze-making emerged, carried out at the village level.

UPLANDS AND LOWLANDS

From about 1000 B.C., major changes began to take place. With the expansion of the Han Chinese in person and influence, small towns began to emerge. In addition, entrepôts, to facilitate trade between India and China, were established along the coasts of the Malay peninsula and of what is today southern Vietnam. These entrepôts became centers of diffusion of Indian culture throughout Southeast Asia.

As these events ensued, primitive and tribal people began to lose their dominant positions. The new civilizations centered around the major river valleys and Cambodia's Tonle Sap basin (Chapter 14). These lowlands are separated by sparsely populated mountainous areas—the uplands. Hill people became increasingly differentiated from lowland people, yet they retained part of their cultural realm as "holders of the wild" or "people of the upland fields." Ceremonial exchange relationships validated the bonds between hill people and lowlanders. Tangible trade between the two groups was important as well. Lowlanders also viewed hill people as potential laborers and slavery was not unusual. Hill tribes often raided each other's settlements and sold the captives as slaves to lowlanders. In some regions such as Vietnam, efforts were made to reduce lowlanders' contact with the hill people to simple tribute payments and restricted trade.

The mountains have acted as barriers between larger settlements. Over time, successive immigrant groups have settled in the lowlands, but more recent and more populous groups have driven earlier, less sophisticated groups into the upland regions.

Lowland areas are inhabited by the dominant ethnic group of each country—Thais in Thailand, Khmers in Cambodia, Javanese in Java, and so forth. Upland, forested areas are inhabited by minority groups such as the Shans in Burma and the Yao and Miao in Thailand.

In these contrasting settings, different ways of life have evolved. Lowlanders primarily practice wet rice cultivation while uplanders focus on some form of dry field agriculture or shifting cultivation. These patterns continue today and affect patterns of development and the resolution of environmental issues.

The Impact of Indian Culture

Mountain barriers channeled the diffusion of Indian culture into Southeast Asia's coastal regions. New ideas, religions, and customs came almost entirely by sea in sailing ships driven by the monsoon across the Bay of Bengal. As early as the sixth century B.C. Indian traders worked the coasts of Myanmar, the Malay peninsula, and western Indonesia. The Indians impressed local leaders with lavish gifts, their knowledge of medicine, and their frequent portrayal of themselves as royalty. They also married into Southeast Asian society. Temple forms, styles of art, music, dance, and tales of the *Ramayana* are woven into the cultural tapestries of most Southeast Asian countries to this day. The impact of Indian culture was greatest during the Gupta Empire—India's Golden Age (320–535 A.D.).

Hinduism and Buddhism were the most salient aspects of Indianization. Missions were sent back and forth to Ceylon (Sri Lanka), the core area of Theravada Buddhism. Mahayana Buddhism was accepted in central Java during the eighth century and in Cambodia's Angkor in the twelfth century. Sanskrit was the first written language of the region. Its related Pali provided the basis for developing local writing systems.

The Hindu concept of kingship, law codes, economic treatises, and other canon were absorbed by Southeast Asian rulers. However, the caste system and inferior standing of women were not adopted. Although there is a patriarchal veneer to present-day Southeast Asian society, couples tend to live with the wife's parents, or at least relate most closely to them. Women have much more social and economic power than in other Asian societies.

The earliest Indian states emerged along the coasts, with their nuclei at river mouths. The first of these, called Funan by the Chinese, became an imperial power in the third century A.D. This was followed by the Khmer kingdom in Cambodia. The kingdom of Champa (1192–1471) was established along the east coast of Vietnam. Another early sea power to accept Indian influence was Srivijaya. Its capital at

Figure 13–5

Built by Buddhists on Java in the eighth century at the height of the Sailendra Dynasty, Borobudur's construction took about 75 years. It comprises 1,600,000 volcanic, andesite stones and is ornamented with at least 1,500 bas reliefs and 500 statues of Buddha. No mortar was used in its construction. The large base platform was added at a later date and hides a panel of reliefs known as the "hidden foot." Borobudur stands as a stupa, a replica of the cosmic Mount Meru, center of the universe, and as a mandala to assist meditation.
Photograph courtesy of B.A. Weightman.

Palembang on Sumatra became a center of Mahayana Buddhism, as was central Java at the time. The temple complex of Borobudur (built in the eighth century) stands as a testament to the strength of Buddhism (Figure 13–5). All of Southeast Asia, with the exception of the Tonkin region of northern Vietnam and the Philippines, was impacted by Indian culture.

The Chinese Influence

The earliest contact with China can be traced to the Qin Dynasty (221–207 B.C.). Chinese influence diffused overland; the region mainly affected was Vietnam. In fact, northern Vietnam remained a Chinese territory until it broke away in 907 A.D. China referred to the region of the southern seas as **Nanyang**. Its cultural impact was the direct result of colonization and conquest. Chinese administration style, writing system, and Confucian classics were all superimposed on indigenous culture. Many Chinese cultural influences have persisted in Vietnam.

By the sixteenth century, Chinese trading communities were permanently established throughout the region. However, large-scale Chinese migration did not take place until the turmoil of China's nineteenth century, when thousands of Chinese migrated from southern China to work in mines, on plantations, and in other European-operated enterprises. Others came as farmers, fishers, traders, retailers, or skilled craftsmen. Increasing numbers of Chinese came to play a critical role in the region's commerce.

With the extant plethora of ethnic groups and the influx of Europeans, Indians, and Chinese, each functioning in its own realm of expertise, Southeast Asia became what is referred to as a **plural society**.

Overseas Chinese in Southeast Asia

Overseas Chinese are prominent in Southeast Asia far beyond their numbers. For example, in Malaysia ethnic Chinese make up 29 percent of the population but control 69 percent of share capital by investing in businesses. In Thailand, ethnic Chinese make up 10 percent of the population, yet they control 81 percent of listed firms by market capitalization. In the Philippines, the Chinese comprise 2 percent of the population but control half of the share market, and in Indonesia 3.5 percent ethnic Chinese control 73 percent of the listed firms. Clearly, the Chinese have clout in Southeast Asian economies. In Singapore, Chinese comprise 75 percent of the total population and control 90 percent of the economy.

Southeast Asia's Chinese have a reputation of being rich and influential but loyal to their own group. They are said to do business with each other and hire their kin rather than the local people. Suspicion and tension are rife in several countries such as Indonesia and Malaysia. The least tension is in Thailand and Singapore.

Chinese family businesses take advantage of their kin networks and regional connections to stay afloat in the worst of times. This situation leads to more jealousy and even hatred among non-Chinese. Chinese have frequently been the scapegoats for any sort of trouble. Thousands have been killed, raped, and tortured and their homes and their businesses destroyed. The last attack against the Chinese was in Indonesia during the economic crisis of 1997.

Linguistic Pluralism

Hundreds of distinct languages and dialects are spoken in Southeast Asia. These can be classified into four main language families: Malayo-Polynesian (e.g., Malay and Javanese); Sino-Tibetan (e.g., Thai, Burmese, and Cantonese); Austro-Asiatic (e.g., Khmer and Annamese, or Vietnamese); and the various languages of West Papua. Although the insular region's languages harken mainly to the Malayan division of the Malayo-Polynesian group, the mainland's languages form a complicated mosaic. There are multiple Sino-Tibetan and Austro-Asiatic languages and dialects spoken throughout the region. For example, there are at least six versions of Chinese. Hill regions, with their isolated pockets, have developed a myriad of dialects. Since the eighteenth century, English has

been employed widely in the fields of education and administration in many areas such as the Philippines, Singapore, Malaysia, Thailand, and Myanmar. In Indochina (Cambodia, Laos, and Vietnam), French was the lingua franca. Dutch served Indonesia until after World War II, when English came into favor. The prevalence of English as a means of general communication is readily apparent on signage in many places.

Religious Pluralism

The mainland is largely Buddhist with the exception of the Vietnamese lowlands, where "Chinese religions" and Roman Catholicism are practiced. Chinese religions are a mixture of Mahayana Buddhism, Daoism, and ancestor worship, a social practice affiliated with Confucianism. The island world is largely Islamic with the exception of Bali, which retained Hinduism after the last Hindu dynasty in eastern Java had been overthrown in the seventeenth century. Christians predominate in the Philippines except on the island of Mindanao, where Islam prevails, and in East Timor, where Christianity remains a vestige of Portuguese colonialism.

Islamic fundamentalism is on the rise in some areas and is sometimes associated with acts of terrorism. Indonesia, southern Thailand, and Mindanao have been the most affected by this religious revival. Read Chapters 14 and 15 for more details.

Angkor Wat

Angkor was the capital of the Khmer empire, a civilization that flourished from about 800 until about 1370 and included Lower Burma (Myanmar), the upper Malay peninsula, central Thailand, Cambodia, and southern Vietnam. A mixture of sandstone, brick, and laterite was used to build an immense complex of temples (the Khmer word *wat* means temple), each one a symbolic representation of Hindu or Buddhist cosmology. The whole of the Angkor complex

Figure 13–6

The main Buddhist wat at Angkor is reflected in a moat surrounding the complex. This temple, built of laterite, covers one square mile (2.5 sq km) and is one of the largest edifices in the world. Photograph courtesy of B.A. Weightman.

covers an area of more than 15.5 miles (25 km) east to west and almost 6.2 miles (10 km) north to south (Figure 13–6).

Angkor Wat, dedicated to the Hindu god Vishnu, is just one temple. Erected as the funerary memorial of King Suryavarman II in the twelfth century, it covers 1 square mile (2.5 square km) and is one of the largest religious edifices in the world. Angkor Thom is slightly larger. It is entered by four axially

aligned gateways that define a square almost 0.62 mile (1 km) across.

Angkor has largely escaped the ravages of time but it is not unscathed. Certainly the jungle has taken its toll. Trees have split walls and lichens etch away illustrious carvings. French and Indian archaeologists have cleared the forest from much of the monument but they can do nothing about the robbers who have absconded with carvings, statue heads, and even whole statues. More recently, Cambodia's war has resulted in further damage to the complex in the form of bullet holes, smashed artifacts, and the carving of political slogans into bas reliefs.

Fortunately, international assistance is supporting archaeologists and the local people in their restoration efforts. Treasure hunting raids still occur however, and certain sections can only be visited in the company of armed guards.

The Coming of the Europeans

European political contacts with Southeast Asia commenced with the Portuguese conquest of Malacca in 1511. Still, there was no major European political or cultural impress on the region until well into the nineteenth century. The Philippines is an exception in that European penetration here was greater than anywhere else in Southeast Asia. Indeed, there were a string of fortified ports in the region (Chapter 1), but even the Dutch, who controlled the western third of Java centered on Batavia (Jakarta), were unaware of Borobudur, perhaps the greatest Buddhist monument in the world.

The Spice Islands

Long known as the Moluccas and now Maluku, the Spice Islands are a volcanic mini-archipelago between Sulawesi and West Papua north of Timor. Ambon, Banda, and Ceram are only 3 of the 1,000 islands. Some of the islands were known for their spices, especially nutmeg and cloves. Spices, important for preserving and flavoring meat, were traded for Indian cotton, Chinese porcelain, and Javanese rice for at least 2,000 years before the Europeans came on the scene.

The arrival of the Portuguese in 1511 initiated decades of conflict among the local sultans and the competing colonial powers. The Dutch established settlements in 1599 and by 1667 had completed their conquest of the islands. Their first treaty with Ambon gave the Dutch monopoly over all the cloves. The Dutch East India Company, founded in 1602, put an end to individual profiteering by ousting all other traders by force and intrigue. At one point, the company eliminated the entire population of Banda.

To guarantee control and keep prices high, the Dutch destroyed all the nutmeg and clove plants except in those areas approved for cultivation. Even the number of trees was restricted. Cloves were removed from all of the islands except Ambon. This was a particularly cruel move because the indigenous people planted a clove tree at the birth of a child and to destroy the tree portended ill-will for the child. The penalty for cultivating non-regulated trees was death.

Sago palms were eliminated from Ceram, then known as the "mother island" because of its supply of this food staple. The area soon became uninhabitable. Today, it is said that Ceram is populated by ghosts.

Restrictive policies were enforced until 1824. By that time the Mollucans had been reduced to subsistence existence. The Company's monopoly was broken when a Frenchman smuggled nutmeg and clove seedlings to Zanzibar. The spice trade also declined with the availability of fresh meat and changing dietary habits in Europe.

The nineteenth century ushered in the Industrial Revolution, creating increased demand for raw materials and markets. The Caribbean, focus of eighteenth-century trade, fell into decline with the abolition of slavery and the slave-labor system. The introduction of steamships and the opening of the Suez Canal in 1869 shortened the journey to Asia, and increasing interest in China put Southeast Asia once more at the crossroads of activity. European missionary zeal led to additional ventures to the region. From about 1825 to 1870, most of Southeast Asia came formally under European control (Figure 1–11).

EUROPEAN IMPRESS

The various powers had their own agendas, which diversified their influences on people and landscapes. In 1896, Britain and France decided to preserve Siam (Thailand) as a buffer state to protect their valuable colonial dependencies of Burma (Myanmar) and Indochina. Britain treated Burma as an extension of its Indian Empire and pursued a policy of development of resources and building the economy. France's doctrinaire policy of assimilation of Indochina into France left most of that region undeveloped. The Dutch commercial exploitation policies left most of Indonesia undeveloped and unprepared, economically or politically, for self-government. In the Philippines, Spanish rule gave way to American control in 1898. American anti-colonial attitudes assured that country's direction toward self-rule.

The Dutch Culture System

The **Dutch Culture System** (*cultuur stelsel*) in Indonesia exemplifies the penetration of capitalism and commercialization into the landscape. First established by the Dutch East India Company, the system required that every farm household allocate one-fifth of its land to the production of a cash or export crop. The system worked well in the sugar-growing areas of Java but proved burdensome where land was badly needed for food crops. The system was exploited by local officials who received commissions for the products they delivered to authorities. When rice was added to the list of export crops in the 1840s, famine struck Java causing widespread starvation and emigration.

Figure 13–7
This photo shows the traditional Chinese/European shophouses that once dominated Kuala Lumpur. These stand in sharp contrast to today's modern architecture. Photograph courtesy of B.A. Weightman.

The Dutch relinquished their monopoly control to private capital after 1877 but compulsory labor was still required. Farmers had to pay substantial taxes and were unable to get reasonable-rate credit. The powerful position of European and Chinese entrepreneurs added to the exploitation and disruption of traditional Indonesian society.

The British, who ruled Burma from Calcutta, brought in Indian money-lender castes to bank and fund the expansion of rice cultivation in the Ayeyarwady Delta. Indians often foreclosed on the Burmese, who were hit especially hard during the world depression. Anti-Indian sentiments in Burma derive from the tactics employed by the British.

Economic concerns also stimulated population movements. For instance, the cultivation of new crops and the opening up of new lands attracted masses of migrants. The rapid expansion of rice cultivation into the Mekong, Ayeyarwady and Chao Praya deltas are testament to increased global demand for rice in the late nineteenth century. In lower Burma, where the rice area burgeoned from 988,000 acres (400,000 ha) to 9,880,000 acres (4 million ha), populations increased from 1.5 million in 1855 to 8 million in 1930.

Populations were also drawn to other commodity regions such as rubber and tin. Rubber was planted over large areas of Malaya, Sumatra, and Java, literally transforming these landscapes. Malay and Chinese smallholders accounted for 39 percent of the rubber industry by 1939, yet it was the capital- and labor-intensive plantations that dominated production. As Malays were not keen on wage labor, companies imported large numbers of indentured Indian and Ceylonese (from Sri Lanka) workers. More than 700,000 Indian workers entered Malaya between 1907 and 1917. The tin industry in Malaya, Indonesia, and Thailand was dominated by Chinese mining enterprises employing cheap Chinese immigrant labor. Between 1890 and 1899, almost 165,000 Chinese laborers a year were entering the Straits Settlement of Singapore and Penang alone.

The influx of Europeans, Indians, and Chinese created plural societies; each group occupied their own residential quarters but amalgamated in the marketplace. Immigrants concentrated in urban places and zones of commercial agriculture or mining.

Cities were now founded on commerce and, with the exception of Bangkok, were dominated by external forces. Cities such as Jakarta, Manila, and Kuala Lumpur were largely foreign places, alienated from the circumstances of the rural masses who made up 90 percent of the population (Figure 13–7).

Transportation (river, road, rail) was immensely improved to serve the colonial presence. Routes linked raw materials with coastal ports, which served as entrepôts for outgoing and incoming goods. This selective transport development meant many areas were devoid of communication systems, isolated from the larger economy. Dual economies emerged, only to be exacerbated as development increased. The rural-urban divide widened as education and health care were made available to urban elites. Only the Philippines had widespread educational facilities.

POLITICAL PATTERNS

Politically, territorial units had little relevance to historical and ethnic circumstances. Territorial forms were derived from European considerations: the need to access raw materials; the need to avoid conflict; and administrative conveniences. Boundaries did not coincide with settlement patterns of ethnic groups that were often divided by boundary constructs. One of the few instances where boundaries were drawn according to ethnic differences is between parts of Laos and Vietnam. Villages where the majority of houses were built on piles with entrances facing east were assigned to Laos. If the majority of houses were built on the ground facing south, that village went to Vietnam.

Burma and Thailand each have long tails trailing down the Malay Peninsula. Laos is landlocked, and what eventually became Malaysia is fragmented between the mainland and island realms.

Although colonials frequently turned internal hostilities to advantage, no attempts were made to integrate territorially based ethnic identities into unified wholes. In particular, Chinese and Indian migrants remained largely unassimilated and resented.

Portuguese, British, and Dutch introduced changes that shredded the fabric of indigenous societies; the net result was the creation of social systems that could not survive without the paternalism of colonial governments. France, Spain, and the United States also played roles in the colonial period but their activities did not as directly alter regional systems.

Rising Nationalism and Independence

Geographer Jonathan Rigg (1991) suggests, "The uneven pattern of economic development, and the subordinate position of the local populations, engendered a great deal of bitterness. In addition, the paternalistic attitude of many colonial administrators

and their occasionally harsh treatment of the local inhabitants created a reservoir of resentment that nationalist leaders were only too happy to exploit in order to achieve their aims."

Nationalistic leaders emerged because certain elites had been given the opportunity to study abroad. These individuals, such as Sukarno of Indonesia, Aung San of Burma, and Ho Chi Minh of Vietnam, spearheaded independence movements. New ideologies were framed in context of the Boxer Rebellion in China (1899), the Chinese Revolution of 1911, and the rise of Japan and its defeat of Russia in 1905, events that proved that the Western powers were not invincible.

After the Japanese invasion of 1941, Japan promoted "Asia for the Asians." It also encouraged emerging nationalist movements and even transferred some powers to local leaders such as the sultans of the Malay States. Although the Japanese occupation proved to be no better than that of the former colonial powers, the period between 1941 and 1945 was important in laying the groundwork for subsequent independence.

After the defeat of Japan in World War II, the colonial powers returned to discover they had to negotiate with well-organized and informed nationalists with popular support. Not all efforts were peaceful, with much acrimony between the negotiating parties. The Philippines gained independence in 1946, Indonesia in 1949, and Malaya, along with British North Borneo and Sarawak (today's Malaysia), in 1957. The process in some of the other states was prolonged. For example, Singapore did not gain independence until 1963 and Brunei in 1984 (Figure 1–11). Whatever the case, the colonial period left what Rigg (1991) calls "an institutional and ideological legacy" of Westernization that became superimposed on or integrated with former modes of governance.

A Bitter Battle: French Indochina

French Indochina—Cambodia, Laos, and Vietnam—became nominally independent in 1949, although political and economic power was retained by France. Various nationalistic leaders including Ho Chi Minh resisted the French with aid from Communist China and the Soviet Union. A guerrilla war ensued in Vietnam, and the French were finally defeated at the Battle of Dien Bien Phu in 1954.

The United States, panicked over the further spread of communism, stepped into what became known in America as the "Vietnam War" and in Vietnam as the "American War." The American-supported non-Communist government of South Vietnam waged battle against the Communist forces of North Vietnam. After years of terrible conflict, loss of life, and increasing corruption in South Vietnam, public opinion in America forced U.S. withdrawal in 1973. The war finally ended in 1975. Divided Vietnam was then unified by the Communists.

The Vietnam/American war was a human and environmental tragedy. More than 4.5 million died in Indochina, including more than 58,000 Americans. Bombs, napalm, and chemical defoliants destroyed millions of acres of forests and cropland as well as people's health. Thousands have been killed or maimed by unexploded ordnance throughout Vietnam, Cambodia, and Laos. As you will see in the next chapter, the ramifications of this war were severe, widespread, and ongoing.

Changing Urban and Rural Worlds

Although there were a number of very large cities in 1900, most of mainland and insular Southeast Asia remained sparsely populated. However, the twentieth and twenty-first centuries witnessed unprecedented population growth in concert with economic, social, and political transformation (Figure 13–8).

Historically, settlement patterns were geared to coasts and rivers and based on seaborne interchange. Inland transportation was difficult until the road and railroad building of the 1800s.

Settlement has been tied to rice cultivation, which has, with Green Revolution improvements, supported ever-larger populations. Centralized

policies, in Angkor and the Red River Delta, were required to command and control the expansion of rice production. Given the fractured terrain, for most of history, small societies and local economies dominated the Southeast Asian world. This situation changed in the late nineteenth century as the tentacles of European imperialism reached beyond port cities into the interior. Large-scale plantations, mines, and administrative cities were constructed in any potentially profitable niche.

With the growth of colonial enterprises, urban centers including traditional entrepôts and new colonial cities exhibited spurts of growth. By 1910, there were eleven Southeast Asian cities of more than 100,000 people: Mandalay, Rangoon (Yangon), Bangkok, Hanoi, Saigon, Georgetown (Pinang), Singapore, Batavia (Jakarta), Surakarta (Solo), Surabaja, and Manila. These cities were primarily administrative and commercial centers without large industrial bases. Moreover, the colonial economy did little to stimulate economic development beyond the export sector.

The rural sector exhibited three categories. The largest was the peasantry tied to a subsistence economy. Second was the commercialized peasantry who participated in the growing market economy based on urban needs. Third was the enclave economies of plantations and mines opened up with Western and Chinese capital and imported wage labor.

A demographic picture of Southeast Asia at the turn of the twentieth century reveals two primary settlement patterns. At one extreme were the dense populations engaged in irrigated rice cultivation. Java, with its population of 30 million in 1901, stands as an example. At the other end of the spectrum were vast insular and mainland areas sparsely settled by shifting cultivators. Around 1900, irrigated rice began diffusing, and several areas such as the remaining frontiers of Java, lower Burma, central Siam (Thailand), and the Mekong Delta transitioned from low to high population densities. The first three decades of the twentieth century continued this pattern of opening up the region to export industries and tightening the links between the colonies and the imperial powers overseas.

The end of the Japanese interregnum in 1945 ushered in a period of transition to independence. During this time, international trade withered, urban and industrial employment fell, and many were forced

Figure 13–8

*Note that the majority of cities in Southeast Asia are either coastal or on major rivers.
Why is this?*

to revert to a subsistence existence. Nevertheless, there were factors of continuity as well as those of change.

The transition to national independence increased population mobility, only part of which was rural to urban migration. Many moved from overcrowded rural areas to less crowded ones. Others moved to cities as squatters, and still others populated the first suburban and even exurban areas outside large cities.

Out-migration of Western colonial administrators and business people—along with ethnic Chinese from several areas, and all Indians from Burma, plus Vietnamese from Cambodia—were other significant movements. Plantation and commercial rice regions were the areas mainly affected by these out-migrations. Many foreign managers and overseers were replaced by indigenous people.

METROPOLITAN EXPANSION

In Southeast Asia, it is the largest metropolitan areas that are growing much faster than other cities. For example, Kuala Lumpur has grown at almost double the rate of Singapore in the post-war period, reflecting its position as the capital of the new state of Malaysia. Once French Indochina split into Vietnam, Laos, and Cambodia, Phnom Penh grew rapidly in contrast to the former Indochina capital of Hanoi. Primate cities such as Bangkok, Kuala Lumpur, Jakarta, Manila, and Ho Chi Minh City (Saigon) continue to dominate the urban hierarchy.

Urbanization in general is on the increase. For example, the proportion of Indonesians living in urban areas of more than one million rose from 32.6 percent in 1971 to 40 percent in 2003. In the Philippines, the Manila Metropolitan Region has expanded to include 10.5 million people, 12 cities and 5 municipalities.

Many of these increases are related to processes of the transactional revolution, whereby flows of people, commodities, capital, and information are no longer hampered by the limits of space. While cities continue to grow, they do so over larger areas. This is called **region-based urbanization** and it produces **extended metropolitan regions** (EMRs).

It is in the areas adjacent to currently defined urban areas that urbanization is occurring most rapidly, and this is reflected in the growing proportion of populations engaged in nonagricultural activities (refer to *kotadesasi* settlements in Chapter 4). EMRs have two main components: the central city and the metropolitan region containing adjacent urban settlements. With the exception of Singapore, it is these outer rings that are experiencing the most rapid growth. Meanwhile, city cores have maintained substantial shares of metropolitan populations. In fact, city populations continue to grow with increasing density and congestion (Figure 13–9). EMRs also

Figure 13-9

More than 10 percent of Thailand's population lives in Bangkok, the capital city. Bangkok is also known as Krung Thep, *meaning "City of Angels." Here, the crowded old city is surrounded by new office and apartment towers. Hundreds of people pour into the city every day seeking employment. Densities are as high as 5,800 people per square mile (3,600 per sq km), and an estimated 400 new vehicles are registered every day. Bangkok is one of the most congested cities in the world.* Photograph courtesy of B.A. Weightman.

continue to attract rural-urban migrants. At the social level these areas are extraordinarily heterogeneous with multiple types of land use existing side by side. You will learn more about these trends and individual cities in the next two chapters.

NEW RURAL WORLDS

There have always been market dependencies and interactions between rural and urban worlds but the rural world of today is different in many ways. While *village* may describe life for the majority of Southeast Asians, millions now spend some of their time in urban areas both living and working and millions more spend time working in urban-type activities that have come to the countryside.

There is an assumption that people who live in villages are farmers (i.e., that "rural" is synonymous with "agriculture"). But this is not necessarily the case. Urban residents from rural areas often still identify themselves as villagers and farmers. In other words, *domicile* does not always mean "home" and *rural* and *urban* are frequently melded.

Recent studies of village economies consistently indicate that between 30 and 50 percent of total household income is derived from non-farm or off-farm sources. Many households have different family members participating in either rural or urban livelihoods, making it problematic to classify the household as rural or urban.

The transport revolution has been instrumental in expanding networks of contact and communication. The construction of new roads, the easing of restrictions on travel, and the advent of the bicycle, motor scooter, and other cheap forms of transport encourage rural and urban interchange in terms of both job transfers and industrial location (Figure 13–10). This increases economic opportunities and supplements incomes. There is a definite link between poor infrastructure and high rates of poverty.

The merging of rural and urban worlds has had other impacts. The use of chemical fertilizers, wage labor, and mechanization has increased. More emphasis is placed on raising animal stock and cash crops for the urban market. More people engage in non-farm activities on the farm, such as mat and basket weaving, pottery, and furniture making. Off-season, full-time employment in factories or public sector jobs has increased, and there is increasing migration for employment beyond local areas.

There are also age and gender considerations. Agriculture remains the dominant occupation among older age groups. Younger groups are more likely to engage in wage-labor, which points to the availability of such opportunities in or near rural areas. A second change is the growing importance of

Figure 13–10

I met these two students in Viangchan, Laos. Both were very proud of their new motorbike. Photograph courtesy of B.A. Weightman.

female employment. Rising female labor is pronounced in export-oriented activities such as the garment industry, electronics, and footwear. Declining fertility, later marriages, education, and increased mobility have fostered conditions for women to work beyond the confines of the village.

Rural and urban changes function in tandem. There is more to rural life than agriculture, and at any time in the year different members of different households may be engaged in rural- or urban-type occupations. Although better accessibility allows rural dwellers to live and work in cities, governments are encouraging factories and other businesses to locate in non-urban settings where a pool of workers is ready to take up non-farm employment.

Population Dynamics

Prior to the intensification of the European role, Southeast Asia was a comparatively healthy region. But, by the nineteenth century, conditions had deteriorated. Western advances in health provision and medical techniques had spurred population growth, meaning that increased concentration and contact and influx of new groups reduced public health effectiveness. Epidemics became a problem. Paradoxically, populations increased with growth rates climbing from 0.2 percent to well over 1 percent in many areas.

The Spread of Disease

When the rain forest was cleared over vast expanses to plant rubber trees, water bodies were exposed to sunlight. This caused the *Anopheles maculatus* to breed. This mosquito is one of the most dangerous epidemic vectors of malaria in the world. Epidemics were so severe that plantation owners had to carry Tamil workers off in droves and replace them with new ones. These deadly mosquitoes continue to plague settlement schemes, logging operations, agricultural clearance, and the like.

In another case, the British accidentally introduced the *Aedes aegypti*, another mosquito originally from Africa. As a super vector of dangerous viruses, it caused dengue fever to spread explosively in urban areas. Today, it is responsible for the emergence and spread of the deadly dengue hemorrhagic fever.

Interregional connections and land conversion policies of the Europeans changed the biotic ecology of disease, especially vectored ones. Millions of victims have suffered or died as a result.

This growth is thought to be attributable to colonial activities. For instance, the spread of Christianity in the Philippines increased birth rates considerably. The expansion of settled agriculture and general peace provided an atmosphere of stability and security in which to raise families. Improvements in agriculture meant improvement in nutrition encouraging longer life expectancies and lower mortality rates. By the 1960s, fertility rates had increased dramatically. For instance, Indonesia's total fertility rate (TFR) was 5.67 and Thailand's TFR was 6.4. However, with family planning (beginning in 1971 in Indonesia) and social pressure, these rates had fallen to 2.6 and 1.7 respectively by 2003. The achievements of these two countries in reducing fertility are to be noted for their noncoercive nature (unlike China).

From 1950 to 2000, population growth was impressive. Thailand's population increased from 20 million to 56 million. In the Philippines, population increased from 19.9 million to 50 million. Indonesia's population grew from 83 million to 212 million.

At the beginning of the twenty-first century, Southeast Asia's population reached more than half a billion. In 1900, the population was a mere 80 million. Even with birth control programs on the part of such countries as Thailand, Indonesia, and Malaysia, population dynamics are uneven. On the one hand, richer countries such as Singapore and Thailand have low fertility rates. On the other hand, poorer countries such as Cambodia and Laos have high fertility rates (Table 3-1). Further, different groups within countries have different rates. For example, Malaysian women of Chinese ancestry (38 percent of the population) tend to have about two children each, while Malays and other indigenous groups average about four children each. In Southeast Asia, the use of birth control is at 55 percent, with contraceptive use rising with educational level. A major problem is access to contraception, as in Vietnam and the Philippines.

Environmental Concerns

The most severe environmental problem in Southeast Asia is deforestation. While slash-and-burn cultivation, clearing for migrant settlement, and the development of tree crop plantations are important causes, perhaps the most important is commercial logging. The forests of Southeast Asia have high commercial value in addition to high yields. Much of today's imported wood products originate in these forests.

Commercial logging is a lucrative activity. The entire process has involved considerable corruption, especially in Indonesia, Malaysia, the Philippines, and Thailand. Much forest is logged illegally. In other cases, timber exports are under-invoiced and the logs smuggled overseas. Most logs from the Philippines are exported illegally. Finally, timber concessions are granted for political or money favors. Who has access to and control of forests is one of the most salient aspects of their use today.

Southeast Asian sources of tropical timber products are concentrated in the Philippines, the Malaysian states of Sarawak and Sabah, and in Indonesia's outer islands. The primary destination of logs and related products is Japan, which has been the world's largest importer of tropical timber since the 1960s. Forest devastation has been so great that today only Sarawak continues to export raw logs legally.

Timber extraction in Southeast Asia is related to Japan's investment and development plans in the region. For example, Japan's early role in the timber industry in Kalimantan began with technical assistance in exchange for timber export agreements. After 1967, most of the major Japanese trading companies such as Mitsubishi, Mitsui, and Sumitomo were involved in loaning money for infrastructural development, including access roads into timber areas. In 1971, Indonesia replaced the Philippines as Japan's first source of raw logs, which were largely processed into plywood in Japan.

By 1980, after rampant forest destruction, Indonesia shifted its efforts to develop its own forest products industry. It also began a phased ban on the export of raw logs, which was fully implemented by 1988. Most Japanese companies withdrew from the Indonesian timber industry but Japan remains the largest importer of Indonesian plywood. This example shows the relationship between one country's political and economic activities and environmental consequences in another.

Although several countries have taken steps to control wholesale forest destruction, illegal logging continues. Deforestation, with its accompanying soil erosion and habitat elimination (Chapter 2), continues as one of the major issues in the region.

Conflicts at Sea

Territorial conflicts have permeated the history of Southeast Asia. Given this ocean-bound world, it is no surprise that some conflicts would be over ocean spaces. Control over ocean space becomes critical when natural resources are involved as in the case of the Paracel and Spratly islands (Figure 2–5). Furthermore, control over ocean space becomes paramount when piracy is involved.

DISPUTED ISLANDS

The Paracels are disputed islands and reefs approximately 16 degrees north of the equator and about 233 miles (375 km) east of Da Nang in Vietnam. They were a part of French Indochina until the Japanese held them in World War II. Although these islands are closer to Vietnam than they are to China, the Chinese took the islands from Vietnam in 1974, claiming they were always a part of China. The Paracels are also claimed by Vietnam and Taiwan. The interest centers on potential oil reserves beneath the ocean.

The Spratly Islands are another group of a hundred or so islands and reefs in the South China Sea lying about two-thirds of the way from southern Vietnam to the Philippines. They were used as a submarine base by the Japanese in World War II. Their natural resources are fish, guano, and undetermined natural gas and oil potential. They are also situated near several primary shipping lanes. About 45 of the islands are claimed and occupied by China, Malaysia, the Philippines, Taiwan, and Vietnam. In 1984, Brunei established an exclusive fishing zone in the southern Spratlys.

In 1994, the American company Exxon signed a multibillion dollar deal with Indonesia to develop the Natuna gas fields in the southern Spratlys, an area partly claimed by China. There were clashes between China and the Philippines in 1995 because China occupied one island within 124 miles (200 km) of the Philippine island of Palawan. That same year, Indonesia expressed concern over China's claim to the natural gas fields. Some have suggested joint development of this region, but no multi-lateral agreements have yet been reached.

PIRACY

Traders have been harassed in the Southeast Asian waters for centuries. The nomadic seafaring Bugis from Sulawesi were once infamous as pirates and continue to ply goods throughout Indonesia. But piracy is not just a thing of the past. Pirates continue to attack from the Straits of Malacca to the Celebes Sea. Cargo ships, fishing vessels, passenger liners, and refugee boats are all victims. There were 445 incidences of piracy in 2003, 20 percent higher than in the previous year.

The situation was particularly tough during the Vietnam War when thousands of Vietnamese and Cambodians were trying to escape persecution. From 1980 to 1986 alone it is estimated that around 1,500 "boat people" (escapees) were killed, and hundreds of women were raped, abducted, and sold to brothels in Thailand (see Chapter 3 for more information on the sex trade in Thailand).

The Law of the Sea has put piratical victims beyond the reach of international law. Piracy is only piracy when attacks occur within jurisdictional waters of coastal states, and most states lack the resources to police their vast marine areas effectively. Many countries now claim straight base lines along their coasts within which the marine areas become coastal waters and pirates are subject to arrest.

Commercial piracy occurs in several regions. It is most common in the Straits of Malacca and the Strait of Singapore. Some 6,000 vessels pass through these passageways every month carrying a quarter of the world's trade and 80 percent of Japan's energy supplies. Such attacks irritate relations among Indonesia, Singapore, and Malaysia. Endemic overfishing in the Gulf of Thailand and poverty levels also foster piracy in that region. The Sulu and Celebes seas, between Malaysia and the Philippines, are frequent scenes of attacks on coastal freighters, passenger ships, and fishermen. Some freighters have taken to repelling the pirates with fire hoses. The Moro Islamic Liberation Front, an insurgency in Mindanao, has hindered suppression efforts and increased tensions between Malaysia and the Philippines.

Because modern pirates in Southeast Asia are land-based, and not rovers like the ancient Bugis, coastal states are expected to take on the responsibility of stopping them. Thus far, it seems that maritime frontiers are virtually impossible to police effectively.

Shapes of Countries

You have probably noticed that Southeast Asian countries take on many different shapes, some rather odd. Country shapes are significant in that they may affect a country's ability to consolidate its territory and control trans-border interactions. The most desirable shape is circular, because that provides the shortest possible border in relation to territory and allows equal access to all places from the center. Although this implies stability, stability is not necessarily the case. Cambodia has a relatively circular or *compact* shape but Cambodia does not have a history of stability. *Elongated* countries such as Vietnam and Laos tend to have unity and control problems.

Prorupt states are almost compact but have at least one extension of territory. This makes control problematic, especially in the prorupted regions. Thailand and Myanmar, with their long territorial tails, are examples of prorupt states and both have problems with their proruptions—especially Thailand with its minority Muslim population.

Some countries are broken up into pieces. Archipelagoes are therefore referred to as *fragmented* states. Both the Philippines and Indonesia are examples of fragmented states. Both have trouble with unity and both have separatist movements. However, there are exceptions. Brunei is a fragmented state yet does not suffer from instability. Why do you suppose this is the case?

Country shapes provide grist for discussion but do not determine a country's political or economic circumstances. Other factors such as resource distribution, physical features, and historic events play even more important roles.

ASEAN

The Association of Southeast Asian Nations (ASEAN) was formed in 1967 to promote social, economic, and political cooperation within the region. After 1945, the newly independent countries had little to bind them together in terms of interstate cooperation, and since economic development was the foremost concern, an international organization seemed prudent. The original members were Thailand, Malaysia, Singapore, Indonesia, and the Philippines. Brunei joined after its independence in 1984, and Vietnam, Laos, Myanmar, and Cambodia joined in the 1990s. East Timor joined after gaining independence in 2002.

ASEAN's main goals are to promote and integrate economic development, encourage social and cultural progress, and guarantee regional peace and stability. Actually, member countries trade more with outside nations than with each other, as so many of them produce the same things. Moreover, international trade activity is dominated by Singapore, Malaysia, Indonesia, and Thailand. Intra-ASEAN trade is also dominated by Singapore. Minerals

and fuels account for more than half the volume of exchange.

Three obstacles stand in the way of increased intra-ASEAN trade. The economies of the members are competitive rather than complementary, which supports trade outside the organization. Second, non-tariff barriers that are employed to protect domestic industries serve as barriers to intraregional exchange. Third, the productivity of countries such as Vietnam, with its low level of development, or Brunei, with its small size, cannot be expected to match the volume and diversity of output of more developed countries such as Singapore or Thailand.

Facing competition from India and China, ASEAN is being forced to confront its failure to realize a long-held goal: integration into a single market that is attractive to foreign investors. Singapore and Thailand are pushing for change and Brunei and Malaysia are somewhat supportive. However, the newer members—Cambodia, Laos, Burma, Vietnam, and East Timor—are reluctant to open their economies fully because they fear they cannot compete with the stronger countries. Some have backslid from original commitments to cut tariffs by 60 percent. The backsliders include Malaysia with its car industry, Indonesia with rice and sugar, and the Philippines with rice and petrochemicals.

India and China sent representatives to the ASEAN summit in Bali, Indonesia in 2003. They boasted about their investments in Southeast Asia and pointed out the importance of their economic roles in the region. These countries want to forge links with ASEAN in a larger economic community.

Pursuing consolidation, ASEAN leaders endorsed a blueprint which envisions a single common market and a single production base with the free flow of goods, services, investment, and capital. The year 2020 is the target date but Singapore and Thailand want to accomplish it sooner. ASEAN leaders did agree to remove non-tariff barriers, and harmonize customs procedures and product standards. Targets were set from 2004 to 2010 for 100 priority sectors ranging from wood products and cars to air travel and tourism.

These economic agreements might change the face of ASEAN which until recently has been most effective in the political arena, curbing hostilities and acting as a forum to solve numerous disputes. Most importantly, a strong regional identity has been created among its members. Geographer Antonia Hussey (1991) wrote, "The sense of identity and cooperation espoused by ASEAN is not found elsewhere in the third world." However, countries have been too hesitant in criticizing blatant human rights violations among their members— such as in Cambodia in the past and in Myanmar (Burma) today.

Economic Crisis of the 1990s

Rapid economic growth in the non-Communist ASEAN countries came to a rapid halt in the late 1990s when over-lending by banks caused a major financial crisis. Thailand, the Philippines, Malaysia, and Indonesia were forced to devalue their currencies in 1997. The problems were caused by high levels of foreign borrowing, government budget deficits, large bank loans to glutted property markets, and slower than expected economic growth. Japan, in its own economic crisis, slowed investment and technical transfers to Southeast Asia.

Thailand suffered most as the crisis led to a change of government and a new constitution. Austerity measures were imposed. Malaysia and Indonesia, long ruled by authoritarian governments, found it more difficult to admit mistakes and launch reforms and restructuring. President Suharto of Indonesia was forced to resign, leaving the country's economic and social life in chaos. Indonesia has had the hardest time recovering from this 1997 crisis, partly because a backlash against the Chinese led to their withdrawal in large numbers together with their money. Indonesia also had to absorb more than a million workers who were deported from Malaysia during the crisis there.

By the early 2000s, China presented new challenges with increased competition and increased markets. The redirection of foreign direct investment to China slowed Southeast Asia's recovery from the 1997 crisis. Moreover, Chinese goods compete in textiles (Vietnam and Indonesia) and higher-value products (Thailand and Malaysia). ASEAN knows that it must attract investors, lower trade barriers to open up markets, and upgrade domestic worker skills to counter China's impact, but inevitably it will face severe competition from China in many areas in the years to come.

🌐 Main Point Summary

1. Southeast Asia is fragmented into mainland and insular regions. Fragmented landscapes have supported the establishment of states and kingdoms either land-based on agriculture or sea-based on maritime trade.

2. Populations have been separated by mountains and tend to concentrate in river valleys and on river deltas. Mountain and lowland people have led a symbiotic existence yet have evolved in separate ways.

3. The Mekong is the longest river in Southeast Asia. It is an artery of human activity but not suitable for international shipping. One-third of the populations of Laos, Thailand, Cambodia, and Vietnam live along the Mekong.

4. Although Southeast Asia is tropical, there are temperature variations, depending on latitude and altitude. Precipitation also varies, with some regions experiencing dry seasons derived from the monsoon and prevailing wind directions. Dry seasons increase in length with increasing latitude and proximity to Australia as in the case of Indonesia.

5. Water and control of ocean passageways have been most important in the historical development of the region. Arab and Indian ocean traders, Hinduism, Buddhism, Islam, the Chinese, the European colonials, and the Japanese all arrived primarily by sea. Commercial interaction ensured diffusion of products and ideas and linked the people of the region with each other.

6. Water is essential to agriculture in its amount and distribution. Elaborate irrigation systems have been developed in many areas. Bali exemplifies how water control has been woven into the tapestry of life in a place.

7. Mountain barriers caused Indian culture to come almost entirely by sea. Southeast Asian societies adopted many aspects of Indian culture, including religion, architectural styles, and concepts of kingship. All of Southeast Asia except the Tonkin region was affected by Indian culture.

8. Buddhism that diffused from Sri Lanka also exerted a powerful influence. Borobudur on Java and Angkor in Cambodia testify to the strength of Buddhism in the region.

9. The Chinese referred to Southeast Asia as *Nanyang*. The Chinese impact was the direct result of colonization and conquest. Vietnam remained a Chinese territory until 907 A.D. By the sixteenth century, Chinese colonies were established throughout the region but large-scale immigration commenced only with China's internal turmoil in the nineteenth century.

10. With the coming together of so many different ethnic groups, languages, religions, and lifeways, most of Southeast Asia evolved as plural societies.

11. European political contacts began with the Portuguese capture of Malacca in 1511. With the exception of the Philippines, there was no major impress on the region until the nineteenth century.

12. One hot spot of activity was the Spice Islands (Maluku), where various European powers contested for spices such as cloves and nutmeg. Eventually, the Dutch wrested control from others and tightened its grip by destroying plants except for in specific regulated areas, devastating the native populations and their economies.

13. Southeast Asia became the crossroads of activity with the demands of the Industrial Revolution, opening of the Suez Canal, and the decline of Caribbean commerce. From 1825 to 1870, most of the region fell under European control.

14. Although the various European powers had their own agendas, Thailand was preserved as a buffer state between British and French interests on the mainland. The strongest rule was exerted by the Dutch Culture System, which forced Indonesians to grow specific amounts of cash crops.

15. The opening of new lands such as the Chao Praya and the Mekong deltas attracted thousands of new migrants. Indians and Chinese were drawn to tin mines and rubber plantations. Plural societies emerged whereby different ethnic groups had their own residential quarters but came into contact in the marketplace. Cities were largely foreign places, alienated from the circumstances of the rural masses.

16. Transportation and other infrastructure components were designed for foreign interests. Coastal entrepôts serving shipping left much of the interiors to their own devices. Dual economies emerged and became exaggerated as uneven development increased.

17. Political units, designed for the needs of Europeans, had little to do with historical or ethnic circumstances. Many groups were divided by arbitrary boundaries. Myanmar and Thailand are protruded and Laos is landlocked. Indonesia and the Philippines are fragmented archipelagoes.

18. In the late nineteenth and early twentieth centuries, revolutions proved that colonial powers were not invincible. Moreover, Japan's occupation, while as exploitative as the Europeans, promoted Asia for the Asians. After 1945, nationalistic leaders, educated abroad, orchestrated independence movements. They were not typically peaceful, and many colonies did not achieve independence until the 1960s and 1970s. Vietnam was one of the most prolonged conflicts, drawing in even the United States. East Timor was the last country to become independent in 2002.

19. Under colonial dominance, infrastructure and settlement patterns changed with more primary activity in the interiors. Large-scale plantations, mines, and administrative cities were constructed in many potentially profitable niches. Increasing percentages of the peasantry

participated in a growing market economy based on urban needs.

20. Two population patterns existed at the turn of the twentieth century: a dense one linked with intensive rice cultivation and a sparse one associated with vast insular and mainland areas inhabited by shifting cultivators.

21. An array of migration streams were encouraged by new-found independence and related events.

22. Larger metropolitan areas are growing much faster than other cities. Urbanization is on the increase. This region-based urbanization is fostering extended metropolitan regions that incorporate growing numbers of rural populations occupied in nonagricultural activities. While *village* may describe life for the majority of Southeast Asians, millions now spend time living and working in urban areas.

23. Rural and urban activities exhibit age and gender differences with older people generally sticking to agriculture and younger people seeking other forms of employment. Females are especially prominent in export-oriented activities.

24. Even in the context of poor sanitation, Western advances in medical techniques and health provision, along with improvements in agriculture, spurred population growth. Population growth was high between 1950 and 1990, causing giant leaps in national populations. Population dynamics remain uneven—countries such as Cambodia and Laos have much higher growth rates than others where family planning programs have been more effective.

25. Environmental concerns are dominated by deforestation, in which Japan continues to play a major role. Increasing pollution is another major concern.

26. Territorial conflicts and piracy continue to disturb Southeast Asia. The Paracel and Spratly islands, with their fisheries and undersea oil potential, are the source of territorial claims of several nations including China. Piracy plagues the South China Sea, the Straits of Malacca, and other critical ocean passageways.

27. The Association of Southeast Asian Nations (ASEAN) is an intraregional organization intended to promote social and economic development and maintain regional stability.

28. The late 1990s saw an economic crisis generated by overlending by banks, investment in already glutted property markets, government budget deficits and slower than expected economic growth. Political and economic reforms are helping hard-hit countries such as Thailand and Indonesia to recover.

⊕ Bibliography

Brierley, Joanna Hall. 1994. *Spices: The Story of Indonesia's Spice Trade.* New York: Oxford.

Dauvergne, Peter. 1997. *Shadows in the Forest: Japan and the Politics of Timber in Southeast Asia.* Cambridge: MIT.

Drake, Christine. 1989. *National Integration in Indonesia.* Honolulu: University of Hawaii.

Duchin, Dian. 1986. *The Subak Agriculture of Bali.* The New York Botanical Garden 10: 6–10.

Dutt, Ashok K, ed. 1985. *Southeast Asia: Realm of Contrasts.* Boulder: Westview.

Dzurek, Daniel J. 1989/90. "Piracy in Southeast Asia." *Oceanus* 32: 65–70.

Ginsburg, Norton. 1972. "The Political Dimension." In *Focus on Southeast Asia,* ed. Alice Taylor, pp. 3–12. New York: Praeger.

Hirschman, Charles. 1994. "Population and Society in Twentieth-Century Southeast Asia." *Journal of Southeast Asian Studies* 25: 381–416.

Hussey, Antonia. 1991. "Regional Development and Cooperation Through ASEAN." *The Geographical Review* 81: 87–98.

Keyes, Charles F. 1995. *The Golden Peninsula: Culture and Adaptation in Mainland Southeast Asia.* Honolulu: University of Hawaii.

Kummer, David M. "The Physical Environment." In *Southeast Asia: Diversity and Development,* ed. Thomas R. Leinbach and Richard Ulack, pp. 6–34. Upper Saddle River, NJ: Prentice Hall.

Ma, Laurence J. 1985. "Cultural Diversity." In *Southeast Asia: Realm of Contrasts,* ed. Ashok K. Dutt, pp. 53–64. Boulder: Westview.

Mastny, Lisa. 2003. "Messing With the Mekong." *World Watch* (November/December): 21–28.

McBeth, John. 2003. "ASEAN Summit: Taking the Helm." *Far Eastern Economic Review* (October): 38–39.

McCawley, Tom. 2004. "Sea of Trouble (Piracy)." *Far Eastern Economic Review* (May): 50–52.

McCloud, Donald G. 1995. *Southeast Asia: Tradition and Modernity in the Contemporary World.* Boulder: Westview.

McGee, T. G. 1995. "Metrofitting the Emerging Mega-Urban Regions of ASEAN: An Overview." In *The Mega-Urban Regions of Southeast Asia,* ed. T. G. McGee and Ira M. Robinson, pp. 3–26. Vancouver: UBC.

McGee, T. G. 1967. *The Southeast Asian City.* London: G.F.Bell.

Michell, George, and John Fritz. 1993. "Angkor Regained." *Geographical* 65: 33–38.

Minority Rights Group. 1992. *The Chinese of Southeast Asia.* Manchester: Free Press.

Noble, Allen G. 1985. "The Physical Environment." In *Southeast Asia: Realm of Contrasts,* eds. Ashok K. Dutt, pp. 36–51. Boulder: Westview.

Rigg, Jonathan. 1991. *Southeast Asia: A Region in Transition.* London: Cambridge.

Rigg, Jonathan. 1997. *Southeast Asia: The Human Landscape of Modernization and Development.* New York: Routledge.

Sanderson, Warren C., and Jee-ping Tan. 1995. *Population in Asia.* Washington, D.C.: World Bank.

Sar Desai, D. R. 1989. *Southeast Asia: Past and Present.* Boulder: Westview.

Shaffer, Lynda Norene. 1996. *Maritime Southeast Asia to 1500.* London: M. E. Sharpe.

Vatikiotis, Michael, and Prangtip Daorueng. 1998. "Survival Tactics (Overseas Chinese)." *Far Eastern Economic Review* 26 (February): 42–45.

Chapter 14

Mainland Southeast Asia: Turmoil and Peace

*"The waves of human passion rise
and fall; worldly things are as
transitory as passing clouds."*

CAO BA-QUAT (1809–1854)

In this chapter you will learn about the five countries of mainland Southeast Asia: Burma/Myanmar; Thailand; Vietnam; Laos; and Cambodia (Figure 14–1). As you will discover, the history of each unfolds in context of the history of the others. Landscapes, with their impress of rice cultivation, Buddhism, and battlegrounds are remarkably similar. Yet, each country exhibits its own character in terms of ethnicity, language, political system, level of development, and various other cultural aspects (Figure 14–2). The complexity of mainland Southeast Asia makes it a geographic conundrum leaving us with more questions than answers.

Burma/Myanmar: Paradise Lost

Imagine living in a country where your universities are closed most of the time. Where you cannot stay at a relative's or friend's house without government permission. Where you voted for democracy but your party is outlawed and your leader is under house arrest. Where barbed wire barricades are stacked on streets ready to block any march or protest. Where people meet in secret to tell you about persecution of minorities and slave labor. This is Myanmar today—oppressive for most within and despised by most without.

Officially, the country is called Myanmar. But, many people refer to it as Burma. The country's original name was Myanmar but this was anglicized as "Burma" by the British who invaded in 1885. In 1989, an unelected military dictatorship changed the name back to Myanmar and this is the name recognized by the United Nations. However, there is a "democracy movement" in Myanmar that insists on the name "Burma." Those who support this movement and who do not recognize the military dictatorship as a legitimate government support the name "Burma" as well. The problem with "Burma" is that it refers to the majority Burman people and does not represent the 8 major nationalities or the 135 ethnic minorities who also reside in the country. Some

Figure 14–1

Defining the Realm: Mainland and Insular Southeast Asia. Note the location of the core regions.
From H. J. de Blij and P. O. Muller, *Concepts and Regions in Geography*, Second Edition, 2005, p. 342.
Originally rendered in color. © H. J. de Blij and P. O. Muller. Reprinted with permission of John Wiley
& Sons Inc.

Figure 14–2

This region is a dramatic ethnic-cultural complex. Note the presence of Chinese and Indian communities.

think that Myanmar/Burma should have a new name, one that represents all the people. I will use Burma and Myanmar interchangeably in the text.

From its prewar position as a productive agricultural colony, a major rice exporter and rich source of minerals, Burma has slumped into being the poorest country in Southeast Asia. How did this happen?

POLITICAL BACKGROUND

Burma is the largest and most ethnically diverse state in mainland Southeast Asia (Figure 14–3). Burmans make up 68 percent of the population but there are 11 major and many more ethnic groups in the country. The Burmans dominate the alluvial plains and cities. Tribal groups, flanking the borders of India, Bangladesh, China, Laos, and Thailand, have long resisted Burman dominance. The Burmans are mainly Buddhist. The Shan and Karen are predominantly Christian, and the Arakanese are split between Christian and Muslim. These religious differences have deepened the rifts among Myanmar's ethnic groups.

Burma has always been concerned about the ambitions of its neighbors. War, competition, and animosity between Burma and Thailand are centuries old. The Mongols once invaded Burma, and the Chinese are still seeking access to the Indian Ocean. Myanmar's economy was once dominated by Indian migrants. Thus, historic acrimony shapes the country's attitudes to outside powers.

The British East India Company, headquartered in Calcutta, envisioned Burma as a buffer zone. Clashes in Arakan adjacent to Bengal mushroomed into the first Anglo-Burmese War from 1824 to 1826. Superior arms gave victory to the British, and Burma was forced to cede territory along the Bay of Bengal. A second conflict erupted in 1850, whereby the British acquired more territory in lower Burma. In 1885, the Burman king was exiled to Calcutta (even the throne was removed), and Britain annexed Burma in 1886.

Ruling from Calcutta, the British imposed alien Indian forms of government. Lower Myanmar—the Burman region—fell under direct rule, and the power of local elites was squashed. In Upper Myanmar, indirect rule left the hill tribes pretty much to their own devices. Consequently, the minorities developed

Figure 14–3

Myanmar (Burma) and its core areas. Note the importance of rivers in the location of urban places and core areas.

a stronger sense of identity in opposition to the British-run region.

British colonial rule was founded on an Anglo-Burmese and Indian elite. Indians were free to migrate to Burma. By 1931, about 7 percent of the country was Indian, and Rangoon was 53 percent Indian. Chinese were recruited from British Malaya and Singapore. They soon added to the population diversity, working in Shan state mines and operating small businesses and rice mills. Shan state is the largest province of Myanmar.

British efforts to develop the Burma delta regions in the 1850s encouraged large-scale migrations of Burmans from the arid northern regions. The Sittang and Ayeyarwady deltas became prominent rice-producing regions. The area under rice cultivation expanded from 800,000 acres (320,000 ha) in 1852 to around 6 million acres (24 million ha) in 1901. The associated population increase went from 1 million to 4 million. By the 1920s, population pressure had become a major problem: many farmers were indebted, and tensions among Burmans, Indians, and Chinese were spiraling into violence.

Britain was content to have Burma as a rice, teak, and mineral exporter. There was no attempt to industrialize the country. This plural society rested singularly on primary economic activities. Moreover, economic position was coincident with ethnicity.

Economic development under colonial rule was accompanied by the spread of Western education and the rise of nationalist leaders. These were not only dominated by ethnic Burmans but also were strongly anti-Chinese and anti-Indian. The fact that Buddhism was at the core of Burman life further alienated the non-Burman minorities, especially those who were Christian.

Three of the prominent nationalists and independence seekers were Aung San, U Nu, and Ne Win. Aung San was assassinated in 1947, U Nu became the first prime minister in 1948, and Ne Win led a coup in 1962 placing the military in power. In 2005, the military was still in power.

The anti-Western Japanese occupation in the 1940s was welcomed by many Burmans, seduced by the Greater East Asia Co-Prosperity Sphere idea of "Asia for Asians." But reality did not meet the vision, and the disillusioned Burman nationalists worked toward independence. Burma was terribly damaged in World War II, with most of its infrastructure destroyed.

After the war, independence came quickly (in 1948). Minorities found themselves under a govern-

ment relentless in its efforts to Burmanize them. A Karen nationalist movement launched several insurrections. These continue into the 2000s. Burma was a democratic state between 1948 and 1962, but its failings were critical. A failing economy, regional insurrections, social urban and rural unrest, and rampant corruption led to Ne Win's overthrow of U Nu. Political and ethnic minority leaders were arrested and opposition from university students and Buddhist monks was ruthlessly suppressed. The Westernized, often Anglo-Burman elite that had run the country under colonial rule and through the 1950s fled, along with large contingents of Indians and Chinese.

The Demise of Teak

To finance his campaign against U Nu, Ne Win's junta sold off much of Burma's teak. An estimated 80 percent of the world's teak is found within Burma's 900-mile (1,500-km) boundary with Thailand. Thailand is hungry for logs as it placed a moratorium on domestic logging in 1988. Scores of logging and fishing concessions were sold to Thai firms, along with the rights to gems, minerals, and oil to foreign concerns. Leaders of the Karen and other minorities have been and are still being displaced by these concessions and associated road building. Even so, they, too, are selling off teak to finance their insurgencies against the Burmans. However, tribal peoples use elephants for logging, and elephants do not require roads. Consequently, their operations are less damaging than the mechanized ones that call for forest clearance for road development.

Burma is losing at least 2 million acres (800,000 ha) of forest cover a year. Forests have been virtually eliminated in coastal areas and in the Ayerawaddy and Sittang valleys.

In 1974 the country was partitioned into administrative divisions: seven minority states and seven Burman divisions based on Western notions of nation and state. The regions were not designed for any level of autonomy but rather to facilitate further centralization. Delineations are not consistent with ethnic or residential integrity; groups and villages are splintered and mixed up. This aggravates the existing tensions among different linguistic and religious groups. Moreover, as geographer Curtis Thomson (1995) explains, traditional mapping with bounded concepts "contrasts sharply with the ephemerally bordered, hierarchically centered territories of mainland Southeast Asia." Boundary definitions remain an important focus of rebel movements.

In 1987, the United Nations gave Myanmar "Least Developed Nation" status. However, there are two economies: one legal and the other illegal. The illegal economy focuses on opium, of which Myanmar is one of the world's major producers.

In 1988, a series of protests and strikes called for an end to military rule. In response, a new organization, the State Law and Order Restoration Council (SLORC), took control and killed thousands of protestors. SLORC decided to hold elections in 1990. Surprisingly, **Aung San Suu Kyi,** daughter of the nationalist hero Aung San, returned from London to lead the National League for Democracy (NLD). She was placed under house arrest in 1989. Nevertheless, the NLD won the election convincingly. But SLORC declared the elections null and void and arrested NLD members. Aung San Suu Kyi remains under house arrest and was awarded the Nobel Peace Prize in 1991. SLORC still rules but under a new label: The State Peace and Development Council (SPDC).

Rules of SLORC/SPDC

The rules of SLORC are known as the "People's Desire." They are printed in daily newspapers and on billboards and include such orders as: Oppose those relying on external elements acting as stooges; Oppose those trying to jeopardize stability of the state and programs of the state; Oppose foreign nations interfering in internal affairs of the state; and, Crush all internal and external destructive elements.

SCRAPING THE RICE BOWL

Burma was once known as the "Rice Bowl" of Southeast Asia. Rice production dwindled during decades of socialist rule but is picking up under SPDC's liberalized economic policies.

Rice trade is one of the regime's most important sources of foreign revenue. The government monopolizes rice exports. It buys rice from farmers at low prices, sells it on the world market, and absconds with the profits. Quotas are set for each cultivated area—15 to 20 percent of each farmer's crop is forcibly purchased for less than half the market price. Many farmers cannot make ends meet and wind up losing their land to the government or to other, richer farmers.

There are 7,000 agricultural supervisors posted around the country to encourage harder work. Signs such as "Lazy People Lose Their Land" and "Be Confident in Your Ability" are intended to stimulate productivity.

Paddy production increased around 30 percent in the 1990s, mainly because of expansion of irrigated areas and increased supplies of subsidized fertilizer, pesticide, and diesel fuel. Low-cost tractors have also been imported from China though few are visible in the landscape.

Pests are a serious problem. Insects devour rice because storage techniques are poor. Moreover, more double-cropping with fast-growing seeds means that much rice is harvested in the rainy season. Without pesticides and adequate storage facilities, the wet grain is even more susceptible to pests.

In border regions, quotas and inadequate technology have compelled many farmers to leave their land altogether and migrate to Thailand. Others have been forced off their land by the military. About a million Burmans now work in factories and farms in Thailand.

Inle Lake's Floating Agriculture

Inle Lake, located in one of the Shan states, is Burma's largest lake. The lake is home to vast

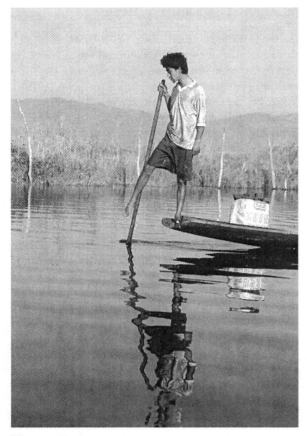

Figure 14-4

Inle Lake. This man is tending his gardens from his dugout boat. Note that the paddle is operated by hand and foot. Increasing numbers of motorboats are displacing the polers. Photograph courtesy of B.A. Weightman.

areas of floating gardens anchored in place with long bamboo poles driven into the lake bottom (Figure 14–4). Adjacent houses are built on stilts and stand in 6 to 10 feet (2 to 3 m) of water.

The gardens are strips roughly 82 feet (25 m) long and 3 feet (1 m) wide. They are cut from floating vegetation at the lake's edge and towed to the garden site. Sediment from the lake bottom is scooped up and placed on top of these mats to provide topsoil for the

planting of vegetables. Plants, especially tomatoes, with their roots close to the water, absorb nutrients from the lake, which is a repository for all the residential waste in addition to runoff from the surrounding hills. The gardens are tended by farmers who move from strip to strip in dugout canoes. Much of the produce is sold at markets around the lake. Fish farming is also practiced, with the floating gardens acting as nurseries.

AUTHORITARIAN DEVELOPMENT

To make way for development projects such as roads and tourist facilities, SPDC has forcibly moved people off their land. Tens of thousands have been moved to satellite towns around Yangon (called "Rangoon" by the British and during the period before SLORC) and Mandalay. The entire population of Bagan was displaced to eliminate dwellings from among the temple complexes. The displacement was part of the government's tourism promotion. When the government wanted to widen a road in Mandalay, it simply sliced the fronts off all the shops lining the route.

One of numerous persecuted groups is the Rohingyas. Burmese Muslims, the Rohingyas began to flee across the border into Muslim Bangladesh in 1978. This exodus, which continues in the 2000s, has strained relations between Burma and Bangladesh. There are an estimated 15,000 Rohingyas living in Bangladesh.

The Karen are pressing for an autonomous state within a federated Burma. The government has brutalized the Karen, arrested their leaders, and done everything it could to squash their ambitions. However, in 2004, five decades of fighting were halted with an informal cease fire. Delegations from the military junta and the Karen met in Yangon to discuss their positions. Only time will tell what the results will be.

Most displaced people work in the cities but find transport costly in both time and money. Many have moved to squatter settlements closer to their place of employment. In any case, community and family relationships and village economic systems are broken or strained.

The country has a serious shortage of electricity. In Mandalay, power is rationed. Many homes have electricity on alternate days, and factories can only operate part time. However, government-supported tourist hotels always have power.

The government relies on forced labor for public works and eco-tourism development. In 1996, when people protested, at least 300 villages were destroyed and many people were killed. There are numerous accounts of people "worked to death." Many eco-tourism projects are in minority regions, as are other developments such as new roads and oil pipeline construction. These projects are controlled by the SPDC military, whose forces have doubled since 1988. Recruitment of 13- to 15-year-olds has boosted numbers to 400,000.

One project of interest is an oil pipeline built from the Yadana oil field in the Andaman Sea to Ratchaburi, southwest of Bangkok. Much of this project used slave labor. Every village along the way had to donate labor, including women and children. On such projects, skilled workers are paid about 50 cents a day; the rest get nothing. Foreign oil companies are heavily involved: Total (partly owned by the French government), the American company UNOCAL, and companies from Myanmar and Thailand. The deal is supposed to give the Yangon generals about US $400 million a year for 30 years. A new railway into the pipeline region is also being constructed with slave labor. UNOCAL denies this, but 5,000 troops have been sent into the region to protect Total personnel.

THE CHINA CONNECTION

The Mekong River and the Burma Road from Kunming, Yunnan in China, to Mandalay now function as major arteries for the influx of Chinese into Burma and elsewhere in Southeast Asia. Thousands of Yunnanese have poured into Mandalay, where they now dominate commercial life.

China refers to the region as the "Great Golden Peninsula." Trade between Yunnan and Burma increased from US $15 million to well over US $800 million annually between 1984 and 1994. Trade has been growing at a rate of 10 percent a year since then. Manufactured goods that once came from Thailand now come from China. In addition, the Burma Road has become the main route for timber into China and arms shipments from China. It is also a major conduit for drugs.

The Chinese are anxious to gain connections to the Indian Ocean. It is surmised that China already has access to Andaman Sea islands for spy operations. Fishing boats with sophisticated monitoring equipment have been detected in the area as well. India has protested to Yangon and Indonesia has expressed concerns over potential Chinese threats to commerce in the Straits of Malacca.

LANDSCAPE OF REPRESSION

For decades a tyrannical military regime has kept Burma in political and economic turmoil. Infrastructure is pathetic at best, and thousands of Burmans and minorities are unemployed, engaged in forced labor, or working under oppressive conditions.

I have a vivid memory of Mandalay. I went to a place where they make gold paper to be used for Buddhist temple offerings. Perched at the edge of a muddy lane by a stream was a thatched, wooden structure with a cellar carved out of the earth. The cellar was filled with the stream's murky water. Floating on the water was a platform, and on the platform squatted four women. The only light was that which crept past a tarpaulin shielding the pit's doorway from the sun's intensity. It was 87°F (30°C). Both heat and humidity were tortuous. Each woman had a substantial piece of wood in each hand with which she rhythmically beat the small stack of gold sheets at her feet. The pounding resonated in the enclosure; the noise was deafening. These women work in this manner for eight hours a day, seven days a week. Think about it. This could be your life!

Unfortunately, Myanmar has been shoved so far into life's cellar that even a change in government will take years to drag the country out of its current state of oppression, corruption, and impoverishment.

Figure 14-5

Opium poppies in the Golden Triangle in northern Laos. Photograph courtesy of B.A. Weightman.

Drugs and the Golden Triangle

The production of illicit narcotics has more than doubled over the past few years in the Burmese section of the Golden Triangle, a place where the borders of Burma, Laos, and Thailand intersect. The area under opium poppy cultivation increased from 228,000 acres (92,300 ha) in 1987 to 380,542 acres (154,000 ha) in 1995 (Figure 14-5). As of 2003, poppy acreage had been reduced by half but more intensive efforts on the remaining land means that opium production has not declined. Heroin is derived from opium. The potential heroin output rose from 54 tons in 1987 to 166 tons in 1995 and to 825 tons in 2002. Close to 400,000 households earn the bulk of their income from opium production.

For years the drug trade was controlled by the infamous Khun Sa, but in the 1980s he lost ground to better equipped heroin barons near the Yunnan frontier. Khun Sa surrendered to the Burmese government in 1996 and moved with his heroin-generated fortune to Yangon, where he lives in luxury. Much of his money was invested in perfectly legal businesses such as hotels, real estate, supermarkets, and construction, generating a mini-boom in Myanmar's economy.

Khun Sa's turf has been taken over by the United Wa State Army, a splinter group of the Burmese Communist Party in northern Shan state. This group maintains a cease fire with the Yangon regime.

Since the 1980s, a new string of refineries has been established in the Shan state close to the Chinese drug market and with links to new markets abroad. One new market is Taiwan, which in 1989 was virtually drug free. Now drug abuse is a major concern. Drug use is increasing in Vietnam, Bangladesh, and India. There are more than 700,000 Burmese heroin addicts, 70 percent of whom have HIV. There are at least half a million addicts in Thailand, and 14 percent of Malaysian teenagers are now using hard drugs. China counts at least 250,000 addicts, with Yunnan having the highest rate of addiction.

Some 70 percent of all heroin in the United States comes from the Golden Triangle, and most of of that originates in Burma. In Australia, with 30,000 addicts, 75 percent of all heroin comes from Burma, with 5 percent from Laos and Thailand.

For years, Thailand was the main conduit for drugs, but stepped up law enforcement and poppy production alternative programs have altered this pattern. Cambodia and Laos have emerged as major transit routes. Cambodia is also a key center for money laundering. Drug laws are vague in Cambodia, and marijuana is freely available in markets. Cambodia is also a major shipper of marijuana. In addition, after years of war, Cambodia become a major source of arms for the Golden Triangle's drug lords. Cambodia's deep-water ports at Koh Kong and Kompong Som, with their hundreds of fishing trawlers, make it easy for drug runners to smuggle the drugs to larger ships in the international distribution web. Cambodia handles about 15 percent of Burma's heroin production. In Laos, drug smuggling is protected by the Mountainous Area Development Corporation, a company controlled by the Laotian army.

Another new trail is from Burma and northern Thailand via northern Laos into Vietnam via Hanoi and the port of Haiphong. Danang is another distribution point. Vietnam, like its neighbors, has a long history of opium smoking among ethnic minorities in mountainous areas. Vietnam provides an alternative, and less-policed shipping route. Addiction is becoming problematic in Hanoi among wealthy teens and the disadvantaged.

Profits are huge. A poppy farmer earns US $330 for 7,000 grams of raw opium, which is then converted into a 700-gram heroin brick, worth US $4,250, at a Burmese refinery. In Sydney, the brick will be worth US $53,050. In New York, the brick will be valued at US $80,000 and will be converted into 28,000 packets of adulterated heroin with a total retail value of US $280,000.

Clearly, drug production in the Golden Triangle has widespread international implications. Chinese-organized crime families from Hong Kong, China, Taiwan, and Singapore show great interest in Cambodia. Various African groups ferry drugs to Nigeria by sea and then on to Europe. The Tamils deal in arms and heroin to finance their war in Sri Lanka.

Democracy alone will not alleviate the drug crisis in northern Burma. Even with a new government, present and past drug lords are unlikely to listen to calls for reconciliation and stop their lucrative activities. In the absence of a comprehensive policy, the fracturing of northern Burma into small, well-armed, and violent drug-trafficking states will continue to create instability in the region and serious drug problems abroad.

Thailand: Tiger in the Metropolis

Like an elephant with its trunk drooping down the Malay Peninsula, Thailand sits in the heartland of mainland Southeast Asia. Because the north-south distance stretches 16 latitudinal degrees, Thailand's climate exhibits great diversity. Mountains in the north give way to limestone-encrusted tropical islands in the south. Rivers and tributaries drain into the Gulf of Thailand via the Chao Praya Delta near Bangkok (Figure 14–6).

The country is divided into four main zones: the fertile central region dominated by the Chao Praya; the drought- and flood-prone, poor, northeast plateau; the rugged northern region dominated by mountains and fertile valleys; and the southern peninsular region characterized by rain forest.

Thailand's population is relatively homogeneous. It is unique in Southeast Asia in that it avoided the disruptions of Western colonial rule and the upheavals of decolonization. After World War II, which produced minimal effects on the landscape, Thailand pursued capitalist development. By the 1990s it was deemed an NIC and now is one of the tigers of Asia. A brief look at Thailand's history will set the scene for the discussion of this Bangkok-centered economy.

THE TAIS

The **Tais** were the principle ancestors of the Thai, Lao, and Shan peoples. *Tai* is a cultural and linguistic term used to denote the various Tai people who became differentiated into a larger number of separate identities. Over many centuries, the Tai migrated southward from western China.

The Tai were wet-rice farmers organized in *muang*—one or more villages under a chieftain. By the thirteenth century, further expansion and alliances transformed *muang* into kingdoms that adapted many of the beliefs of peoples they came into contact with. For example, the Tai probably adopted Theravada Buddhism from the Mon people of central Thailand and the Burmans at Pagan. Angkor served as a blueprint for Tai state builders. From Angkor came the

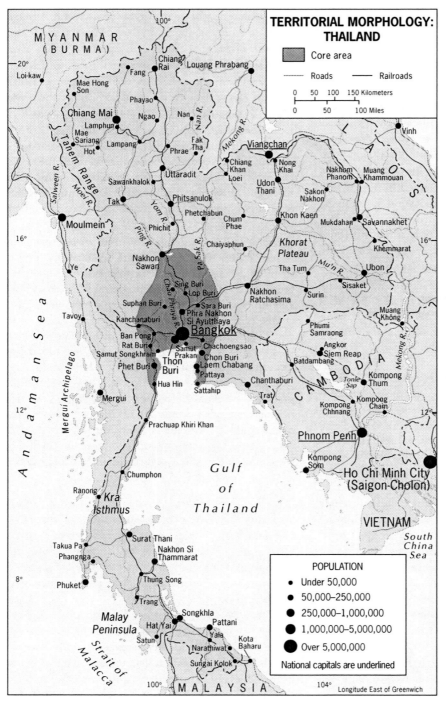

Figure 14-6

Thailand has many cities and towns but the core area focuses on the primate city of Bangkok. Why do you think this is the case? From H. J. de Blij and P. O. Muller, *Concepts and Regions in Geography,* Second Edition, 2005, p. 355. Originally rendered in color. © H. J. de Blij and P. O. Muller.

concept of society as a divinely ordained hierarchy under a ruler incarnate of a Hindu deity or Buddhist *boddhisattva*. Angkor also provided a model for administering large and scattered populations as well as a range of arts and technologies. Cultural borrowings increased as a result of Tai attacks on Angkor in the fourteenth and fifteenth centuries.

Sukothai

Meanwhile, in the thirteenth century the Tai kingdom of Sukothai had arisen. Modern Thais regard Sukothai as the birthplace of the Thai nation. The people of Sukothai adopted the name "Thai" to distinguish themselves from other Tai-speaking people. The Khmer system of writing was adopted and modified for a Tai written language. This is the predecessor of today's Thai writing system. Ultimately, Sukothai faded. In 1351 a new kingdom emerged to the south, Ayuthaya, which eventually became known as Siam.

Ayuthaya

Ayuthaya lasted until 1757, prospering from its strategic position only 43 miles (70 km) up the Chao Praya from the sea. Commanding the vast Chao Praya plain, Ayuthaya provided rice for export. Ayuthaya was seen by the Europeans as a great trading power. However, Ayuthaya had to face a challenge from the Burmans who crushed the empire by 1767. Ayuthaya's ruling class was decimated. Thousands of people and their wealth were taken away. The city, which had been one of the greatest and wealthiest in Asia, was burned and vast tracts subjected to a scorched earth policy.

The Rise of Bangkok

By 1785, new Thai rulers had emerged to smash the Burmese and establish new kingdoms. These eventually consolidated into an area larger than Ayuthaya had ever known. The new empire included all of mainland Southeast Asia excluding Burma and Vietnam. It also included some of the northern Malay states. A new capital was established at Bangkok in 1782.

In 1896 an Anglo-French agreement recognized Thailand as an independent buffer state between British Burma and French Indochina. Not experiencing any intense Western pressures, Bangkok remained traditional until the mid-nineteenth century. At this time, many Thai nobles studied Western languages, science, and mathematics. Bangkok was prepared to orient itself toward the West. In 1855, the Bowring Treaty was signed with Britain, giving Britain extraterritorial rights. Treaties with other countries followed.

Eventually the Thai empire was stripped of some of its holdings. It had to abandon its Cambodian and Laotian territory to the French and the Malay states to the British. In 1896, a treaty between England and France guaranteed the independence of most of the territory that today forms Thailand. Siam became Thailand in 1939, although it was still referred to as Siam during World War II.

Modernization proceeded with the building of railways and Western-style education for royal and upper-class children. Rice exports boomed, as did teak and rubber. But there was no significant industrialization. Western and Chinese interests controlled the economy.

The government stimulated Chinese assimilation into Thai society through business partnerships and intermarriage. The Chinese also adopted the Thai language, education, and culture.

European predominance in Southeast Asia was challenged by the Japanese, who invaded Thailand in 1941. When Western forces collapsed in the area, Thailand nominally allied itself with Japan, although effectively, it was an occupied country.

The defeat of Japan in 1945 was followed by an era of close relations with the United States in an alignment against the spread of communism in the region. Thailand has served as a refugee center for hundreds of thousands fleeing the conflicts of former French Indochina.

CHANGE, FAITH, AND ORDER

America's involvement in Vietnam was a powerful factor in opening up Thai society. In the 1970s, international tourism struck the country in a big way and labor migration began in earnest. This stimulated consumer culture and the emergence of a plethora of "lifestyles." What does this mean for Thai culture?

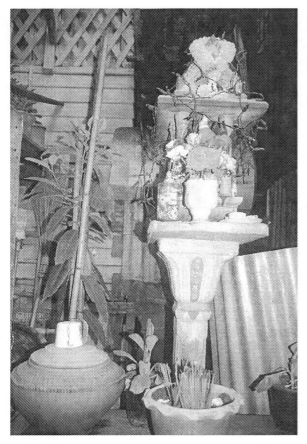

Figure 14–7

This is a phrapbuum, a Thai spirit house. Thais believe that if it is well-treated and respected, it will protect homes and businesses. Photograph courtesy of B.A. Weightman.

The Thais are confident that future developments will be adapted to the Thai way of life. One important aspect of Thai culture is order. Thais like a well-regulated life with a bent for aestheticism and appreciation for harmonious social order. Implicit in this is respect for interpersonal relationships and hierarchy in concert with ensuing rights and obligations. Even with modernization and social transformation, the expectation that society should be orderly remains.

The role of the *sangha* in the knowledge of *Theravada*—original teachings—is predominant. The monks are fed twice a day by the common people before noon. This gives people a chance to acquire merit. Almost every Thai male earns merit for his family by going into the monkhood as a novice for at least a rainy season. Traditionally, few Thai women would consider marrying a man who had not spent time thinking about ethics, learning to read, meditating on the Four Noble Truths, or focusing on how to be an upstanding human being. Even in the whirling commercial world of Bangkok, the role of monks and *wats* and the belief system of Theravada Buddhism are central.

In Thai thought, Buddhism encompasses virtue and wisdom, which can liberate one from the common order of life. It transcends and relates to the ultimate order of morality and goodness symbolized by the home, the mother, and the female symbols of Mother Earth and Mother Rice. Buddha and these female entities constitute the domain of moral goodness.

Next comes the less tenuous order of the realm of supernatural powers outside the home. This area of *saksit* is the focus of religious preoccupation, with rules for dealing with it.

Beyond the area of *saksit* lies the area of chaos represented by wickedness and immorality. Forces of goodness can intermingle with forces of evil. Pure order can descend into chaos; Buddha can emerge into evil; pure virtue can become deeply immoral; and safety can transform into danger.

Each Thai house and business possesses a spirit shrine of the **phrapbuum**—"lord of the place" (Figure 14–7). If respected, well-treated, and given offerings, the spirit shrine will protect and care for the place. If things do not go well, it is not the fault of the *phrapbuum*; the cause is always elsewhere.

Just as individual households have lords of the place, so do temples, villages, and provinces, where the lord resides at the city pillar. These rulers have no power outside their local territories and must be vested with appropriate rites and offerings.

What's a Wat?

Wat is a Thai word from a Pali-Sanskrit word meaning "dwelling for pupils and ascetics." A *wat* is a Buddhist compound where men or women can be ordained as monks or nuns

Figure 14–8

Wat Pho, *originating in the sixteenth century, is the oldest and largest* wat *in Bangkok. It is also the earliest center of public education.* Photograph courtesy of B.A. Weightman.

(Figure 14–8). Virtually every village has a *wat*, while towns and cities may have several. Each *wat* has a number of buildings such as a *bot* where ordinations are held; a *sala* for community meetings and lectures; a *tripitaka* or library for Buddhist scriptures; and various *chedis* (mounds or monuments) and *stupas* (large pagodas) housing the ashes of worshippers. *Wats* are multifunctional and are fundamental in villages. They operate schools, clinics, community centers, old age homes, and so forth. The typical *wat* is a locus of social activity. Especially important are the temple fairs that take place on auspicious dates such as Buddha's birthday.

In the five southernmost provinces, 85 percent of the inhabitants are Muslim. These are not Thai but rather ethnic Malays with ties to the population of Malaysia to the south. For more than a century, the southern provinces have had stronger ties with Malaysia than with Bangkok 1,000 km (600 mi) away. With the rise of Islamic fundamentalism and incidents of terrorism in recent years, southern Thailand has come to international attention.

The Pattani United Liberation Organization, named after the capital of the Muslim south, was active in the 1960s and 1970s but has been dormant since. Nevertheless, there has been a series of bombings of hotels, schools, a Buddhist temple, and a Chinese shrine in the 2000s. Bomb-making materials have been found in the home of a Muslim cleric and an al-Qaeda operative has been in the region. Thai police have arrested a number of suspects and are watching this region very closely. The government is particularly concerned about the Andaman Sea coast. This is an important tourist area and one that might become a target of terrorism directed against non-Muslim tourists. The Thailand-Malaysia border is another area of concern. This area is extremely porous. Muslim Malays can cross into Thailand with ease.

THE ECONOMY

Agriculture

Sixty percent of Thailand's exports are agricultural. Thailand ranks sixth in the world in rice production but number one in rice exports. It ranks third in tapioca (manioc), fourth in sugar, and sixth in coconut production. It is the world's largest producer of natural rubber. Other exports include corn, jute, pineapple, palm oil, and canned or frozen shrimp, tuna, and chicken. Chickens are grown American-style in large, industrial feed houses, and they are exported frozen. Pigs are also raised in industrial fashion. One of the risks of this kind of industrial food production is the outbreak of epidemics such as Japanese encephalitis. Freshwater aquaculture is also important. Thailand exports mainly to the United States, Japan, and Singapore.

About 50 percent of Thailand's work force is engaged in agriculture, but the sector is fraught with problems. One of these involves the availability of water, which has dwindled in the dry season. Some worry that decreased flow in the Chao Praya will draw salt water from the Gulf of Thailand to foul waterworks north of Bangkok. Climatic change has resulted in less rainfall. Water shortage is also attributed to deforestation. Thailand has lost 50 percent of its forests over the last 30 years, and there is now a moratorium on logging. Thailand is procuring its wood from Burma and Laos.

Farmers are used to expanding space rather than intensifying inputs to increase production. This is no longer possible. The industrial boom has sent harvest wages soaring, and many people have left the farm for work in industry.

Rising competition is another problem. Sugar, the only crop subsidized by the government, faces stiff competition in the newly established ASEAN Free Trade Area. Vietnamese rice exports, especially lower grades, are hurting thousands of subsistence farmers. Chicken and shrimp exporters face competition from China. The rubber industry is threatened by cheaper labor in Indonesia. Thailand is especially vulnerable to world market prices as 70 percent of its land is planted in only six commodities: rice, sugar, rubber, maize, cassava, and sorghum.

Some changes are taking place. Improvements are being made in mechanization and intensification, along with land consolidation and better merchandising. A dairy industry is taking off as cows from New Zealand and the Netherlands are being crossbred with other breeds to withstand Thailand's climate. Agribusiness firms are entering into new areas such as fresh and processed fruit and cut flowers. The country has become a center of tissue culture; laboratories commercially mass-produce plant clones for customers around the world. Further, Thai firms have been exporting agribusiness technology to other developing countries (e.g., China, Indonesia, and Vietnam).

New Theory Farming

New Theory Farming is a form of land and agricultural management considered by various government agencies as a means of introducing a community-based economy. In the initial stage, farmers divide their 6-acre (2.4-ha) plot of land into three 30 percent and one 10 percent portion allotted to rice, field crop or fruit, water supply, and residence. This initial stage should enable farmers to become self-sufficient and to pay off their debts.

In subsequent stages, farmers join to form cooperatives and then link up with financial and energy sources such as commercial banks and oil companies. In the third stage, farmers will manage their own mills and co-ops and take on joint ventures.

In the first year of the project, 1998, thousands of villages participated in various aspects of the program. Community markets, networks, water source development, agricultural service centers, and civil society activities were put into place. More than 5,000 families have participated in the New Theory self-sufficiency project with positive results. It is the hope of the Thai government that these programs will help farmers pay off their debts and ease themselves out of poverty.

Tourism

Since 1987, tourism has become a leading earner of foreign exchange, outdistancing even textiles. Ecotourism has increased substantially among the northern hill tribes. Sex tourism, with its accompanying risk of AIDS, continues in Bangkok and some of the southern resorts such as Pattaya (Chapter 3). Unfortunately, many of the hill tribe traditions have been compromised to suit tourist desires, leaving substantial numbers of indigenous peoples culturally marginalized. Now, much of hill tribe life is "staged" for the tourists. The timing of work or festivals is directed to fit in with tourist schedules. Traditional lives are disrupted as the tribes are reduced to being "objects" of tourist attention.

Figure 14–9

Recent years have seen a dramatic increase in construction in Bangkok. Although money is made in construction, vacancy rates are very high. Photograph courtesy of B.A. Weightman.

Industry

Manufactured goods have become increasingly important as a source of foreign revenue. Textiles, cement, and electronics lead the way. Car and truck manufacturing are on the rise since several international manufacturers such as Toyota, Mitsubishi, Mazda, Ford, and GM have established themselves in Thailand. Thailand now has the second largest pickup truck market in the world after the United States.

The 1980s and 1990s have witnessed a surge of foreign investment from such countries as Japan, South Korea, and Taiwan, where labor costs had risen. Thailand had numerous attractions for relocating industrial operations from these countries. Unskilled, trainable, and cheap labor was available. It had a decent infrastructure of roads, airlines, and port facilities, as well as sufficient power and water supplies for industry. Tax incentives, infrastructure provision, and other favorable policies encouraged foreign investment. Bureaucratic procedures were streamlined. Moreover, there was political stability and lack of racial conflicts in most of the country. Finally, there was no adversity toward Chinese business, an important consideration for firms coming from Hong Kong and Taiwan.

Much foreign capital has flowed toward primary industries such as rubber, tiger prawns, livestock, nonferrous metals, and marble. It has also gone to resource-based industries such as processed foods, palm oil, sugar products, wood-furniture parts, rubber gloves, and paper products.

Thailand remains an attractive destination for foreign involvement. Japan, the major investor, and other countries view Thailand as the hub of a golden pentagon comprising Thailand, Myanmar, and Indochina. Thailand is seen as an ideal site to establish satellite companies into a surrounding hinterland of cheap labor and a potential market of more than 120 million consumers. Thai companies, offshoots of joint ventures with the Japanese, are already trading in Vietnam and have a strong presence in Laos, where they dominate extractive industries.

Thailand is unique among developing countries because industrialization is concentrated in and around the primate city of Bangkok (Figure 14–9).

Ya Baa

Coming from clandestine laboratories in Myanmar and the Golden Triangle, little pink methamphetamine pills are supplanting heroin as the drug of choice in Thailand. The number of drug offenders in Thailand rose from just over 10,000 in 1992 to 100,000 in 2001, thereby swamping prisons and courts with more than 80,000 cases a year. Most of the drugs are manufactured in Myanmar, particularly in the states controlled by the United Wa State Army. This pro-Yangon ethnic group operates at least 80 permanent and countless mobile production labs near the Thai border and is said to be in collusion with the Yangon military government.

Ya Baa or "crazy medicine" as it is known in Thailand, has broad usage across all segments of society. Construction workers, sailors, farmers, and taxi drivers take the drug to help them through long and arduous days. Students use it to stay awake to study and get through exam periods. Night club devotees take it to dance the night away. Children as young as six become hooked and hospitals treat more *Ya Baa* addicts than alcoholics. The worst symptoms, after years of addiction, are paranoia and hallucinations leading to uncontrolled violence. Cures take several months and relapse is common.

In 2003, Prime Minister Thaksin decided on all-out war against drug trafficking. More than 50,000 drug dealers were arrested including 2,750 large-scale retailers. Forty million pills were seized as well. Anti-drug programs have succeeded in 82,500 villages that are currently drug free. Nevertheless, about five percent of Thais still abuse methamphetamines—*Ya Baa.*

Future Plans

Thailand is pursuing plans to replace Singapore as Asia's fuel-trading and transport hub. This includes the construction of an energy land bridge between the Indian Ocean and the Gulf of Thailand. In January 2004, a state-owned energy company opened a new petroleum trading center at Sriracha, a deepwater port on Thailand's eastern seaboard just southeast of Bangkok. The government plans to build a "Strategic Energy Land Bridge" which will include two deep-sea oil terminals (the one at Sriracha and another at Laem Chabang) and a pipeline for transporting oil across the southern Thai isthmus from the Andaman Sea to the Gulf of Thailand.

This project is of particular interest to the oil-importing countries of China, South Korea, and Japan. Almost 70 percent of East Asia's oil currently passes through the Straits of Malacca in tankers. Of growing concern is the security situation in Sumatra where Indonesia is trying to control an insurgency in the province of Aceh. Then there are old and current worries about piracy as well as new ones regarding possible terrorist attacks. Some analysts speculate that China is concerned about a potential American blockade of the Straits of Malacca that would cut off China's oil imports, although this is highly unlikely. It is estimated that the 22 million to 25 million barrels of crude oil that East Asia imports every day will more than double by 2020 led by China's growing appetite for fuel. Oil security is a top priority for East Asia.

An especially grand plan is to cut a canal through the Isthmus of Kra to shorten shipping routes between Europe and East Asia. Investors from Australia and Japan have expressed interest in the project. Some economists argue that the relatively small savings for shippers will not be enough to recoup the estimated cost of US $35 billion, but at least the canal would allow ships to avoid the dangerous Straits of Malacca.

Another plan involves transforming the northern city of Chiang Mai into an international aviation hub. Prime Minister Thaksin inaugurated a new air route to Chittagong in Bangladesh although planes fly virtually empty. Ongoing discussions with Singapore Airlines are intended to establish a discount-fare route between Chiang Mai and Singapore. The whole project could fail because the surrounding economy is agricultural and short on expertise and facilities for manufacturing and service industries.

The Thai government also plans to turn the resort island of Phuket into a laboratory for high-tech research and development. This will prove difficult as most high-tech workers live around Bangkok and will be reluctant to leave.

These grandiose schemes are, in large part, in response to China's emergence as a rival low-cost manufacturer and a magnet for foreign investment. China's cheap manufactured goods are eroding key Thai export markets in the United States and Japan. The Thai Textile Association estimates that Chinese rivals will drive 30 percent of Thai textile producers out of business by 2005. Furthermore, foreign investment in the Thai economy has dropped dramatically since 2001. It is hoped that these new developments will attract more foreign investment.

BANGKOK METROPOLITAN REGION

Why is Thailand a tiger in a metropolis? Consider these facts. Over half of Thais are employed in agriculture, yet agriculture contributes only 16 percent of the GDP. Although only 16 percent of Thais live in Bangkok, it produces more than half of the GDP. Bangkok is Southeast Asia's largest city and is the economic pulse of the region.

A city of more than 10 million, the historic and current primacy of Bangkok relates to its excellent location. It has been the capital since its founding in 1792 and lies in the rich Chao Praya rice region. The government is highly centralized, the city is the spiritual center of the nation, and it is the residence of the highly esteemed royal family.

Bangkok is a newly industrializing city. Economies of agglomeration are supported by a significant informal sector. Urbanization has spread from the Bangkok Metropolitan Area (BMA) into the Bangkok Metropolitan Region (BMR), which comprises five provinces circumscribing the city. The BMR contains 15 million people. Transportation, water, and utility costs are lower in Bangkok than elsewhere in the country. Labor is more productive because of the higher educational levels and greater experience of workers. The city attracts labor-intensive, export activities. Smaller firms are concentrated in the BMA where land is more expensive. Larger firms prefer the cheaper land of the BMR. More than a third of Thailand's manufacturing enterprises are located in the BMR.

Ribbon development has extruded onto agricultural land. Rice-producing and other agricultural areas are being displaced by industry, housing estates, and recreational facilities such as golf courses. Thailand has more golf courses than any other Southeast Asian country. Although development is clustered in the BMR, numerous benefits flow from this wealth, including upgraded health and education systems and improved social and physical infrastructure throughout the country.

Rapid industrialization has taken its toll on the environment. Air and water pollution are severe and traffic congestion is among the worst in Asia. These seemingly unmanageable problems are constraints on development. Yet construction continues as profits are made in the development process rather than from the final product. Many construction projects are joint ventures with Japanese firms.

The population of Bangkok fluctuates daily and seasonally. An estimated one million people commute into the city every day. In February, a slack agricultural period, many workers move to the city in search of temporary jobs in the informal sector of the economy.

Gridlock

In Bangkok, an estimated 4 million cars and 1.5 million motorcycles are in the city every day. The number of cars is rising by 1,000 a month. Nearly half the traffic police are undergoing treatment for respiratory disease. A commuter spends an average of 44 days a year in traffic. Gas stations sell Comfort 100, a portable toilet for when a motorist is stuck in traffic jams.

Road surfaces account for only 8.5 percent of urban space; they should make up 20 percent. Motorcycles, buses, taxis, and trucks comprise the bulk of the traffic. Buses, which account for a mere 1 percent of the traffic, contribute more than half of the pollution.

Several mass transit systems are either in the planning stages or under construction. However, lack of coordination, interdepartmental quarrels, and problems with contractors have rendered many of these stagnant.

Bangkok is called the Venice of Asia because of its *khlong* (canal) system linking various parts of the city with the major artery—the Chao Praya. Now many of the *khlongs* have been paved over and quicker boat traffic has given way to slower vehicle traffic.

With development so concentrated in the BMR, the question is this: Is Thailand a newly industrializing country or is Bangkok a newly industrializing city? It does appear that Thailand is more of a metropolitan tiger than a national one.

Vietnam: Baby Tiger

"Two baskets of rice separated by a bamboo pole." This is Vietnam—the densely populated Red River Delta in the north strung by 1,000 miles (1,610 km) of mountains and a narrow coastal strip to the watery world of the Mekong Delta in the south. On these two rich, alluvial plains live two-thirds of the Vietnamese population, now numbering 80.8 million (Figure 14–10).

Eighty percent of the people in Vietnam are ethnic Vietnamese. In addition, there are at least 50 minority groups. Some minorities are identifiable by their religious affiliation. Roman Catholics are the largest group. At independence in 1954, there were about 2 million. When Vietnam was divided later into two zones, about two-thirds of the 900,000 who left the north were Catholic. Another group is the **Cao Dai**, with its headquarters in Tay Ninh Province near Ho Chi Minh City. Cao Dai means "high tower" and is a mix of Buddhist, Christian, Confucian, and Taoist beliefs.

Although the population is dominated by the ethnic Vietnamese, the presence of these other groups has made it difficult for any government to create a centralized and highly integrated society. Moreover, regional differences are strong among the northerners, southerners, and the central Vietnamese. Currently, the Communist regime is attempting to reduce these dichotomies by integrating southerners into the more regimented system of the north. Moreover, modernization is moderating differences throughout the country.

ORIGINS AND CHINESE INFLUENCE

It is possible that the ancestors of the Vietnamese originated in southern China, but they first appear in history as *homo erectus* in the Red River Delta (Tonkin region) at the same time as Peking and Java Man, about 500,000 years ago. Available evidence supports the fact that they had mastered primitive agricultural techniques as early as 9,000 years ago. If this is so, the Vietnamese are among the first people to practice settled agriculture. By the second millenium B.C., the inhabitants of the Red River delta had created an advanced civilization based on foreign trade and wet rice cultivation.

Then in the second century B.C., the region was absorbed into the Han Empire, the Chinese calling it *Nan-Yueh* or *Nan-Viet*. Thus began over 1,000 years of Chinese rule. Chinese writing was introduced along with the Confucian classics, Chinese art and architecture, and Mahayana Buddhism. This differentiated Vietnam from the rest of mainland Southeast Asia with its Indian and Theravada Buddhist ideals.

The peak of Chinese influence was during the Tang Dynasty, whose rulers termed Vietnam *Annam*, or "the pacified south." In 939 the Vietnamese asserted their independence and remained free from Chinese dominance except for a 20-year stint under the Ming Dynasty from 1407 to 1428. Today, Chinese influence is stronger in Vietnam than in any other Southeast Asian country, with the exception of Singapore.

EXPANSION AND EUROPEAN INFLUENCE

Vietnam was constantly under pressure to find more farm land for its growing population. Between the eleventh and the seventeenth centuries, it gradually extinguished the kingdom of Champa in what is today central Vietnam. Subsequently, it acquired the Mekong Delta from the Khmers. During the nineteenth century Vietnam would probably have gotten all of Cambodia had the Thais not challenged their

Figure 14–10

Vietnam is part of a region known as Indochina, including Laos and Cambodia. All were French colonies. Note how the Annamese Cordillera dominates the physical geography of Vietnam. With rice-producing core areas at either end of the country, Vietnam is indeed "two baskets of rice separated by a bamboo pole." From H. J. de Blij and P. O. Muller, Concepts and Regions in Geography, *Second Edition, 2005, p. 352. Originally rendered in color. © H. J. de Blij and P. O. Muller. Reprinted with permission of John Wiley & Sons, Inc.*

advance and the French not established Cambodia as a "protectorate" in 1863. This was the origin of the anti-Vietnamese sentiments in Cambodia today.

Expansion of the Vietnamese state southward coincided with the appearance of the Portuguese and other European traders in the region. Vietnam's first exposure was in 1535, when a Portuguese ship entered the port of Da Nang on the central coast. Within a few years the Portuguese set up a trading post at Faifo (Hoi An). By the seventeenth century several European powers had posts along Vietnam's coast.

The first Catholic mission arrived in 1615, when Jesuits established a mission at Hoi An. Another mission was set up at Thang Long (Hanoi). The Jesuit priest Alexander de Rhodes organized a society to train missionaries to propagate the faith. However, he was known more for his transliteration of the Vietnamese spoken language into the Roman alphabet. Known as *quoc ngu* or "national language," it did not come into general use immediately but was used to translate the Bible into Vietnamese. Consequently, thousands of new converts were gained.

Missionary activities eventually antagonized the authorities, who feared a subversion of Confucianism. Christianity was ultimately barred from both the north and south, and European missionaries were either expelled or executed. In 1697, the French closed down their small trading post at Hoi An.

FRANCE AND THE *MISSION CIVILISATRICE*

French interests changed the history of the region forever. In France, religious organizations demanded protection for their colleagues overseas. Commercial interests feared the expansion of Britain in Burma and the potential loss of the "China market." The French attacked Vietnam and in 1862 acquired three provinces in the south, the opening of port cities, and the freedom to propagate their religion. In 1867, French units seized the remainder of the south and transformed the region into the colony of Cochin China. Further, they assumed Vietnamese rights in Cambodia and turned it into a protectorate. The

French finished their conquest of all of Vietnam in the mid-1880s. Shortly thereafter, France completed its "balcony on the Pacific" by making Laos a protectorate as well.

France's civilizing mission was applied to an Indochinese Union comprising five territories or states: north Vietnam, or **Tonkin** (meaning "eastern capital"); central Vietnam, known as **Annam** (from the Chinese "the pacified south"); the colony of **Cochin China** in south Vietnam; Laos; and Cambodia. A hierarchy of French "residents" was installed to run the country.

The primary goals were economic—to exploit natural resources and open up markets for goods produced at home. The French also emphasized the moral aspect of their deeds, and the French citizenry believed that this *mission civilisatrice* (civilizing mission) would bring the benefits of modern civilization to the "primitive peoples" of the world. In fact, the civilizing mission was subsumed by the greater goal of commercial profit. A small colonial elite was educated in Western ways, but the masses of people had only rudimentary exposure to Western culture.

With all its faults, there was economic progress under the French regime. A vigorous commercial and manufacturing system emerged, although it was dominated by foreigners including Chinese and Indians. Coal mining north of Haiphong was boosted; Vietnam remains a coal exporting country to this day. Bridges were built over major waterways, and railroads were laid from Saigon to Hanoi and on to Kunming in China. It is said that Vietnam, under the French, had the best system of paved roads in all of Asia.

The French also expanded agricultural land. In the Mekong Delta, they drained marshlands and constructed canals, putting thousands of acres under cultivation for the first time (Figure 14–11). The development of the Mekong Delta enabled Vietnam to become one of the world's leading rice exporters by the 1920s. The French also established coffee, tea, and rubber plantations along the hill slopes toward Cambodia. However, much of the Mekong Delta lands was held by absentee landlords who charged exorbitant rents for the land. Many farmers

Figure 14-11

Women sifting and cleaning rice in the Mekong Delta.
Photograph courtesy of B.A. Weightman.

lacked security as sharecroppers on these vast estates and suffered from poor living and working conditions. Further, the commercialization of agriculture forced thousands into the cities looking for jobs.

By the 1930s, there was substantial opposition to the French. However, there was also a new class of native "collaborators" working in tandem with the colonial regime. These French-speaking individuals wore French clothing, ate French food, and lived in French-style houses. They were the middlemen between the traditional masses and the foreign regime and to outsiders served as evidence of the "success" of *mission civilisatrice*. Nevertheless, Vietnam was a cauldron of social unrest.

THE ROAD TO COMMUNISM

The collapse of Confucian government and the triumph of Western colonial aims put traditional Vietnamese beliefs into question. What path should be taken into the future? Some opted for Western-style democracy; some looked to China (itself in turmoil) for a restoration of Confucian values. Others became interested in the new Indochina Communist Party (ICP) founded in 1927 by young idealists including **Ho Chi Minh**. But in fact, the most imposing popular movements were *Cao Dai,* with over a million adherents by the 1930s, and a Buddhist sect known as *Hoa Hao,* also popular in the south. Christianity had also grown, claiming 10 percent of the population.

The Japanese entered French Indochina in 1940 and set up a nominal government under the emperor Bao Dai. Wartime disruptions in the economy and disastrous weather created terrible famine in Tonkin and Annam. More than one million people died. When the war ended, there was a power vacuum in Vietnam, and the stage was set for the August revolution of the **Viet Minh,** a front organization of the International Communist Party.

The Viet Minh took control of most of northern and central Vietnam. They were less successful in the south, where they were opposed by politicians, business people, and religious groups. Ho Chi Minh set up his government in Hanoi. Meanwhile, by late 1945, the French had regained control of the south. War ensued, with the French pitting their conventional tactics against the guerrilla tactics of the Viet Minh. With assistance from China and the Soviet Union, Ho Chi Minh's forces were able to subdue the French at **Dien Bien Phu** in 1954. The power of the Viet Minh ended French efforts to hold Vietnam.

Partition and War

As Dien Bien Phu fell, the great powers were meeting in Geneva to plan Vietnam's future. The decision was made to divide the country at the seventeenth parallel, with the northern Democratic Republic of Vietnam (DRV) to be ruled by Ho Chi Minh and the southern Republic of Vietnam to be run by Bao Dai. Elections were to be held in 1956 to reunite the country. These never took place. Ultimately, the Americans gave support to Ngo Dinh Diem, an administrator in the Bao Dai government, and Bao Dai left the country.

While President Diem spent his aid money on military acquisitions, the north pressed ahead with socialism and agricultural collectivization. Even through the "rich" suffered, popular support for Ho Chi Minh's government remained high. By 1959, the National Liberation Front (NLF) or the **Viet Cong** had organized to oppose Diem's oppressive and ineffective rule. By the early 1960s, the Viet Cong were in control of vast areas of the southern countryside, which compelled the Americans to increase their aid to Diem. By 1963, there were 17,500 American military advisors in Vietnam.

By now Diem and his followers had antagonized all groups in South Vietnam. Buddhist monks were burning themselves alive to protest his regime. Ultimately Diem was assassinated. Several years of unstable government followed, but meanwhile the Americans decided to confront the NLF. The United States then began to bomb targets in both the north and the south and, in 1965, landed troops in the south. Thus began the Vietnam or American war (Chapter 13), which did not end until 1975, with a combined DRV/NLF victory.

The environmental damage of the war on Vietnam was appalling. From 1965 to 1971, Indochina, an area slightly larger than Texas, received twice as many bombs as were dropped in all the combat theaters of World War II combined. In a typical B-52 mission, seven planes delivered 756,500 pounds of bombs in a pattern that saturated nearly 1,000 acres. The B-52s alone created about 100,000 new craters a month. The area had as many as 26 million bomb craters. Soon the craters filled with weedy grasses and became breeding grounds for mosquitoes. About 10 percent of the agricultural land of southern Vietnam had to be abandoned. In addition, thousands of tons of chemical defoliants were dropped. Vast tracts of forests, rubber plantations, and mangroves were obliterated or seriously damaged. In these areas, there was a significantly higher rate of pregnancies that resulted in still-births and congenitally abnormal babies. The effects of the wartime bombing are still felt today.

Thousands fled the country to refugee camps in adjacent countries, especially Thailand. The first migrants used a map torn out of a school atlas to navigate the Gulf of Thailand. More than a million refugees have been resettled since 1975. Of those who remained in refugee camps in Southeast Asia, most were "screened out" as economic rather than political refugees. For them, there was no hope of resettlement. These people were repatriated to Vietnam voluntarily or involuntarily.

Aftermath of War

Since the 1950s, the northerners had lived under an austere form of socialism that reemphasized the traditional regard for social hierarchy and community obligation. The southerners had been introduced to a quasi-capitalistic consumer economy and the trappings of American popular culture. In 1975, the north moved swiftly to subordinate the south to direct rule from Hanoi. In 1976, the country was renamed the Socialist Republic of Vietnam; Saigon became Ho Chi Minh City. Socialization of the southern economy proceeded quickly over the next two years.

However, things did not go as well as planned. There was an acute economic crisis, worsened by

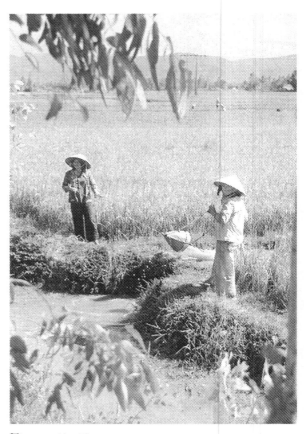

Figure 14-12

I came across this scene in the Red River Delta. The two women are operating a bucket to lift water from a lower to higher level in the rice paddies. Courtesy of B.A. Weightman.

Figure 14–13

This scene in Saigon shows new hotel construction, as well as evidence of foreign investment. Photograph courtesy of B.A. Weightman.

floods and other natural disasters in 1977 and 1978. Ambitious industrial targets were not achieved, rice production diminished, and food had to be rationed. In 1978, Vietnam attacked Cambodia to oust the genocidal, anti-Vietnamese regime of Pol Pot. In the 1980s, thousands of Chinese and Sino-Vietnamese fled as *boat people.* These refugees went to Australia, Canada, the United States, France, and many other countries.

DOI MOI

A hard-line communist approach was held in Vietnam until 1985 when reforms in the Soviet Union gave impetus to reforms in Vietnam. In 1986, the Vietnam Communist Party (VCP) instituted the policy of *doi moi,* meaning "renovation." Measures were introduced to encourage private-sector production, make prices more flexible, and give more discretion to managers on the use of their profits. This was followed by the introduction of a peasant agricultural contract system in 1981. The system allowed farmers to keep anything over their set quotas. State enterprises were permitted to sell production exceeding their quotas on the open market.

Reforms were given further impetus with the collapse of the Soviet Union and the European Eastern Bloc's abandonment of socialism. (This resulted in a 30 percent cut in Vietnam's budget.) Support for reform was also bolstered by China's new policy of market socialism. Today, Vietnam is a hybrid of a one-party government, faltering state enterprises, and burgeoning free enterprise.

TODAY'S ECONOMY

Vietnam is anchored by two cities: the capital of Hanoi in the north and Ho Chi Minh City in the south. Hanoi, a quiet city of bicycles in 1995, is now a bustling city of motorcycles and automobiles. Hanoi dominates the Tonkin Basin and its productive agricultural hinterland. Agricultural methods remain traditional—dependent on human and animal labor (Figure 14–12). The port of Haiphong, a mere 60 miles (97 km) away, is linked to Hanoi by a new four-lane highway that passes through a developing *kotadesasi* landscape.

Ho Chi Minh City (Saigon) and Cholon (Chinatown), with its 8.5 million people, is an urban agglomeration on the Saigon River. It contains about 10 percent of the country's population. The name Saigon is commonly used. Unlike Hanoi, Saigon can be accessed by ocean-going vessels and a Special Economic Zone has been developed downstream from the port. Saigon is a bustling city of every form of transportation imaginable, with traditional districts selling a wide array of consumer goods. High-rises punctuate the skyline everywhere (Figure 14–13). Modernizing Saigon is distant in form and fact from the more traditional capital of Hanoi. With the lifting of a United States embargo in 1994, foreign investors began investing in Vietnam. Modern high rises and construction cranes mark the landscapes of Saigon and now, even Hanoi.

Now about 65 percent of the labor force works in the primary sector in agriculture, forestry, and fisheries. The primary sector includes rice, corn, potatoes, tea, pepper, coffee, rubber, and fish and animal products. In 2002, the government announced that Vietnam was the world's number one exporter of coffee. Land is being taken from hill people in large swaths and planted with coffee and other export crops such as pepper. Mining is important as is the manufacture of cement and fertilizers. Food processing and the textile and apparel industries are also important factors in the economy. Exports to foreign countries grew by 73 percent in the 1990s. However, a heavy reliance on two commodities—rice and crude oil—makes this situation tenuous. Tourism is the main emphasis in the service sector and numerous hotels have been constructed. In fact, the scale of development is such that Vietnam refers to itself as "Asia's baby tiger."

Laos: Land of a Million Elephants

The Lao People's Democratic Republic, now often referred to as simply *Lao* or *Laos* is a mountainous, landlocked, sparsely populated country sharing borders with Thailand, Myanmar, China, Vietnam, and Cambodia. Laos, as we know it today, is a creation of the French, but much of its history has involved conflict with its surrounding countries.

LAN XANG

There are no early documentary records of early Lao history but we know that the region had links to Chiang Mai in the eleventh century and that it was a satellite of the Khmer empire in the twelfth century.

The first kingdom of Laos emerged in 1353. This was known as **Lan Xang**: "a million elephants."

Lan Xang stretched from China to Cambodia and from the Khorat Plateau in present Thailand to the Annamite Mountains in the east. Louang Phrabang, on the Mekong, became the capital of Lan Xang, and the king made Theravada Buddhism the official religion. A census in 1376 indicates that there were 300,000 Thais living in the kingdom. *Muang* administrative districts were organized, and these lasted until abolished by the Communists in 1975.

In the sixteenth century, Vientiane (Viangchan) on the Mekong, became prominent as a religious and trading center. Lan Xang weathered wars with the Vietnamese, the Thais, and eventually the Burmese. Nevertheless, Vientiane reached its zenith in the seventeenth century. The first European visitors reported on the city's religious structures and imposing buildings. For lack of a royal heir, the kingdom broke up in 1694 and Vientiane and Louang Phrabang became centers of rival states. The south fell under Thai patronage.

FRENCH TAKEOVER

The nineteenth century was a period of devastation for a Laos caught in regional rivalries. The surviving kingdom at Louang Phrabang acknowledged the overlordship of both the Vietnamese and the Thais, but was really within the Thai orbit of control.

The French takeover between the 1860s and 1885 resulted from several factors. They perceived the Mekong as a route to China, although this was an incorrect assumption, as the Mekong is broken by impassable rapids. They feared Thai interests in their territories. They were concerned about the intentions of armed bands of renegade Chinese who were attacking northern Laos and Vietnam. And, they were worried about British encroachment from Burma. By 1885, the French controlled the Vietnamese emperor's claims to overlordship of the Lao territories. Still, Laos remained the center of rivalries among the surrounding states in addition to the competing colonial powers.

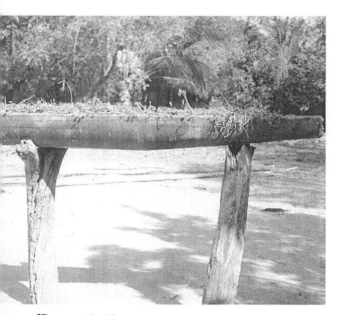

Figure 14–14

In Laos, a bomb casing makes a perfect container for a herb garden. Photograph courtesy of B.A. Weightman.

The French's attitude to Laos was one of "benign neglect." Few French actually went there, and Vietnamese were placed in administrative positions. Trade was left to the Vietnamese and growing numbers of Chinese. Most people in the region continued as subsistence farmers, the lowlanders growing rice and the uplanders pursuing slash-and-burn cultivation. The colony's most important products were tin, mined by the Vietnamese, and opium grown by the Hmong and other mountain dwellers. The French had a monopoly on the opium trade throughout Indochina.

Prior to World War II, modernization was extremely limited. A mere 8,050 miles (5,000 km) of mostly unpaved roads eased communication, but most services, including health care, were limited to the towns, and no Western-style education was available beyond the primary level. However, a small Vietnamese or French-educated elite did emerge in Viangchan, and by the 1940s it became the core of a Lao nationalist movement.

THE WAR YEARS

Nationalist politicization was a feature of the war years in Laos. After 1945, there were attempts at independence, but the French recaptured Laos in 1946. In the late 1940s, Lao guerrilla groups developed along the mountainous Lao-Vietnam border, aided by Viet Minh know-how and supplies. Meanwhile, the French declared the king in Louang Phrabang king of all Laos, held elections for a national assembly, and declared the country independent in 1949. Nevertheless, the guerrilla movement expanded, known by then as the **Pathet Lao**. By 1954, it controlled large areas of northern Laos.

In 1954, the same year as the French were defeated at Dien Bien Phu, a conference was held in Geneva to settle Indochina's disputes. The Pathet Lao were not invited. By 1959, guerrilla warfare was in full swing, with the United States involved in supporting the Royal Lao Forces. At the same time, the CIA was clandestinely forming links with the opium-growing Hmong, whom they assisted in selling drugs to new markets in Southeast Asia.

Unfortunately, Laos became deeply involved in the Vietnam War. About one-third of the bombs dropped in Indochina fell on Laos, especially on the Plain of Jars, part of the route of the Ho Chi Minh Trail. Tens of thousands of unexploded bombs still mar the landscape of northern Laos. When I was in Laos in 1998, seven children were killed by an unexploded bomb. It will take many years for these bombs to be removed. Meanwhile, the people collect bomb casings, aircraft fuel tanks, and other materials and incorporate them as practical items in their landscape (Figure 14–14).

By 1972, the Pathet Lao, backed by the well-armed DRV, was gaining ground. In 1973, the two sides reached an "Agreement on the Restoration of Peace and Reconciliation in Laos." When, in 1975, communist organizations toppled the power structures in Saigon and Phnom Penh, the Pathet Lao supported a "popular revolution." The king abdicated. A total of 300,000 Lao—10 percent of the population—left the country between 1973 and

1975. These were mostly middle class people. Many ended up in the United States and France. Then, in December 1975, a National Congress of Peoples Representatives voted for the establishment of the Lao People's Democratic Republic.

The Plain of Jars

The Plain of Jars is a plateau 3,280 feet (1,000 m) high, covering an area of 390 square miles (1,000 sq km). It is also known as *Plaine des Jarres*. Upon this bomb-crater pocked plateau lie more than 300 stone jars ranging from 3 feet (1 m) to 6 feet (2.5 m) high and around 3 feet (1 m) in diameter. Each 2,000-year-old jar equals the weight of three small cars (Figure 14–15).

Generations of archaeologists, geographers, and others have pondered the origin and use of the jars. Were they used to brew *Lau-lao*, a wicked local drink? Perhaps they were used to store grain or water. The most likely theory holds that they were burial urns with the larger jars being for the elite.

Tools, bronze, and ceramic objects have been found in the jars, indicating that they were used by a sophisticated society. But no one can relate them to any previous civilization. The stone that they are hewn from is not found locally.

During the war, the Pathet Lao set up their headquarters in a cave near the jars and the region was heavily bombed. Miraculously, the majority of jars remain unscathed.

RURAL SOCIETY

Over 85 percent of Laos is rural, and most people live in villages ranging from 10 to 200 households, or up to approximately 1,200 persons. Towns are mainly administrative and market centers. Most district centers have little more than a middle school

Figure 14–15

The Plain of Jars in Laos is an amazing place. I'm sitting on one of the jars to give you an idea of the scale of these objects. Their origin remains a mystery.
Photograph courtesy of B.A. Weightman.

and a few officials. Some places have minimally supplied clinics. Since 1975, the government has spent a lot of effort at national unification and instilling a sense of belonging to a Lao state.

One priority has been to improve infrastructure. However, both mountainous topography and torrential rains make access, work, and maintenance problematic. Lao's low population density means that the cost per capita for road construction is often prohibitive. War damage such as bomb craters and the presence of mines and cluster bombs inhibit most development.

Lowland Society

Lao Loum (Laotians of the valleys) has been the dominant group since the founding of Lan Xang in the fourteenth century. The Lao Loum make up half of the population of 5.8 million. They live in stilt houses constructed of wood, bamboo, and thatch. Progressive village houses sport a corrugated metal roof. Steamed glutinous rice, or "sticky rice," is the staple food. Eaten with the fingers, it is dipped into a soup or stewed dish.

Lao Loum live in stable villages along rivers or streams and indulge primarily in paddy rice cultivation. Where terrain is not level, they practice swidden cultivation. Other crops are cotton, tobacco, or sugarcane, but these are grown for personal use. Villagers keep chickens, pigs, and draft animals. Hunting, gathering, and fishing supplement subsistence.

Household tasks are usually divided by gender, but divisions are not rigid. For instance, both men and women cut and carry firewood. Women and children carry water, tend gardens, look after small animals, and are the main marketers for produce. Men look after and market draft animals. In field work, men plow and harrow the rice paddies, women plant seedlings, and both sexes transplant, weed, and harvest the rice.

Lowland Lao are mostly Buddhist. Most villages have a *wat*, which serves as a religious and social center. The *wat* also serves as the village meeting house. *Wats* are populated by two to six monks and a few novices, depending on the size of the village.

Midland Society

Lao Theung (Laotians of the mountain slopes) comprise about a quarter of the population and consist of 37 different ethnic groups. They speak languages of the Austro-Asiatic family. Many are mutually incomprehensible. Characterized as swidden farmers, the people of the mountain slopes are semi-migratory, moving their villages on occasion as swidden plots are exhausted. Many of these villagers were disrupted in the war, and many have moved closer to the lowlands. But traditionally, lands were managed in a way to maintain sedentary communities.

Lao Theung villages are generally smaller than *Lao Loum* villages. Some are divided into two segments with a men's common house in the middle. Villages of stilt houses are situated near a stream to access water that may be diverted via bamboo aqueducts.

Swidden rice economy is the basis for most communities. Both regular and glutinous rice are cultivated. Corn, cassava, and wild tubers supplement this diet. Hunting and fishing are also important.

Gender role differentiation is much greater among the hill people. Men clear and burn the swidden fields and punch holes for seeds. Women follow, dropping seeds in the holes and covering them with topsoil. Both sexes weed, but this is regarded primarily as women's work. Harvesting is a joint activity. In the house women cook, gather wood and water, and look after children. Men weave baskets, repair farm tools, hunt for small game, and are much more likely than women to manage finances and engage in trade. Where villages have schools, both boys and girls attend but girls have significantly higher dropout rates.

Upland Society

Lao Sung (Laotians of the mountain tops) include six ethnic groups. The Hmong are among the most numerous. They also live in Vietnam and Thailand. The *Lao Sung* are among the most recent migrants to Laos, having initiated a series of moves from China in the nineteenth century. All *Lao Sung* villages are located in the north. Only the Hmong live as far south as Viangchan.

Lao Sung typically live above 3,280 feet (1,000 meters) on steep slopes and mountain tops. As swidden cultivators, most groups are semi-migratory. Yet some villages have remained sedentary for over 100 years. Hmong houses are built directly on the ground, with walls of vertical wood planks and roofs of thatch or split bamboo.

Hmong swidden farming is based on white (nonglutinous) rice, corn, and a variety of tubers, vegetables, and squash. They also raise pigs and chickens in large numbers. Opium poppies, a cold-season crop, are planted after the corn harvest.

Opium, the sap drawn from the poppies, is a low-bulk, high-value item that is easy to transport. The Laotian government has outlawed opium production and instigated a program of alternative crops. Nevertheless, the practice continues in the remoter regions.

Hmong gender roles are strongly differentiated. Task divisions are similar to those of the *Lao Theung*.

Most of the *Lao Sung* are animists, but some have converted to Christianity after contact with missionaries. Most believe that illness is caused by spirits, and shamans play a powerful role in the communities. Shamans are "chosen" by the spirits and may be either male or female.

Lao Sung swidden cultivation is endangered. In the late 1980s, the Thai government imposed a ban on logging, so Thailand began looking to surrounding countries for timber. Logging in Laos is causing deforestation and soil erosion. However, the practice is to blame this on the swidden farmers. The government wants to see the disappearance of swidden farming early in the twenty-first century and the resettlement of *Lao Sung* in the lowlands.

URBAN SOCIETY

With a population of 532,000 in 2004, Viangchan is the only city of any significant size (Figure 14–16). Other cities such as Louang Phrabang and Pakse are

Figure 14–16
This is known as Victory Gate in Viangchan. It was constructed using American aid money designated for airport runway construction. Photograph courtesy of B.A. Weightman.

only around 20,000. Laos is 17 percent urbanized, including district centers of 2,000 of 3,000 people. The fundamental village nature of Laotian society is evident in the cities. For example, most residents of Viangchan are rice farmers. Very little rural-to-urban migration occurs as there is not much to migrate to.

NEW THINKING

In recent years, the government has shied away from the centrally planned economy and embarked upon a policy of *Chin Thanakan Mai,* or New Thinking. The hammer and sickle were removed from the state crest in 1991 and a series of economic reforms was instigated. With the demise of handouts from Moscow, Viangchan has turned to the West for investment. At first the pace of change was *koi koi bai*—"slowly, slowly." But in 1991, everything became privatized. Limited tourism is being promoted. Nongovernmental organizations are ubiquitous. Thai television, with its materialist message, is beamed across the Mekong. Likewise, more consumer goods, such as children's toys and motor scooters, are appearing in the markets of at least the largest cities. Foreign investment, mainly Japanese, is evident in the capital. Many Thai businessmen have appeared on the scene as well and new businesses are opening every day.

Infrastructure is still very poor. Only a few of the larger towns have municipal water systems. Electrification is in progress but as of 2003, only 20 percent of the country had electricity. Most roads are not paved, and many are impassable in the wet season. The Friendship Bridge (built by the Australians in 1994) links Laos and Thailand by road across the Mekong. With upgrading of roads in Laos, soon it will be possible to drive all the way from Singapore to Beijing!

Many Lao see their country's future as a transport hub linking northeast Thailand with the port of Da Nang in Vietnam, as well as southern China with Thailand, and, from north to south, Malaysia and Singapore. Many fear the effect of rapid development and influences of the surrounding states.

Laos remains one of the poorest countries in Asia. More than 75 percent of the population lives on less than two dollars a day. Subsistence agriculture contributes 53 percent of the GDP and employs 80 percent of the workforce. Even worse, inequality between various parts of the country is increasing. Development has been concentrated in Viangchan and other towns on the Mekong River plain, adjacent to Thailand, while remote provinces in the mountainous north have seen little to no change.

Cambodia: Beyond the Killing Fields

Cambodia is famous for two things. It is highly noted for the wonders of Angkor and it is highly notorious for the genocide that took place from 1975 to 1978. What happened in this small and compact country of Southeast Asia?

This land of the Khmer people is the third smallest nation in the region after Brunei and Singapore. About 5 percent of Cambodia is rivers and lakes. The plains of Cambodia are ringed by the last foothills of the Himalayas, the Cardamom range in the west, the Elephant Mountains to the southwest, and the lengthy Dangrek chain to the north. The eastern frontier with Vietnam runs up into the high and inaccessible Moi Mountains, which are crossed by only one highway. Rising out of the plains are ranges of hills called *phnom*. These were worshipped and shelter numerous settlements. The Mekong crosses the Cambodian Plain and south of Phnom Penh, splits into the Bassac to the southwest and the lower Mekong to the east. Inland, 340 miles (550 km) from the sea, the Mekong is still navigable year round in Cambodia.

Many thousands of years ago, Cambodia lay under the waters of the Pacific Ocean. The action of the Mekong River filled in the submerged land to form a fertile plain. This plain, watered by rivers and tributaries, formed natural irrigation systems. Where the land was not uniform, large lakes such as Tonle Sap were created. These lakes are relics of the former ocean gulf. The civilization of Cambodia developed on this moist plain.

The history of Cambodia is one of rivalries between outside forces and internal power struggles. Although the country has experienced periods of peace, the peace has often been enforced. This is the situation today.

THE EARLY PERIOD

By the early centuries of the Christian era, the Khmer and the related Mon peoples occupied a broad swath of land stretching from Burma to Vietnam. This was Funan, which flourished from the third to the seventh centuries. Early records also mention the states of "water Chenla" in the Mekong Delta and "land Chenla" further inland. Chenla brought down Funan.

The consolidation of Khmer society is clearer by the ninth century when Jayavarman II ruled over Angkor and established the concept of *Devaraj*: the god-king. Jayavarman II built temples at widely spaced sites across Cambodia. Eventually, Angkor, near Siem Reap, became the core region with an irrigated area of 13,585 acres (5.5 million ha) and a large population. Then, King Suryavarman II initiated the construction of Angkor Wat, the largest religious complex in the world. The complex was enlarged and enhanced by subsequent rulers (Figure 12–7).

The focus of Khmer society slowly altered. Varieties of Buddhism had long existed with Hindu cults, but during the thirteenth century, Theravada Buddhism won general allegiance. Now rulers demonstrated their power through the building of temples, schools, and monasteries.

In the 1440s, Angkor was abandoned in favor of capital sites in the region of Phnom Penh. The abandonment could have been the result of the silting up of the irrigation system or even malaria. The rise of the Thai state of Ayuthaya was probably the main reason. The capital sites were closer to the sea and the booming sea trade of the fifteenth century.

Until the late sixteenth century, the Khmer kingdom appears to have been as strong as its rivals Ayuthaya and Lan Xang. But thereafter, the Khmers became embroiled in regional machinations of the Thais and the Vietnamese. Vietnamese and Chinese began to dominate Cambodian ports. Lively trade ensued even as Cambodia was caught in the pincers of Thai and Vietnamese expansionism. Repeated incursions left the Thais dominant in the eighteenth century.

In the nineteenth century, the Vietnamese rose to the fore and Vietnamese people were encouraged to colonize the region. Vietnamese language, law, and modes of dress were imposed. A countrywide rebellion broke out in 1840; for five years Thais, Vietnamese, and Cambodian factions fought an inconclusive war.

COMING OF THE FRENCH

The French, having established their colony of Cochin China by 1862, were interested in Cambodia. In 1863 France signed a "treaty of protection" with the king that placed a French resident at Phnom Penh. This gave the French control of foreign policy and opened the door to commercial interests. However, the French soon realized that there was little economic return to be had from Cambodia and so concentrated on Cochin China.

By 1904, the French, with the intent of expanding their presence in Southeast Asia and countering British interests, achieved complete authority over the protectorate. In 1921, the population of Cambodia was assessed at 2.5 million. A Chinese rice export business had developed with rice purchased from the Khmer farmers. However, Cambodian rice was regarded as inferior to the rice of Cochin China so did not fare as well in the market. There were also small Chinese timber and pepper industries and French-financed rubber plantations using Vietnamese labor. Maize, kapok, and fish were gleaned from the region of Tonle Sap. The Mekong remained Cambodia's main trade artery but the port of Saigon handled the shipping.

Around 95 percent of Cambodians survived as subsistence farmers. They were characterized by the French, Vietnamese, and others as "lazy," "ignorant," "fatalistic," and "child-like." Western observers condemned them as a "decadent race." As the interlopers took in their profits, no industries of consequence were developed. Towns remained small, with Phnom Penh housing only 100,000 by 1930. Towns were dominated by aliens, and the first Khmer language newspaper only appeared in 1938.

WORLD WAR II AND INDEPENDENCE

During World War II, the French reached an agreement with Japan to retain control of Indochina in return for allowing free movement of Japanese troops. In order to deflect Cambodian fascination with Japanese rhetoric, the French stimulated a nationalist movement. They tried to institute a new form of writing as in Vietnam but the Buddhist community rebelled against what they saw as an attack on Cambodian culture and heritage. When the war ended, Cambodia had no clear nationalist group as existed elsewhere in Southeast Asia. The country drifted and the French returned to power.

While keeping control of key elements of government, the French permitted the formation of political parties. By the 1950s however, lack of progress and cooperation produced acute strains. In 1951, the Khmer People's Revolutionary Party (KPRP) was founded and began to organize guerrilla operations in outlying areas. Meanwhile, Prince Sihanouk toured France and the United States demanding independence. The French agreed and granted it in 1953.

SIHANOUK, WAR, AND REVOLUTION

Sihanouk, who came to power under rigged elections, promoted "Buddhist socialism." He was intolerant of opposition and forced hundreds of "dissidents" to disappear. He greatly expanded education but expressed little interest in economic affairs. The Cambodian economy went into decline. Meanwhile, the Vietnam conflict was omnipresent and Sihanouk permitted the northern Viet Minh and the southern Viet Cong to use Cambodian territory. He rejected the United States and opened relations with China. This eliminated United States aid and angered conservatives. Communist-led revolts over government seizures of rice were brutally crushed. In 1970, while Sihanouk was overseas, the National Assembly withdrew its support and installed Lon Nol as the head of the new government of the "Khmer Republic."

The anti-Sihanouk coup polarized the Cambodian population. The United States supported Lon Nol and his totally inept government. Aid programs funded by the U.S. fostered gross corruption in both government and military. At the same time, the United States was carpet bombing Cambodia with more than half a million tons of bombs as part of its offensive against North Vietnam. About 500,000 soldiers and civilians were killed. The Communist insurgency was slowed by this devastation but not defeated. The Lon Nol regime ultimately collapsed and the insurgents took Phnom Penh in April 1975.

The Khmer Rouge (Red Khmer) set up the state of Democratic **Kampuchea**. This name derived from *Kambuja*, meaning "the sons of Kambu"—the ascetic who married a celestial nymph and founded the kingdom of Chenla, forerunner to the great Khmer empire. Their communist leadership was made explicit in 1977 when they announced the existence of the Communist Party of Kampuchea (CPK), which had been founded in 1968. A Paris-educated school teacher rose to take the helm. His name was Saloth Sar, later known as **Pol Pot**. One of his first acts was to declare 1975 as the year zero.

THE KILLING FIELDS

In April 1975 the Khmer Rouge marched triumphantly into Phnom Penh. The city was filled with refugees and its population had jumped from 600,000 to more than 2 million. Pol Pot immediately imposed his radical Maoist-style agrarian society on to the Cambodian people. Within four days, most of the inhabitants of Phnom Penh had been forcibly relocated to the countryside. Later, hundreds of thousands of people were forced from the southeast of the country to the northwest.

Pol Pot's goal was pure socialism. Everything modern was to be eliminated. A typical slogan was "We will burn the old grass and new will grow." Money, newspapers, education, and technology were outlawed. Intellectuals were killed. This was to show the strength of the people who, "though bare handed . . . can do everything."

Food was scarce under Pol Pot's inefficient system of farming and administration was built on fear,

Figure 14–17

This skull map of Cambodia symbolizes the millions of people who died at the hands of the Khmer Rouge. Photograph courtesy of B.A. Weightman.

torture, and execution (Figure 14–17). Little was known of the Khmer Rouge atrocities until a few refugees trickled over the Thai border. During the Khmer Rouge reign of terror, more than a million people died; some estimates are as high as 2 to 3 million. The Khmer Rouge turned Cambodia into a terrifying work camp in which families were abolished and murder was used as a tool of discipline. Pregnant women were disemboweled, babies were torn apart limb from limb, people were buried alive or bashed to death with axe handles. Of 64,000 Buddhist monks, only 2,000 survived. Half of the Cambodian Vietnamese were eliminated, along with thousands from other groups. The CPK even turned upon itself, purging those who had received training in Vietnam.

In 1978, Vietnamese troops entered Cambodia. The Khmer Rouge collapsed before them abandoning their centers of power. The Vietnamese took Phnom Penh in 1979 but failed to capture Pol Pot and his close comrades. The Vietnamese quickly established the People's Republic of Kampuchea (PRK) with a Cambodian government. This included Hun Sen, who became premier in 1985 and won in elections held much later in 1998.

The country suffered a devastating famine and it was not until the mid-1980s that subsistence farming reached an equilibrium and shops and markets began to function normally. Meanwhile, the PRK became an international pariah supported by only a few countries, including the Soviet Union. In 1989, the Vietnamese withdrew from Cambodia. Kampuchea was renamed the State of Cambodia and committed itself to a private enterprise economy and the restoration of Buddhism.

The United Nations became involved in Cambodia's internal affairs and elections were held in 1993. Multiple parties contested and a coalition government was formed. King Sihanouk was reinstalled, although he later died of cancer. The coalition government was heavily financed by foreign aid, but internal dissension and corruption meant that little progress was made toward rebuilding the shattered economy and infrastructure. Phnom Penh University opened in 1988. Only 5,000 teachers had survived of the original 20,000 in 1975. A sex ratio imbalance was evident in employment. For example, 70 percent of workers in the textile and tobacco industries are women, mostly widows of pre-1975 workers. A former school in Phnom Penh, where 20,000 died, has been turned into a museum to Pol Pot's genocide.

In disputed elections in 1998, Hun Sen became the head of government. Pol Pot committed suicide in the same year and other Khmer Rouge leaders have either defected or live clandestinely in various parts of the country.

Did you know that Cambodia is the world's biggest minefield?

Death Awaits

Because of the influx of military hardware to various factions during the 1980s, Cambodia has become the most heavily mined place in the world today. There are more than 3 million mines, mostly concentrated in the northwest province of Battembang, a Khmer Rouge stronghold. In addition, there are countless thousands planted along the main roads. When resettlement of refugees began in 1992, there were 7,000 amputees in the camps. Now there are about 300 new amputees every month. One in every 236 Cambodians is an amputee.

Mine clearing is a slow process. A British-based group has been involved since 1990, although fewer than 6,000 mines have been removed. Every week animals or people are maimed or killed by land mines.

CAMBODIA TODAY

Cambodia continues as a very poor country. Forty percent of children under age five are malnourished. Only 13 percent of primary school students complete grade five in five years. Only 19 percent of students in upper secondary school are female. More than 90 percent of births in rural areas are at home and attended by traditional midwives. Of the rural population, only 26 percent have access to safe water. A mere 6 percent of rural people have access to improved toilet facilities. Clearly, Cambodia remains one of the least developed countries in the world.

Rice production accounts for nearly 88 percent of the available cultivated area. Farms tend to be very small—around 2.5 acres (1 ha). A farm of this size supports a family of five. Rice is supplemented by fruit and vegetable cultivation. Although rice yields are low, Cambodia became self-sufficient in rice in 1996 and had a surplus to export.

Cambodia's rubber plantations are second only to timber as a source of export earnings. In 1995, US $3 million worth of natural rubber was exported to the United States. Expansion of rubber production is planned for the eastern provinces. Cambodia should be able to produce 600,000 tons of rubber annually.

Fishing is important as it provides the main source of protein in the Cambodian diet. Fisheries are both inland and marine, and fish are also raised in ponds, pens, and cages where water is available around the year. Some frozen fish are exported.

Teak and rosewood are among Cambodia's most valuable assets. However, more than 70 percent of Cambodia is deforested and the government has taken measures to stop timber exports. Illegal cutting and exporting continues despite the prohibitions. In 2003, the government expelled Global Witness, an environmental group that was monitoring efforts to combat illegal logging. As a consequence, aid to the country's forestry program from the International Monetary Fund and the World Bank has been stalled amid concerns over high-level government involvement in illegal logging.

Cambodia is rich in minerals, including precious stones such as sapphires and rubies. It has oil in the Gulf of Thailand, which is now being exploited by both domestic and foreign concerns.

Foreign aid and investment is critical to Cambodia's development. Light industry is more developed than heavy industry. The more developed industries include household goods, textiles, soft drinks, alcohol, tires, pharmaceuticals, and cigarettes. The extent of foreign involvement can be demonstrated by the textile industry. Of 78 companies, Hong Kong runs 25, Taiwan 15, Cambodia 11, and China 7. Malaysia and Korea each run 6, and the rest are divided up among Singapore, France, Britain, and the United States.

DANGEROUS LIAISONS

After decades of civil war and isolation, Cambodia has become a favorite destination for foreign tourists. Some 790,000 foreigners visited the country in 2002 and this number is expected to rise to 2.2 million in 2006. If that materializes, the tourism industry will bring in revenues of a billion dollars. Many tourists want to visit Angkor Wat. On the more sinister side, the tourists also include thousands of pedophiles who take advantage of Cambodia's weak law enforcement. Prostitution and drug abuse have also led to the highest HIV/AIDS infection rate in Asia. According the United Nations, 160,000 Cambodians between the ages of 15 and 49 have HIV.

AIDS so far has killed about 90,000 people since it was discovered in the country in 1991. Another 200,000 people are expected to develop the disease in the next 10 years. This will overwhelm Cambodia's grossly inadequate health-care system. Moreover, the economy will be impacted because most of the sick and dying will be young people—potential workers. The epidemic is already being called "the new killing fields."

⊕ Main Point Summary

1. Mainland Southeast Asia (MSA) is a complex region with an intertwined history.
2. Burma, or Myanmar, is the largest and most ethnically diverse state in MSA.
3. Although the boundaries of Burma were created by the British, its territory originally centered on Bagan and Mandalay.
4. British colonial rule was founded on an Anglo-Burman and Indian elite. Efforts to develop the delta regions caused a large southward migration.
5. A plural society rested on the fact that the British were only interested in primary products such as teak and not industry.
6. World War II was accompanied by the rise of nationalism, but regional insurrections and rampant corruption led to the downfall of the government.
7. The country was divided into regions not coincident with minority territories.
8. In the 1980s, SLORC took power in Myanmar and imposed a military dictatorship even though elections had been held. The winner of the elections, Aung San Suu Kyi, was put under house arrest where she remains.
9. Rice trade is Myanmar's major form of revenue, although quotas and inadequate technology have driven many farmers off the land.
10. The government of Burma has forced thousands of people to move elsewhere to depopulate cities, and to get out of the way of new development projects. Forced labor is widespread.
11. Most heroin (a derivative of opium) in the United States originates in the Golden Triangle shared by Burma, Thailand, and Laos. Burma is one of the world's largest producers of opium.
12. Thailand has four main zones dominated by the fertile valley of the Chao Praya.
13. The principle ancestors of the Thai were the Tai people who were wet-rice farmers. The Tai founded the kingdom of Sukothai—the birthplace of the Thai nation.
14. Bangkok became prominent in the eighteenth century while Thailand lost territory to colonial powers. Thailand served as a buffer state between the territorial ambitions of Britain and France.
15. Theravada Buddhism and the *sangha* are critical elements of Thai society. A sense of order is central to Thai culture. Shrines for "lords of place" along with *wats* are basic elements of the Thai landscape.
16. More than half of Thai exports are agricultural, and about half the population are farmers. Deforestation has contributed to a shortage of water for irrigation.
17. New Theory Farming is community-based farming depending on the cooperation of multiple sectors of agricultural production.
18. Tourism is a leading earner of foreign exchange. Eco-tourism and sex tourism both compromise the cultures of native peoples.
19. Industrialization is focused around the city of Bangkok, where economics of agglomeration

are supported by an informal sector. Ribbon development has extruded onto agricultural land. Bangkok has an array of severe environmental problems.

20. Vietnam's population is concentrated in the Tonkin and Mekong delta regions.

21. Although Vietnam's population is dominated by the Vietnamese, the presence of other groups such as Cao Dai and Catholics makes it difficult to create a fully integrated society.

22. In the second century B.C., the region was absorbed into China's Han Empire. This began a period of a thousand years of Chinese domination. Chinese culture had a great influence on Vietnam and differentiated it from other parts of Southeast Asia.

23. The Jesuits created *quoc ngu*, a writing system using the Roman alphabet.

24. The French acquired a colony in the south, which they named Cochin China. Eventually, their *mission civilisatrice* was applied to the Vietnamese regions known as Tonkin and Annam, in addition to Cambodia and Laos.

25. The collapse of the Confucian government and occupation by the Japanese fostered nationalist and popular movements within Vietnam. The Viet Minh, under Ho Chi Minh, took control of most of northern and central Vietnam.

26. The French were defeated at Dien Bien Phu in 1954, and U.S.-led efforts to prop up a democratic government in the south failed. The Viet Cong emerged to control vast areas of the southern countryside.

27. The Vietnam War took an enormous human and environmental toll. Even after the war, north-south differences prevailed.

28. A hard-line Communist approach eased in the 1980s when the government adopted *doi moi*, economic reforms leading to market socialism.

29. Laos or Lao is a landlocked, sparsely populated country, created by the French as a part of Indochina. It first existed as Lan Xang—"a million elephants."

30. The French attitude to Laos was one of benign neglect. Administration was controlled by the Vietnamese. There was very little modernization.

31. During World War II, Communist-inspired guerrilla movements emerged, known as the Pathet Lao. They controlled a large area of northern Laos.

32. During the Vietnam War, the American CIA established links with the opium-growing Hmong people to sell drugs in new markets in Southeast Asia to help finance the war.

33. Laos was heavily bombed with enormous damage to the landscape. Many bombs still exist unexploded, and people are killed by them on a weekly basis.

34. Rural Laos can be understood according to lowland, midland, and upland ways of life. All the people of these areas have differentiated gender roles.

35. Viangchan is the only city of any significant size. There is very little rural to urban migration.

36. The government is now supporting a policy of "New Thinking," a kind of market socialism. Foreign investment, mainly Japanese, is taking advantage of the more liberal economic policies.

37. Infrastructure is poor but Laos has possibilities as a transport hub, given its crossroads position in MSA.

38. Cambodia is noted for the temple complex at Angkor, as well as for the genocide that took place under Pol Pot in the 1970s.

39. Cambodia is ringed by mountains and dotted with hills known as *phnom*. The great lake of Tonle Sap is a remnant of when Cambodia was under the sea millions of years ago.

40. Consolidation of Khmer society occurred in the ninth century with the institution of a god-king and concentration of development of religious sites at Angkor. Hindu influence gave way to Theravada Buddhism.

41. Until the sixteenth century, the Cambodian kingdom rivaled both Ayuthaya and Lan Xang. Then it became embroiled in the expansionist machinations of both the Thai and Vietnamese.

42. The French fostered industries such as rice production, timber, pepper, and rubber, and the products were shipped via the Mekong and the port of Saigon.

43. In order to deflect Cambodian interest in the Japanese, the French encouraged nationalist movements. In 1951, the Khmer People's Revolutionary Party was founded. Independence was granted in 1953.

44. The Khmer Rouge set up the state of Democratic Kampuchea in 1975. Pol Pot headed the government.

45. Pol Pot's efforts to create a completely peasant agrarian society resulted in millions of intellectuals and other people being executed. Most Buddhist monks were eliminated.

46. The Vietnamese withdrew in1989 and Kampuchea was renamed Cambodia. Cambodia committed itself to a private enterprise economy and the restoration of Buddhism.

47. In 1998, Hun Sen was elected to head the government and most of the Khmer Rouge defected. Pol Pot committed suicide.

48. Foreign aid is critical to Cambodia's development, and Cambodians are counting on tourism to earn foreign exchange.

49. As tourism increases, so does the sex trade. There has been a substantial increase in cases of HIV/AIDS in Cambodia since the disease was discovered in 1991.

⊕ Bibliography

Birsel, Robert. 1990. "Military Regime Minority Insurgence." *Geographical Magazine* 62: 28–32.

Bradshaw, Adrian. 1989. "Sequel to the "Killing Fields"?" *Geographical Magazine* 61: 22–27.

Church, Peter, ed. 1995. *Focus on Southeast Asia.* ASEAN Focus Group. Sydney: Allen and Unwin.

Cohen, Eric. 1992. "The Growing Gap: Hill Tribe Image and Reality." *Pacific Viewpoint* 33: 165–69.

Crispin, Shawn, W. 2003. "Thailand: Big Risks in Big Plans." *Far Eastern Economic Review* (February 27): 16–17.

Crispin, Shawn, W. 2004. "Thailand: Pipe of Prosperity." *Far Eastern Economic Review* (January 19): 12–16.

Cummings, Joe. 1997. *Thailand: Lonely Planet Survival Kit.* Singapore: Lonely Planet.

Duiker, William, J. 1995. *Vietnam: Revolution in Transition.* Boulder: Westview.

Eliot, Joshua et al., eds. 1995. *Vietnam, Laos, and Cambodia Handbook.* Chicago: Passport.

Fairclough, Gordon. 1996. "Rice to the Occasion: Burma Exhorts Farmers to Grow More Faster." *Far Eastern Economic Review* (August 29): 55–56.

Freeman, Donald B. 1996. "Doi Moi Policy and the Small-Enterprise Boom in Ho Chi Minh City, Vietnam." *The Geographical Review* 86: 178–97.

Forbes, Dean. 1996. *Asian Metropolis.* New York: Oxford.

n.a. 1998. "Granting A New Life to the Thai People." *Kinaree* 15: 26–27.

Handley, Paul. 1993. "Thought For Food: Thai Farm Policy Faces Long-Term Choices." *Far Eastern Economic Review* (April 29): 46–49.

Huke, Robert E. 1998. "Myanmar Visited." *Focus* 45: 18–25.

Hussey, Antonia. 1993. "Rapid Industrialization in Thailand, 1986–1991." *The Geographical Review* 83: 14–28.

Jackson, Karl D., ed. 1989. *Cambodia 1975–1978: Rendezvous with Death.* Princeton: University Press.

Lintner, Bertil. 2003. "Cambodia: Dangerous Liaisons." *Far Eastern Economic Review* (March 13): 47.

Lintner, Bertil. 1997. "Drugs, Insurgency and Counterinsurgency in Burma." In *Burma: Myanmar in the Twenty-First Century*, ed. John J. Brandon, pp. 207–45. Bangkok: Open Society Institute.

Lintner, Bertil. 2003. "Laos: Aid Development." *Far Eastern Economic Review* (July 3): 45.

Lintner, Bertil. 1994. "Burma: Enter the Dragon." *Far Eastern Economic Review* (December 22): 22–23.

Lintner, Bertil. 1998. "The Dream Merchants (Narcotics)." *Far Eastern Economic Review* (April 16): 26–27.

Mabbett, Ian, and David Chandler. 1995. *The Khmers.* Cambridge: Blackwell.

Mulder, Niels. 1996. *Inside Thai Society: Interpretations of Everyday Life.* Kuala Lumpur: Pepin.

Norman, Colin. 1983. "Vietnam's Herbicide Legacy." *Science* 219: 1196–97.

Reardon, Douglas. 2004. "The Qualms of a Geographer in a Pariah State." *Focus* 48: 7–11.

Rigg, Jonathan. 1995. "In the Fields There is Dust." *Geography* 80: 23–32.

Rigg, Jonathan. 1995. "Managing Dependency in a Reforming Economy: the Lao PDR." *Contemporary Southeast Asia* 17: 147–72.

Ryan, John C. 1990. "War and Teaks in Burma." *World Watch* 3: 8–9.

Savada, Andrea, ed. 1995. *Laos: A Country Study*. Washington, D.C.: U.S. Government.

Shawcross, William. 1994. *Cambodia's New Deal*. Washington, D.C.: Carnegie.

Tesoro, Jose, M. 1998. "The Young and the Restless." *Far Eastern Economic Review* (February 13): 23–24.

Thayer, Nate. 1993. "Diverted Traffic (Narcotics)." *Far Eastern Economic Review* (March 18): 24–25.

Thomson, Curtis N. 1995. "Political Stability and Minority Groups in Burma." *The Geographical Review* 85: 269–85.

Tyner, James. 2003. "Cities of Southeast Asia." In *Cities of the World*, eds. S.D. Brunn, J.F. Williams, and D.J. Zeigler, pp. 373–411. New York: Rowman and Littlefield.

Wellner, Pamela. 1994. "A Pipeline Killing Field: Exploitation of Burma's Natural Gas." *The Ecologist* 24: 189–93.

Westing, Arthur H., and E. W. Pfeiffer. 1972. "The Cratering of Indochina." *Scientific American* 226: 21–29.

Chapter 15

Insular Southeast Asia

"There is not one state truly alive
if it is not as if a cauldron burns
and boils in its representative
body, and if there is no clash of
convictions in it."

SUKARNO (1901–1970)

Welcome to the peninsular and island realm of Southeast Asia (*see* Figure 14–1). With six nations including Malaysia, the Sultanate of Brunei, Singapore, Indonesia, East Timor, and the Philippines, this is one of Asia's most complex regions. Like the rest of Southeast Asia, the region's history is bound up with ocean interactions, foreign intervention, and struggles for identity in the modern world. In this chapter, starting with Malaysia, you will read some of the most intriguing stories of geographic interrelationships in the world today.

Malaysia 2020

The course of Malaysia's history has been influenced by its strategic position, which made it a natural meeting place for traders from both east and west. The lush tropical forests and abundance of water sustained numerous small, self-supporting communities. But,

unlike mainland Southeast Asia, the presence of mountains and the absence of broad plains precluded the development of elaborate systems of water control so fundamental to mainland societies. Sea trade bred contacts from the outside, and this was the foundation of Malaysia's history.

Whence "Malay"?

Many Malay areas were colonized by Sumatrans long ago. It is conceivable that the word *Melayu* derived from the *Sungei Melayu* (Melayu River) in Sumatra. Then there is the possibility that it came from a Tamil word *malai*, meaning hill.

The geographer Ptolemy called the Malay Peninsula *Aurea Chersonesus*, or "The Golden Chersonese." This was the fabled land of gold.

During the Portuguese and Dutch colonial periods, and prior to the British colonial period, the whole peninsula was labeled *Malacca*. There was very little mapping of the peninsula until the early nineteenth century. As you will read shortly, many name changes would occur before reaching the name "Malaysia."

EARLY SETTLEMENT

The earliest of the current inhabitants of Malaysia are the *Orang Asli*, meaning "Original People" who inhabited the Malay peninsula (Figure 15–1). Other groups include the Dayak and Penan of Sarawak and the Rungus of Sabah, people whose nomadic way of life has been almost completely destroyed in the wake of development.

The next arrivals were the Malays, the Proto-Malays establishing themselves around 1000 B.C. These were followed by other immigrants, including the Deutero-Malays, streaming down the peninsula from mainland regions, including China. Many migrated even further to the island realms of Indonesia and elsewhere. The peninsular Malays had their closest links with the Malays of Sumatra, and for centuries the Straits of Malacca served as a corridor for interaction between the two regions.

Together with the *Orang Asli*, all these people share a common culture and make up the *Bumiputera* (*Bumiputra*)—"Sons of the Soil." Cultural characteristics were rooted in an agrarian-maritime economy and were reflected in village society, where leadership was largely consensual and attitudes were impregnated by a belief in an all-pervasive spiritual world. Although this culture came to be overlaid by Hinduism, Islam, and Western ideologies, elements of the basic culture persist.

Trade with China and India began around the first century B.C., although Hindu and Buddhist elements from India were more influential in terms of impacts on language, literature, and social customs. Numerous temples were constructed throughout the region. For the greater part of this time, Malaysia was subject to either Javanese or Sumatran power structures.

The Peranakan

By 1400, Malacca had become an important outpost for Chinese traders. They arrived from November to March on the northeast monsoon and departed in June with the southwest monsoon. Malacca's sultans even paid tribute to the Ming court. Many Chinese settled in Malacca,

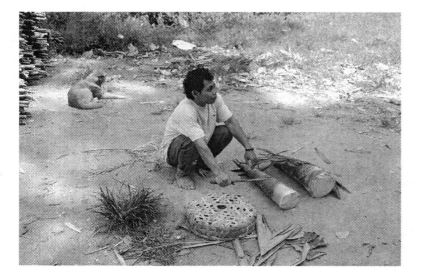

Figure 15–1

This man is an Orang Asli, *one of the original inhabitants of the Malay peninsula. He is cutting a sago palm, one of the most useful palms in the forest. The carbohydrate-rich pith is dried as "pearl sugar" and is eaten with fish and fruit. The palm provides fiber for rope, matting, or cloth; the leaves can be used for thatch; and the sap ferments into a liquor.* Photograph courtesy of B.A. Weightman.

especially around *Bukit Cina*—Chinese Hill. Today, *Bukit Cina* is a historical park, housing the world's largest traditional Chinese cemetery.

Subsequent generations of Straits Chinese became known as *Peranakan*. The Peranakans were known for their business acumen and lavish lifestyles. They built classic townhouses and developed all the trappings of a "high society." They also developed their own cuisine, a blend of Malay and Chinese, which is noted for its meticulous preparation and pungent flavors.

Blending easily into British society, the Peranakan remained divorced from other Chinese. Many became doctors, lawyers, and entered public office. In Pinang they were known as "the Queen's Chinese."

The Hindu-Buddhist period of Malaysian history ended with the arrival of Islam. Introduced primarily by Indian and Arab traders, Islam became a major influence after 1400 with the conversion of formerly Hindu rulers of Malacca. Islam diffused along the trade routes into the rest of the insular region. The Sultanate of Malacca and the Kingdom of Brunei became major proponents of the faith.

Both Malacca and Brunei were shattered by the influx of Europeans. The Portuguese, Dutch and Spanish made their presence felt through force of arms. Despite their technological superiority, the European's had limited influence until the coming of the British at the end of the eighteenth century and their introduction of the resources and organization of the Industrial Revolution.

MAKING MALAYSIA

Under the British, Pulau Pinang (1786), Singapore (1819), and Malacca (1824) became known collectively as the Straits Settlements. Garnering other Malay states under their wing of control, the Straits Settlements were later expanded to include the island of Labuan off the northern coast of Brunei. On the peninsula, a system of Federated Malay States was established with the capital at Kuala Lumpur. Several other peninsular states were given British advisors and became known as the Unfederated Malay States. Sarawak on the island of Borneo became a British protectorate ruled by the Brooke family. Sabah also became a protectorate ruled by the Chartered Company of British North Borneo.

This confusing arrangement satisfied few and opposition frequently turned violent. In 1948, the British established the Federation of Malaya, including all the nine Malay states of the peninsula. They also committed to the region's independence. Local elections were held in 1951 and federal elections in 1955. The British released their sovereignty in 1957.

The idea of the formation of Malaysia was introduced in 1961. The new nation was to include the peninsular states, the Borneo states, and Singapore. The name "Malaysia" is derived from "Malay"—the people, "si"—Singapore, and "a"—Malaya. Elections were held shortly thereafter, with only Brunei declining to join. Tension was apparent with regard to Singapore, however. Its large contingent of Chinese (with their potential political power aligned with Chinese elsewhere in Malaysia) concerned the Malay majority. Singapore was pressured out of the union in 1965 (Figure 15–2).

TIN AND RUBBER

At the beginning of the nineteenth century the population of peninsular Malaysia was no more than 300,000. Most were Malay rice cultivators. A heavily forested terrain deterred the British, who were more interested in their ventures in Pulau Pinang and Singapore. However, the success of the British entrepreneurs stimulated migrations from India and China. The Indians and Chinese would eventually become intermediaries between the British and the Malays.

Chinese miners revolutionized tin mining, which had previously been on a very small scale. Given a low-cost means of extraction and the location of tin-bearing gravels along the western foot of the Main Range, a mere 25 to 30 miles (40 to 50 km) from the coast, production surged. In the 1890s, a

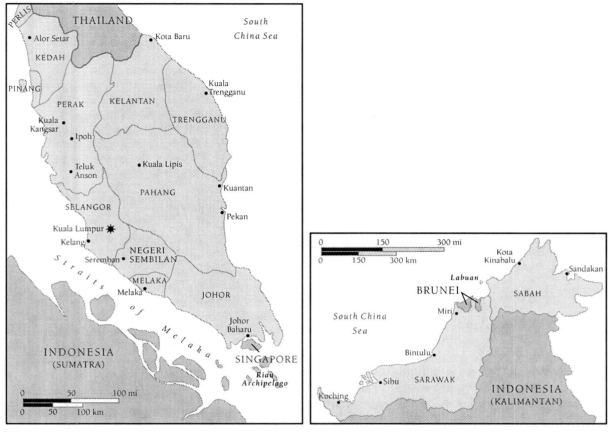

Figure 15–2

Malaysia. Like Indonesia and the Philippines, Malaysia is a fragmented country.

"tin rush" ensued in the three western states of Perak, Selangor, and Negeri Sembilan.

Relations between the Chinese and Malays were not good, so the British decided to provide a system of law and order in the region. This took the form of an elaborate infrastructure of road, rail, and port facilities. Smelting and exporting of tin were centered at Pinang and Singapore.

In the early 1900s, an increasingly motorized world was demanding rubber. The climate of Malaysia was perfect for the raising of this commodity. Consequently, new rubber estates were established within the area already possessing an infrastructure. Indian rubber tappers were brought to work the

trees, and the rubber industry boomed until the Great Depression of 1930. When the global economy recovered, so did rubber. Malaysia remains the world's largest rubber producer.

The peninsula's development was severely unbalanced. A rubber and tin belt occupied the southwest. It covered only a quarter of the land area but accounted for 85 percent of its economic activity. Here, too, were the ten largest towns and a large but diverse population of Chinese, Malays, and Indians. The east was largely undeveloped and contained a much lower density of predominantly Malay peasants. When Malaysia became independent, a major goal was to rectify this imbalance.

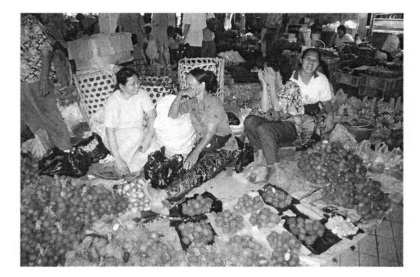

Figure 15–3

Here the Iban people of Sarawak are selling "jungle products," which they have collected from the forest. The fruit in the foreground is "rambutan." The Iban are traditional shifting cultivators but many have moved to coastal towns to participate in the oil, lumber, and plantation economy. They make up nearly one-third of the population of Sarawak. Photograph courtesy of B.A. Weightman.

EAST MALAYSIA

East Malaysia consists of Sarawak and Sabah. Prior to the economic intervention of the Europeans, northern Borneo's trading economy was dominated by indigenous, unprocessed, and luxury products. Jungle products such as bezoar stones (concretions from the stomachs of animals such as goats), hornbill beaks, camphor, and birds' nests were the most important. Other products included rattan, bamboo, other woods, and gutta percha (a resin from the gutta tree used to make cable coverings and glue).

This was an economy separate from the world market. Traditional networks of collecting and marketing channeled jungle products to coastal ports where Chinese or Malay traders sold them on the regional market (Figure 15–3).

By the late nineteenth century, the East Malaysian economies had begun to shift toward meeting the needs of world markets. Production systems were increasingly controlled from abroad as plantation farming, rubber, timber, and oil took precedence over jungle products. Indigenous groups, such as the Penan and the Dayak (engaged directly in accessing these products), fell increasingly under the influence of others, particularly the Chinese. Javanese workers were also recruited to work in the region. However,

as jungle produce continued to deliver significant income, indigenous people clung to those activities while outsiders involved themselves in tobacco, pepper, and other new production enterprises. Oil was very important in British North Borneo, where it was the stable backbone of revenue.

Pre-colonial and colonial trading economies had become intermeshed. Whatever the case, the focus was on environmental exploitation. Native populations became ever-more marginalized as commodity trade increasingly fulfilled the requirements of the growing world economy. This heritage affects the developmental patterns of East Malaysia and Brunei to this day.

ETHNICITY AND DEVELOPMENT

Malaysia has a plural society and development policies have favored indigenous Malays. However, the relationship between ethnicity and status is difficult to define as there are not viable definitions of ethnicity. Ethnic pluralism is a historical construct. For example, in Malaysia the British perceived the Malay as farmers unwilling to do wage labor and the Chinese and Indians inherently suited for mining and plantation work. Thus, Malays were shoved to the margins of development by both British practice and policy. As geographers Mark Cleary and Brian Shaw (1994) observed: "Ethnic pluralism in Malaysia might be more accurately represented as a series of overlapping Malay, Chinese or Indian ethnic dimensions (language, business practice, occupation, religion) rather than lines of rigid demarcation. They represent socially constructed rather than 'natural' difference."

Just as economic and infrastructural factors were imbalanced, so was population distribution. By the late 1960s, some 15 percent of the Malay population was classified as rural. About 45 percent of Chinese and 35 percent of Indians were classified as urban. Malays in the cities were heavily concentrated in government employment. Very few participated in the commercial and industrial sectors.

Modern development in Malaysia has been defined by a New Economic Policy (NEP) and subsequent "Malaysia Plans." A fundamental goal of the NEP was to restructure Malay society by raising incomes for all groups and eliminating the association of ethnicity and economic function. This meant an increase in the production of export commodities such as timber and oil as well as export-oriented industrialization. Explicit within the NEP was the intention to create a Malay (*Bumiputera*) business community that would own or manage at least 30 percent of all these activities by 1990.

DEVELOPMENT SCHEMES

One widely acclaimed land development scheme was the Federal Land Development Authority (FELDA) settlement plan. Commencing with a 2,470 to 4,940 acre (1,000 to 2,000 ha) block of land, contractors were brought in to clear the jungle, plant the main crop—rubber in the early years and palm oil, cocoa, and coffee in later years—and build basic infrastructure such as roads, houses, and schools. Settlers were selected, moved in, and employed as salaried workers until the crop began to produce. Each family was given title to a plot of land once the mortgage was paid off. By 1987 there were 422 FELDA schemes in all states except Pinang (Figure 15–4).

As palm oil replaced rubber, some of the FELDA schemes became highly integrated, including their own harvesting, milling, marketing, and shipping facilities. These organizations, with their economies of scale and integrated management, contributed to the increasing commercialization of all FELDA operations.

Figure 15–4

This is the fruit of the oil palm, the economic focus of many FELDA schemes. The fruit is crushed for oil which is used for cooking and in such products as margarine and soap. Malaysia is the world's primary palm oil producer. Photograph courtesy of B.A. Weightman.

Today, FELDA schemes are more likely intensively run "agro-businesses" than traditional Malaysian farming communities. Moreover, the schemes increasingly are becoming part of the Western, mechanized, uniform production system. FELDA plans were cut to zero in the Sixth Malaysia Plan of 1991.

The Eighth Malaysia Plan of 2001 promotes a shift from those mono-crop, low-technology, and small-scale operations that still exist, to integrated, high-tech, and large-scale enterprises. Emphasis on palm oil and rubber is to be reduced while the production of fruit and vegetables is to be promoted. The plan also proposes to intensify land use by integrating more food crops with plantation crops. Furthermore, the government is encouraging urban horticulture and agro-technology parks. Land settlement is now seen as the concern of the private sector in context of regional planning frameworks.

Rural Poverty

For all its impressive growth rates and higher per-capita incomes, Malaysia ranks among the most unequal societies in Southeast Asia in terms of income distribution. While it is true that absolute poverty levels have declined—from 29 percent in 1980 to 6 percent in 2000—critics complain that these figures are misleading. The poverty line is at US $134 a month and this cannot keep up with the rising costs of living. The World Bank reports that the trend of inequality reduction has reversed itself since 1990. Another report that measures Malaysia's Gini coefficient—where 0 indicates perfect equality and 1.0 indicates perfect inequality—finds that the country's 0.49 score is one of the highest in the region. Indonesia is at 0.32 and Thailand is at 0.41.

According to another research study, intra-ethnic inequality appears to be higher among Malays. Two decades of affirmative action policies did increase the stake in the economy of the *Bumiputeras* and led to the emergence of a Malay middle class. However, a huge amount of this stake is held by state-backed investment agencies holding shares in trust.

Another disparity concerns the rural-urban divide. By 1999, rural incomes were only 55 percent of urban incomes. Many of the poor rural households can be found in Kelantan, Terengannu, Kedah, and Perlis states. It is in this "heartland" that support for conservative Islam is the strongest. In East Malaysia, indigenous groups in the interior of Sarawak and Sabah are very poor as are the Orang Asli of peninsular Malaysia.

A POPULATION ANOMALY

Malaysia's population growth is relatively high, given its level of development. It is also higher than the average rate for Southeast Asia. Malays have significantly higher total fertility rates (TFRs) than the Chinese and the Indian populations. The national TFR in 2003 was 3.3 of which Malays had a two-thirds contribution.

High Malay TFRs can be related to Islamic fundamentalism and associated decline in the use of contraception. Furthermore, population policy in 1984 promoted five-child families. Malay families receive preferential treatment in university education, government employment, and the like. Malaysia 2020 has pronounced a goal of 70 million people as the optimum population size. This would substantially enlarge the consumer market in light of industrialization ambitions.

The structure of Malaysian population has been altered with Malays now forming a dominant majority. Many Chinese, seeing no future for their children, left for Singapore, Canada, and Australia. Malays are now 58 percent of the population as compared to 56 percent in 1970. The Chinese make up 25 percent and the Indians 7 percent. Multiple non-Malay ethnic groups, mostly in East Malaysia, make up the final 20 percent.

In East Malaysia, populations are concentrated along the coast and in surrounding urban centers. This contrasts with the sparsely populated interior where settlements are along river arteries. Sabah's population has been boosted by the in-migration of nearly half a million Filipinos during the 1980s.

Up until 1997, an estimated two million Indonesians lived and worked in Malaysia. After the financial crisis, about one million of these people were deported.

URBANIZATION

Migration has taken place in two ways. In the 1960s and 1970s, government efforts were concentrated in redistributing rural populations to other rural areas via land development schemes. The 1980s witnessed a dramatic change, with the largest share of migration being rural to urban. The rural sector was less able to absorb more people due to a fall in commodity prices, and the NEP pointedly tried to reduce rural poverty and dissociate Malays from strictly agricultural pursuits.

The result has been the cultural diversification of urban places once dominated by Chinese. More than one-half of Malays now live in urban areas. Moreover, the cultural landscapes of the largest urban centers have been noticeably "Islamicized" in recent years. More mosques, the high-rise Islamic Bank, Islamic archways over roads leading into the cities, and the presence of more women wearing head coverings are some of the more noticeable landscape elements (Figure 15–5).

Urban population distribution is still unbalanced. The west coast states accounted for more than 60 percent of the country's urban population in

2000. Historical inertia plays a role in this phenomenon. In East Malaysia, rates of urbanization have been rapid, yet the percentage of urban dwellers is low (37 percent in Sarawak and 33 percent in Sabah). This can be attributed to the absence of an urban-based manufacturing economy.

The paramount urban conurbation is that of Kuala Lumpur and the Kelang Valley (Figure 15–6). Because of the domination of Singapore and Pinang in the

Figure 15–6

The Kuala Lumpur-Kelang development corridor.

Figure 15–7

Selangor's new capital building at Shah Alam is part of the Kuala Lumpur-Kelang development corridor. Photograph courtesy of B.A. Weightman.

colonial period, Kuala Lumpur's population exceeded 100,000 only by 1930. Kuala Lumpur was designated a federal capital in 1974. Rapid growth ensued in the 1970s under NEP policies and today the city is the economic, political, and cultural core of the country.

The city incorporates a number of satellite cities in the state of Selangor, including an east-west corridor between it and the port of Kelang. Petaling Jaya was established in 1952 as a "new town," and by 1991, Petaling Jaya and Kelang were the nation's fourth and fifth most populous urban centers. In the 1970s, a free trade zone was established at Sungei Way. Further west is Shah Alam, the newly planned Selangor state capital that houses one of Southeast Asia's largest mosques (Figure 15–7). All these are joined by superhighways.

Kuala Lumpur is a modern landscape with a mix of modern and colonial buildings, Chinese shops, shopping malls, a fabulous Moorish-style railroad station (to become a hotel), and the 88-story, 1,483 feet (452 m) Petronas "Twin Towers"—the world's tallest buildings when completed in 1997.

Housing ranges from squatter settlements (fewer than in most of Asia) to condominiums, to suburban homes, to mansions. Because of a housing shortage, the government has built high-rise public housing (Figure 15–8). Squatters do not like these structures because it removes them from ground level, where

they can garden and watch their children play. Moreover, it is difficult to operate a home-based industry in a high rise. In addition, these projects are frequently distant from employment and services.

Since this area does not have vast expanses of paddy, it does not fit the *kotadesasi* model exactly, although it has some elements of it. *Kampungs* (Malay villages) and agricultural land do exist in the interstices of urban places, highways, industrial parks, and the like. Many original inhabitants supplement their income with off-farm jobs.

A Multimedia Super Corridor (MSC) was launched in 1996. This 15 by 50 km (9 by 30 mi) corridor stretches from the Petronas Towers in Kuala Lumpur southward to the international airport and includes the new cities of Cyberjaya and Putrajaya. Kuala Lumpur and the airport are joined by a high-speed rail system that opened in 2002. There is a multimedia university in Cyberjaya and Putrajaya is designed to be the new administrative capital of Malaysia with paperless, electronic government. As of 2004, over 1,000 high-tech firms had located in the MSC.

INDUSTRIALIZATION

Malaysia's national economy is now anchored by high-value manufacturing, including electronics and textiles for domestic consumption and steel and auto assembly for domestic consumption. In the 1980s, the main catalyst in economic growth was an outward-looking industrialization strategy. Rapid industrialization was encouraged by free trade zones and other strategies. Pinang has become Malaysia's "Silicon Island," attracting both foreign and domestic industrial concerns (Figure 15–9). However, Pinang will face competition from the MSC. The manufacturing sector is the largest generator of employment. The Eighth Malaysia Plan (2001) points to placing more emphasis on manufacturing and exports to foreign markets. The plan stresses the need for diversification and the acquisition of more knowledge-intensive industries.

The Proton Saga

The Proton Saga automobile company is one of several government-owned public enterprises managed by Malays. Its assembly plant is in

Figure 15–9
Georgetown, once a British fort, on the island of Pinang, is rapidly industrializing with high-tech industries. Pinang, joined to the mainland by one of Asia's longest bridges, is becoming Malaysia's "Silicon Island." Photograph courtesy of B.A. Weightman.

Shah Alam. A joint venture with Japan's Mitsubishi, Proton Saga automobile company produces a car replete with the star and crescent symbols of Islam. Initially, the Proton Saga captured over half of the domestic market by targeting middle and upper class consumers. However, this is a low-volume market and the company has looked overseas for more potential customers. Britain is the largest thus far. Targeting foreign markets is difficult because of competition and inability to meet Western safety and emission standards.

Expansion of foreign connections is seen as necessary for the success of Malaysia's auto industry. HICOM, the Proton holding company, is lining up with other foreign companies in Australia and Britain to manufacture certain components such as wheels and catalytic converters. The Malaysian government has turned to Daihatsu to build another car—the Kancil—that is being geared for export markets. Citroen of France is contributing diesel technology to lower the vehicles' overall costs.

Foreign investment—to provide capital, technology, and access to foreign distribution networks—was critical to success in industrialization. As a result of changing government policies, foreign investment increased tenfold in the 1980s decade. Nearly 50 countries were involved, the major sources being Japan, Taiwan, the United Kingdom, the United States, Hong Kong, Indonesia, and South Korea. The growth of manufacturing has had a tangible effect on urbanization as it generates urban employment.

With NEP policies, foreign ownership in the corporate arena fell to 21 percent by 1995. Malay involvement increased to 21 percent. Moreover, the Malay employment share in the manufacturing sector rose from 26 percent to 51 percent and in the service sector from 31 to 45 percent. Chinese and Indians experienced parallel share declines.

Beginning in the 1970s, the government targeted a mangrove swamp and an area of aging rubber trees directly east of Johor Baharu for economic development. Much of the early investment was in palm oil refineries and metal-processing plants. By the 1980s the region supported a more diverse industrial base in response to the establishment of a Singapore-centered growth triangle. In 1988, nearly half of investors were from Singapore, and activities had expanded to include such enterprises as chemicals, textiles, rubber products, and low-value electronic components.

With cheaper land prices, lower labor costs, and access to the Malaysian system of tariffs, Singapore ventures are able to establish the labor-intensive stages of production in Johor prior to adding higher value enhancements in Singapore. The Malaysian government expresses concerns that this evolving core will pull activity away from the Kuala Lumpur-centered development corridor and that too much Chinese investment will be detrimental to the stated goals of economic equality.

DEFORESTATION AND DEVASTATION

Deforestation is one of Malaysia's most pressing environmental problems. From its origins in early swidden agriculture and colonial plantation, mining, and timber operations, deforestation has reached a zenith with land development schemes and logging over the past two decades. Although devastated, forest cover is greater in East Malaysia than in West Malaysia.

One-quarter of the deforestation is attributable to logging. I recall driving from Kuching, Sarawak, to Brunei along a beautiful, modern road, lined with barren hillsides, stacks of enormous logs, and fires burning off debris. The route is punctuated by Malay *kampungs*, squeezed on the edges of coastal towns that are dominated by Chinese shops and fronted with docks piled high with logs and cut lumber. Rivers run brown with eroded material from clearcut plots deep in the once magnificent rain forest. I thought of the polluted and debris-ridden water, the damage to fish, and the resultant flash floods. All along the route, signs announced the coming of palm oil, pepper, or other estate enterprises, along with new housing and industrial sites.

As I looked upon government-built, tin-roofed long houses lining parts of the highway, I thought of the once-proud Dayak hunters and gatherers whose lives have been disrupted and homes displaced

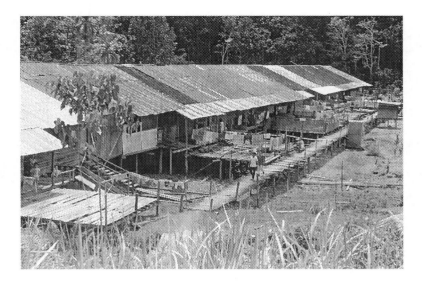

Figure 15–10

An Iban longhouse along the main highway in Sarawak. As deforestation and modernization ensue, many Iban are forced to seek work in the coastal towns. Although they retain their nuclear family structures and unified housing, the influx of income enables their acquisition of such luxuries as metal roofing and televisions. Photograph courtesy of B.A. Weightman.

(Figure 15–10). I also thought of the Penan, another group of hunters and gatherers who unsuccessfully protested the removal of their forest home. Many of these displaced people now work for the logging operations or on plantations.

A striking note is that women play a leading role in direct action protests. While most men are forced into the cash economy, women are left behind to care for villages and families. Women often work part-time in local industries but still must maintain their subsistence agricultural plots. With their diverse experiences and interests, women are driven to participate in protest events against the logging companies.

A sign of the times is a cultural center near Kuching, which has model villages of all the major tribes of the region. This has been developed for tourists. If you want to see the old ways of doing things, you can go there. Traditional ways of life in Malaysia are sadly disappearing in the wake of development. Both environments and peoples are being devastated.

Brunei: Micro-State

Brunei Darussalam—the country's official name—means "Abode of Peace." Territorially, Brunei is about the size of Luxembourg, a mere 3,577 square miles (5,769 sq km). In 1981, the government purchased a cattle ranch in Australia's Northern Territory. The ranch is larger than Brunei. Brunei is a fragmented state, and people go from one part to the other by boats referred to as "flying coffins." These long boats are totally enclosed. Passengers are entertained by programs of Kung Fu genre films.

In the early 1900s, Brunei was known for its exports of *cutch* (or gambier), a dye produced from mangrove trees and used for dying leather. Today, the country is known for its oil, with Royal Dutch Shell as the major company. Oil accounts for half of Brunei's GDP and 90 percent of its exports. So much wealth derives from oil that Bruneians are the wealthiest people in Asia after the Japanese. Brunei is sometimes called the "Shellfare State," as its citizens do not pay income tax, enjoy free education, medical care, and old-age pensions. Bruneians seldom do manual labor; this is done by Filipino and Thai immigrants. More than half of Brunei citizens hold government jobs. Most Malays are well-housed, although there is a long-lived stilt village in the primate city capital of Bandar Seri Begawan

Figure 15–11

Note the contrast between Kampung Ayer and the Omar Ali Saifuddin Mosque in Bandar Seri Begawan, the capital of Brunei. Kampung Ayer is 400 years old and is the world's largest stilt village, housing around 30,000 people, mostly Malays. In 1987, it was declared a national monument and remains Brunei's most popular tourist attraction. Photograph courtesy of B.A. Weightman.

(Figure 15–11). This more traditional village has electricity and running water.

The Sultan of Brunei is one of the world's richest men, reputedly earning $100 a minute from oil. He lives an extravagant lifestyle and has no problem reconciling his lavish spending and partying with the strictures of Islam. Rock star concerts and alcohol consumption in the palace are generally ignored by the populace. After all, "he is the Sultan." To ensure his safety, the sultan maintains a contingent of (British) Gurkha troops in addition to his own army.

In 2000, Malays made up 65 percent of the total population of Brunei. Most of the rest are Chinese and a handful of indigenous groups. About 11 percent are Asian and Western expatriates connected to the gas and oil industries. Brunei has one of the fastest-growing populations in the world. Women continue to have large families, as they are not integrated into industry.

Aside from oil, other primary sectors are not well developed. Brunei is one part of the island of Borneo where most of the forests remain intact. Oil and gas are concentrated at Seria and Kuala Belait. Production began at an onshore field at Seria in 1932, but as of the 1990s, most production is from offshore fields. Reserves of oil and natural gas are expected to last for 40 years. There is little diversification of the oil industry. Downstream processing takes place in facilities at Miri and Bintulu in Sarawak. In terms of patterns of development, Brunei has emerged as a tertiary-based economy with a small primary sector and no secondary sector.

Brunei must import about three-quarters of its food needs. Modern grocery stores are dominated by *halal* foods—foods that are processed according to Islamic law—but the main ones have small "non-*halal*" sections for the benefit of the expatriate community. McDonald's and other fast-food chains are prominent in Bandar.

Singapore: The "Intelligent Island"

Geography is central to Singapore's evolution. This 240-square-mile (600 sq km) island, holding 4 million people, is situated at the tip of the Malay peninsula, separated from the mainland by a narrow stretch of shallow water. It is a pivotal, highly developed

Figure 15–12

Singapore. Note how the urbanized area is oriented toward the sea. From H. J. de Blij and P. O. Muller, *Concepts and Regions in Geography,* Second Edition, 2005, p. 359. Originally rendered in color. © H. J. de Blij and P. O. Muller. Reprinted with permission of John Wiley & Sons, Inc.

island in the Straits of Malacca (Figure 15–12). This island city-state's success is contingent on its strategic location, as well as its relationships with its closest neighbors—Malaysia and Indonesia.

ORIGINS

The British sought strategic ports to counter Dutch power in insular Southeast Asia. They acquired Pinang in 1786, and, in 1819 the British flag was hoisted over Singapore by Sir Stamford Raffles. At that time Singapore, with no more than a thousand people, was a small fishing village

known as Temasek (Sea Town) and later as Singapura (Lion City). Control of Singapore meant that the British East India Company (BEIC) controlled both the northern and southern entries to the Straits of Malacca.

By the 1830s, Singapore had become the major trading port in Southeast Asia. It was challenged by both Manila and Batavia (Jakarta), but the island had certain advantages. Most ships trading between China, India, and Europe had to pass Singapore. Unlike its competitors, Singapore was a free port. Furthermore, it was enmeshed in the powerful British commercial and industrial empire.

Tin played a major role in Singapore's development. Tin miners in Malaya and Thailand imported their supplies from Singapore and used the port to ship their tin to the world. The island attracted traders of all kinds: British, Indians, Malays, Arabs, and Chinese.

The Chinese were the labor force upon which Singapore was built, and the city was the conduit for thousands of Chinese laborers going to Malaya and the Netherlands East Indies. Most Chinese came to Singapore as poor indentured laborers hoping to make a fortune and send money to their ancestral home. The forced opening of the Treaty Ports in China and the annexation of Hong Kong in 1842 (Chapter 9) accelerated the migration of Chinese—not only to Singapore, but worldwide. Singapore remains a predominantly Chinese country with a 78 percent Chinese population.

By the late nineteenth century, Singapore had become a major transshipment point as well as a commercial and financial center. Here, products from Southeast Asia, such as tin and rubber, were received, packaged, and re-exported. The largest commercial firms were British. Growing numbers of Chinese-owned enterprises were linked to the business web of the **Chinese Diaspora**. There was little manufacturing in Singapore before 1960; three-quarters of the population were involved in the service sector.

There was also a significant Indian contingent, fluctuating from 6 to 12 percent of the population. The minority was large enough to form its own community known as "Little India." The Indian community was far from united, however. There were regional, religious, and caste divisions. During colonial times, thousands were imported from India as convict laborers to build Singapore's infrastructure. Free Indians were primarily merchants or worked in public employment as teachers, policemen, and the like. There were strong links between the Indian communities of Singapore and Malaya. By the time of World War II, Singapore was a multiracial, multireligious, multilingual society governed by a British elite.

Sea-oriented Singapore was ill-prepared for a land invasion and fell to the Japanese in February 1942. All communities suffered greatly at the hands of the Japanese during World War II, a fact that has contributed to a legacy of distrust of Japan.

After the war, the British reoccupied the island. Although its vision for Malaya was independence, it looked upon Singapore quite differently. The British feared the Chinese majority in light of the Communist takeover in China in 1949 and the insurgence of communism in the region. It was believed that an independent Singapore would come under Communist control.

Limited self-government was introduced in 1955. In 1959, the People's Action Party (PAP) began its dominance of Singapore's politics. Led by Lee Kuan Yew, the PAP was the party of a new English-educated elite intent on creating a new, industrialized Singapore.

The creation of Malaysia in 1963 included Singapore. However, unresolved issues surrounding the combined power potential of the Chinese in both Singapore and Malaya led to the creation in 1965 of a separate Republic of Singapore. Under the tutelage of Lee Kwan Yew and the PAP, the island transformed itself from an entrepôt economy to an industrial and post-industrial economy in the space of 30 years. Under the strong, stable, and corruption-free PAP, Singapore is a model of planned development.

PLANNING OFFENSIVES

In less than two generations, Singapore has transformed itself from a swampy, overcrowded slum to a modern, gleaming metropolis where rules and regulations control activities and behavior. For example, littering brings a fine of US $500 and chewing gum was forbidden until 2003.

The city is a traveler's paradise with magnificent botanical gardens, a fine zoo, and a plethora of shopping opportunities. As early as 1975, the government introduced a licensing system costing US $3 for cars entering the business district between 7:30 A.M. and 6:30 P.M. on work days. Now there is also an electronic payment system to control traffic on the busiest streets. Moreover, Singapore has an extensive and efficient subway system.

The Land Acquisition Act of 1967 allowed the government to acquire land anywhere, making serious urban planning possible. Before than, the city was plagued with a soaring population and housing shortages. PAP set up a Housing Development Board

(HDB), which cleared slums and built new housing along the Singapore River. Hundreds of Chinese shops (with living quarters above) gave way to multi-unit apartment blocks. In the case of older housing, residents paid only 10 to 20 percent of the cost of refurbishing. The government paid the rest.

The HDB then proceeded to build new towns, complete with high-rise apartment blocks, town centers, markets, schools, and light industries to provide jobs. With government loans, apartments (flats) are sold far below free market prices and now house about 90 percent of the city's population.

Population planning is also part of the grand plan. Prior to the 1980s, strict family size controls and incentives were introduced, with a resultant drop in fertility rates, especially among the Chinese majority. Suddenly, the government realized that there were not going to be enough workers for the growing economy. In 1987, new policies were introduced encouraging people to have more children. While Malays and Indians have higher fertility rates, the Chinese remain the majority at 78 percent of the population. Indians comprise 7 percent and the Malays 14 percent.

Many more plans are in the making. A new mass transit line will connect more new towns. Deep tunnel sewage systems will be placed 65 to 164 feet (20–50 m) below ground to avoid contamination of usable aquifers. Water desalination plants are intended to reduce the reliance on Malaysia for about half of the city's water. The city has also developed drinking water from sewage water. This bottled water is called "NEWater." Producing water is of paramount importance because the contract with Malaysia runs out in 2011. Developers are creating land to join five small islands of the southwest coast into one big island. The new Jurong Island is earmarked for petrochemical industries.

CYBERCITY

Already one of the world's largest hubs for shipping and commerce, this economic tiger is staking its future on becoming a global nerve center for media and communications. The National Computer Board's report titled "IT2000: Vision of an Intelligent Island" proposes an infrastructure that includes both a fiber-optic network and the information services that re-

quire this form of digital communication. Singapore is second to Japan in terms of the number of computers it manufactures. In addition, the island makes more than half of the world's hard disc drives and almost all the sound cards for personal computers.

A new emphasis is being placed on bio-medicine. The island wants to make the biomedical sciences, which include the research and development of pharmaceuticals, biotechnology, medical equipment, and health-care products and services, the "fourth pillar" of its economy. The other three pillars are electronics, engineering, and chemicals.

More than 7,000 multinationals operate out of Singapore, employing 60 percent of the island's work force and producing 80 percent of its exports. Companies regard Singapore's infrastructure as superb. It has the world's largest container port, state-of-the-art utilities, and one of the world's most advanced telecommunication networks.

CREATING A HINTERLAND

Building an external economy is considered a national imperative. Foreigners already make up 20 percent of the labor force, and an overseas push is the only place for expansion. Many companies have already relocated in such countries as India, Thailand, Vietnam, Russia, Hungary, and Belgium. Twenty-two Singapore companies are constructing a huge industrial park near Suzhou (west of Shanghai) in China. In 2000, nearly 60 percent of Singapore's investment was in Asian countries, with 10 percent in Malaysia. In Malaysia, more than half went into financial-services projects and the rest into manufacturing. Cheaper labor is not the only reason for these investments. Singapore is hoping to tap into large potential markets as well.

Singapore's goal is for greater interdependence with its neighbors, especially with Malaysia and Indonesia. The adjacent Malay peninsula and Indonesia's Riau Islands linking Singapore to Sumatra have potential for a development triangle. Water can be obtained from Sumatra, granite from an island west of Batam, and both Batam and Johor can offer cheap labor. All these areas provide land for expansion. Batam has been developed as an industrial estate, agribusiness site, and tourist haven. It is

Figure 15–13
Singapore's modern skyline rises above Chinese shops that have been recently "boutiqued" and now contain trendy restaurants and shops. This is one of the city's deliberately designed heritage landscapes. Photograph courtesy of B.A. Weightman.

the third most visited island in Indonesia after Java and Bali. A second causeway has already been built to Johor where spatial patterns and linkages have effected an extended metropolitan region.

The islands of Bintan and Bulan are also being developed. Bintan is becoming a playground for Singaporeans. It is destined to have 20 hotels and 10 golf courses. Bulan is touted as an agribusiness center. Pig production supplies 50 percent of Singapore's demand for pork. Chicken production is expected to supply all Kentucky Fried Chicken outlets. The island will also have the world's largest orchid farm.

Singapore's growth triangle is neither complete nor proven entirely successful. While these areas are not yet fully integrated, it is still argued that there is more to be gained from cooperation than from competition. This growth triangle is still in the making. The lack of formalization is partly because the growth triangle means different things to each of the participants. According to geographers Scott Macleod and T. G. McGee (1996), "In many ways the strength of the Growth Triangle as a concept lies in its vagueness; it can mean all things to all sides."

FINDING AN IDENTITY

One way in which Singapore is forging a sense of place is through heritage tourism. Certain districts of the city are being refurbished to display their cultural ori-

gins (Figure 15–13). "Little India" and "Chinatown" are examples of this phenomenon. However, localism is combined with globalism, as many buildings now host international business chains such as upscale shops and restaurants. In advertising, the city is being portrayed as having, "its head in the future and its soul in the past." As geographer T. C. Chang (1999) points out, "On the conceptual front, the power relations between the 'global' and the 'local' in Singapore demonstrate that 'globalization' and 'localization' are mutually constitutive." These relations are revealed in the city's carefully orchestrated heritage landscapes.

Language, as the context for thought, is vital to forging identity. In the 1960s, the PAP stressed the importance of Malay and English. As Singapore's prosperity grew and the economy became more internationalized, PAP shifted its emphasis to Mandarin and English at the expense of Malay and other Chinese dialects. Mandarin and English are now the official state languages.

Singlish

Singlish is a bewildering mixture of English, Mandarin, Chinese dialects, and Malay. There is concern that the rise of this dialect will demote the official languages, which are intended to internationalize the state. The artistic community has embraced the dialect enthusiastically, and Singlish is even popularized on television shows. An example is: "Don't pray pray *lah*," which means "Don't fool around." Many people feel that Singlish is their own personal language for everyday communication and "*So shiok*" ("That's great"). New words and expressions are invented almost on a daily basis fueled by the Internet experience. As the debate over Singlish and English rages, it is clear that Singlish is the language spoken on the street.

In the twenty-first century Singapore will be among the few prosperous countries in the world. Its geographic situation may lessen in importance as it becomes enmeshed in the global economy's computer revolution. In regional terms, Singapore regards itself as the economic engine of the region.

The growth triangle is evidence of this view. Nevertheless, Singapore worries about its image and is constantly advocating campaigns extolling what it means to be Singaporean. Its 1990s drive to reveal historic aspects is only part of this continuing search for national identity. However the future is charted, it will be in the context of "Okay can!"

Indonesia: Integration vs. Disintegration

With more than 17,000 islands, Indonesia is the world's largest archipelago (Figure 15–14). With a population of more than 220 million, it is the fourth most populous country in the world. There are more than 350 ethnic groups speaking an estimated 583 languages and dialects. Indonesia also has the world's largest Muslim population.

Indonesia's geography is just as varied. There are vast expanses of tropical swamps in Kalimantan

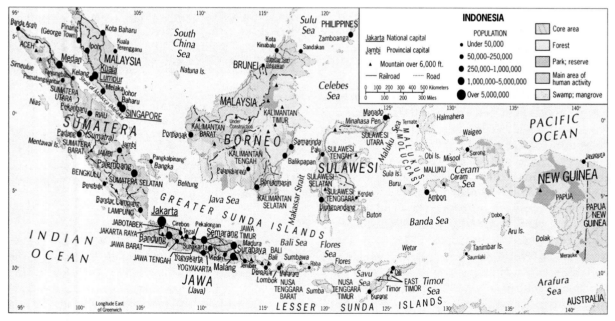

Figure 15–14

Indonesia. Note the concentration of cities on the island of Java as well as the distances between Java and the outer islands. From H. J. de Blij and P. O. Muller, *Concepts and Regions in Geography,* Second Edition, 2005, p. 361. Originally rendered in color. © H. J. de Blij and P. O. Muller. Reprinted with permission of John Wiley & Sons, Inc.

(part of the island of Borneo) and Sumatra and glaciers in the central mountains of West Papua (the western part of the island of New Guinea formerly known as Irian Jaya). There are more than 300 volcanoes, 200 of which have been active in historical time. Indonesia's motto *Bhinneka Tunggal Ika*—"Unity in Diversity"—is fitting indeed.

Because of Indonesia's cultural diversity, most people first develop a regional identity, only learning the national language *Bahasa Indonesia* when they attend school, and with it a national identity. The Indonesian government is fully aware that national unity and national cultural identity must be created.

Indonesia can be divided into the inner islands and outer islands. The inner islands are Java, Bali, and Madura. The outer islands are the rest, such as Sumatra, Kalimantan, Sulawesi, and the islands of Maluku (Figure 15–15). The political, historical, and economic heart of the country are the islands of Java and Bali. These two islands contain 67 percent of the population but only 7 percent of the land area.

THE EARLY YEARS

As you already know, Southeast Asia lies across the great trading routes from India to China. Chinese, Indian, and Arab traders were common sights in the ports that dotted the region.

In the Indonesian archipelago, there were two types of states. First were the coastal states located at river mouths with their secure harbors which were dependent on regional and international trade. Second were the inland states. These were based on agricultural production derived from the rich volcanic soils and alluvial plains. The most prominent states were in central and eastern Java and Bali.

The earliest kingdoms were Hindu/Buddhist states. The religions were adopted by local rulers who were attracted by religious and philosophical ideas as well as court rituals. Yogyakarta is the center of this history and culture. Within 48 miles (30 km) of this city are the great monuments of Borobudur (Buddhist) and Prambanan (Hindu) (Figures 13–5 and 15–16). The island of Bali remains predominantly Hindu.

Founded in 650 A.D., Sri Vijaya was one of several sea-oriented states. Its capital was at Palembang

Figure 15–15

This woman is a Dayak who lives in the interior of eastern Kalimantan. The Dayaks were once head-hunters. Their traditional way of life in the jungle is rapidly changing with resource exploitation, deforestation, influx of migrants, and government impress from Java. Note the filter cigarette. Clove cigarettes are commonly smoked in Indonesia.
Photograph courtesy of B.A. Weightman.

in southeastern Sumatra and its location gave it access to the region's two strategic waterways: the Straits of Malacca and the Sunda Strait. Palembang provided an excellent port for ships riding the monsoon winds from India to trade with China. The city became a center of Buddhist learning and had more than 1,000 Buddhist monks in its monasteries. Sri Vijaya expanded its territory so that it could monopolize trade in the region. At its peak, the empire

Figure 15–16

Evidence of early Hindu kingdoms on Java, these temples at Prambanan have suffered earthquake damage. The 232 temples of Prambanan were built between 900 and 930 A.D. Like Borobudur, they replicate the cosmic Mount Meru. The largest three temples are dedicated to Brahma, Shiva, and Vishnu. Photograph courtesy of B.A. Weightman.

controlled parts of southern Thailand, peninsular Malaysia, western Java, and parts of western Borneo. Sri Vijaya ruled the region's sea lanes and trade until it was defeated by the Javanese in 1290.

Muslim traders arrived as early as the seventh century but the Islamicization of Indonesia began in the thirteenth century with the conversion of the ruler of Aceh in northwestern Sumatra. The nature of Islam in Indonesia varies greatly, ranging from strict adherence among the Acehnese to elsewhere where people have a more relaxed attitude to the faith.

When the Europeans arrived in the middle of the sixteenth century, they found a series of well-established states across the archipelago. The major ones were at Aceh, Solo (Surakarta) and Yogyakarta in central Java, Bali, the Moluccas (Maluku), and Makassar. There was a constant flow of people and goods across the region. Traders used Malay as their means of communication. The Indonesian and Malay languages are similar in the same way that British and American English are similar.

THE NETHERLANDS EAST INDIES

The Netherlands United East India Company (VOC) began to trade with local kingdoms in the early seventeenth century. In 1619, they attacked Jayakarta (Jakarta), then a major fort in the Javan kingdom of Banten. Jayakarta was renamed **Batavia**. It remained the capital of the Netherlands East Indies until independence in 1945, when it was renamed Jakarta.

Total control of Java was not achieved until 1756 after the bloody Java War whereby the Dutch defeated the Yogyakarta sultanate. Several years later, in 1796, the Dutch Crown took over from the corrupt Company and ruled what by then was the largest state in the archipelago. The Dutch extended their control gradually through the nineteenth century. Sumatra and eastern Indonesia came under Dutch rule and, with the destruction of the Balinese kingdom in 1905 and Aceh in 1911, the colony of the Netherlands East Indies was complete.

The Dutch maintained a centralized state. Power was vested in Batavia, and the colonial government kept a close eye on religious leaders. Any stirrings of resistance were quickly quashed. The Dutch promoted a Western secular elite by educating the children of the pre-colonial elites and making every effort to prevent the formation of an Islamic elite.

Regional trading networks were gradually destroyed. External trade became the preserve of European companies and internal trade became dominated by the Chinese, who were encouraged to migrate from southern China.

Javanese agriculture was transformed by the *Culture System* (Chapter 13). Islanders were compelled to produce designated crops and by the end of the nineteenth century Java was the world's largest sugar producer. Village Java's subsistence economies faded to commercialization, and cities and towns emerged to handle the export trade. By the beginning of the twentieth century, most Javanese were tenant farmers, sharecroppers, or wage laborers.

In Sumatra, vast areas of virgin forest were removed to make room for tobacco and rubber plantations. When oil was discovered in the 1920s, it became the nucleus of what would become the Royal Dutch Shell Oil Company. Much of the labor was Chinese, and although they never made up a large portion of the colony's population, they became dominant in local trade and urban commerce.

The economic transformation of Indonesia led to increased urbanization, and by 1910 Java's cities were unable to cope with the flow of rural to urban migration. Urban places were crowded, without piped water or sanitation systems, and people were suffering from malaria and other water-borne diseases. Living conditions for most Indonesians worsened into the 1970s.

The Dutch introduced Western education to provide a needed pool of skilled labor but entry to these schools was very difficult for ordinary Indonesians. The best schools used Dutch as the medium of instruction. Graduates entered the professions or even attended universities in the Netherlands. Most who benefited were children of the elite or government officials. By the end of the Dutch colonial era, literacy rates were lower than any other European colony in Asia, with the exception of the Portuguese colony of East Timor.

NATIONALISM AND SUKARNO

The term *Indonesia* was first used in the early 1920s. By 1928 there were enough educated youth with determination to create a modern Indonesia free from Dutch rule. Although several political parties emerged, the most prominent nationalist was a man named **Sukarno**, who charmed the masses with his brilliant oratory. Sukarno succeeded in spreading the idea of freedom across all levels of society. His most important contribution was the popularization of nationalist ideology. An important theme was that people should set aside their ethnic and religious differences to unite in opposition to colonial rule. It was generally agreed that Indonesia should be a secular state. However, none of this pleased the Dutch who expelled or jailed many nationalist leaders. Sukarno was imprisoned.

THE JAPANESE AND INDEPENDENCE

The Japanese occupied Indonesia in March 1942 and quickly alienated the Indonesians by requiring all males to work for the war effort and treating people with disdain and brutality. However, they removed the Dutch from administrative functions and replaced them with Japanese and Indonesians. Thousands of Dutch were thrown into concentration camps where they perished from torture, disease, and starvation.

Certain Japanese policies eventually served the cause of freedom for Indonesia. The Japanese prohibited the use of the Dutch language and encouraged the use of Japanese and Indonesian. Various military training programs were initiated for young people. Nationalist leaders, including Sukarno, were freed from jail. As the Japanese realized they were losing the war, they promoted moves toward independence. The Republic of Indonesia was declared on August 17, 1945, in Jakarta. The Dutch rejected this notion and a four-year guerrilla and diplomatic war ensued. An agreement was reached to end Dutch colonial rule in 1949.

AFTER THE REVOLUTION

The revolution witnessed the rise of a strong army intent on taking the initiative in leading the country to national unity and modernization. The army continues to play a strong role in Indonesian politics to this day. It insists on its *dwifungsi*, "dual function," to defend the nation from external threat and internal subversion and to promulgate economic development.

Despite the perceived power of the army, it did not come into full power until after 1965. Numerous political parties vied for control of Indonesia, and by 1960 competition was intense. Many were concerned about the influence of the Indonesian Communist Party (PKI). Moreover, in 1965, there were rumors that Sukarno was terminally ill and that the PKI was preparing a coup. On September 30, a group of army officers overthrew the government and took power under the leadership of General Suharto, who proclaimed a government of the "New Order."

Over the next six months, the army vigorously sought out PKI members. At least 400,000 were killed, mostly in Java and Bali. Many of the victims were Chinese. The PKI was destroyed.

Pancasila

With the demise of the PKI, there arose a fear of the power of Islam. The consensus political system has embraced what is called *Pancasila* democracy. *Pancasila* is the five principles adopted by Sukarno in 1945 as the basis for Indonesian ideology and development: Belief in one God; national unity; humanitarianism; democracy based on consensus and representation; and social justice. Its vagueness allows for many interpretations. All students, members of the civil service, and the armed forces must pass exams based on the principles of *Pancasila*.

Despite its great mineral, timber, and agricultural wealth, Indonesia remains a poor country with the largest labor surplus in Southeast Asia. An estimated 40 percent of its workers are unemployed or underemployed. Half of its workers are employed in agriculture. The mid-1997 financial crisis dealt the country a devastating blow, with Indonesia's currency falling in value lower than in any other Asian country. This economic crisis generated political turmoil that saw the fall of General Suharto in 1998. This was followed by a period of *reformasi*, "reform" that encouraged liberalization. Elections were held in 1999 and a new and more liberal government was put into place. Also exacerbated by the economic turmoil were Indonesian-Chinese tensions and criticisms regarding the country's population distribution and transmigration program. After 1997, thousands of Chinese fled the country taking more than one billion dollars of investment with them. Peaceful elections were held in 2004, with a new president—Susilo Yudhoyono—defeating Sukarno's daughter, Megawati Sukarnoputri.

POPULATION DYNAMICS

In the case of Indonesia, economic growth has far outpaced population growth. Women of the 1990s had only half as many children as women did in the 1960s. In part, this is due to the changing status of women and the decline of the extended family (Figure 15–17).

Figure 15–17

These residents of Jakarta have one and two children, respectively. Smaller family size reflects the success of family planning in Indonesia, as well as modernizing trends in terms of the status of women and growing preference for nuclear families. Photograph courtesy of B.A. Weightman.

Figure 15–18

This is the home of a transmigrasi in Kalimantan, the destination of thousands of settlers in Indonesia's transmigration programs. The owner derives a living from growing cloves and collecting rubber and other products from the remaining forest. Although typical housing has open windows with shutters these glass windows reflect outside influence and are a sign of relative affluence, unusual for most migrants. Photograph courtesy of B.A. Weightman.

However, strong family planning programs have been in place since 1971.

The program was started in Java and Bali and was progressively extended to the larger outer islands, where it gained support from local governments as well as Muslim leaders. The program involved hospitals and family planning services, a battery of field workers, and village leaders. It focused on both reducing family size and improving quality of life.

In the last quarter century, there has been nearly a 50 percent decline in fertility. Mortality levels have also been reduced, a function of the improved nutrition of children. Life expectancy has risen but still is not at the same level as in other Asian countries. Maternal mortality remains a problem. Furthermore, there has been an increase in chronic and degenerative diseases such as cancer and heart disease.

Gaining access to medical care is difficult for most Indonesians, especially outside Java. More than a third of Indonesians still do not have access to safe drinking water, and nearly half do not have access to sanitation facilities.

TRANSMIGRATION

Population distribution has long been a concern of Indonesian governments. Even the Dutch had colonization programs to provide pepper estates with la-

bor. *Transmigrasi* started up again in 1969 with the first five-year plan of development. In the next three decades, some 6.5 million people were moved to try to correct population imbalances among the islands (Figure 15–18). The proportion of Indonesians living in Java declined from 66 percent at the time of independence to nearly 60 percent in 1995. This relatively small change is due to the fact that a number of people from the outer islands moved to Java. Java continued to be one of the most densely populated islands in the world.

The aims of *transmigrasi* were to relieve population pressure in crowded areas, to open up new lands, and to increase the presence of Javanese in sensitive areas such as Aceh, West Papua, and Kalimantan. People with farming experience were usually selected for the program, but it was rife with problems.

Lack of funds to prepare areas and assist settlers was one problem. Another was the unsuitability of the land to be developed. For instance, attempts were made to grow wet rice on sandy soil. Some migrants to Sumatra ended up in mangrove swamps. In many cases there were cultural clashes between the immigrants and the original settlers. For example, pig farmers from Bali insulted the sensitivities of Muslims—who regard pigs as unclean—in parts of Sulawesi.

Figure 15–19

In Jakarta, modern buildings form the backdrop for part of the Medan Merdeka *or Liberty Square, one of the largest squares in the world.* Photograph courtesy of B.A. Weightman.

From 1997, there were violent outbreaks between tribal groups and settlers who were cutting down forests for palm oil plantings. More problems arose as illegal loggers and miners chopped down more sections of rain forest. Many migrants tried to move back to their place of origin but no longer had homes or land to return to. A number of these people became part of the 1.5 million internally displaced people in the country. The Indonesian government ended the *transmigrasi* program in March 2001.

URBANIZATION

The dominant population movement in Indonesia today is rural to urban migration. The proportion of Indonesians living in urban areas has doubled during the last 25 years. By 2004, one in three lived in cities. Many people living in rural areas actually work in urban areas. Improved transportation has fostered commuting on a daily basis. Other people spend part of the year working in the city while maintaining a family in a rural community. These are **circular migrants.** Both commuting and circular migration have increased with expanded transportation systems, advances in education, changes in the role of women, and greater industrial development.

Jakarta is Indonesia's megacity, a sprawling metropolis situated on the north coast of Java. Jakarta originated as a fortified port town called Sunda Kelapa. Today, this Special Capital Region of Jakarta has more than 13 million people. As with many primate cities in Southeast Asia, urban growth has spread into the hinterland. In recognition of this sprawl, the government began to refer to the entire urban region as **JABOTABEK** in the mid-1970s. JABOTABEK stands for Jakarta and the contiguous administrative units of Bogor, Tangerang, and Bekasi (Figure 15–19). Another urban agglomeration includes Bandung and its surrounding administrative regions. The Bandung complex is blurring into JABOTABEK.

JABOTABEK

This is the fastest growing area in Indonesia and is becoming a single, integrated urban region with Bandung. JABOTABEK and Bandung consist of nine administrative units. In 2000 the combined population of this region was nearly 25 million people. Of the 17.1 million in JABOTABEK, 13.1 million were urban and 4 million were rural. Increased spatial unification is expressed in socioeconomic corridors of activities including agriculture, residences, and retail trade.

An important part of this mega-urban region is housing development. New housing developments have been stimulated by the high cost of city center land; the lack of affordable housing for middle-income families in the main cities; increasing employment opportunities in urban fringe areas; improved transportation facilities; and an increase in the social and economic facilities provided by real estate developments outside the main cities.

Housing development is in the form of large-scale subdivisions. Towns associated with industrial developments are also on the drawing board. Much of this development is associated with foreign investment. Both government and private enterprises are participating in development. However, the government is losing control to private enterprises which dominate construction projects. The result is the emergence of numerous problems.

Prime agricultural land has been lost to developers. Farmers are reluctant not to sell as demand increases and prices rise. Water supply has become critical—both groundwater and surface water are in short supply. Seawater is encroaching inland as groundwater is depleted. Mangrove habitats are giving way to grand tourism schemes. Displacement of local people is a serious problem as well. While local communities are the last to be told about forthcoming development, they are the first to be removed. So far, the government has not adequately addressed these and other problems of rapid growth in the JABOTABEK region of Java.

Jakarta remains Indonesia's most important metropolitan area. It is both the national capital and the principal administrative and commercial center of the archipelago. The greater part of foreign investment goes to Jakarta and focuses primarily on manufacturing, construction, and service sectors. The city remains an important manufacturing center but current and future growth does and will derive from tertiary and quaternary activities.

In the past decades, the central core of the city has been transformed from residential to higher-intensity commercial and office land use. There are also many new high rise apartments there. However, Jakarta reflects the urban woes characteristic of many primate cities. There is a lack of adequate public housing. Traffic congestion, air and water pollution, and sewage disposal present additional problems. The newly elected government (2004) must move on these issues as the metropolitan population continues to increase.

ECONOMIC TRANSFORMATION

The Sukarno years were ones of stagnation. Under General Suharto, five-year plans were initiated and development ensued. Beginning in 1969, economic development actually outpaced population growth. From 1970 to the early 1980s, economic growth was sustained by global increases in oil prices. However, with the fall in oil and gas prices during the 1980s, the government was forced to shift its strategy. It proceeded to adopt a market-oriented approach. Liberalization of controls and processes of deregulation allowed the economy to become more diversified, and growth picked up again. Indonesia is now much less reliant on its oil and gas sector. Indonesia also has the world's largest gold mine and the second largest copper mine, both in West Papua.

Although agriculture has decreased as a contributor to Indonesia's GNP, rice is still an important commodity. Paddy land has expanded in most regions, but a substantial increase in rice production has been achieved through Green Revolution technology— better hybrid seeds, more fertilizer and pesticides, and better use of irrigation—rather than with increased labor. In the 1990s, the agricultural work force shrank to less than half the overall work force for the first time.

Change in village life is endemic. In the traditional way, all villagers participated in rice cultivation for a share of the harvest. But with commercialization, agricultural work was put on a pay basis, which broke the patron-client relationships between landed and landless peasants.

Growing at more than 10 percent a year, the industrial sector rose to account for 40 percent of the

Unemployment is concentrated in the young, educated group. On the other hand, underemployment could be as high as 40 percent of all workers. Another important aspect not to be ignored is the informal sector, which absorbs labor and redistributes income within the larger economic scheme of things.

The Chainsaw Massacre

Indonesia is home to 10 percent of the world's rain forests. Products derived from these forests such as rattan, plywood, sawn timber, and wood furniture are its most profitable export after oil and gas. The biological diversity of these forests is astonishing—more than 10,000 species of trees, 1,500 varieties of birds, and 500 types of mammals. Conservationists claim that the forests are being decimated by logging companies and transmigration settlers.

Of 356 million acres (144 million ha), 74 million acres (30 million ha) are designated as protected; 49 million acres (20 million ha) have been set aside for national parks and wildlife preserves; 77 million acres (31 million ha) are "conversion forests" for settlers; and the remaining 158 million acres (64 million ha) are earmarked for logging. However, millions of acres (ha) not so designated have already been deforested or converted for use by small land owners. Others claim that the logging sections are not always suitable for logging.

Indonesian loggers are granted 20-year concessions to log in areas ranging in size from 49,000 acres (20,000 ha) to 247 million acres (100 million ha). Many of these concessions are connected to Chinese or military interests. Selective cutting is supposed to be practiced in order to ensure a regenerative growth to allow another cutting in 35 years. But the realities are often different.

The World Bank claims that Indonesia's forests are disappearing at the rate of 247 million acres (100 million ha) a year. About a quarter of that amount is attributable to poor

Figure 15–20

Here in a Dayak village in Kalimantan, a chief wears traditional garb specifically for the benefit of tourists. Tourism is an important source of Indonesia's foreign exchange.
Photograph courtesy of B.A. Weightman.

economy in 2000. Manufacturing plays a large role in this increase. Industries include petroleum refineries, textiles, food products, tobacco, chemicals, and wood products. Indonesia's service sector is also growing, employing 39 percent of the labor force in 2000. Tourism, once one of the country's principal sources of foreign exchange, has fallen off dramatically since September 11, 2001. (Figure 15–20).

Much of the manufacturing growth is in and around Jakarta and Surabaya. Nevertheless, in terms of numbers of firms, increases were greater in the outer islands such as Sulawesi.

Figure 15-21

This Kalimantan forest fire has been burning since 1983. More than 9 million acres (80,000 ha) of forest and its resident wildlife have been destroyed.
Photograph courtesy of B.A. Weightman.

logging practices and over-logging. Slash-and-burn cultivators, forest fires, and a variety of development projects such as palm oil plantations have also contributed to forest losses. There are an estimated 2 million shifting cultivators, although their methods are sustainable when land is freely available. Illegal logging is rampant in all designated forest environments. Many experts have said that Indonesian logging practices are not conducive to sustaining its forests.

Fire has been a major destructive force. As of 1997, according to official estimates, fire had swept through more than 198,000 acres (80,000 ha) in Kalimantan, Sumatra, Sulawesi, and West Papua. In 1982 to 1983 a blaze eradicated more than 9 million acres (3.6 million ha). Vast areas of Kalimantan, with its underground coal deposits, still burn. A blackened wasteland is the result (Figure 15–21). The blazes reach temperatures that roast the subsoil and kill off root systems and microorganisms to a depth of 6.6 feet (2 m). You can walk on this smoldering landscape and have the soles of your shoes melt. This is what happened to me! Forest fires in Indonesia often affect the air as far north as mainland Southeast Asia and the Philippines where visibility has been cut to 50 feet (15 m)

and the sun seems to disappear. Satellite photos reveal that many of these fires are deliberately set by logging and plantation companies.

One well-known endangered animal is the orangutan—"man of the forest." This animal is native to the forests of Sumatra and Borneo. They are largely vegetarian, so destruction of the forest is destruction of their habitat and food source. Adult orangutans are frequently shot and their offspring are taken as pets or to be sold to zoos. Conservation efforts exist in both Malaysia and Indonesia, but these are small in relation to the actual devastation of these and other precious animals.

NATIONAL INTEGRATION

National integration is of primary concern in Indonesia. After Suharto's downfall in 1998, there was a power vacuum and a fear of disintegration. In many outlying areas, local autonomy or independence is seen as the panacea for long-felt ills, and *reformasi* has allowed long-seething discord to surface.

In northern Sumatra, there have been violent clashes over land rights between farmers and security forces protecting the land rights of plantation owners. In southern Sumatra, there has been fighting between the locals and transmigrants. In western Kalimantan, the Dayaks have fought with Madurese migrants. On the island of Ambon in Maluku, Christians have clashed with Muslims. Christians and Javanese Muslims have also clashed on the island of Sulawesi.

Since 1998, the government has recognized the need for decentralization and more autonomy for Indonesia's different regions. Autonomy would allow outlying regions to make their own arrangements with foreign investors and set their own development priorities. This could spur voluntary migration from overcrowded Java. On the other hand, these areas are typically rich in natural resources and the revenue they generate is needed by the central government in Jakarta. Autonomy is not likely.

Take Kalimantan as an example. This province is rich in coal resources, most of which lie under primary forest. Indonesia's oil is expected to run out in the next 20 or 30 years and coal will be an essential fuel. What will happen to the people and the environment of Kalimantan? Kalimantan will remain an essential part of a developing Indonesia.

Two very important regions of dissent are Aceh and West Papua. Aceh, in northern Sumatra, and West Papua, in the western half of New Guinea, both have had long-running separatist movements.

Aceh

About 3.5 million Acehnese live in the northern Sumatran province of Aceh. The region is important for oil, gas, gold, tin, and coal. It also produces coffee, pepper, rice, tobacco, rubber, and timber.

Aceh was where Islam first gained a foothold in Indonesia, and it was the last part of the archipelago to fall to Dutch rule. After a difficult colonial rule lasting 69 years, the Dutch withdrew from Aceh in 1942. Aceh was persuaded (bribed) to become a part of an independent Indonesia in 1949. Since then, many Javanese have been moved to Aceh in an attempt to integrate the region into the larger Indonesia. Strict Acehnese Muslims resent the more religiously relaxed and alien Javanese.

Complaining that Indonesia was sucking the resources out of Aceh without fair revenue return, the Aceh-Sumatra National Liberation Front was formed in 1976. The movement is also known as *Gerakan Aceh Merdeka* or "Free Aceh Movement." In that same year, Aceh declared itself independent. Since then, thousands of Acehnese have been killed by the Indonesian army. Since 1992, thousands more have simply disappeared. There are more than 16,000 recorded orphans in Aceh. Furthermore, about 15,000 Javanese migrants have been driven out of areas where they have lived since the 1970s.

Human rights are systematically violated in Aceh by the Indonesian army. The people are forbidden to use their own language in written form. There is no freedom of expression. Workers are prevented from forming trade unions. They work for foreign companies at below-subsistence wages. Hundreds of schools and other public buildings have been burned and 70,000 refugees driven into impoverished camps along the northern coast. Villagers remain poor, even though Aceh accounts for 15 percent of Indonesia's exports.

Reformasi has made some difference in the situation. The army has backed off but the Aceh *Merdeka* movement remains strong. Weapons are being smuggled in from mainland Southeast Asia and the movement is well-funded from outside sources. Although independence for Aceh is "nonnegotiable" according to the government, many Acehnese do not recognize the legitimacy of Indonesia. In mid-2003, a negotiated settlement was brokered by a Swiss team. The Achenese were to give up seeking independence and the Indonesian government would consider their demands. However, the situation remains unresolved and violence continues even after the devastating impact of the tsunami in 2004.

West Papua

In West Papua, there is a 13,500-foot (4,050 m) mountain thrusting out of the crocodile and malaria infested swamps of the glacier-capped Sundirman Range. Here is a hunk of rock called Grasberg that holds the world's richest copper and gold deposits. The Grasberg (estimated worth of US $50 billion) is owned by the mining company of Freeport McMoRan Copper and Gold of Louisiana. The mine was one of Suharto's first development projects and is Indonesia's single largest source of tax revenue.

Since the Freeport mine opened, the lives of the indigenous inhabitants have been radically altered. Tribes have received little if any compensation for their lands. Indonesian law requires indigenous peoples to relinquish customary rights over their land in the national interest. After riots in 1996, the mine implemented a scheme to parcel out 1 percent of their revenues to the 20,000-strong Melanesian population. But the scheme collapsed due to corruption and mismanagement among provincial officials.

The Freeport mine has provided 18,000 jobs, but only 1,500 of these are filled by West Papuans, and only 400 are filled by local people. In addition, locals are paid only one-seventh as much as other employees. There have been intertribal conflicts. Alcoholism and prostitution abound in the nearby town.

On the mountain's eastern border lies the Lorentz National Park, a 6-million acre (2.5 million ha) expanse whose biotic richness is incomparable. The park's ecological integrity is endangered. Lava flows containing the mineral deposits also exist in the national park. The mountains are rich in timber and oil and gas deposits lie offshore. How is this area to be protected from exploitation?

Freeport has one of the world's worst environmental records. It dumps tailings and other chemical and toxic wastes into local rivers, and the mining operations leave behind a bare patch large enough to be viewed from space. Tailings devastate lowland vegetation by choking the life out of the plants. Trees are killed by tailings-laden floodwaters. Ninety-seven percent of ore ends up as tailings, which contain arsenic, lead, mercury, and other dangerous metals. These have killed fish, poisoned sago forests (a traditional food source), and made the water unsafe to drink.

Some scientists claim that the vegetation will regenerate on its own. Experiments are taking place to find out what crops might grow on the tailings. World pressure has forced Freeport to put more resources into environmental considerations.

West Papua is an Indonesian province but it lies in the Pacific geographic realm. Along with the ethnic Papuans, about 200,000 Indonesians from other parts of the archipelago, especially Java, inhabit the province. Most of the non-Papuans are concentrated in the capital of Jayapura (a non-Papuan name). Local opposition against Indonesia's rule has continued for decades. For example, the *Organisasi Papua Merdeka* (Free Papua Movement or FPM) held rallies in Jayapura in 1999 and, in 2000 demanded independence. They went so far as to display a Papuan flag—the Morning Star. Later that year, Indonesian troops killed demonstrators on West Papua's southern coast. Papuan animosity towards other Indonesians and the Indonesian government is deepening and there is no solution in sight.

RADICAL ISLAM

For centuries, the Indonesian version of Islam was a moderate and tolerant one. Many Indonesians were "nominal" Muslims holding only loosely to Islam's precepts. In recent years, however, radical Islamic movements have arisen. As the West's war on terror reached into Indonesia, the government was pressed to investigate possible al-Qaeda links. Terrorists responded in 2002 by bombing a nightclub in Bali. More than 200 people, mostly vacationing Australians, were killed. A Marriott hotel and the Australian Embassy in Jakarta have also been bombed.

An organization called Jemaah Islamiah is said to be responsible for these and other terrorist acts. Jemaah Islamiah adheres to the Wahabi sect of Islam, a strictly fundamentalist form of the faith imported from Saudi Arabia. This and other fundamentalist groups want *sharia* (Islamic law) applied as the law of the land, as it already is in Aceh. Mosques and religious schools promote hatred of the United States and Israel. More than 30 people linked to Jemaah Islamiah have been convicted and sentenced to prison for their part in the Bali and Marriott bombings. The group's suspected leader, Abu Bakar Bashir, is in jail in Jakarta awaiting trial on terrorism charges.

Once regarded as a haven for moderate Muslims, Indonesia is in the throes of a struggle between moderates and conservatives. The moderates seek pluralism and accommodation, a form of Islam that steers away from the sort of orthodoxy that discriminates against women and non-Muslims, and embraces democracy. The conservatives want Islam to control all aspects of all citizens' lives.

Orthodox Islam appeals to the poor and also to the middle class, many of whom have a western education. The transition to democracy since 1998 has done little to improve the lives of the vast majority of Indonesians. The government claims to have stabilized the economy and secured an annual growth of almost four percent. But critics point to rising prices and rampant corruption. At the same time, political freedom has allowed proponents of conservative Islam to promote their views.

The outcome of the struggle has profound implications, not just for security, but also for economic progress and political stability in the region. Should the conservatives prevail, it could provoke conflict with the country's sizeable non-Muslim community (more than 20 million Christians for example), produce a foreign policy more critical of the West, and an investment environment less friendly to non-Muslim investors. It would also ensure a steady stream of potential recruits for radical groups like Jemaah Islamiah.

PROGNOSIS

As Indonesia's resources are increasingly eroded and as population growth continues, it will be difficult, if not impossible, to reign in the outer resource regions to the will of the industrial core of Java. It also will be difficult, if not impossible, to suppress the many self-determination movements across this vast region. It also will be problematic to control the progress of conservative Islam. These are only three of many problems facing the newly elected government and President Susilo Bambang Yudhoyono.

East Timor: New Nation

Timor is the easternmost of the lesser Sunda Islands. The eastern part of Timor was a Portuguese colony until it was overrun by Indonesia in 1975 and annexed in 1976. Subsequently, East Timor became the scene of a bitter struggle for independence. Tens of thousands of Timorese were killed by the Indonesian army. The infrastructure was destroyed and thou-

sands of people were dislocated. Foreign intervention, led by Australia (once Indonesia's only supporter of its claim to East Timor), brought some semblance of order. East Timor became the independent nation of Timor-Leste in 2002.

As Figure 15–22 reveals, sovereignty has created geo-political complications. The political entity of East Timor comprises a main territory containing the capital of Dili, along with a small exclave on the north coast of Indonesian West Timor. This exclave is called Ocussi. Although there is a road from Ocussi to the main territory, relations between Indonesia and East Timor may not allow this road to be used, so that the two parts of the state would have to be connected by boat traffic only.

A second problem has to do with oil. South of East Timor is an area called the *Timor Gap* that contains major oil and gas reserves. While Indonesia ruled East Timor, Indonesia and Australia agreed on a maritime boundary greatly favoring Australia in 1972. Once East Timor became independent, it wanted to get rid of this boundary and replace it with one based on United Nations recommendations. The United Nations boundary would be midway between East Timor and Australia and give the bulk of the resources to East Timor. Australia rejected this proposal. However, in 2004, the two countries hammered out a proposal to develop the Greater Sunrise gas field. Revenues will be shared with East Timor gaining US $3.5 billion in tax and royalty payments. The problem of the maritime boundary has yet to be solved.

East Timor is one of Asia's poorest countries with 42 percent of its people living below the official poverty line in 2004. The population is growing at a rate of 2.1 percent. The total fertility rate is 3.7 and 38 percent of the population is under 14 years of age. In 1999, anti-independence militias, along with the Indonesian army, destroyed the country's entire infrastructure. Homes, schools, irrigation and water supply systems, and the electrical grid must be rebuilt from scratch. East Timor has an array of natural resources but few are developed. A quarter of East Timorese are involved in agriculture and coffee is the main export. More than half of those employed work in the service sector. Industry accounted for only 17 percent of the economy in

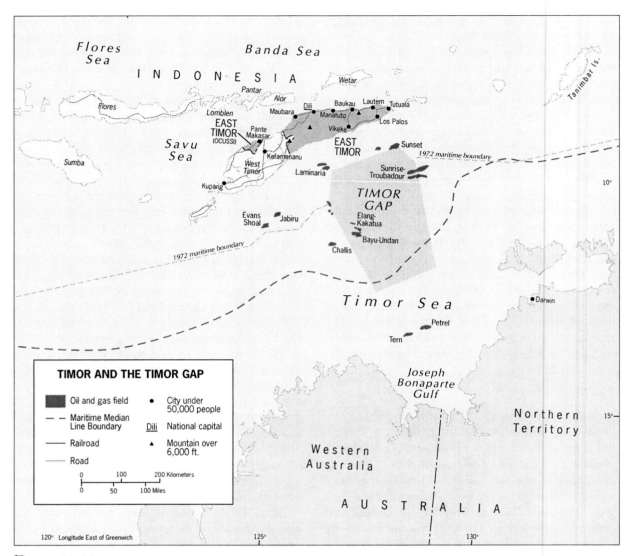

Figure 15–22

East Timor and the Timor Gap. From H. J. de Blij and P. O. Muller, *Concepts and Regions in Geography,* Second Edition, 2005, p. 363. Originally rendered in color. © H. J. de Blij and P. O. Muller. Reprinted with permission of John Wiley & Sons, Inc.

2004. Unemployment and underemployment are widespread—as much as 50 percent. However, there is the possibility of new jobs in the production of coffee and vanilla, and in the oil industry. Problems with the development of East Timor are extremely challenging. As of 2005 the country was relying on international assistance.

The Philippines: Pearl of the Orient Seas

The Philippine archipelago began to take shape about 50 million years ago, the result of volcanic eruptions and the buckling of the earth's crust. The archipelago, comprising more than 7,000 islands,

Figure 15–23

The Philippines. Note the focus of activity on Luzon and the Visayan Islands. From H. J. de Blij
and P. O. Muller, *Concepts and Regions in Geography,* Second Edition, 2005, p. 364. Originally rendered in
color. © H. J. de Blij and P. O. Muller. Reprinted with permission of John Wiley & Sons, Inc.

stretches 1,100 miles (1,770 km) forming a land
chain between the Pacific Ocean and the South
China Sea. Only 154 of these islands exceed 5 square
miles (14 sq km) in area, and 11 of them contain

about 95 percent of the total population. Luzon
and Mindanao are the largest islands. These
bracket a regional grouping of islands, known as
the Visayas (Figure 15–23).

Rainfall is variable because of the mountainous nature of the islands and shifting wind directions. On Luzon, for instance, rainfall ranges from 35 to 216 inches (84-549 cm) a year, Manila is wettest during the southwest monsoon, which delivers 82 inches (208 cm) from June to November. The islands also straddle the typhoon belt and experience about 15 cyclonic storms yearly which are increasingly violent and destructive. Active volcanoes and earthquakes can also wreak havoc.

In June 1991, Mount Pinatubo erupted and devastated an area of 154 square miles (400 sq km) with volcanic ash. Fortunately, predictions allowed precautions to be taken, and only 250 people died in the event.

Knowledge of the Philippines' prehistory is sketchy, but Hominid remains estimated to be 30,000 years old have been discovered on Palawan. Malay-related people of Mongoloid descent are predominant today. Other groups are Negritos of obscure origin. Chinese trace their origins back to traders of the tenth century. Five centuries later, Islam from Borneo and the Sulu Islands diffused as far as Mindanao, only to be stopped by the arrival of the Spanish Christians in the sixteenth century.

Philippine society is complex, with at least 60 ethnic groups among the Malay majority. Chinese, Americans, other Asians, and indigenous people are also part of the population mix. Several languages are spoken, with **Tagalog**—the language of the people around Manila Bay—being the official language. However, English is the lingua franca throughout the islands.

THE SPANISH PHILIPPINES

Ferdinand Magellan claimed the Philippines for Spain in 1521. Magellan came upon the islands during his circumnavigation of the world. He never completed that journey, however, and was murdered in the Philippines. Only one of his ships managed to complete the entire voyage.

Spain controlled the Philippines for the next 377 years, with the familiar twin goals of spreading the Christian faith and setting up trade. The first settlement was on the island of Cebu in 1565, but headquarters were moved to Manila in 1571. The islands were named after the Spanish Crown Prince Felipe.

The economic base of Spanish occupation of the islands was the galleon trade between Macao on the south China coast, Manila, and Acapulco in Mexico. Great galleons left Acapulco for Manila laden with silver. In Manila the silver was exchanged for silk brought from Portuguese Macao. The silk was then taken to Acapulco and on to Europe.

In the seventeenth and eighteenth centuries, Manila became a rival of the major colonial port of Batavia (Jakarta). By this time Manila was in many ways a Chinese city. Chinese traders and *mestizos* (Chinese-Filipino or Mexican-Filipino mix) organized the entrepôt's trade and soon became the most influential political and economic force in the Philippines.

Power was vested in Luzon in the north. Efforts to control the Muslim-controlled Sulu archipelago and Mindanao were repeatedly rebuffed by the Islamic sultanate. These regions were not incorporated into the Philippines until the American takeover at the end of the nineteenth century.

Spanish control was based on a close relationship between church and state. Jesuit, Dominican, and Colombian friars were sent into the countryside to proselytize and at the same time establish plantations. Today, more than 80 percent of Filipinos are Roman Catholic. Eventually, a state-controlled tobacco monopoly was established in northern Luzon with local people forced to provide the labor.

As the galleon trading monopoly came to a close, the islands were open to private investors and traders. In 1785, the Spanish established the Royal Philippine Company to invest in export crops such as sugar, coffee, indigo, and pepper. By exporting these crops, the Philippines came into direct competition with the Netherlands East Indies. Furthermore, British and American interests had become prominent. By the nineteenth century, British-American export houses dominated the export economy. The volume of trade increased 15 times between 1825 and 1875, with major exports being sugar, coffee, tobacco, and abaca—the strongest vegetable fiber obtained from trees in the banana family.

The incorporation of the Philippines into the world economy had two chief results. First, it saw the rise of Philippine nationalism, as well as the emergence of a modern nation state. Second, it created regional economic, political, and social forces that would ultimately weaken the country. Powerful regional elites became the critical social forces of the twentieth century.

Philippine nationalism was the earliest of its kind in Southeast Asia but it had little impact on the other colonies. The Chinese *mestizos* dominated ownership of export plantations. The haciendas (large estates) developed by powerful regional families were worked by tenant farmers. The landowners became richer and the tenants became poorer. From this base arose the powerful landed families that controlled politics (a situation that lasted even into the 1990s). This elite educated their children in Spanish schools, seminaries, and universities. These Spanish-educated children, known as *ilustrados,* were influenced by liberalized policies in Spain. They were anti-clerical and demanded separation of church and state. As their demands were ignored, they began to call themselves **Filipinos**. The name *Filipino* became a symbol of nationalism. In 1896, a rebellion erupted.

THE AMERICAN PHILIPPINES

At the same time Spain was dealing with open rebellion in the Philippines, it was at war in Central America and Cuba. U.S. intervention in Cuba resulted in the Spanish-American War. The Americans sailed to Manila, smashed the Spanish fleet, and laid siege to Manila. Philippine opportunists seized the opportunity to declare independence in 1898. The Filipinos were the first people in Asia to fight their colonial oppressors and declare a nation-state. However, the Americans were not done with the Philippines.

In 1899, war broke out between America and the Philippines. Hostilities ended in 1901, when the Americans effectively bought off the *ilustrado* elite, promising to maintain their wealth and power in return for cooperation with American colonial rule.

The United States occupied the Philippines with a "democratic mission" of leading the Philippines into self-rule. The Tydings-McDuffie Act of 1934 promised independence in ten years. The Americans stressed the development of health, education, and democratic processes. In 1941, the literacy rate was

five times that of Indonesia. In 1935, a Democratic Republic was established with Manuel Quezo as its first president. Government leaders were from the landed elite. While espousing complete independence, their political and economic interests were well served by American rule.

World War II interrupted the Philippines' march to independence. Fighting in the Philippines was more intense than elsewhere in Southeast Asia. However, Japanese slogans such as "Japan, the light of Asia" had little attraction among Filipinos, and resistance to the Japanese was strong. Filipino nationalists welcomed the American liberators, who promised independence by 1946.

INDEPENDENT PHILIPPINES

Independence changed very little in that the landed elite maintained control. Patron-client relations and vested links between rich landowners and poor peasants have continued to underpin the political process. Although rebellions have been crushed, rural discontent and unrest remain problems.

In the 1950s and 1960s, a new industrial class emerged. Some were from the landed elite who were diversifying their capital. Others were professional traders who had merged interests with American or local Chinese businessmen. By the 1960s, the Philippines was the most successful manufacturing country in Southeast Asia. Agribusiness, real estate, and banking were the basis of several multinational conglomerates. Unfortunately, subsequent government policies spoiled the record and the economy became stagnant.

During the rule of Ferdinand Marcos, who proclaimed martial law in 1972, per capita income fell, foreign indebtedness rose to serious levels, and the middle class protested corruption and the inability of the government to handle the country's social ills. Moreover, Marcos, his friends and relatives, and the military practiced *crony* capitalism—absorbing private businesses and state enterprises—on an unprecedented scale. Family conglomerates controlled 78 percent of all corporate wealth in the country. Land reform and the end of rural poverty remained sheer rhetoric.

By the 1980s, the nation was fragmenting. The Moro Nationalist Liberation Front (MNLF) had 50,000 to 60,000 guerrillas fighting for an independent Muslim state in Mindanao. Half the Philippines' army was fighting that battle. Most of the others were fighting communist-controlled guerrilla attacks elsewhere in the country. Opposition to Marcos became strident. He called an election in 1986 and lost to Corazon Aquino, wife of assassinated popular politician Benigno Aquino. Marcos ultimately fled the Philippines.

Although the government returned to its 1970s' democratic form, little changed in the social structure. The landed elite retained their riches, land reform was minimal, and political loyalties remained personal rather than institutional. However, there was an impressive economic turnaround.

ECONOMIC PROGRESS

Ferdinand Ramos took power in 1992, and by the mid-decade the Philippines had reached growth levels approaching those of other ASEAN countries. However, the country has a long way to go. Although the agricultural sector has remained relatively constant, with 17 percent of the total economy, it employs 35 percent of the labor force. Industry and manufacturing account for 20 percent of employment while services account for 45 percent. But still, a third of the population is unemployed or underemployed.

Subic Bay

In 1992, under pressure from the Philippine government and people, the United States gave up the Subic Bay Naval Base and Clark Air Base, both located on the island of Luzon. Subic Bay alone had employed 32,000 local people and indirectly created 200,000 jobs. However, it also imported some negative aspects of American culture, such as sex work and rampant consumerism. Soon Subic Bay was turned into a free trade zone, and 200 companies—including many from Taiwan—pledged investments in the zone. In 1995, the American company Federal Express established Subic Bay as its air hub for overnight parcel deliveries in the region.

All has not gone as planned for Subic Bay. With the economic crisis in the late 1990s, many Taiwanese, Japanese, and other potential investors backed away from their commitments. Subic's largest investments are tourism-related but poor transport from Manila has limited domestic use. Subic's development is also hampered by a lack of space. The complex is blocked by jungle and the ocean. The airport runways, designed for fighter aircraft, cannot handle large passenger planes, as the runways are lined on three sides by water. Current plans involve expanding the harbor to provide a more convenient international port than Manila.

The traditional industrial base was the processing of agricultural products, but this was altered in the 1990s with a policy of import substitution. For example, the Philippines' largest company, the San Miguel Corporation (of beer fame) diversified into food and soft-drink processing and invested in coconut products, banking, and other commodities.

Recent emphasis has been placed on export-oriented industrialization. Export-processing zones (EPZs) are located near Manila, Cebu, and in Baguio. In addition to Subic's free port, Clark Air Base is now a special economic zone. Numerous private enterprises have been established in other areas, mainly in Metro Manila. Today, the leading exports of the Philippines are apparel and electronics. Gateway Electronics has developed a business park, and numerous chip manufacturers have gravitated to the site. Texas Instruments, Motorola, and Philips have also established industrial sites near Manila.

Manufacturing is highly concentrated in Metro Manila. Its metropolitan area accounts for more than half the total Philippine manufacturing employment. The next largest centers are Cebu and Davao.

Cebu is a world center for the manufacture of rattan products. Nearby locations such as Mactan

Island—where Ferdinand Magellan was killed during the first circumnavigation of the globe—include an international airport and an export-processing zone. Mactan has also been developed for tourism, with fine hotels and beach resorts. One-third of all foreign tourists, mostly Japanese and Taiwanese, visit Cebu and Mactan. The Japanese are the major foreign investors in the EPZ. One of the zone's first factories was Timex. Now 80 percent of all Timex watches are made in the Mactan plant. Other companies produce semiconductors, cameras, furniture, apparel, and so forth.

Another important manufacturing center is General Santos City on the southern coast of Mindanao. The major industries here are food processing, especially of fish. It is the largest producer of tuna, most of which is exported to Japan as sashimi. The agricultural region surrounding the city accounts for 30 percent of the nation's maize production, 10 percent of its coconuts and copra, 20 percent of its rice, 99 percent of its asparagus, and 40 percent of its pineapples. One of the largest employers is the Dole Corporation, involved in everything from pineapples and bananas to vegetables.

Another economic scheme is an effort to link Brunei, Indonesia, Malaysia, and the Philippines into an East Asian Growth Area (BIMP-EAGA). These countries are examining ways to enhance trade in the region with the support of the Manila-based Asian Development Bank. A new international airport, expanded harbor facilities, and a new EPZ are part of the future of this new economic growth triangle.

OVERSEAS FILIPINOS

Foreign exchange revenue inflows are an important element of Philippine earnings. With millions of workers deployed in over 190 countries and territories, billions of dollars come into the country in the form of remittances every year.

The first significant outflow of Philippine workers was to Hawaii to work on plantations. Others have worked in U.S. defense industries in Guam and Vietnam. These *balikbayans* (overseas Filipinos) remain important sources of revenue for the Philippines. More than 90 percent (around 1.4 million) live in the United States, Canada, and Australia.

Many go to work in the Middle Eastern oil fields of Saudi Arabia, Kuwait, and the United Arab Emirates.

Adjacent Asian countries also offer opportunities for employment but these are for low-paid domestics and "entertainers." Geographer James A. Tyner (1996) has found that the marketing and recruiting strategies of the Philippine government, especially in promotional literature, present Filipinos as loyal, disciplined, and obedient and that promotional materials draw attention to gender, associated roles and occupations. For example, women are portrayed as nurses and men as construction workers. Tyner notes that, "Global marketing of workers is organized around specific gendered assumptions of male versus female occupations. The use of gender by government and private institutions as an organizing device provides a gendered context in which migrants and their households must subsequently operate." Gendering of jobs is also most pronounced in Japanese firms. Unfortunately, many Filipino foreign workers find themselves turning to prostitution to pay labor recruitment fees, and otherwise suffer abuse, unable to save up enough money to go home.

ONGOING PROBLEMS

The Philippines has many problems to overcome. Environmental damage is severe, with diminishing forests and damaged coral reefs. Little has changed for the poor in either urban or rural areas. Manila is marred by vast squatter settlements and lack of adequate housing. Wealthy families still control the land and tenant farmers remain impoverished. Income differentials are huge—some of the largest in Asia—and the middle class remains small. According to the WB, nearly 40 percent of Filipinos still live in poverty.

Another pressing problem focuses on the Philippines' small Muslim population that is concentrated in the southeastern part of the archipelago (Figure 15–23). Muslims feel that they are marginalized in this predominantly Christian country. Over the past three decades, several Muslim organizations, including the Moro National Liberation Front and the Moro Islamic Liberation Front, have promoted the Muslim cause with tactics ranging from peaceful negotiation to violence. Basilan Island has become the base for the extreme Abu Sayyaf which receives support through al-Qaeda. The United States has sent troops to help the Filipinos pursue these terrorists. Deadly bombings in southern towns have forced the Philippine government to spend a large share of its budget on crushing the insurrection. In addition, a communist uprising in 2004 required a response in difficult, remote mountain terrain where military operations are very expensive.

Conclusion

Clearly, the countries of insular Southeast Asia are moving along the path toward modernization. Singapore has achieved this end and is well entrenched as a fully developed country in the world of transnationals and cyberspace. However, it is now challenged to preserve its historic landscapes. Malaysia is challenged by the problem of uneven development exacerbated by its fragmented nature. Environmental destruction is taking its toll, especially in Sarawak and Sabah, where traditional lifestyles are being destroyed. Both Indonesia and the Philippines are fraught with problems deriving from their fragmented nature and their ethnic and religious complexities. Intergroup rivalries are detrimental to a smooth transition to more even economic development in addition to political freedom and self-determination for all concerned. Furthermore, environmental problems overshadow all forms of progress.

Main Point Summary

1. With five nations, this is one of Asia's most complex regions.
2. The course of Malaysia's history has been influenced by its strategic position, which makes it a meeting place between east and west. Sea trade was the basis of Malaysia's history.
3. The earliest inhabitants were the Orang Asli to be followed by the Malays. The Straits of Malacca served as a corridor of interaction between mainland and insular regions.
4. Trade with China had an early history in the region. The Peranakan remind us of the importance of Chinese settlers in the region's commerce.
5. Islam was introduced by Arab traders and became a major influence after 1400.
6. European influence was limited until the coming of the British at the end of the eighteenth century.
7. The British established themselves throughout Malaysia although this political entity did not come into being until the twentieth century.
8. Tin mining and rubber production dominated the economy. Both Indians and Chinese came to work these enterprises. These activities laid the basis for Malaysia's core region on the southwest of the peninsula.
9. East Malaysia's economy was for a long while dominated by indigenous people who dealt in jungle products. By the late nineteenth century, they were transferring these goods to Europeans who were participating in a growing world economy. Development was coastal.
10. Although Malaysia has a plural society, pluralism is a historical construct. Different groups were pigeonholed in specific functions. Modern development in Malaysia stresses the development of all groups, especially the native Malays known as *Bumiputera*.
11. FELDA plans have sought to redistribute population in rural areas. The focus has been on *Bumiputeras* and associated with land distribution and plantation crops such as rubber and palm oil. Modern FELDA schemes are more like agro-business operations. The program no longer exists as of 1991.
12. Malay populations grow at faster rates than other populations such as Indians and Chinese. Malays are now a dominant majority, and more Malays are moving to urban areas.

13. Urban populations are still unbalanced with west coast states accounting for the most people. In East Malaysia, populations are mainly coastal yet urbanization is relatively low given the lack of an urban-based manufacturing economy.

14. The paramount urban conurbation includes urban areas from Kuala Lumpur to the port of Kelang. This is the core of the country.

15. Kuala Lumpur has a modern landscape with a mix of modern and colonial buildings. Housing is varied and includes many government apartment buildings and few squatter settlements. The city shows many signs of the country's Islamic culture.

16. Malaysia's national economy is now anchored by high-value manufacturing. It has an outward-looking industrialization strategy. Pinang has become the country's "Silicon Island." Foreign investment, especially from Japan and Taiwan, are critical to continued growth.

17. Malaysia is involved in the development of a "growth triangle" involving the state of Johor and Singapore whereby Singapore can garner less expensive land and employ Malaysia's cheap labor.

18. Deforestation is a pressing problem. Much primary forest has been lost to plantations, logging, and land settlement schemes. Logging in East Malaysia has removed one-third of its forests and displaced many indigenous peoples.

19. Brunei is a fragmented micro-state that derives its wealth from oil. A Muslim country, it is ruled by a sultan who is one of the world's richest men. Aside from oil, other industrial sectors are not well developed. Brunei must import most of its food needs.

20. Singapore is a pivotal island in the Straits of Malacca. It is second only to Japan in development in Asia. Its success is contingent in part on its strategic location relative to other Southeast Asian states, India, and China.

21. By the 1830s Singapore was a major trading post attracting a variety of traders. The Chinese were the labor force upon which the country was built, and they dominate the population today.

22. Singapore had become a major trans-shipment point and financial center by the late nineteenth century. By the time of World War II, Singapore was a multi-racial, multi-religious, and multi-lingual society governed by a British elite.

23. Singapore was part of Malaysia from 1963 to 1965, when it became an independent republic. It has transformed itself from an entrepôt economy to a post-industrial economy in 30 years. It is a model of planned development.

24. Singapore is one of the world's largest hubs for shipping and commerce, and is staking its future on becoming a global nerve center for media and communications.

25. Building an external economy is a national imperative, and the country has expanded overseas seeking both cheap labor and potential markets. It is also seeking greater interdependence with its neighbors Malaysia and Indonesia. It is central to the growth triangle of Johor and Indonesian islands offshore.

26. Singapore is forging a sense of place through redevelopment and heritage tourism. High-tech industry and medical technology are at the forefront of economic development. Both Mandarin and English are official languages, although a linguistic mix called "Singlish" is very popular among young people.

27. Indonesia is the world's largest archipelago and is one of the most complex ethnically and linguistically. It has the world's largest Muslim population. It can be divided into inner and outer islands with Java as the most populated and urbanized island.

28. Because the region is so diverse, people tend to focus on local identities before national ones. The Indonesian language helps inculcate a national identity.

29. The earliest kingdoms were Hindu and Buddhist, but Islam started to take over in the thirteenth century. The Kingdom of Sri Vijaya in Sumatra controlled trade from Thailand to Indonesia.

30. When the Europeans arrived they found a series of well-established states such as in central Java and Aceh. The Netherlands United East India Company (VOC) began to trade with local

kingdoms in the early seventeenth century. Not long after, they set up their headquarters at Jayakarta which they renamed Batavia. The Dutch completed their control of the islands in 1911.

31. The Dutch promoted a Western secular elite at the expense of the Indonesians and encouraged the Chinese to engage in trade.

32. Under the Culture System the islands were compelled to grow designated crops. Cities emerged to handle the export trade but most Javanese were tenant farmers, sharecroppers, or wage laborers. Plantation crops were important to the economy.

33. Oil, discovered in the 1920s, became the nucleus for the development of a modern economy. Indonesia is a member of OPEC.

34. Java's cities were unable to cope with large flows of in-migrants and living conditions for most were grim. Substantial improvements were not made until the 1970s.

35. Western education benefited the children of the elite but spawned nationalism and a desire to throw off their Dutch rulers. Sukarno became the leader of an independent Indonesia after 1949. He promoted the concept of *Pancasila* with its five goals to make a better and more unified society.

36. The revolution for independence witnessed the rise of the army. The army has a dual function: to defend the nation and to lead the way in development.

37. In 1965, Sukarno was overthrown and General Suharto took over. He established a government of a "New Order." Many so-called communists were killed after the coup, especially Chinese.

38. Indonesia remains a poor country with high levels of unemployment and underemployment.

39. General Suharto was ousted in 1998, and the government elected in 1999 stressed *reformasi*—reform toward liberalization. Economic and social turmoil is problematic with different ethic and other groups expressing dissatisfaction with the current state of affairs. A new democratic government was elected in 2004.

40. Indonesia has instituted a successful family planning program and birth rates have fallen dramatically as a result. This has happened in concert with the changing status of women. The extended family is less common.

41. Transmigration programs, introduced by the Dutch, continued in the 1960s to redistribute population from crowded Java to the outer islands. This was also an effort to Javanize these regions. Some of these *transmigrasi* schemes were ill thought-out and failed. Others are structurally successful, but the Javanese are resented in the presence of different ethnic groups.

42. The dominant population movement in Indonesia today is rural to urban migration. Improved transportation and other factors have led to people moving or commuting to cities to work. Circular migrants spend part of the year working in cities and the rest at home in rural communities.

43. There has been a dramatic lateral extension of urban areas. JABOTABEK centered on Jakarta is one of these urban conurbations. Housing development is an important element of urban sprawl. Foreign investment plays an important role as well. Prime agricultural land is being lost to developers.

44. Indonesia has adopted a market-oriented approach to development and is attempting to rely less on its oil and gas sector for economic growth. Manufacturing is becoming increasingly important.

45. Fewer people are working in agriculture. Rice remains significant with gains mostly due to increased inputs and mechanization. With commercialization, change in village life is endemic.

46. Indonesia is home to 10 percent of the world's rain forests and these forests have the world's greatest biodiversity. Forests are being decimated by logging, plantation and settlement schemes, and fire.

47. National integration is of primary concern in Indonesia. It cannot afford to lose the outer islands, which hold most of the natural resources such as coal, oil, and gas. Many of the outer islands desire independence from a Java-controlled Indonesia, however. Aceh in northern Sumatra and West Papua are two prominent examples of trouble spots.

48. Radical Islam is on the rise in Indonesia which historically has been moderate in its religious views. The organization Jemaah Islamiah, with suspected ties to al-Qaeda, is responsible for a series of bombings and other terrorist acts.

49. The Philippines archipelago consists of thousands of volcanic islands. Luzon and Mindanao are the largest. The islands straddle a typhoon belt and experience a monsoon climate.

50. The population is predominantly Malay-related people of Mongoloid descent. There are at least 60 ethnic groups.

51. *Tagalog* is the official language of the Philippines but English is the lingua franca.

52. Spain controlled the islands for more than 300 years after 1521, when Magellan claimed them for Spain. The Spanish were interested in trade and spreading Christianity. The first settlement was on Cebu, but headquarters were established at Manila.

53. The economic base of the Spanish in the Philippines was the silk- or silver-laden galleon trade between southern China and Mexico.

54. Chinese traders and *mestizos* became the most dominant economic forces in the islands, organizing the entrepôt trade. Manila rivaled Jakarta as a port in the seventeenth and eighteenth centuries.

55. Mindanao and the Sulu archipelago, dominated by Muslims, were not brought under Spanish control until the nineteenth century.

56. Spanish control was based on a close relationship between state and church. Today, more than 80 percent of Filipinos are Roman Catholic.

57. After the galleon trade ended, the Spanish invested in export crops such as tobacco and pepper. By the nineteenth century, British export houses were the most prominent.

58. The incorporation of the Philippines into the world economy saw the rise of nationalism. It also created economic, social, and political forces that saw the rise of regional elites and fostered division.

59. The Spanish-educated *ilustrados* demanded separation of church and state. They began to call themselves Filipinos, and the term became a symbol of nationalism. The Filipinos were the first people in Asia to fight their colonial oppressors.

60. In connection with the Spanish-American War, the Americans fought the Filipinos and the country became an American colony in 1901. The United States had a "democratic mission" to lead the Philippines to independence.

61. Independence still left the landed elite in control. Rural rebellions have been crushed but rural discontent remains.

62. A new industrial class emerged in the 1950s and 1960s, and by the 1960s, the Philippines was the most successful manufacturing country in Southeast Asia. However, subsequent government policies spoiled this record and the economy stagnated. Martial law was proclaimed in 1972 and "crony capitalism" under Ferdinand Marcos became the rule. The country was fragmenting with unrest, and Marcos was defeated in an election in 1986 and fled the country.

63. Even with a new government little reform in terms of land redistribution and reducing the power of the elite was instituted. However, there was an economic turnaround.

64. In the 1990s industry accounted for a third of the economic productivity. Agriculture still employs more than half the country's workers but contributes about a quarter of the economy. A third of the population is unemployed or underemployed.

65. In 1992, the United States gave up the Subic Bay Naval Base and Clark Air Base. These areas are targeted for development based on foreign investment. Problems with infrastructure have caused many potential investors to shy away from the projects.

66. A policy of import substitution replaced processing of agricultural products in the 1990s. Recent emphasis has been placed on export-oriented industrialization.

67. Manufacturing is highly concentrated in Metro Manila. Cebu, Mactan, and General Santos City are other important nodes of economic activity.

68. The Philippines is planning to participate in a growth triangle with other countries in the

region including Brunei, Indonesia, and Malaysia.

69. Many Filipinos live overseas and are important for foreign exchange earnings for the Philippines. Frequently they are "marketed" for specific job types according to gender. Exploitation is common.

70. The country faces many social, economic, and environmental problems. Wealthy families still control the land, and nearly 40 percent of Filipinos continue to live in poverty.

71. There is a Muslim insurgency in the southern islands.

⊕ Bibliography

Airriess, Christopher. 2000. Malaysia and Brunei. In Southeast Asia: Diversity and Development, eds. Thomas R. Leinbach and Richard Ulack, pp. 341–78. Upper Saddle River, NJ: Prentice-Hall.

Asian Strategy and Leadership Institute. 1997. Malaysia Today: Towards the New Millennium. London: ASEAN.

Balakrishnan, N. 1991. "Single-Minded: Census Fails To Allay Concern on Marriage Patterns." Far Eastern Economic Review (June 20): 17.

Bunnell, Tim. 2004. Malaysia, Modernity and the Multimedia Corridor. New York: Routledge.

Cartier, Carolyn L. 1993. "Creating Historic Open Space in Melaka." The Geographical Review 83: 359–73.

Chai, Michael. 1992. "The "Other Indigenous" Peoples of Sarawak." Development 2: 56–61.

Chang, T. C. 1999. "Local Uniqueness in the Global Village: Heritage Tourism in Singapore." The Professional Geographer 51: 91–103.

Church, Peter, ed. 1995. Focus on Southeast Asia. Hong Kong: Allen and Unwin.

Cleary, M. C. 1996. "Indigenous Trade and European Economic Intervention in North-West Borneo c. 1860–1930." Modern Asian Studies 30: 301–24.

Cleary, Mark, and Brian Shaw. 1994. "Ethnicity, Development and the New Economic Policy: The Experience of Malaysia, 1971–1990." Pacific Viewpoint 35: 83–107.

Dharmapatni, Ida Ayu Indira, and Tommy Firman. 1995. "Problems and Challenges of Mega-Urban Regions in Indonesia: The Case of Jabotabek and the Bandung Metropolitan Area." In The Mega-Urban Regions of Southeast Asia, eds. T. G. McGee and Ira M. Robinson, pp. 296–314. Vancouver: UBC.

Dolven, Ben. 1999 a. "Forward March (Singapore)." Far Eastern Economic Review (March 25): 45.

Dolven, Ben. 1999 b. "Singlish: Can or Not?" Far Eastern Economic Review (September 2): 32–33.

Drake, Christine. 1989. National Integration in Indonesia. Honolulu: University of Hawaii Press.

———. 1992. "National Integration in China and Indonesia." The Geographical Review 82: 295–312.

Heyzer, Noeleen. 1992. "Rainforest Management and Indigenous Livelihoods: A Malaysian Case Study." Development 4: 14–17.

Hiebert, Murray. 1996. "It's A Jungle Out There (Singapore)." Far Eastern Economic Review (April 25): 58–61.

———. 1997. "On the Offensive: Singapore's Rules May be Strict—But They Do Curb Urban Blight." Far Eastern Economic Review (February 6): 38–39.

Hugo, Graeme. 2000. "Indonesia." In Southeast Asia: Diversity and Development, eds. Thomas R. Leinbach and Richard Ulack, pp. 304–40. Upper Saddle River: Prentice-Hall.

n.a. 1997. Information Malaysia 1997 Yearbook. Kuala Lumpur: Berita.

Jardine, Matthew. 1995. East Timor: Genocide in Paradise. Tucson: Odonian.

Jayasankaran, S. 2004. "Brunei: Buoyed By Oil." Far Eastern Economic Review (August 12): 41.

Johnstone, Bob. 1995. "Culture Clash in Cyberspace." New Scientist (March 25): 38–41.

Lee, Boon Thong. 1995. "Challenges of Superinduced Development: The Mega-Urban Region of the Kuala Lumpur-Klang Valley." In The Mega-Urban Regions of Southeast Asia, eds. T. G. McGee and Ira M. Robinson, pp. 315–27. Vancouver: UBC.

Leinbach, Thomas R. et al. 1992. "Employment Behavior and the Family in Indonesian Transmigration." Annals of the Association of American Geographers 82: 23–47.

Macleod, Scott, and T. G. McGee. 1996. "The Singapore-Johore-Riau Growth Triangle: An Emerging Extended Metropolitan Region." In Emerging World Cities in Pacific Asia, eds. Fu-chen Lo and Yue-man Yeung, pp. 417–64. New York: United Nations.

Mapes, Timothy. 2003. "No Solution in Sight for Aceh." *Far Eastern Economic Review* (September 25): 21–22.

McBeth, John et al. 1999. "Worse to Come (Indonesia)." *Far Eastern Economic Review,* July 29: 16–19.

McBeth, John. 2004. "East Timor: Not Yet Ready To Go It Alone." *Far Eastern Economic Review* (May 20): 12–14.

Monbiot, George. 1989. "The Transmigration Fiasco." *Geographical Magazine* 61: 26–30.

Murphy, Dan. 1999. "The Next Headache (Indonesia)." *Far Eastern Economic Review* (April 29): 20–21.

Parsonage, James. 1992. "Southeast Asia's "Growth Triangle": A Subregional Response to Global Transformation." *International Journal of Urban and Regional Research* 16: 307–17.

Netto, Anil. 2004. "Rural Malaysia Mired in Poverty." *World Press Review* (January): 14.

Perry, Martin. 1991. "The Singapore Growth Triangle: State, Capital and Labour at a New Frontier in the World Economy." *Singapore Journal of Tropical Geography* 12: 138–51.

Rodrigue, Jean-Paul. 1994. "Transportation and Territorial Development in the Singapore Extended Metropolitan Region." *Singapore Journal of Tropical Geography* 15: 56–74.

Runyan, Curtis. 1998. "Indonesia's Discontent." *World Watch* 11: 12–23.

Schwarz, Adam. 1992. "A Saw Point for Ecology (Indonesia)." *Far Eastern Economic Review* (April 19): 60–61.

Sutton, Keith, and Amriah Buang. 1995. "A New Role for Malaysia's FELDA: From Land Settlement Agency to Plantation Company." *Geography* 80: 125–37.

Taylor, D. M. et al. 1994. "The Degradation of Rainforests in Sarawak, East Malaysia, and its Implications for Future Management Policies." *Geoforum* 25: 351–69.

Tiglao, Rigoberto. 1999. "About-Face: New Chief Changes Game Plan for Subic Bay." *Far Eastern Economic Review* (February 18): 44–45.

Tyner, James A. 1996. "The Gendering of Philippine International Labor Migration." *Professional Geographer* 48: 405–16.

Ulack, Richard. 2000. "The Philippines." In *Southeast Asia: Diversity and Development,* eds. T. R. Leinbach and R. Ulack, pp. 408–33. Upper Saddle River: Prentice-Hall.

Vatikiotis, Michael. 2003. "The Struggle For Islam." *Far Eastern Economic Review* (December 11): 54–58.

Glossary

Activity space   The cumulative space employed in the activities of daily life.

Adi Granth   The Sikh holy book containing the teachings Guru Nanak, the founder of Sikhism.

Adivasi   Means "original inhabitant." Applied to scheduled tribes in India.

Aeolian   Referring to wind.

Age of Exploration   The period of European exploration of the world beginning in the fifteenth century.

Age of Imperialism (1870–1914)   The period during which competing European powers acquired, controlled, and exploited colonies.

Ainu   Aboriginal people of Japan. Now confined mainly on Hokkaido.

Allah   In Islam, "God."

Alluvial soils   Fertile, water-deposited soil.

Altitudinal zonation   The variation in plant communities caused by climatic change associated with increasing altitude.

Amenity-migration   The movement of people to places offering amenities such as rest and recreation or a more comfortable climate.

Americanization   The promotion and acceptance of American culture.

Ammonite   A flat, spiral fossil shell from the ancient Tethys Sea.

Analects   The ideas of the Chinese philosopher and teacher Confucius as recorded by his students.

Annam   A Chinese word meaning "the pacified south." The Chinese name for Vietnam during the Tang Dynasty. During the period of French rule, one of the five territories of the Indochinese Union.

Aquaculture   The cultivation of sea products such as shrimp, oysters, seaweed, and pearls.

Aquifer   A permeable subsurface rock layer that can store, transmit, and supply water.

Archipelagoes   Collections of islands. Japan, the Philippines, and Indonesia are all archipelagoes.

Arithmetic density   The number of people existing per square mile or kilometer of land in a particular country.

Artesian basin   An aquifer in which water is held under pressure. Drilling a well would allow water to flow freely to the surface without artificial pumping.

Aryans   Relatively light-skinned nomadic herders who invaded India from Central Asia around 1500 B.C. The Aryans spoke Sanskrit.

Asia-Pacific   Similar to the economic concept of the Pacific Rim. The countries of the Asia–Pacific region trade primarily with one another.

Asia Pacific Economic Cooperation (APEC)   A regional trading block similar to the European Union (EU) or the North American Free Trade Association (NAFTA).

Asoka   The most important Mauryan ruler in India. Asoka adopted Buddhism as the State religion.

Aung San Suu Kyi   Leader of the National League for Democracy (NLD) in Burma. She has been under house arrest since 1989.

Azad Kashmir   A revolutionary government in northern Kashmir.

Bagasse   Sugarcane fiber.

Bangla   The Bangladeshi name for Bengali, the language of Bangladesh.

Balikbayans   Overseas Filipinos. They are important sources of revenue for the Philippines.

Bao Dao   The name for Taiwan as portrayed on ancient Chinese maps.

Barani   Dry, non-irrigated regions of Pakistan.

Bari   The basic spatial unit in a Bangladeshi village consisting of six houses built around a square or rectangular courtyard.

Base population   The actual population from which growth or decline occurs.

Batavia   The Dutch name for Jakarta, the capital of Indonesia. Batavia was renamed Jakarta in 1945.

Benazir Bhutto   The youngest and first female leader of a Muslim state. Served as prime minister of Pakistan from 1988 to 1990 and again from 1993 to 1997.

Bilateral aid   Government to government transfers of cash, low-interest loans, food, and the like.

Biodiversity   The range of living organisms defined in terms of genes, species, and ecosystems.

Biological oxygen demand (BOD) The measure of the amount of oxygen required to support microorganisms that break down organic wastes such as those found in raw sewage.

Bodhisattva In Buddhism, a divine being such as Avalokitesvara in Sanskrit, who postpones entry into *nirvana* in order to save other beings; meaning "the lord who looks in every direction."

Bollywood India's movie industry located in Mumbai (Bombay).

Brahman In Hinduism, the one ultimate reality.

Brahmins Hindu priests in India's ancient social hierarchy. In most recent times, any one of this class.

Buddha In Sanskrit, the enlightened one. The name of Siddhartha Gautama upon his enlightenment.

Buffer state A country situated between two other opposing countries.

Bumiputera Means "sons of the soil." The Malay peoples of Malaysia.

Bungalow A low-lying, thick-walled, one-story house found in the hill stations of northern India.

Burakumin Social outcasts in Japan.

Burghers Descendants of Portuguese or Dutch in Sri Lanka.

Buttresses Giant, supportive blade-like structures of a shallow-rooted tree.

Canopy The upper layer of the rain forest. The forest's biomass.

Cao Dai A Vietnamese word meaning "high tower." A Vietnamese religion that is a mixture of Buddhism, Christianity, Confucianism, and Daoism.

Cardamom A highly valued spice commonly used in South Asian cuisine.

Caste system The division of Hindu society according to social ranking and occupation.

Castle town In early Japan, a fortified town along a transportation corridor.

Ch'i The vital force or cosmic breath that gives life to all things. The most important force of *feng shui.*

Chaebol A giant, corporate conglomerate in South Korea.

Chars In Bangladesh, the name given to small, sandy islands formed by rivers as they sway between valley walls.

Chawl Slum in Mumbai (Bombay).

Chiang Kai-shek The leader of the Nationalist forces during China's civil war and World War II. Chiang opposed Mao Zedong and the Communists and was defeated by them in 1949. He fled to Taiwan in 1949 and ruled the island as the Republic of China until his death in 1975.

China Frontier The rugged, arid, and sparsely populated environment of western China.

China Proper The relatively low relief, humid, and densely populated environment of eastern China.

Chinese Diaspora Dispersal of Chinese from the mainland to all parts of Southeast Asia.

Chipko Movement A grass roots, women's movement to save trees in the Indian Himalayas.

Christianity A monotheistic faith proclaiming the idea of a single deity, God, and Jesus Christ, the manifestation of God on earth.

Cilao Portuguese name for Ceylon (Sri Lanka).

Cinchona Source of quinine extracted from tree bark; it was the only treatment for malaria in historic times.

Circular migrant A person who lives and works in the city for part of the year and spends the remainder of the year in his or her village.

Civil line British civilian district socially and spatially separated from Indian society.

Cloud forests In the Himalayas, forests existing in fog and mist at elevations between 8,000 and 13,000 feet (2,400–39,000 m).

Cochin China A colony in southern Vietnam. One of the five territories of France's Indochinese Union.

Collectivization The end to land ownership in China as instituted by Mao Zedong in 1950. Land was worked collectively.

Colonizer's model The idea that civilization and progress diffuse from Europe to the rest of a culturally barren world.

Commoditization The division of a particular market into two sectors: big-name brand marketing and sales, and hundreds of no-name component suppliers. The garnering of the no-name component sector is known as the Original Equipment Manufacture market.

Confucius A Chinese teacher of ethics, ritual, and philosophy.

Congress of Vienna 1815 A meeting at which European boundaries were defined and Europe was demarcated from Asia by the Ural Mountains, through the Caspian and Black seas, and along the eastern end of the Mediterranean Sea to the African continent.

Congress Party The Indian National Congress was formed in 1865 to provide a forum for nationalist sentiments. The Congress Party was run by brahmin Hindus.

Continentality The size of a landmass.

Continental drift A process that took place millions of years ago in which a continent known as Pangaea broke into pieces forming separate continents.

Convectional rainfall A feature of equatorial regions in which precipitation is brought about by the immediate rising, cooling, and condensing of moisture-laden air.

Coral reefs Biological entities formed in sea water by small marine animals living in symbiotic relationships with algae. Coral reefs occur along continental shelves and around islands.

Core-periphery The concept that there are powerful core regions of development surrounded by less powerful regions of lesser development. The core-periphery concept is applicable in any context and at any scale.

Cultural convergence Takes place when lifeways and belief systems blend in a landscape.

Cyberspace The sphere of information technology.

Cyclone A low pressure center. Can be associated with cyclonic storms having high velocity winds and copious precipitation.

Dalits The name used by India's untouchables in reference to themselves. It means "oppressed."

Dao de Jing (Tao-te-ching) Means "classic of the way." A book of philosophy concerned with the Dao, or the way the universe works.

Dasht Expanses of black pebbles common to Baluchistan, Pakistan.

Deciduous forests Refers to trees that lose all their leaves in a cold or dry season.

Decolonization A nation's process of gaining of its independence from a colonial power.

Demographic transition model Sequential changes in population growth or decline beginning with high births and high deaths and ending with low births and low deaths.

Deng Xiaoping Ruler of China from 1976 to 1989. A colleague of Mao Zedong and survivor of the Long March. He began to open up China to a market economy.

Dependency theory The idea that core regions can grow only through exploitation of their dependent peripheries.

Deposition The depositing of material through the process of erosion.

Deserts Regions receiving less than 10 inches (254 mm) of precipitation a year.

Developing countries (DCs) Countries that are beginning the industrialization process. India and Indonesia are examples.

Dien Bien Phu The battle in 1954 between the Communist Viet Minh and the French. The French loss effectively ended their efforts to hold Vietnam.

Diffusion The spatial spread of a phenomenon.

Dipterocarps Trees with two-winged seeds that dominate the upper canopy of the rain forest.

Doab The level and fertile land between the Ganga and Yamuna rivers. It means "two rivers." Also, any such land between two rivers.

Doi Moi A Vietnamese word meaning "renovation." Reform policies instituted by the Vietnamese Communist Party in 1986.

Domestication The process of learning how to grow plants and manage animals for food and related products. The invention of 'agriculture.'

Dowry deaths The death of a new bride at the hands of her husband's family who are displeased over the size of the young woman's dowry and harass or kill her.

Dragon and Tiger Economies Newly industrializing economies of East and Southeast Asia.

Dravidians Dark-skinned people of the Indus Civilization in India.

Drignam Namzha A traditional values and etiquette policy launched by the king of Bhutan in order to salvage Bhutanese culture.

Drukpas The Bhote tribe of northern Bhutan, they were originally from Tibet. They revere the king of Bhutan as their spiritual leader.

Drukyul Original name of Bhutan.

Dual economy The economy characterized by the existance of two levels of economic activity; a formal sector functioning at a large scale in the realm of national and international trade, and an informal sector, functioning at a smaller scale at local levels.

Dutch Culture System During the colonial period, the Dutch requirement that every Indonesian farm household allocate one-fifth of its land to the production of a cash or export crop.

Dwifungsi Means "dual function." Refers to the Indonesian army's dual role of defending the country and promoting economic development.

Dynasties In ancient China, periods of rule by one family.

Dynastic cycle The rise and fall of a particular dynasty in China. A dynasty would begin as virtuous, peaceful, and productive but over time would become corrupt, warring, and a victim of natural disasters, thereby losing the Mandate of Heaven.

Dzongs Fortified monasteries in Bhutan.

Economic Cooperation Organization (ECO) Established by Pakistan, Iran, and Turkey in 1985. Now includes Afghanistan and the Central Asian Republics.

Economic development zones (EDZs) Part of Deng Xiaoping's new economic policies. Areas of China designated as special zones for foreign investment.

Ecumene The inhabited world.

Edo Traditional name for Tokyo.

Eightfold Path The Buddhist way to overcome desire and achieve the ultimate state of *nirvana*.

Emergents Trees rising above the rain forest canopy.

Endemic   Plants or animals occurring in a given area only.

Endogamy   The practice of marrying within one's caste or *jati* (occupation).

Entrepôt   A transshipment point, usually a port city such as Malacca.

Environmental criticality   Environmental degradation associated with socioeconomic deterioration such as the kind that takes place when forests are destroyed.

Environmental refugee   A person who has lost his or her home and land due to environmental disaster or deliberately planned environmental change.

Environmental transformation   Changes in the physical environment wrought by values, attitudes, policies, and practices.

Ephemerals   Desert plants that bloom sporadically in response to rain.

Epiphytes   Plants (such as orchids and ferns) physically but not nutritionally supported by other plants. These are found in the middle level of the rain forest canopy.

Erosion   The wearing away of the earth's surface by the action of wind and water.

Estate Tamils   Indian, Tamil laborers imported by the British to Sri Lanka to work on plantations or estates.

Ethnocentrism   Judging other countries or cultures by one's own which is considered superior.

Eurasia   The geographical landmass comprising Europe and Asia.

Eurocentric   A worldview that puts Europe at the central and superior position relative to the rest of the world.

Exclusive Economic Zone (EEZ)   The 200-nautical-mile zone belonging to coastal states as defined by the 1982 Law of the Sea Treaty issued by the United Nations.

Extended Metropolitan Regions (EMRs)   Regions that have evolved through the continual growth of cities over a large area.

Extraterritoriality   A consular jurisdiction whereby a foreigner is subject to the laws of his or her own country.

Exudates   Resins (such as copals and dammars) that ooze out of a tree.

Factories   Trading posts established by the Dutch and British in South and Southeast Asia.

Fa-hsien   The first Chinese Buddhist pilgrim to visit India. His journey took place from 399 to 412 A.D.

FELDA   Federal Land Development Authority in Malaysia. FELDA develops schemes to open up jungle and virgin forests to settlement and agriculture.

Female infanticide   The deliberate killing or neglect of female children.

Feng shui   A Chinese term meaning "wind and water." Forces of nature that must be dealt with appropriately in behavior and landscape design.

Filial piety   Confucian concept of hierarchical loyalties of ruler to heaven; subject to ruler; son to father; younger brother to older brother; and wife to husband.

Filipinos   People of the Philippines.

Five Pillars of Islam   Rules for proper adherence to Islam.

Floating population   The movement of more than a million individuals in China from city to city seeking work opportunities.

Fluvial   River-related.

Flying geese model   An economic model used in the 1960s to describe the industrial life-cycle in recently industrialized nations.

Foot-binding   In China during the dynasties, the practice of binding women's feet with the toes folded under the arch. Known as the Golden Lotus, bound feet were considered sensual. The practice began among the upper classes and court circles of the Southern Song Dynasty (1127–1278) and continued into the early 1900s.

Forward-thrust capital   A new capital city built in a frontier region to decentralize population from elsewhere and serve as a growth pole. Islamabad, in Pakistan is an example.

Four Noble Truths   The Buddhist notion that the world is filled with suffering which is caused by desire and that desire can be overcome by following a set of rules.

Frontline state   A state which serves as a barrier to the progress of an undesirable political system.

Functional interactions   Relationships between people and the natural world that are expressed in political, social, and economic landscapes.

Ganga Ma   Mother Ganges. A river in India that is considered sacred by Hindus.

Gender   A culturally constructed social institution. Gender roles are expected and assumed in cultural contexts and society is organized around gender.

Genghis Khan   Lived 1155–1227. A military genius who amassed his nomadic hordes into the world's most formidable cavalry. Occupied northern China, Mongolia, and the Far East.

Geographic realms   Based on physical and cultural characteristics, the largest units into which we can divide the ecumene, or the inhabited world.

Geomancer   An expert in *feng shui*. Often called upon in building orientation and design.

Glaciation   The action of ice in the process of erosion and deposition.

Global feminization of poverty   The fact that increasing numbers of women are becoming poor and that the majority of the world's poor are women.

Globalism   The concept that countries are linked in various ways on a global basis.

Globalization   The process by which regions become increasingly linked with global capitalist enterprises and economic systems.

Grameen Bank   A non-government organization in Bangladesh. The bank offers small loans to the poor, mostly women, to get a business started.

Great Leap Forward   The communization of China's workers beginning in 1958. Agricultural cooperatives were reorganized into People's communes holding several cooperatives and thousands of individuals. Communization was expanded to include urban and industrial regions.

Great Proletarian Cultural Revolution   A period of giant social, political, and economic upheaval in China. Beginning in 1966, an estimated 100 million people were targeted for reeducation and forcible movement to different regions of the country.

Greater East Asia Co-Prosperity Sphere (GEACPS)   Japan's ambition to lend an empire composed of Asian countries minus European imperialists. The plan was halted when the United States was drawn into World War II by the Japanese attack on Pearl Harbor in 1941.

Green Revolution   The use of high-yielding, pest-resistant seeds in agriculture.

Growing season   The number of consecutive, frost-free days in a year.

Growth triangle   A triangle of economic growth regions often involving more than one country.

Growth pole   A center designated to attract industry and development.

Gun-salute states   Indian princely states and estates where the number of gun salutes—21, 15, 10—indicated the importance of the ruler.

Guptas   Rulers of India who united northern India around 320 A.D. Art, mathematics, science, and literature flowered during the Gupta period, also known as India's Golden Age.

Gurkhas   British-trained Nepalese troops.

Guru   A spiritual advisor in Hinduism.

Guru Nanak   Founder of Sikhism in the fifteenth century.

Hajj   A pilgrimage to the holy cities of Mecca and Medina. One of the Five Pillars of Islam.

Hamun   Dry lakes or *playas* found in Baluchistan, Pakistan.

Hangul   Korean lettering system developed in the fifteenth century under King Sejong.

Hartal   A labor strike in Bangladesh.

Hearth areas   Places of origin of cultural phenomena.

Hill stations   Communities at higher elevations used by the British as refuges from the summer heat of the lowlands. During the Raj, they were socially and spatially segregated from Indian areas.

Hindutva   "Hinduness," a desire to make India a country where Hindu principles prevail.

Hiragama   Japanese syllabary that permits the expression of words and parts of speech not easily represented by Chinese characters.

Historic inertia   A phenomenon in which people continue to live in and migrate to places where other people already live and opportunities are perceived to exist. As a consequence, densely populated areas are most likely to continue to be densely populated.

Ho Chi Minh   The founder of the Indochina Communist Party. Leader of North Vietnam from 1954 until his death in 1969.

Hollowed-out industry   In Japan, an industry that has lost its competitive edge due to rising labor costs and is forced to seek cheaper labor elsewhere, typically overseas.

Holy Bible   The holy scriptures of Jews and Christians.

Household responsibility system   A return to family farming in China whereby farmers are to meet government quotas but have the remainder of the crops to keep or sell on the open market.

Hsia-fang (downward transfer) **movement**   The transfer of millions of youths and others to the countryside during China's Cultural Revolution.

Hsuan-tsang   A Chinese pilgrim who spent 13 years in India from 630 to 643 A.D. He furthered the spread of Buddhism in China.

Hu Jintao   General Secretary of the Chinese Communist Party since 2002. Also president of the People's Republic of China since 2003.

Hukou   Household registration system in China. Every Chinese citizen is registered as nonagricultural or agricultural. Registration includes one's particular geographic location, and it is very difficult to alter one's classification.

Hydrological   Referring to water.

Hyperurbanization   Extremely rapid city growth.

Ilustrados   Spanish-educated descendants of powerful landed families in the Philippines.

Imam   Islamic religious leader.

Indentured servitude   A state of being poor farmers who borrow from landlords, become victims of usurious interest charges, and remain indebted for the rest of their lives.

India's Golden Age   A period during the fourth–sixth centuries when art, science, and mathematics flourished under the Guptas.

India's Silicon Valley   A region of high-tech industries centered on Bangalore.

Indus Waters Treaty   In 1960, divided the tributaries of the Indus River between India and Pakistan.

Industrial crops   Non-food crops such as cotton or rubber.

Infidels   According to Muslims, people who do not believe in Islam.

Intermediate technology   Technology that lies between the most primitive and the most modern.

Intertropical Convergence Zone (ITCZ)   The equatorial low-pressure trough that dominates the central area of the Indian Ocean.

JABOTABEK   The fastest growing area in Indonesia. Comprises the metropolitan regions of Jakarta, Bogor, Tangerang, and Bekasi.

Jainism   A religion focusing on non-violence against any life form. Founded in the sixth century by Mahavira.

***Jajmani* system**   A Punjabi custom for Harijan laborers to work for their Jat landlords in exchange for grain and fodder. The system is both an economic system and a social contract rooted in caste relationships.

Jarkhand   Name for a new independent state in the Jamshedpur industrial region in India. Jarkhand came into existence in 2000.

Jati   A subdivision of *varna* defined by and named by occupation.

Jawaharlal Nehru   Leader of the Congress Party and the first prime minister of India.

Jesus Christ   Believed by Christians to be the Son of God and the manifestation of God on earth.

***Juche* philosophy**   Kim Il Song's ideas of self-reliance and self-sufficiency for the people of North Korea.

Jungle   Dense vegetation of the tropics that occurs where light penetrates the rain forest canopy.

Jute   A fiber crop used to make sacking, twine, ropes, carpet, car interiors, lining for asphalt roads, and many other products. Jute is one of the major crops in Bangladesh.

Kamikaze   Japanese term meaning "the divine wind." A typhoon which was responsible for the defeat of Kublai Khan in his attack on Japan in 1281.

Kampuchea   Cambodia's name under the Communist Khmer Rouge and Pol Pot.

Kampung   A village in Malaysia and Indonesia.

Kanji   The Japanese writing system which was borrowed from the Chinese character system.

Kansai Science City   Research and development center focusing on Kyoto, Osaka, and Nara in Japan.

Karez   Underground irrigation systems consisting of hand-dug, gravity-flow tunnels. Karez are found in northwest China and in southwest Asia. The tunnels carry water from aquifers to surface channels and ponds.

Karma   In Hinduism, the accumulation of all of a person's deeds in a lifetime.

Karst topography   Pitted and lumpy limestone formations caused by chemical weathering.

katchi abadis   Squatter settlements on the periphery of Karachi, Pakistan.

keiretsu   Japan's modern industrial conglomerates. A hierarchical organization of companies.

Khalistan   Name for the independent state desired by many Sikhs in Punjab.

Kharif   In South Asia, crops (such as rice) sown after the onset of the wet monsoon. Kharif crops are harvested in autumn.

King Harsha   Controlled an empire from Punjab to the Narmada River during the 600s A.D. The last north Indian ruler to encroach south of the Narmada for 600 years.

Koran   The holy book of Islam; from the Arabic word *qur'an,* meaning "recitation."

Koryo   A Korean kingdom that emerged in 918 A.D. in the central region of the peninsula. Origin of the name "Korea."

Kotadesasi   From the Indonesian words *kota* (town) and *desa* (village) meaning the coincidence of urban and rural activity in the same spatial territory.

Kshatriyas   Members of the warrior class in India's caste system.

Kublai Khan   Grandson of the Mongol warrior Genghis Khan. Kublai Khan dominated all of China, Korea, and northern Vietnam. The Mongols ruled China as the Yuan Dynasty, 1279–1368.

Lan Xang   A Laotian word meaning "a million elephants." The first kingdom of Laos that emerged in 1353.

Landscape   The world that we see and experience.

Lao Loum   A Laotian term meaning "Laotians of the valley." Lao Loum make up one-half the population of Laos.

Lao Sung   A Laotian term meaning "Laotians of the mountain tops." Lao Sung are among the most recent migrants to Laos.

Lao Theung   A Laotian term meaning "Laotians of the mountain slopes." Lao Theung make up about a quarter of the population of Laos.

Laozi   In Chinese, "the old one." Laozi was a contemporary of Confucius. The philosopher most closely associated with Daoism, the study and contemplation of the way universe works.

Laterite   Rock-like, lower layer of tropical soils formed by the presence of aluminum and iron oxides.

Lianas   Woody climbers or vines found in the middle level of the rain forest canopy.

Link-language   A language used to allow communication between speakers of other languages. English is a link-language in Sri Lanka.

Loess   Fine, wind-blown soil lacking horizontal stratification. Loess can stand in vertical cliffs.

Long March   The name given to the 6,000 mile-long trek (9,660 km) by the Communist Army in 1934 and 1935, as it fled the forces of the Nationalists. The Long March was led by Mao Zedong and took the Communists from southern China to Yenan in the loess hills of Shensi Province.

Madrassas   Islamic religious schools.

Maharaja   A word meaning "great ruler." Maharajas ruled various states in India following the breakdown of the Mughal empire.

Mahavira   Founder of Jainism.

Mahayana Buddhism   The school of Buddhism practiced in China, Korea, Japan, and Vietnam.

Mainlanders   Those who moved to Taiwan from the Chinese mainland between 1945 and 1949. They are associated with Chiang Kai-shek and the Nationalist forces.

Malesia   Malaysia and Indonesia.

Manchukuo   A Japanese puppet state in Manchuria established in 1931.

Mandate of Heaven   In dynastic China, the belief that Heaven, bestowed the right for Chinese rulers to rule. The mandate demanded virtue and the integrity of proper relationships.

Mangal   A mangrove community of tropical and subtropical coasts.

Mao Zedong   Leader of the Chinese Communist Party whose forces won the revolution and control of China in 1949. Mao ruled China until his death in 1976.

Mappaemundi   Religiously-inspired maps of the world which portrayed the earth as flat.

Market gardening   The production of fresh agricultural products for urban markets.

Mega-cities   Giant cities often incorporating several cities in a cluster.

Mega-urban regions   Urban agglomerations that extend beyond metropolitan boundaries.

Mercantilism   An economic system whereby raw materials are extracted from a colony and brought to the colonial power for manufacture. The manufactured goods are shipped back to the colony for purchase.

Mestizos   In the Philippines, people of Chinese-Filipino or Mexican-Filipino descent.

Mission civilisatrice   A French term meaning "civilizing mission." The French belief that colonialism would bring the benefits of modern European civilization to the primitive peoples of the world.

Mohandas Gandhi   He believed in non-violent protest and was a key figure in gaining India's independence.

Monotheistic   Having one god.

Monsoon   A climatic system based on intense seasonal changes in pressure and wind patterns. A realm affected by the monsoon has wet and dry seasons.

Monsoon Asia   Those parts of Asia under the influence of the monsoon climatic system.

Muhajirs   Urdu-speaking, post-partition, Muslim migrants to Pakistan from India.

Muhammed   The prophet of Islam.

Muhammed Ali Jinnah   Leader of the Muslim League and first prime minister of Pakistan.

Mujibur Rahman (Mujib)   Leader of the Awami League in Pakistan.

Multilateral aid   Aid deriving from international organizations such as the United Nations, World Bank, or Organization for Economic Cooperation and Development.

Multinational enterprise   An economic, social, or political entity with ties to several nations.

Multiregional hypothesis   Proposes that humankind originated in several locations simultaneously.

Mullah   Islamic religious leader.

Mura   Traditional rural village in Japan.

Muslims   Those who submit to God; adherents of Islam.

Myth of Asia   The fact that there is no cultural or historical entity that equals Asia.

Nagas   A Sanskrit word meaning "serpents." The *naga* is an aquatic symbol that appears in art form as a dragon, snake, or other bodily form. Part of cosmologies in Southeast Asia.

Nanyang   China's name for the southern sea regions (Southeast Asia) in early times.

Native Taiwanese   Immigrants from southeastern China who arrived from the fourteenth century into the twentieth century. They speak the Taiwanese dialect and make up about 70 percent of Taiwan's population.

Natural boundary   A boundary demarcated by physical features such as mountain ridges or rivers.

Natural vegetation   Undisturbed, mature plant communities.

New Theory Farming   In Thailand, a form of land and agricultural management employed as a means to introduce community-based economies.

New Thinking   A policy aimed at liberalizing the centrally planned economy of Laos.

Newly industrializing countries (NICs)   Countries that have recently begun to industrialize. Used in reference to several East Asian countries such as South Korea and Taiwan.

Nirvana   In Buddhism, a state of desirelessness and peacefulness.

Non-Governmental Organizations (NGOs) Non-governmental aid organizations. NGOs have an agenda, be it political, humanitarian, or religious.

North-South dichotomy The difference between the rich, industrialized countries of the Northern Hemisphere and the poorer, developing countries of the Southern Hemisphere.

Nucleated Clustered, as in human settlements.

Nutritional density The number of people existing per square mile or kilometer of cultivated land.

Occident The West.

Oracle Bones Bones, especially scapulae, heated and their cracks interpreted by ancient Chinese rulers and sages. Also inscribed with questions and interpretations.

Orang Asli Means "original man." Earliest of the current inhabitants of the Malay Peninsula.

Orient The East.

Orographic precipitation Precipitation brought about by the presence of a topographic barrier which forces a moisture-carrying air mass to rise, cool, and condense.

Oustees Uprooted persons who flee from war and violence, oppression, disasters, economic stagnation, and human-induced environmental displacement. About half of all oustees are landless or sharecroppers.

Out of Africa Theory that postulates that humankind originated in Africa.

Overseas Chinese Chinese—both individuals and corporations—living outside of mainland China.

Pacific Rim An economic concept referring to all countries situated around the Pacific Ocean. These countries' primary trading partners are each other.

Pancasila The five principles adopted by Sukarno in 1945 as the basis for Indonesian ideology and development.

Pan-gu In Chinese mythology, a primeval being whose body parts transmuted into various parts of the cosmos.

Pat Clay-desert in the Sind province of Pakistan.

Pathet Lao Communist guerrilla fighters in Laos.

Pax Mongolica A period of peace and prosperity when China was ruled by the Mongols from 1280 to 1300.

People's communes A collection of cooperatives housing thousands of people organized under China's Great Leap Forward of 1958.

Percentage urbanized The percentage of a country's population living in cities.

Phrapbuum A Thai word meaning "lord of the place." It refers to the spirit shrine exhibited in houses, office buildings, and other private and commercial establishments in Thailand.

Pinyin A simplified character and transliteration system for the Chinese language. Pinyin closely approximates the sound of the word in the Mandarin dialect of Chinese.

Plate tectonics A theory that postulates that rigid plates in the upper portion of the lithosphere float on the more plastic upper layer of the earth's mantle. Movement along the plates' fracture zones cause seismic activity.

Playas Salt-encrusted dry lakes found in Baluchistan, Pakistan.

Plural society A society in which two or more culture groups live adjacent to each other in the same country without mixing.

Pol Pot Leader of the Khmer Rouge in Cambodia from 1975 to 1978. Responsible for the death of millions.

Polytheistic Having many gods.

Popular culture Usually refers to American popular culture that is diffusing via transnationals and media into all corners of the world.

Population composition The age and sex structure of a particular population.

Population pyramids Pyramidal graphs showing the percentage of a country's males and females in various age cohorts.

Power relationships The fact that men and women are accorded unequal roles in virtually every society.

Primate city A nation's largest city, within the context of urban primacy.

Purdah The Muslim custom of secluding women.

Quality of life The state of human well-being.

Quoc Ngu Means "national language." Transliteration of Vietnamese spoken language into the Roman alphabet.

Rabi In South Asia, crops (such as wheat) sown at the end of the wet monsoon. Rabi crops are harvested in spring.

Rainshadow The protected, leeward side of a topographic barrier which is the dry side in the process of orographic precipitation.

Raj A word meaning "the rule." The Raj refers to the period of British rule in India.

Ramadan A month of daytime fasting; one of the Five Pillars of Islam.

Reformasi A period of economic reform and liberalization in Indonesia following the end of General Suharto's rule in 1998.

Regional state A natural economic zone shaped by the global economy in which it participates. Guangdong Province in China is an example.

Regionalism Focusing on the local rather than the global.

Region-based urbanization The continual growth of cities over large areas.

Reincarnation In Hinduism, the notion that a person is reborn into another cycle of existence depending on the Karma, or sum of ideals, of the previous life.

Religion A set of sacred beliefs and practices.

Religious revivalism In Islam, the resurgence of religious fundamentalism. In the early 2000s, characterized as extreme hostility to the West and Western culture.

Restructuring The rapid change in local, national, and international social, political, and economic systems wrought from the dynamics of global interaction.

Rich-poor gap The difference between rich and poor between parts of the world, but also between countries and within individual nations.

Sangha Monastic orders of Theravada Buddhism.

SARS Severe acute respiratory syndrome, a highly contagious pneumonia-like disease that infected thousands of people between November 2002 and July 2003. China was the source of the disease.

Sati In early India, the practice of burning a widow alive on her husband's funeral pyre. *Sati* was outlawed by the British in 1829 but continued to exist for many years.

Scheduled castes and tribes Also called *adivasis,* original inhabitants of India such as the Nagas and Bondas.

Sawah Terraced rice paddies in Indonesia.

Semi-deciduous Refers to trees that lose some of their leaves in the dry months of the monsoon.

Sepoy A British-trained Indian soldier.

Serendip Arab name for Ceylon (Sri Lanka).

Sex ratio The numerical ratio between men and women in a population.

Sex-selective abortion The deliberate abortion of males or females (typically females) after sex determination of the unborn fetus.

Shadow economy Illegal economic activities, such as black market enterprises, that are unregulated, untaxed, and unseen.

Shatter-zone A region of multiple, independent cultural groups often generated by fragmented terrain.

Sherpas Herders and traders of Nepal's Himalayas who are now dependent on tourism. Many serve as guides for climbing expeditions.

Shifting cultivation A migratory, field rotation system of agriculture found mainly in forested regions with relatively sparse populations.

Shiites A breakaway Islamic group with roots in Persia (Iran).

Shinto Traditional animist and nature-centered religion of Japan.

Shoguns Military leaders who ruled Japan from the medieval period to 1868.

Sikh A follower of Sikhism.

Sikhism Belief system founded in fifteenth century by Guru Nanak in India. Sikhs believe that all people are the children of God. They also oppose the caste system.

Simhala Dvipa Island of the Lion Tribe (Sri Lanka).

Sinhala Language of the Aryan, Buddhist, Sinhalese in Sri Lanka.

Sinicize To make Chinese.

Sinocentric A world view that places China at the central and superior position relative to the rest of the world.

Slash-and-burn method The cutting down of all but the largest trees in an area and the burning of the debris to clear the land for planting.

Social construction The construction of landscapes by internal and external political, social, and economic forces.

Sodic land Land where crusts of salt or other alkalis prevent nutrients and water from going below the surface. Reclamation of sodic lands is a goal of the Indian government.

South Asian Association for Regional Cooperation (SAARC) Formed to disassociate political conflicts from economic cooperation. It includes Bangladesh, Bhutan, India, the Maldives, Nepal, Pakistan, and Sri Lanka.

Southward policy Taiwan's policy of increasing investment in Southeast Asian countries to offset investments in China.

Spatial duality The differences between the landscapes of core and peripheral regions.

Special economic zones (SEZs) Zones set aside by China's government to attract foreign technology and investment.

Species diversity Variation in plant or animal species.

Squatter settlements Makeshift settlements in and around cities; associated with rural to urban migration.

State capitalism A system in which the government plays a strong role in economic development such as in South Korea.

Steppe Short grasslands.

Storm surge Water pushed toward shore by the force of the winds swirling around the storm. When combined with a tide in an increasingly narrow channel across low, flat terrain, a storm surge can cause an increase in water level of 15 feet or more.

Subak Agricultural cooperative in Bali.

Sudras Anyone belonging to the class of serfs in India's caste system.

Sukarno Indonesian nationalist who ruled from 1945 until 1965.

Sun Yat-sen Asia's first Asian nationalist. Proclaimed a Chinese Republic in 1912.

Sunderbans An area of ever-shifting mangrove, mudflats, islands, and distributary channels found in the Indus and Brahmaputra delta in the Bay of Bengal.

Sunna The way of life prescribed in Islam, based on the teachings of Muhammed and the Koran.

Sustainable development Development that sustains environmental integrity.

Swidden   Shifting cultivation in Southeast Asia.

Tagalog   The official language of the Philippines.

Taiping Rebellion   An uprising against the corrupt and ineffective Manchu rulers of China. It began in 1850 and ended in 1864 with the loss of 20 million lives.

Tais   Principle ancestors of the Thai, Lao, and Shan peoples. *Tai* is a cultural and linguistic term used to denote the various Tai people who became differentiated into a larger number of separate identities.

Taiwanese   Everyone who lives on the island of Taiwan.

Tamil   Language of the Dravidian Hindus in Sri Lanka.

Tamil Eelam   The proposed name for an independent state being fought for by the Tamil Tigers in northern Sri Lanka.

Tamil Tigers   A militant Tamil group of northern Sri Lankans who are agitating for an independent state called Tamil Eelam.

Tantric Buddhism   Also called Vajrayana Buddhism; it developed in north India in the seventh century and spread to the Himalayas where it was influenced by Tibetan shamanism. Official religion of Bhutan and Sikkim.

Techno-Archipelago   A series of technopolises linked by information and communication channels.

Technobelt   Zone selected for research and development and linked by communication and telecommunication.

Technopoles   Places where information and technology disseminate from. In India, Bangalore's "silicon plateau" is an example of a technopole.

Technopolis   A network of science cities centering on a "mother city" and connected to a major metropolis.

Technoscapes   Vast, intertwined and overlapping channels of communication produced by flows of technology, hardware, and software disseminated by transnational or supranational corporations, organizations, and agencies.

Terai   A marshy plain in southern Nepal situated 900 feet (270 m) above sea level.

Theravada Buddhism   The school of Buddhism practiced in Sri Lanka and most of Southeast Asia. It stresses the responsibility of the individual for his or her own salvation through the accumulation of merit by good behavior and religious activity.

38th parallel   The line of latitude that divides North and South Korea into two nations. The parallel division was instituted in 1945 by the Soviets and Americans.

Three Gorges Dam   World's largest water conservation project on the Yangzi River in China. When completed, Three Gorges will have the world's most powerful hydroelectric dam and improve shipping from the east to the interior. However, thousands of acres of farmland will be lost and countless numbers of residents will need to be relocated. In addition, numerous important cultural sites will be destroyed.

Tied aid   The linking of foreign aid to structural adjustment or specific activities or purchases.

Time-space convergence   More space can be covered in less time due to improvements in transportation and communication.

Timor Gap   The disputed ocean and seabed between Australia and Indonesia.

Tokaido megalopolis   Urbanized, industrialized region from Tokyo to Osaka. Named for the old Tokaido Road.

Tonkin   Region of the Red River delta in north Vietnam. Under French colonial rule, Tonkin was one of the five territories of the Indochinese Union.

Topographic barrier   An area of raised relief (such as the Himalayas) that prevents the passage of wind or other phenomena.

Topography   Landforms.

Toponyms   Place names.

Total Fertility Rate (TFR)   The number of children a woman would have if age-specific birth rates remained constant throughout her childbearing years—age 15 to 45 or 15 to 49.

Trans-border development region   An economic development zone in adjacent political units.

Transgenic plants   Human-engineered plants.

Transmigrasi   People who move from crowded areas of Indonesia to open up and populate frontier regions.

Transmigration of souls   In Hinduism, the changing of people's life status through rebirth according to the merits of their behavior.

Transnational enterprise   An economic, social, or political entity with ties to several nations.

Tree line   The line above which the growing season and the soil depth do not support trees.

Tribal Areas   Mountainous, tribal, and fractious regions, especially those in the western boundary of South Asia, where tribal chiefs and religions leaders rule.

Tropical cyclone   A violent storm created by a deep, circular, low-pressure cell. Called a typhoon in the Pacific.

Tsunamis   Seismic sea waves caused by earthquakes.

Tundra   A sub-arctic environment above the tree line that is a zone of low-lying flora, mosses, and lichens.

Typhoon   In the Pacific, a violent storm created by a deep, circular, low-pressure cell. Widely known as a tropical cyclone.

Underemployment   A Western notion regarding the underutilization of labor.

Unipolar concentration A situation arising from the migration to and growth of one city such as Tokyo at the expense of other places.

Untouchables Non-caste individuals associated with "polluting" occupations in India.

Upanishads A work of Hindu literature, from the Sanskrit phrase meaning "sitting down near" one's spiritual advisor.

Upazilas A subdistrict in Bangladesh created in the 1980s.

Urban agglomerations Clusters of cities.

Urban primacy A situation that exists when the population of a nation's largest city is greater than the combined populations of its next three largest cities.

Vaishyas Peasant farmers in India's caste system.

Varnas "Varna" is a word meaning "color." In Hinduism, classes of people such as priests or warriors.

Vedas The earliest known Hindu literature, from the Sanskrit word *veda,* meaning "knowledge."

Veddas Aboriginals of Sri Lanka.

Vegeculture The cultivation of tuber and tree crops such as taro and bananas. Vegeculture preceded seed cultivation.

Viet Cong National Liberation Front in southern Vietnam. Opposed to the American supported South Vietnamese government.

Viet Minh The front organization of the International Communist Party in Vietnam. After World War II, the Viet Minh took control of most of northern and central Vietnam.

Wallace Line The line of demarcation between the distinct zoogeographic regimes of Malaysia and Indonesia.

We-they syndrome Ethnocentric peoples (we) have a sense of superiority to other cultures (they) that are considered inferior.

White revolution The transformation of India's dairy industry to produce more milk.

Wind-chill factor The role that wind plays in reducing temperature.

World cities Cities at the top of a global urban hierarchy such as London, Tokyo, or New York.

Xenophobia Fear of foreigners.

Xerophytic Drought-resistant.

Yang With *yin,* one of two opposing yet complementary forces that govern the universe.

Yi jing (**I-ching**) A Chinese literary work in which hexagrams depict the harmonies of complementary opposites.

Yin With *yang,* one of two opposing yet complementary forces that govern the universe.

Zaibatsu Japanese military-industrial conglomerates. They were disbanded after World War II.

Zeylan Dutch name for Ceylon (Sri Lanka).

Index

A reference with an italicized number refers to figures, tables, boxes, and captions.

443